Zooarchaeology and Modern Human Origins

Vertebrate Paleobiology and Paleoanthropology Series

Edited by

Eric Delson
Vertebrate Paleontology, American Museum of Natural History,
New York, NY 10024, USA
delson@amnh.org

Eric J. Sargis
Anthropology, Yale University
New Haven, CT 06520, USA
eric.sargis@yale.edu

Focal topics for volumes in the series will include systematic paleontology of all vertebrates (from agnathans to humans), phylogeny reconstruction, functional morphology, Paleolithic archaeology, taphonomy, geochronology, historical biogeography, and biostratigraphy. Other fields (e.g., paleoclimatology, paleoecology, ancient DNA, total organismal community structure) may be considered if the volume theme emphasizes paleobiology (or archaeology). Fields such as modeling of physical processes, genetic methodology, nonvertebrates or neontology are out of our scope.

Volumes in the series may either be monographic treatments (including unpublished but fully revised dissertations) or edited collections, especially those focusing on problem-oriented issues, with multidisciplinary coverage where possible.

For other titles published in this series, go to
www.springer.com/series/6978

Zooarchaeology and Modern Human Origins: Human Hunting Behavior during the Later Pleistocene

Edited by

Jamie L. Clark

Department of Anthropology, University of Alaska Fairbanks, Fairbanks, AK, USA

John D. Speth

Museum of Anthropology, University of Michigan, Ann Arbor, MI, USA

Springer

Editors
Jamie L. Clark
Department of Anthropology
University of Alaska Fairbanks
Fairbanks, AK
USA

John D. Speth
Museum of Anthropology
University of Michigan
Ann Arbor, MI
USA

ISSN 1877-9077 ISSN 1877-9085 (electronic)
ISBN 978-94-007-6765-2 ISBN 978-94-007-6766-9 (eBook)
DOI 10.1007/978-94-007-6766-9
Springer Dordrecht Heidelberg New York London

Library of Congress Control Number: 2013940746

Cover Illustration: The cover image of a fallow deer (*Dama mesopotamica*) was taken by nature photographer Eyal Bartov in 2011 at the Hai Bar Carmel Nature Reserve (Israel), and is reproduced here from the photographer's "Mammals of Israel" series with the permission of Eyal Bartov.

Printed on acid-free paper

Springer is part of Springer Science+Business Media (www.springer.com)

Preface

The period between 71 and 25 ka (Marine Isotope Stages [MIS] 4 and 3) was a critical one in the evolution of our species. This time span witnessed the spread of anatomically modern humans across the Old World, the disappearance of the Neanderthals, the transition from Middle Stone Age (MSA)/Middle Paleolithic (MP) tool traditions to Later Stone Age (LSA)/Upper Paleolithic (UP) technologies, an explosion of evidence for material culture with symbolic significance, and significant global population growth. Climatically speaking, the phase preserves the transition from a glacial period to an interstadial, with records indicating the onset of remarkably high variability in climate during MIS 3. The recent publication of genetic data which indicate that not only were Neanderthals and early modern humans capable of interbreeding, but that Neanderthals did make a genetic contribution to our species, has made it clear that gaining a deeper insight into potential behavioral differences between these populations—particularly in terms of their adaptive strategies/capabilities—will be of key importance in understanding not only the rapid spread and success of early modern humans, but also the ultimate demise of the Neanderthals. The zooarchaeological record provides unique insight into these issues, serving as a valuable source of information on human responses to changes in both climate and demography, and on the ways which variation in technology may have impacted hunting and processing strategies. This volume, which is based on a session that was held at the 11th International Conference of the International Council of Archaeozoology (ICAZ) in Paris in August 2010, brings together a diverse group of scholars who have worked with faunal assemblages dating to MIS 4 and/or 3 from Eurasia and Africa, in order to more fully explore the range and nature of variability in human hunting behavior during this period, and to explore the ways that the zooarchaeological record can contribute to our broader understanding of modern human origins.

We would like to thank the many people who made this volume possible; first, the organizing committee of the 11th International ICAZ Conference, Drs. Jean-Denis Vigne, Christine Lefèvre, and Marylène Patou-Mathis, who helped make the Paris meetings a tremendous success. Thanks also to all the participants in our session—both oral and poster presentations were interesting, informative, and stimulating. Special thanks go to those who prepared their presentations for publication in this volume—their hard work (and patience) is much appreciated. We also owe a debt of gratitude to all those who responded to our requests for reviews—especially to those individuals who graciously agreed to review more than one chapter. Finally, thanks to Judith Terpos and the team at Springer, as well as to Eric Delson and Eric J. Sargis, the Editors of the Vertebrate Paleobiology and Paleoanthropology Series, for their support throughout the publication process.

Contents

Contributors

Madelaine Böhme
Abteilung Paläobiologie und Senckenberg Center for Human Evolution and Quaternary Ecology, Institut für Geowissenschaften, Eberhard-Karls-Universität Tübingen, 72070 Tübingen, Germany
m.boehme@ifg.uni-tuebingen.de

Jamie L. Clark
Department of Anthropology, University of Alaska Fairbanks, 310 Eielson Building, Fairbanks, AK 99775, USA
Institut für Naturwissenschaftliche Archäologie, Universität Tübingen, Rümelinstr. 23, 72070 Tübingen, Germany
Institute for Human Evolution, University of the Witwatersrand, WITS, Johannesburg 2050, South Africa
jlclark7@alaska.edu

Nicholas J. Conard
Abteilung Ältere Urgeschichte und Quatärökologie, und Senckenberg Center for Human Evolution and Quaternary Ecology, Schloss Hohentübingen, Institut für Ur- und Frühgeschichte und Archäologie des Mittelalters, Eberhard-Karls-Universität Tübingen, 72070 Tübingen, Germany
nicholas.conard@uni-tuebingen.de

Sandrine Costamagno
TRACES–UMR 5808, CNRS, PRES Toulouse, Université Toulouse 2-Le Mirail, 5 Allées A. Machado, 31058 Toulouse Cedex 9, France
costamag@univ-tlse2.fr

Gerrit L. Dusseldorp
Institute for Human Evolution, University of the Witwatersrand, WITS, Johannesburg 2050, South Africa
Faculty of Archaeology, Leiden University, P.O. Box 9515, 2300 RA Leiden, The Netherlands
g.dusseldorp@gmail.com

James G. Enloe
Department of Anthropology, University of Iowa, 114 MacBride Hall, Iowa City, IA 52242-1322, USA
james-enloe@uiowa.edu

Ivana Fiore
Soprintendenza al Museo Nazionale Preistorico Etnografico "L. Pigorini", Sezione di Paleontologia del Quaternario e Archeozoologia, P. le Marconi 14, 00144 Rome, Italy
iva_fiore@yahoo.it

Monica Gala
Soprintendenza al Museo Nazionale Preistorico Etnografico "L. Pigorini", Sezione di Paleontologia del Quaternario e Archeozoologia, P. le Marconi 14, 00144 Rome, Italy
monarix@yahoo.it

Keiko Kitagawa
Institut für Naturwissenschaftliche Archäologie, Universität Tübingen, Rümelinstr. 23, 72070 Tübingen, Germany
keiko.kitagawa@ifu.uni-tuebingen.de

Petra Krönneck
Institut für Naturwissenschaftliche Archäologie, Universität Tübingen, Rümelinstr. 23, 72070 Tübingen, Germany
petra.kroenneck@uni-tuebingen.de

Ana B. Marín Arroyo
Instituto Internacional de Investigaciones Prehistóricas de Cantabria, Universidad de Cantabria, Santander, Spain
anabelen.marin@unican.es

Eugène Morin
Department of Anthropology, Trent University, DNA Bldg Block C, 2140 East Bank Drive, Peterborough, ON K9J 7B8, Canada
eugenemorin@trentu.ca

Susanne C. Münzel
Institut für Naturwissenschaftliche Archäologie, Universität Tübingen, Rümelinstr. 23, 72070 Tübingen, Germany
susanne.muenzel@uni-tuebingen.de

Laura Niven
Department of Human Evolution, Max Planck Institute for Evolutionary Anthropology, Deutscher Platz 6, 04105 Leipzig, Germany
laura@lauraniven.com

Marco Peresani
Dipartimento di Biologia ed Evoluzione Sezione di Paleobiologia, Preistoria e Antropologia, Università degli Studi di Ferrara, 32 Corso Ercole I d'Este, 44121 Ferrara, Italy
marco.peresani@unife.it

Elspeth Ready
Department of Anthropology, Stanford University, Main Quad Building 50, 450 Serra Mall, Stanford, CA 94305-2034, USA
eready@stanford.edu

Matteo Romandini
Dipartimento di Biologia ed Evoluzione Sezione di Paleobiologia, Preistoria e Antropologia, Università degli Studi di Ferrara, 32 Corso Ercole I d'Este, 44121 Ferrara, Italy
matteo.romandini@unife.it

John D. Speth
Museum of Anthropology, University of Michigan, 4013 Museums Building, Ann Arbor, MI 48109-1079, USA
jdspeth@umich.edu

Lawrence G. Straus
Department of Anthropology, University of New Mexico, Albuquerque, NM 87131-0001, USA
lstraus@unm.edu

Antonio Tagliacozzo
Soprintendenza al Museo Nazionale Preistorico Etnografico "L. Pigorini", Sezione di Paleontologia del Quaternario e Archeozoologia, P. le Marconi 14, 00144 Rome, Italy
antonio.tagliacozzo@beniculturali.it

Reuven Yeshurun
Zinman Institute of Archaeology, University of Haifa, Mt. Carmel, 31905 Haifa, Israel
ryeshuru@research.haifa.ac.il

José Yravedra Sainz de los Terreros
Departamento de Prehistoria, Universidad Complutense de Madrid, Madrid, Spain
joyravedra@hotmail.com

Chapter 1
Introduction: Zooarchaeology and Modern Human Origins

Jamie L. Clark and John D. Speth

The nature, behavior, and ultimate fate of the Neanderthals have been topics of fascination and mystery for both scholars and the lay public alike, a tradition of focused inquiry and abundant speculation that has persisted since paleoanthropology first emerged as a recognizable discipline. Nor has that interest waned over the years. In fact, quite the contrary—scholarly articles, monographs, and books focused on the Neanderthals and the ever-captivating question of modern human origins have been pouring forth at an unprecedented rate, and media interest has never been greater.

This fascination no doubt derives from the fact that the Neanderthals were among the last "archaic" humans to inhabit the Old World before being replaced by "modern" humans—that is, by people who looked and presumably acted much like "us". There are many things about Neanderthals that are very much like what we see in ourselves—big, presumably intelligent brains; sophisticated and quite complex technologies fashioned from both organic and stone materials; an emphasis on hunting large, often quite formidable prey; intentional burial of their dead; and broadly dispersed populations that colonized a remarkable diversity of environments and successfully coped with the harsh vicissitudes of Ice Age Eurasia for more than 200,000 years.

But the fossil and archaeological records also suggest that Neanderthals were quite different from us, and it is those differences that make them so enigmatic. They buried their dead, but with little or no clear-cut evidence that they did so with any conception of an afterlife. For most of the Middle Paleolithic, we see virtually no sign of art or ornamentation, nor patterned variability in artifact forms that might be construed as stylistic markers of corporateness or status, and over the millennia we see remarkable stasis in their technology and overall way of life.

It is these differences between "them" and "us" that have kept academic opinion sharply divided. On one side are those who see Neanderthals and other quasi-contemporary "archaics" as technologically and culturally impoverished versions of modern humans, people who possessed the mental capacities and potentials that we have, but whose limited technology, low population densities, and/or high mobility stifled innovation and quashed any divisive tendency toward social differentiation. On the other side are those who see Neanderthals as a lesser form of human, one which was smart and very successful at surviving the harsh realities of the Ice Age, but one that nonetheless was separated from "us" by an unbridgeable cognitive gap.

This debate has fueled vast amounts of research aimed at exploring the nature and extent of the behavioral and biological differences between Neanderthals and anatomically modern humans. As recently as three decades ago, research into this topic was almost entirely focused on the European record; while research in Europe continues unabated, the development of large-scale, multidisciplinary projects in Africa has greatly increased the quantity and quality of information at our disposal about hominin behavior during the Late Pleistocene. And yet, the more we know, the more it becomes clear that more work in other, lesser-studied regions such as central and eastern Asia will be equally critical in expanding our knowledge about the lifeways of both late archaic and early modern humans.

J. L. Clark (✉)
Department of Anthropology, University of Alaska Fairbanks, 310 Eielson Building, Fairbanks, AK 99775, USA
e-mail: jlclark7@alaska.edu

and

Institut für Naturwissenschaftliche Archäologie, Eberhard Karls Universität Tübingen, Tübingen, Germany

and

Institute for Human Evolution, University of the Witwatersrand, Johannesburg, South Africa

J. D. Speth
Museum of Anthropology, University of Michigan, Ann Arbor, MI 48109-1079, USA

Jamie L. Clark and John D. Speth (eds.), *Zooarchaeology and Modern Human Origins: Human Hunting Behavior during the Later Pleistocene*, Vertebrate Paleobiology and Paleoanthropology, DOI: 10.1007/978-94-007-6766-9_1,

Developments in geology and climate science have also greatly contributed to our understanding of the timing and context of the Middle Paleolithic/Middle Stone Age and of the spread of anatomically modern humans out of Africa. We now possess a vastly improved understanding of Middle and Late Pleistocene paleoclimates, made possible by the deep-sea oxygen isotope record, together with detailed climatic histories extracted from ice cores in both Greenland and Antarctica. In addition, until quite recently, almost all of the Middle Paleolithic and Middle Stone Age lay beyond the effective range of radiocarbon dating (~40,000 years), making our reconstructions of the temporal placement of Neanderthal and AMH sites and their associated human fossils little more than educated guesses. Now, using thermoluminscence (TL), optically stimulated luminescence (OSL), and a host of other techniques, we are able to date the artifacts and human remains with considerably greater precision, with results that at times have been the reverse of what many had anticipated—for example, the well-known AMH fossils from the Israeli sites of Qafzeh and Skhul turned out to be older, not younger, than the Neanderthal remains from the nearby sites of Kebara and Amud (Valladas and Joron 1989; Valladas and Valladas 1991; Mercier et al. 1993; Valladas et al. 1999). Equally surprising, carefully shaped bone tools, shell beads, and engraved ocher, artifacts that we once thought belonged to the last 35,000 or 40,000 years of human existence (and were the hallmarks of modern humans in Europe), have been dated to greater than 75,000 years ago at sites in North Africa, the Near East, and the southernmost tip of Africa (e.g., d'Errico et al. 2005; Vanhaeren et al. 2006).

Most recently, genetics has entered the picture in a big way, first through the successful extraction and sequencing of mitochondrial DNA (mtDNA) from the original Feldhofer Neanderthal in Germany and now through the sequencing of both nuclear and mtDNA recovered from an ever-expanding number of fossils from sites throughout Europe and Asia. The findings of these genetic studies have profoundly changed our thinking about Neanderthals and modern human origins, challenging many long-held assumptions and raising a myriad of new questions. The synergy among all of these seemingly disparate techniques and approaches has given rise to an unparalleled level of excitement that permeates every facet of the study of modern human origins, and has brought Neanderthals and their contemporaries front and center both in academic circles and in the public eye.

This volume, which brings together 13 contributions by 22 scholars, focuses specifically on the insights that can be gained from the zooarchaeological record concerning the larger debates about modern human origins. This record serves as one of our primary sources of information about how Neanderthals and modern humans made their living. Not only can faunal data provide unique insights into the ways these human groups responded to changes in both climate and demography, but it can also be used to explore how variation in technology may have impacted hunting and processing strategies. Most scholars assume that the Middle to Upper Paleolithic transition should involve major changes in hominid subsistence patterns, as a new, presumably better adapted human form entered Eurasia and eventually replaced the archaic forms they encountered along the way, presumably driving some groups to extinction and forcing others into isolated refugia. In other words, many paleoanthropologists are looking for something tangible that they can argue reflects a genuine forward leap in cognitive wherewithal—for some distinctive behaviors that would have given modern humans some type of competitive edge—and the zooarchaeological record stands as one potential source that could inform of behavioral differences with adaptive significance.

Faunal studies are not without their problems, of course. We have come to appreciate the many natural processes than can alter the composition of archaeological bone assemblages, regardless of how abundant or well-preserved the bones might be. Diagenesis, sediment compaction, and trampling can destroy fragile bones or fragment them beyond recognition. And the specific elements, or parts of elements, that end up being removed from the record (or preserved) is seldom random, but depends instead on their size, shape, and density, whether or not they had been burned, their spatial position in relation to the original mouth of the cave, and many other factors. As a result of these complex processes, the bones that archaeologists dig up cannot be assumed to provide a simple, direct reflection of what the site's inhabitants actually hunted, ate, and discarded there; they may instead be just a remnant, quite possibly a very mixed and biased remnant at that, of the original assemblage. Teasing apart the complex interplay of the many natural and human processes that may have been at work in the formation of Pleistocene faunal assemblages, an area of research collectively known as "taphonomy," has become a central focus of contemporary Paleolithic zooarchaeology. This concern is very evident in many of the papers that follow.

One of the strengths of the current volume is the demonstration of the diversity of approaches that zooarchaeologists have taken when exploring questions relating to modern human origins. Although the papers are presented geographically, beginning with Africa (Clark), moving to three papers on the Near East (Speth, Yeshurun, and Marín-Arroyo), and ending with the European record (Yravedra, Straus, Tagliacozzo et al., Niven, Enloe, Conard et al., Dusseldorp, Costamagno, and Morin and Ready), a number of different themes and approaches are represented.

The role that palaeontological assemblages can play in our understanding of human hunting behavior, particularly as relating to prey choice, is the focus of two papers in the volume: Dusseldorp (Chap. 12) and Yeshurun (Chap. 4). Both examine bone accumulations of Middle Paleolithic age that were formed largely or entirely by natural processes (hyenas in the former case, a natural pit-trap in the latter) in order to generate regional baselines against which to compare assemblages that for the most part are products of human (Neanderthal) hunters.

In his study of French materials, Dusseldorp shows that while hyenas and humans took broadly similar spectra of larger prey, the emphases of the two predators were quite different, hyenas preferentially going after large bovids and equids, as well as megafauna (proboscideans and rhinos), while Neanderthals concentrated more heavily on cervids (especially red deer and reindeer). Interestingly, Dusseldorp's findings do not match the results of recent isotope work, which suggest that Neanderthals relied heavily on meat from mammoth and rhino (Bocherens 2011). It remains to be seen whether this discrepancy stems from selective transport decisions made by Neanderthals who stripped the meat from the carcasses of megafauna prior to transport, discarding most bones at the kill; or instead arises from unresolved issues in the way we interpret the isotope results.

In a similar vein, Yeshurun looks at the ungulates recovered from a natural pit-trap (Rantis Cave) in Israel that dates to the Middle Paleolithic, and compares its contents to more or less contemporary ungulate assemblages from archaeological sites in the same region. Hyenas or other carnivores were not involved to any significant extent in the formation of either the pit-trap's contents or the assemblages from the archaeological sites. Both trap and human occupations are overwhelmingly dominated by the same two animals—Persian fallow deer (*Dama mesopotamica*) and Mountain gazelle (*Gazella gazella*)—yet deer are proportionately more abundant in the natural accumulation while gazelles, despite their smaller size, are more prominent in the archaeological sites. Both Dusseldorp's and Yeshurun's finding underscore the likelihood that Neanderthals were often selective in the prey they targeted, implying that archaeological faunal assemblages dating to the Middle Paleolithic may not reflect in any simple one-to-one fashion the natural abundance of the various prey taxa on the landscape—this is something that must be demonstrated, not assumed.

Two of the papers aimed at exploring behavioral differences among Neanderthals and anatomically modern humans are heavily focused on taphonomic issues: Marín-Arroyo (Chap. 5) conducted a thorough taphonomic analysis of the faunal materials recovered from Tabun Cave, which were then utilized in a broader comparison of Neanderthal and AMH subsistence and mobility, while Yravedra (Chap. 6) discusses the importance of taphonomic analysis for identifying what components of an assemblage were introduced by humans versus being introduced by other predators—information that is critical if one wants to compare the subsistence behavior of Neanderthals and early modern humans.

Marín-Arroyo re-examined an old and extremely biased faunal collection from Dorothy Garrod's classic excavations in the late 1920s and early 1930s at the site of Tabun in Israel. The bias in what was kept in the original excavations is clearly underscored by the fact that over 90 % of the specimens in the collection were taxonomically identifiable. Despite these biases, Marin-Arroyo is still able to find interesting patterning, showing, for example, that the taxonomic make-up of the assemblage is quite similar to what one observes in other roughly contemporary Levantine sites, with a clear emphasis on mid-sized ungulates. She also found that large and extremely large animals such as aurochs and rhino, though never really abundant at Tabun, are most common in the earlier part of the sequence, an observation that echoes results from studies elsewhere in the region. Marín-Arroyo also found that the bones from the uppermost Middle Paleolithic level in Tabun (B), an assemblage dominated by fallow deer, may be a natural pit-trap accumulation with animals falling in through an open "chimney" in the cave's roof. The fact that most of the animals that became victims of the pit-trap were fallow deer is reminiscent of what Yeshurun observed at Rantis Cave.

Based on detailed taphonomic analysis of material from several Spanish sites, Yravedra argues that human subsistence was actually remarkably consistent across the MP-UP transition, with the greatest changes seen in the Solutrean, at which time hunters began focusing more intensively on smaller ungulates such as ibex and chamois. While both of these species have been identified in assemblages dating to the MP and early UP, Yravedra uses the results of his taphonomic analyses to argue that these species were primarily accumulated by carnivores; it is only in the Solutrean that direct evidence for human interaction with ibex and chamois becomes common.

The papers by Straus (Chap. 7) and Conard et al. (Chap. 11) both present regional comparisons of subsistence behavior during the Middle and Upper Paleolithic; Straus focuses on the Iberian peninsula, while Conard focuses on sites in the Swabian Jura in Germany. Both contributions are particularly interesting for their discussions relating to the exploitation of smaller mammals, and, in the case of Conard et al., for the discussion of the use of fish and fowl.

In his discussion of subsistence during the MP and UP in the Iberian Peninsula, Straus highlights the evidence for continuity among the main game species exploited. He finds no marked increase in small game or marine resources in the EUP as compared to the MP; in fact, he points out an increasing body of evidence for the exploitation of rabbits in eastern Spain and Portugal during the Mousterian. As was the case in Yravedra's study, evidence for subsistence intensification seems to primarily occur later in the UP.

The contribution of Conard and colleagues paints a slightly different picture. While they have previously identified continuity across the MP-UP transition in the Swabian Jura in terms of the primary game species exploited, they found a somewhat different pattern when considering the smaller mammal, bird, and fish remains. Smaller mammals and fish seem to have been exploited in both the MP and UP; however, there was an increased focus on these species in the UP. The Aurignacian also showed a sharp increase in the exploitation of birds. The authors argue that, taken as a whole, the data from the Swabian Jura do indicate an expansion of dietary breadth and intensification in the exploitation of animal resources during the UP. And yet, as is the case with the other contributions in this volume, the papers by Straus and Conard et al. make it eminently clear that although MP and early UP hominids may have been capable of exploiting smaller mammals, these hominins were most interested in ungulates, especially small- to mid-sized ones. What is not supported in any of these studies is the supposed heavy focus on megafauna that has been predicted by recent isotope work.

Two papers present comparisons of faunal remains from Mousterian contexts with those associated with the so-called "transitional" industries or techno-complexes: Tagliacozzo et al. (Chap. 8) focus on the Uluzzian, while Enloe (Chap. 10) looks to the Châtelperronian. The question of authorship for these industries has been a source of debate for decades; while some find the Neanderthal burial at Saint Césaire as solid evidence for a Neanderthal origin for at least the Châtelperronian, others have argued that AMH were responsible for producing the transitional technocomplexes (see Bar-Yosef and Bordes 2010), particularly after a recent publication of the results of a study of Uluzzian dental remains (Benazzi et al. 2011); the authors argued that the teeth belonged to AMH, not Neanderthals, and thus that modern humans were responsible for the production of this industry. In this respect, the detailed description of the Uluzzian fauna from Fumane Cave that is presented by Tagliacozzo and colleagues is particularly interesting; their results indicate that the Uluzzian fauna is distinctive compared to the underlying Mousterian, and actually looks more similar to the Aurginacian fauna from the site. These results actually make Fumane distinctive compared to many of the other sites/regions in this volume, wherein subsistence behavior across the MP-UP transition seems quite consistent.

Enloe looks at the Mousterian and Châtelperronian from the Grotte du Bison at the famous site at Arcy-sur-Cure; while he finds that the Châtelperronian occupations at the site appear to be more internally structured, and that there was likely a greater and more systematic focus on marrow exploitation from nutritionally rich elements during the same period, he cautions against reading too much into these results because the Mousterian faunal sample appears to include a substantial component that was not introduced by human activity—again highlighting the importance of detailed taphonomic analyses before attempting broader scale comparisons of subsistence practices.

In addition to papers that make direct comparisons between Middle and Upper Paleolithic assemblages, we also have included three contributions that focus primarily on variability in human hunting behavior *within* the Middle Stone Age/Middle Paleolithic: Clark (Chap. 2) explores the relationships between environmental, technological and subsistence change during the MSA at Sibudu Cave (South Africa), while Speth (Chap. 3) explores evidence for over-exploitation of large game during the MP at Kebara Cave (Israel), and Niven (Chap. 9) looks for evidence of changes in subsistence during the course of the Mousterian occupations at the site of Pech de l'Azé IV (France).

Clark utilizes faunal remains from two consecutive phases of the MSA of southern Africa—the Howieson's Poort and the post-HP MSA—in order to explore the hypothesis that the disappearance of the behavioral innovations evidenced in the HP (including production of geometric backed tools and finely made bone points) was a response to environmental change. The fauna indicates a gradual/continuous shift that crosses the boundary between the two technocomplexes, with the most marked changes in subsistence occurring during the course of the post-HP MSA. She thus argues against the environment as a prime mover for the shifts in technology and material culture witnessed in the sequence, proposing instead that demographic or social variables may have played a more important role.

Speth also addresses the potential role of demography in constraining or encouraging the economic and cultural changes we see over the course of the Late Pleistocene. During the latter part of the Middle Paleolithic, perhaps already by about 100 ka, and certainly after about 70–60 ka, Neanderthals and their archaic quasi-contemporaries had successfully colonized a huge swath of the Eurasian continent, spreading out over a vast array of landscapes and environments. It is hard to imagine that this was accomplished by populations that were demographically static. This is also a time when, in a number of areas, one begins to see a dramatic increase in the number of sites; in sites with deeply stratified deposits, high densities of artifacts and faunal remains, and large numbers of hearths, ash lenses, and middens; and, in some areas, aggregations of human burials that might be construed of as cemeteries (e.g., Qafzeh, Skhul, and Shanidar), a phenomenon that archaeologists working in Holocene contexts would not hesitate to attribute to increasing populations and emerging corporateness or ethnicity (Saxe 1970: 119; Goldstein 1981: 61; Pardoe 1995). Speth's work with the fauna at Kebara Cave indicates an intensification in the exploitation of ungulates over the course of the late MP (and into the EUP)—including evidence for the overhunting of the largest game—that may indicate an early manifestation of the pattern of subsistence intensification one sees in both the plant and animal components of the diet throughout much of the subsequent Upper Paleolithic and Epipaleolithic in the region (e.g., Weiss et al. 2004; Stiner 2009; Stutz et al. 2009).

While not identifying any clear directional trends of the sort highlighted by Speth in the MP at Kebara, Niven's work on material from Pech de l'Azé IV, a site which preserves deposits spanning more than 60,000 years, indicates that the Neanderthals occupying the site were quite behaviorally flexible. The faunal remains suggest shifts in site-use strategies in response to changes in climate and the availability of the primary prey species. Taken together, these contributions indicate that MSA and MP subsistence adaptations were flexible, responding to shifting environmental conditions and perhaps also to demographic pressures.

The final two contributions to the present volume, by Costamagno (Chap. 13) and Morin and Ready (Chap. 14), tackle an important and perplexing issue that is especially near and dear to the heart of JDS, one that has as yet seen too little attention by archaeologists and paleoanthropologists—if Neanderthals were "top predators," right up there with cave lions and hyenas—as suggested by recent stable isotope studies (Bocherens 2011)—how did they avoid the problems that are likely to have arisen from high protein intakes? Nutritional studies, bolstered by literally hundreds of ethnohistoric accounts, indicate that modern humans cannot subsist for extended periods on diets in which protein intake exceeds roughly 300 g per day (which typically translates into about 35–40 % of total calories), the actual amount varying with body mass and allowing for some degree of adaptation to prolonged elevated intakes (Speth 2010). The remainder of the diet must come from non-protein sources, either fat or carbohydrates, and for historically and ethnographically documented foragers in cold, highly seasonal northern latitudes, it is clear that fat, not lean meat (protein) provided by far the largest component of the diet.

Assuming that Neanderthals faced a similar upper limit to the total amount of protein they could safely consume, how did they manage to acquire an adequate supply of fat? Obviously, one way is to target animals with the highest levels of body fat at the time of year when hunting took place—in most ungulates, females in the fall and males in the spring, and prime adults rather than old adults or juveniles (the commonplace underrepresentation of young animals in Paleolithic sites may not always be a consequence of attritional processes). Neanderthals could also significantly augment their fat intake by selecting fat-rich body parts, particularly those that are least likely to be fat-depleted—typically, these include the brain, tongue, internal organ fats, and especially the marrow in the more distal limb bones, feet, and beneath the cheek teeth in the mandibles. Morin and Ready's contribution clearly shows that the marrow index, especially a variant of the index that measures the unsaturated fatty acid content of the marrow, is the best predictor of skeletal element frequencies in a large array of French Paleolithic faunal assemblages, generally outperforming indices that are weighted toward meat yields, such as Binford's (1978) MGUI and Metcalfe and Jones's (1988) FUI. Both Enloe and Costamagno in their contributions also emphasize the prominence of major marrow bones in the assemblages they consider.

Among modern dwellers of the northern latitudes, marrow alone does not seem to provide a sufficient supply of fat to meet their annual non-protein food needs. Most of these groups, at least traditionally, also routinely boiled the lipids ("grease") from spongy bones such as the vertebrae and limb epiphyses (e.g., Stefansson 1956; Eidlitz 1969; Binford 1978). But archaeologists and paleoanthropologists dealing with the Middle Paleolithic commonly assume that Neanderthals lacked the technology to boil, eliminating what otherwise might have been an important, perhaps crucial, source of non-protein calories. Prior to Costamagno's contribution in this volume, we are aware of only one other person—Marean (2005)—who has addressed the issue of bone-grease processing in the MP in detail. Recognizing the likely importance of fat to Neanderthals, and assuming that they had no means of boiling the lipids out of bones, Marean suggested that these denizens of glaciated Eurasia actually ingested comminuted bone and extracted the precious lipids directly from the fragments by digestion.

Costamagno presents the results of a detailed taphonomic analysis which suggests that the under-representation of reindeer epiphyses from the MP site of Noisetier Cave (and likely at other French MP sites) was the result of grease extraction rather than resulting from the use of bone as fuel or some other process of density-mediated attrition. But the item that most archaeologists would accept as the "smoking gun" for boiling technology—fire-cracked rock—is noteworthy for its absence in these sites, as it is in most others as well. So, does this mean that Neanderthals could not boil fats from bones after all? While this remains a possibility, Costamagno, in a position similar to Marean's, suggests that Neanderthals could have been producing and directly ingesting "bone paste".

However, we are convinced that a thorough search of the literature from Classical Antiquity and the Middle Ages, and probably from similar time periods in areas far beyond the bounds of Europe, will reveal that the practice of cooking (boiling and stewing) directly over open flames in skin, paunch, or even bark containers *without the use of heated stones* was a common practice (as examples, see Derricke 1581 [1883]: Plate III; Rawlinson 1859: 52, 53; Ryder 1966, 1969; Skinner 1911: 557). So why not Neanderthals? Perhaps the introduction of heated-stone cooking, which at least in Europe probably occurred sometime during the Upper Paleolithic (perhaps hand-in-hand with the beginning stages of the LGM), rather than being a radically new technology introduced by humans who entered Europe with enhanced mental hardware, instead was another step in the gradual intensification of human ways of acquiring and processing foods—in short, the ratcheting-up of an already existing technology in response to ever increasing imbalances between resources and the number of mouths that needed to be fed. This is clearly an area in need of closer scrutiny, because it may reveal that the success of Neanderthals in penetrating the cold harsh landscapes of glaciated Eurasia was underwritten by their ability to extract precious lipids from the cancellous tissue of their hunted resources.

Taken as a whole, the papers presented herein demonstrate that the MP-UP transition, at least in terms of what we can see through the lens of zooarchaeology, shows few obvious changes that cannot more parsimoniously be attributed to shifts in climate, differences in local topography and environment, seasonality of hunting activities, and intensification stemming from local or regional changes in the size and/or distribution of human populations. Perhaps it is not—or should not be—that surprising that we are not finding dramatic differences between the Middle and Upper Paleolithic (or the comparable transition in Africa) that are clearly attributable to the change in hominin. After all, Neanderthals and other "archaic" humans from the same timeframe appear to have been very accomplished hunters, capable of taking prey across the entire size spectrum; slow-moving prey and prey with remarkably high escape velocities; species inhabiting open steppes, tundra, and savanna and those found only in rugged upland terrain; prey that are likely to be encountered during daylight hours and prey that remain sequestered throughout the day; and obligate drinkers that are best ambushed at waterholes and animals that obtain most of their moisture from their forage and must be hunted elsewhere using other strategies. That's a pretty impressive resume.

It is clear that zooarchaeology has a lot to offer in the way of information and insights into the lifeways of Neanderthals and early modern humans. We are pleased and honored to have been able to work with the many colleagues who contributed to this volume, which we feel makes an important contribution to the literature on modern human origins and will be useful to a broad range of scholars interested in the behavior and biological evolution of our species.

References

Bar-Yosef, O., & Bordes, J.-G. (2010). Who were the makers of the Châtelperronian culture? *Journal of Human Evolution, 59*, 586–593.

Benazzi, S., Douka, K., Fornai, C., Bauer, C. C., Kullmer, O., Svoboda, J., et al. (2011). Early dispersal of modern humans in Europe and implications for Neanderthal behaviour. *Nature, 479*, 525–528.

Binford, L. R. (1978). *Nunamiut ethnoarchaeology*. New York: Academic Press.

Bocherens, H. (2011). Diet and ecology of Neanderthals: Implications from C and N isotopes—insights from bone and tooth biogeochemistry. In N. J. Conard & J. Richter (Eds.), *Neanderthal lifeways, subsistence and technology: One hundred fifty years of Neanderthal study* (pp. 73–85). New York: Springer.

Derricke, J. (1581 [1883]). *The Image of Irelande with a Discoverie of Woodkarne*. J. Small (Ed.). Edinburgh: Adam and Charles Black.

d'Errico, F., Henshilwood, C., Vanhaeren, M., & van Niekerk, K. (2005). *Nassarius kraussianus* shell beads from Blombos Cave: Evidence for symbolic behaviour in the Middle Stone Age. *Journal of Human Evolution, 48*, 3–25.

Eidlitz, K. (1969). *Food and Emergency Food in the Circumpolar Area*. Studia Ethnographica Upsaliensia. Uppsala: Almqvist and Wiksells Boktryckert.

Goldstein, L. (1981). One-dimensional archaeology and multi-dimensional people: Spatial organization and mortuary analysis. In R. W. Chapman, I. Kinnes, & K. Randsborg (Eds.), *The archaeology of death* (pp. 53–69). Cambridge: Cambridge University Press.

Marean, C. W. (2005). From the tropics to the colder climates: Contrasting faunal exploitation adaptations of modern humans and Neanderthals. In F. d'Errico & L. Backwell (Eds.), *From tools to symbols: From early hominins to modern humans* (pp. 333–371). Johannesburg: Witwatersrand University Press.

Mercier, N., Valladas, H., Bar-Yosef, O., Vandermeersch, B., Stringer, C. B., & Joron, J.-L. (1993). Thermoluminescence date for the Mousterian burial site of Es-Skhul, Mt. Carmel. *Journal of Archaeological Science, 20*, 169–174.

Metcalfe, D., & Jones, K. T. (1988). A reconsideration of animal body-part utility indices. *American Antiquity, 53*, 486–504.

Pardoe, C. (1995). Riverine, biological and cultural evolution in southeastern Australia. *Antiquity, 69*, 696–713.

Rawlinson, G. (1859). *The history of Herodotus, Vol. III, Book IV—Melpomene*. London: John Murray.

Ryder, M. L. (1966). Can one cook in a skin? *Antiquity, 40*, 225–227.

Ryder, M. L. (1969). Paunch cooking. *Antiquity, 43*, 218–220.

Saxe, A. A. (1970). *Social dimensions of mortuary practices*. Unpublished PhD dissertation, Department of Anthropology, University of Michigan, Ann Arbor, Michigan.

Skinner, A. (1911). A comparative sketch of the Menomini. *American Anthropologist, 13*, 551–565.

Speth, J. D. (2010). *The paleoanthropology and archaeology of big-game hunting: Protein, fat or politics?* New York: Springer.

Stefansson, V. (1956). *The fat of the land*. New York: Macmillan.

Stiner, M. C. (2009). Prey choice, site occupation intensity and economic diversity in the Middle-early Upper Palaeolithic at the Üçağızlı Caves, Turkey. *Before Farming, 3*, 1–20.

Stutz, A. J., Munro, N. D., & Bar-Oz, G. (2009). Increasing the resolution of the broad spectrum revolution in the Southern Levantine Epipaleolithic (19–12 ka). *Journal of Human Evolution, 56*, 294–306.

Valladas, H., & Joron, J.-L. (1989). Application de la thermoluminescence à la datation des niveaux Moustériens de la Grotte de Kebara (Israel): ages préliminaires des unités XII, XI et VI. In O. Bar-Yosef & B. Vandermeersch (Eds.), *Investigations in South*

Levantine Prehistory (Préhistoire du Sud-Levant) (pp. 97–100). BAR International Series. Oxford: British Archaeological Reports.

Valladas, H., & Valladas, G. (1991). Datation par la thermoluminescence de silex chauffés des grottes de Kébara et de Qafzeh. In O. Bar-Yosef & B. Vandermeersch (Eds.), *Le Squelette Moustérien de Kébara 2* (pp. 43–47). Cahiers de Paléoanthropologie. Paris: CNRS.

Valladas, H., Mercier, N., Froget, L., Hovers, E., Joron, J.-L., Kimbel, W. H., et al. (1999). TL dates for the Neanderthal site of the Amud Cave, Israel. *Journal of Archaeological Science, 26*, 259–268.

Vanhaeren, M., d'Errico, F., Stringer, C., James, S. L., Todd, J. A., & Mienis, H. K. (2006). Middle Paleolithic shell beads in Israel and Algeria. *Science, 312*, 1785–1788.

Weiss, E., Wetterstrom, W., Nadel, D., & Bar-Yosef, O. (2004). The broad spectrum revisited: Evidence from plant remains. *Proceedings of the National Academy of Sciences of the United States of America, 101*, 9551–9555.

Chapter 2
Exploring the Relationship Between Climate Change and the Decline of the Howieson's Poort at Sibudu Cave (South Africa)

Jamie L. Clark

Introduction

The Howieson's Poort (HP; ~65–60,000 years ago) is an enigmatic sub-stage of the southern African MSA which is notable for its technological innovations, including finely made bone points and backed stone tools (both of which may have served as arrow points; see Backwell et al. 2008; Wadley and Mohapi 2008; Lombard and Phillipson 2010) as well as for its evidence for symbolically mediated behavior (engraved ostrich eggshell; Parkington et al. 2005; Texier et al. 2010). One of the more perplexing aspects of the MSA record is the apparent disappearance of the innovative behaviors associated with the HP after ~60,000 years ago. Given the apparent chronological association between the HP and a glacial period—the phase falls within MIS 4, a glacial period, and disappears around the time of the transition to an interstadial (MIS 3, ~59 ka)—the innovations evidenced in the HP have long been modeled as an adaptation to a particular set of environmental conditions (e.g., Ambrose and Lorenz 1990; McCall 2007). In fact, it has been proposed that the appearance (and disappearance) of the HP simply reflects shifts in the adaptation of an indigenous population in response to climate change; some have even argued the episode stands as one of the earliest examples of a highly dynamic (and "distinctly human") response to changing environmental conditions (e.g., Ambrose and Lorenz 1990).

And yet, in comparing a suite of optically stimulated luminescence (OSL) dates from HP and post-HP MSA sites located across southern Africa to ice core data from Antarctica, Jacobs and colleagues (Jacobs et al. 2008; Jacobs and Roberts 2008) argued that the appearance of the HP correlates with a period of climatic amelioration, not decline, and that the ages for the post-HP MSA also line up with a period of warming. Furthermore, because HP sites cross-cut so many distinct ecozones, they argued that the phase cannot be associated with any particular climatic or resource zone. As such, they argued against climate as being a primary catalyst for the HP.

More recently, Chase (2010) has taken issue with Jacob et al.'s use of polar ice core data as a proxy for southern African paleoenvironments, arguing that the region has a distinct climatic regime that does not necessarily align with global trends identified using polar records. Using a variety of paleoclimatic proxies from the southern African subregion, Chase argues that the disappearance of the HP does appear to correlate with an abrupt termination of the MIS 4 humid phase, thus keeping alive the possibility of a link between climate change and the disappearance of the HP.

Others have suggested that the disappearance of the HP reflects a localized population extinction (Bar-Yosef 2002) or even the northward migration of HP populations, who would eventually move out of Africa (Mellars 2006). Given these competing hypotheses, it is clear that gaining a higher resolution picture of the transition from the HP to the post-HP MSA is relevant to larger questions about modern human origins and the adaptive capabilities and strategies of early modern humans. In this paper, I explore the relationship between climate change and the end of the HP using faunal data from the HP and post-HP MSA deposits at Sibudu Cave (KwaZulu-Natal, South Africa).

J. L. Clark (✉)
Department of Anthropology, University of Alaska Fairbanks, 310 Eielson Building, Fairbanks, AK 99775, USA
e-mail: jlclark7@alaska.edu
and
Institut für Naturwissenschaftliche Archäologie, Eberhard Karls Universität Tübingen, Tübingen, Germany
and
Institute for Human Evolution, University of the Witwatersrand, Johannesburg, South Africa

Jamie L. Clark and John D. Speth (eds.), *Zooarchaeology and Modern Human Origins: Human Hunting Behavior during the Later Pleistocene*, Vertebrate Paleobiology and Paleoanthropology, DOI: 10.1007/978-94-007-6766-9_2,

Background

Located approximately 40 km north of Durban and 15 km inland from the Indian Ocean, Sibudu preserves an extensive MSA sequence that spans from the pre-Still Bay (>75 ka) through the final MSA (~39 ka; Fig. 2.1; Table 2.1; see Wadley and Jacobs 2006 for a detailed discussion of stratigraphy and excavation procedures). It is one of a few HP sites with good faunal preservation, and one of the only known sites to preserve the transition from the HP to the post-HP MSA. Current excavations at the site have been ongoing since 1998. More than 20 sq. m of MSA deposits have been excavated, although the deepest layers (including most of the post-HP MSA and everything below) have only been reached in a 6 sq. m unit. Deposits are excavated in 50 cm quadrants and screened through nested 2 and 1 mm sieves; materials from the 2 mm sieve are subject to initial sorting and processing on site.

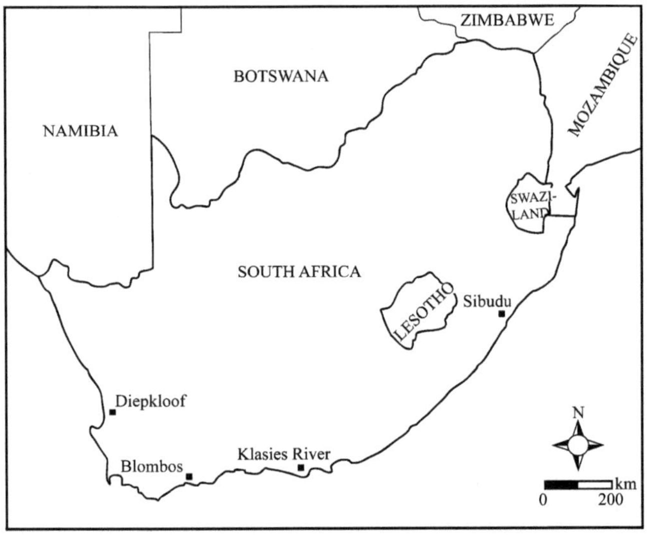

Fig. 2.1 Map showing the location of Sibudu and other MSA sites

Table 2.1 Optically stimulated luminescence dates from Sibudu (Wadley and Jacobs 2006; Jacobs et al. 2008)

Phase	OSL dates (ka)
final MSA	38.6 ± 1.9[a]
late MSA	47.7 ± 1.4[a]
post-HP MSA	58.5 ± 1.4[a]
HP	61.7 ± 2.0
	63.8 ± 2.8
	64.7 ± 2.3
Still Bay	70.5 ± 2.3
pre-Still Bay	72.5 ± 2.5
	73.2 ± 2.7
	77.3 ± 2.7

[a] Weighted mean ages

Figure 2.2 shows a profile drawing of the HP and post-HP MSA deposits at Sibudu. Note that while the post-HP MSA was initially defined as a single unit based on the OSL dates and other shared characteristics (including similarities in the lithic assemblage; Wadley and Jacobs 2006), it has since been divided into an upper and a lower unit, known as the post-HP MSA 1 and post-HP MSA 2, respectively. A number of lines of evidence support this division (see Clark and Plug 2008 for a detailed discussion); most notably, archaeomagnetic data indicated a dramatic shift in mineral magnetic properties during the course of the post-HP MSA, perhaps reflecting the transition from MIS 4, a glacial, to MIS 3, an interstadial (Herries 2006), and lithic data show a marked shift in raw material use at the same time.

A number of features distinguish the HP from the post-HP MSA. First are those related to the archaeological remains themselves. The lithic assemblage from the HP is comprised primarily of the geometric backed tools which define the industry, and typical MSA hunting technology—that is, stone points—are absent. Residue and microwear analyses by Lombard (2007, 2008) demonstrated that the backed tools were likely hafted in composite tools; the presence of impact damage suggests that at least some of these were used for hunting (Lombard and Pargeter 2008; Wurz and Lombard 2007). Metric studies of the backed segments recovered at Sibudu indicate that the dimensions of some of these tools fits within the known range of arrow points, leading Wadley and Mohapi (2008) to propose that bow and arrow technology may have been present in the HP (this argument has recently been expanded in Lombard and Phillipson 2010).

The HP at Sibudu also includes a worked bone industry; bone tools present in the HP layers include a finely made bone point which is virtually indistinguishable from prehistoric and ethnographically documented arrow points, leading Backwell and colleagues (2008) to argue that these tools may stand as an additional line of evidence for the presence of bow and arrow technology during this phase. In addition to these innovative technologies, Wadley and colleagues (Wadley 2006, 2010; Clark and Plug 2008) have proposed that remote capture technology such as snares, traps, or nets, may also have been employed during the HP. Given that there is no direct evidence for these technologies (which are generally made of perishable materials that would not survive to become part of the archaeological record), these arguments were based on the nature of the HP faunal assemblage; a more thorough exploration of this hypothesis is currently underway.

In contrast, the post-HP MSA deposits contain a more typical MSA lithic assemblage comprised primarily of unifacial points and scrapers (Cochrane 2006). Bone points (or other formal bone tools) are also absent. It would thus appear that a very different technological system,

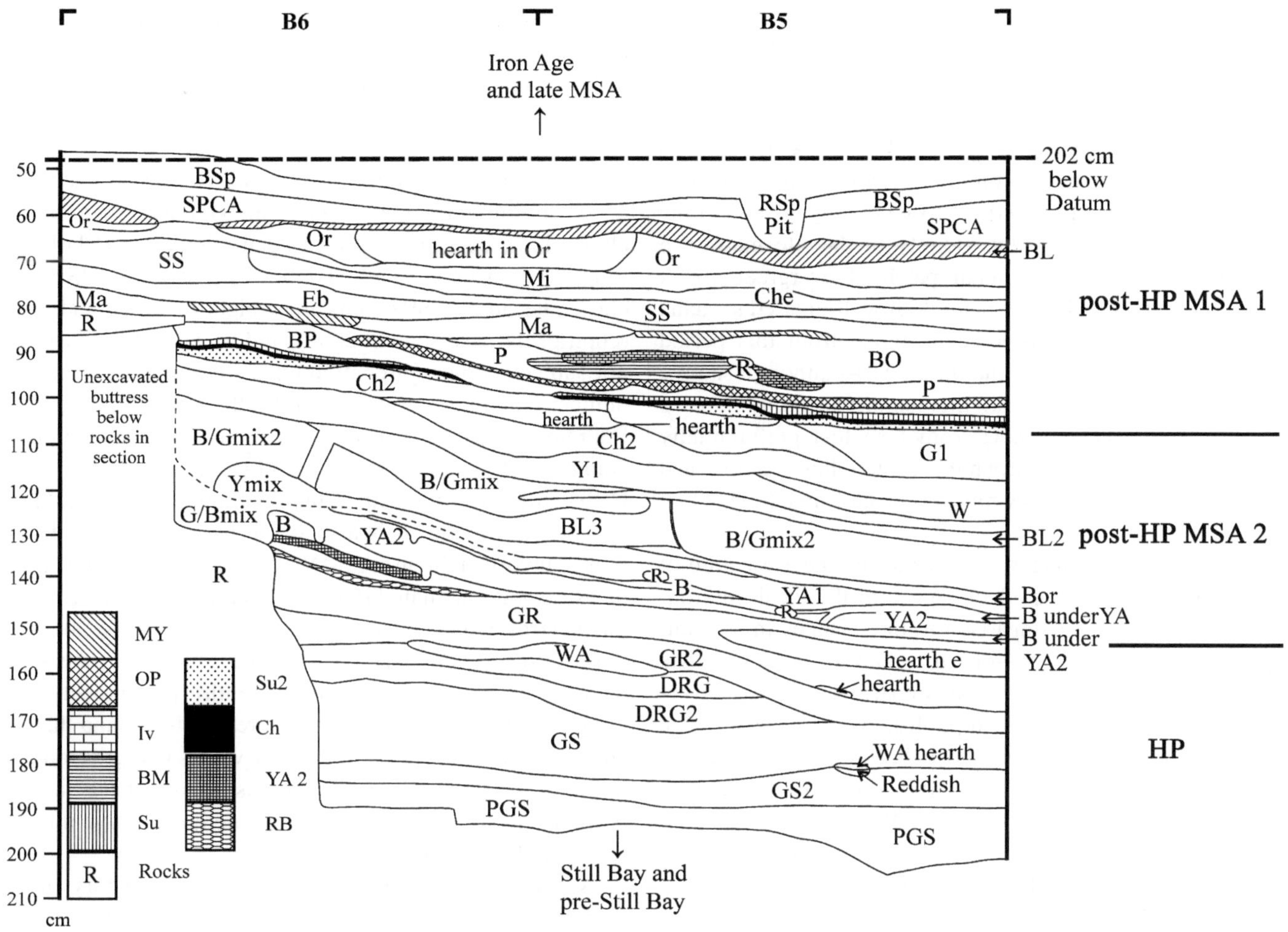

Fig. 2.2 Profile drawing of the HP and post-HP MSA at Sibudu

particularly as relating to hunting weaponry, was in place during the post-HP MSA.

In addition to the changes in material culture evidenced between the HP and the post-HP MSA, there are also a number of characteristics that distinguish the deposits themselves. The HP is comprised of thicker, monotonously colored deposits, while the post-HP MSA consists of thin, brightly colored deposits that contain palimpsests of interfingering combustion features (Pickering 2006; Wadley and Jacobs 2006; Goldberg et al. 2009). This distinction may be due in part to greater intermixture of deposits in the HP caused by more intensive trampling (Goldberg et al. 2009). The post-HP MSA is also more variable in sediment composition than is the HP, and gypsum inclusions, which occur frequently in the post-HP MSA, are largely absent in the HP (Schiegl and Conard 2006).

Furthermore, while Jacobs and colleagues argued that broadly similar conditions prevailed during the HP and post-HP MSA (Jacobs et al. 2008; Jacobs and Roberts 2008), the faunal and botanical data available from Sibudu suggest that the local environment underwent significant changes during this period (Allott 2006; Wadley 2006; Clark and Plug 2008; Hall et al. 2008; Clark 2009, 2011). During the HP, the flora and fauna both indicate a strong evergreen forest component; the evergreen *Podocarpus* was the most frequently identified genus in the charcoal assemblage (Allott 2006), while forest dwelling species such as the blue duiker (*Philantomba monticola*) and bushpig (*Potamochoerus larvatus*) comprised the majority of the faunal remains that could be identified to genus or species (Clark and Plug 2008; Clark 2009, 2011). Several lines of evidence also indicated that conditions were relatively humid during this phase, including the presence of the Gambian giant rat (*Cricetomys gambianus*) and the horseshoe bat (*Rhinolophus clivosus*), both of which primarily occur in more humid environments, as well as the absence of gypsum (which is water soluble) in the HP deposits (Glenny 2006; Schiegl and Conard 2006).

The available data indicated that marked changes in the local environment had occurred by the end of the post-HP MSA, with data from the post-HP MSA 1 suggesting drier conditions and a shift to more deciduous/open vegetation (Allot 2006; Hall et al. 2008). Botanical data indicated that the evergreen forest surrounding the site may have significantly declined in size during this period, with *Podocarpus* absent from the uppermost layers and open species such as *Acacia* becoming more common (Allott 2006). Wadley (2006) and Hall et al. (2008) proposed that by the end of the phase, the vegetation in the vicinity of the site may have been characterized by a small patch of evergreen forest close to the river and larger patches of well-grassed savanna on the neighboring hillsides. The fauna from this period were also distinctive from that in the HP—although sample sizes were relatively small, the majority of the remains identifiable to species came from taxa that preferentially inhabit more open environments, including equids and wildebeest (*Connochaetes taurinus*; Clark and Plug 2008; Clark 2009).

However, questions remained as to the precise nature of the timing and nature of the changes in the local environment—especially in comparison to the changes in technology and material culture—due to a lack of botanical data from the post-HP MSA 2 (charcoal from these layers remains to be fully analyzed/published), and because the initial faunal sample from these layers was quite small (Clark and Plug 2008; Clark 2009). Here I present new faunal data from the HP and post-HP MSA deposits at Sibudu, which allow for a higher resolution reconstruction of environmental change during the HP and post-HP MSA.

Materials and Methods

Table 2.2 provides summary data on the faunal remains analyzed through August 2010. While the site preserves large quantities of fauna, the assemblage is highly fragmentary, with the vast majority of the remains consisting of non-identifiable fragments less than 2 cm in maximum dimensions. Of the more than 800,000 fragments analyzed thus far, the number of identifiable specimens (NISP) is only ~5,600, meaning that less than 1 % of the assemblage was identifiable. Among the identified bones (which are limited to the macromammalian remains; analysis of the remainder of the assemblage, including fish, birds, and microfauna, is not yet complete), the vast majority consisted of bovid remains that could only be identified to size class; Table 2.3 provides a breakdown of these size classes and representative species for each.

The analyses presented herein utilize NISP rather than MNI (minimum number of individuals); the reasons for this choice have been discussed in detail elsewhere (Clark and Plug 2008; Clark 2009). To briefly reiterate, NISP was chosen because NISP counts are straightforward to calculate (contra MNI; see Lyman 2008), and because the measure is not subject to the aggregation effects that plague minimum number counts (Grayson 1984). Furthermore, Marshall and Pilgram (1993) have demonstrated that minimum number counts can be more depressed (and less representative) than NISP when dealing with a highly fragmented assemblage like that at Sibudu. It is also worth noting that despite the issues involved in using MNI data at Sibudu, when the measure was calculated (Clark and Plug 2008; Clark 2009), the patterns evidenced using NISP were still intact, although the resulting sample sizes were quite small. That the same patterns were evidenced was not entirely unexpected, as Grayson (1984) and Grayson and Frey (2004) found that for any given fauna, minimum number values can generally be tightly predicted from NISP counts. Given these factors, NISP seems to be the most appropriate quantitative unit for the assemblage at hand.

A detailed discussion of the taphonomic history of the assemblage is beyond the scope of this paper (see Clark and Plug 2008; Clark 2009; Clark and Ligouis 2010); however, it is relevant to note that all of the available evidence indicates that humans were the primary contributor to the assemblage. More than 8,600 bones were examined under a microscope for signs of human activity, and while there was a low occurrence of surface damage (likely due to poor cortical preservation caused by high degrees of burning, see Clark and Ligouis 2010), signs of human involvement (in the form of cut marks and percussion damage) were always more common than carnivore or raptor damage (evidenced primarily by tooth scores/punctures and signs of digestion)—1.5 % of the assemblage showed evidence for surface damage associated with human activity and 0.05 % showed evidence for carnivore/raptor involvement.

Table 2.2 Faunal summary

Phase	Non-ID < 2 cm*	Non-ID > 2 cm	NISP	ID (%)
post-HP MSA 1	~290,337	28,938	905	0.28
post-HP MSA 2	~275,297	17,191	1,313	0.45
HP	~176,864	10,766	3,577	1.87
Total sample	~742,498	56,895	5,619	0.70

* Full counting of the <2 cm non-ID remains was undertaken in units B5 and B6 only. The full sample of small fraction non-IDs was counted for one quadrant per each level/unit for the rest of the material and this data used to estimate the total count of this material

Table 2.3 Bovid size classes with representative species

Size class	Live weight (kg)	Species (list not inclusive)
Bov I	<23	Blue duiker, common duiker, steenbok
Bov II	23–84	Mountain reedbuck, bushbuck, impala
Bov III	85–295	Red hartebeest, blue wildebeest, kudu
Bov IV	295–950	African buffalo, eland
Bov V	>950	Giant buffalo (extinct)

Furthermore, there was a very low representation of carnivores in the assemblage (less than 2.5 % of the total NISP), most of which were small carnivores such as the mongoose, which would not be contributing significantly to the ungulate assemblage. Finally, the fauna is highly burned and is associated with large quantities of stone artifacts and hearth structures. Taken in concert, all of these lines of evidence indicate that humans were primarily responsible for accumulating the faunal remains. Data also indicate that the relative degree of fragmentation was similar across the three units (Clark 2009), and thus variability in the degree of post-depositional destruction should not be skewing the results.

The faunal data were used to explore the timing and nature of changes in the local environment on two different scales. I first conducted an analysis of the habitat preference data, which is based on that fraction of the faunal remains which could be identified to genus and species, using NISP. The data were divided into three classes; animals that preferentially inhabit closed or semi-closed environments, those that primarily occupy open environments, and those which have no specific habitat requirements (see Table 2.4, which also includes NISP counts for each taxon). Data on habitat preferences were taken from Skinner and Chimimba (2005); "open" habitats include savanna and open woodland or woodland savanna, while "closed" or "semi-closed" habitats include forest (riverine or evergreen), thickets, and underbrush.

I also approached this issue by exploring variation in the ungulate size class data over time. This much larger dataset provides broader information about animal exploitation strategies, but can also serve as a proxy for environmental change over time, as the larger species represented in this assemblage also tend to be those which occupy more open environments. Given that this is predominantly a human produced assemblage, and because Sibudu's inhabitants may not have been selecting prey in direct proportion to their presence on the landscape, this dataset may not provide a direct reflection of the local environment; however, as discussed above, the floral and faunal data do indicate the same trends, indicating that the patterning evidenced in the faunal data is meaningful.

Because the goal for this study was to gain as high a resolution picture of the transition between the HP and post-HP MSA as possible, the three primary units of analysis (HP, post-HP MSA 2, post-HP MSA 1) were broken down as finely as sample sizes would allow. While the HP could be broken down into its three primary layers (Pinkish Grey Sand, Grey Sand, and Grey Rocky), sample sizes for the two post-HP MSA units were small enough that each could only be subdivided into an upper and lower unit (denoted in the figures as "pHP2 low" and "pHP2 upp" for the lower and upper halves of the post-HP MSA 2, etc.).

Results

Figure 2.3 presents the habitat preference data. The patterning evidenced among the initial faunal sample remained intact (c.f. Clark and Plug 2008; Clark 2009, 2011), with a vast majority of the remains from all of the HP layers coming from species that preferentially inhabit closed (and particularly forested) environments. Interestingly, while the difference in the proportion of open versus closed dwelling species was significant between layers Grey Sand and Grey Rocky, both of which are within the HP, the difference between the uppermost-HP layer (Grey Rocky) and the lower post-HP MSA 2 was not (Table 2.5). As a whole, it appears there was a relatively gradual increase in open-dwelling species over time. It is intriguing that a marked increase in open-dwelling species occurs at the boundary between the post-HP MSA 1 and 2, particularly given Herries' (2006) proposal that the transition between MIS 4 and 3 occurs at/around this boundary.

Before moving on, it is worth noting that the sample sizes for the HP are so much higher than those in the post-HP MSA in large part because of the species-level identification of the blue duiker, a small, forest dwelling bovid that is the most commonly identified taxon in the HP. To ensure that the broader pattern is not simply a reflection of changes in the presence of blue duiker across the sequence, the species was removed from the analysis; however, as indicated in Fig. 2.4, the pattern remained intact (see Table 2.5 for a further demonstration of the consistency of the results).

Turning now to the ungulate size class data (Fig. 2.5), which can serve as a rough proxy for environmental change, the data indicate a steady, stepwise decline in the relative proportion of small game that extends from the lowermost HP through the post-HP MSA. This trend continues right across the HP to post-HP MSA transition; once again, the greatest changes occur at the end of the post-HP MSA

Table 2.4 List of species present with habitat preferences (from Skinner and Chimimba 2005) and NISP counts for each species

Habitat preference	Taxa	HP (PGS)	HP (GS)	HP (GR)	post HP-MSA 2 (low)	post-HP MSA 2 (upp)	post-HP MSA 1 (low)	post-HP MSA 1 (upp)
Both	*Atilax paludinosus,* marsh mongoose	1	2	0	0	0	0	0
	Hystrix africaeaustralis, Cape porcupine	0	2	0	0	0	0	2
	Kobus ellipsiprymnus, waterbuck	0	0	4	0	0	0	1
	Oreotragus oreotragus, klipspringer	0	0	0	0	2	0	0
Total both		**1**	**4**	**4**	**0**	**2**	**0**	**3**
Closed or semi-closed	*Cephalophus natalensis,* red duiker	3	0	0	0	0	0	0
	Cercopithecus albogularis, Sykes' monkey	23	0	0	0	0	0	0
	Chlorocebus pygerythrus, vervet monkey	26	34	3	3	7	0	0
	Cricetomys gambianus, Gambian giant rat	29	16	5	0	1	0	0
	Philantomba monticola, blue duiker	662	358	234	23	18	0	2
	Potamochoerus larvatus, bushpig	117	114	68	45	6	2	0
	Redunca fulvorufula, mountain reedbuck	0	0	0	0	0	0	1
	Sylvicapra grimmia, common duiker	0	2	0	0	0	0	0
	Tragelaphus scriptus, bushbuck	4	3	0	2	0	0	0
Total closed/semi-closed		**864**	**527**	**310**	**73**	**32**	**2**	**3**
Open or open w/ shade	*Aepyceros melampus,* impala	0	2	5	0	1	0	0
	Connochaetes taurinus, blue wildebeest	0	0	2	0	0	2	4
	Damaliscus pygargus, blesbok	0	0	0	0	0	0	1
	Equus quagga, plains zebra	0	1	5	0	0	0	22
	Equus sp.	0	0	0	4	3	3	10
	Galerella sanguinea, slender mongoose	2	0	0	0	0	0	0
	Hippotragus equinus, roan antelope	0	4	1	0	0	0	0
	Hippotragus sp.	0	0	0	0	0	0	1
	Orcyteropus afer, aardvark	1	0	0	0	0	0	0
	Papio hamadryas, chacma baboon	10	4	1	0	0	0	0
	Pelea capreolus, grey rhebok	1	0	2	0	0	0	0
	Phacochoerus africanus, common warthog	0	0	5	6	4	4	3
	Procavia capensis, rock hyrax	9	8	11	4	2	1	2
	Pronolagus crassicaudatus, Natal red rock rabbit	0	2	0	2	0	0	0
	Pronolagus sp.	3	2	0	0	0	0	0
	Raphicerus campestris, steenbok	1	3	14	0	0	0	0
	Syncerus caffer, African buffalo	4	3	12	1	0	1	9
	Thyronomys swinderianus, greater canerat	1	0	3	0	3	0	0
	Tragelaphus oryx, eland	2	0	3	0	1	1	0
	Tragelaphus strepsiceros, kudu	0	0	0	0	0	2	6
Total open		**34**	**29**	**64**	**17**	**14**	**14**	**58**

Fig. 2.3 Habitat preference data (NISP)

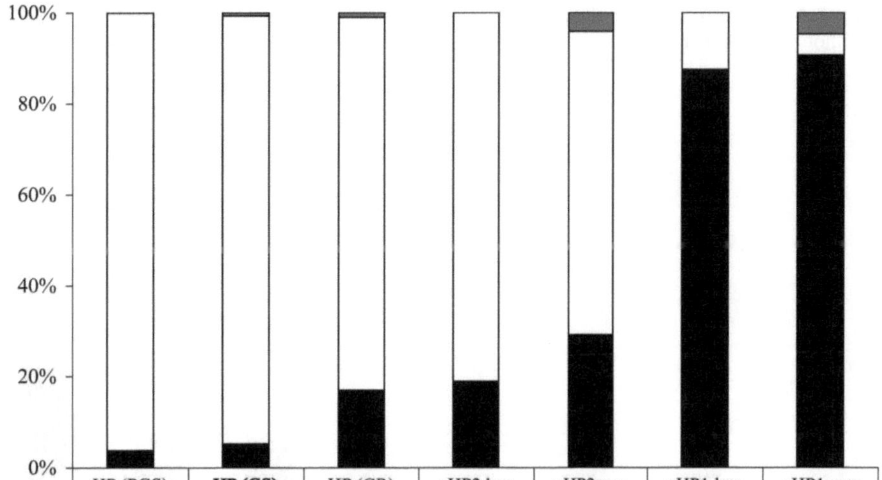

	HP (PGS)	**HP (GS)**	HP (GR)	pHP2 low	pHP2 upp	pHP1 low	pHP1 upp
■ Either	1	**4**	4	0	2	0	3
□ Closed/Semi-closed	864	**527**	310	73	32	2	3
■ Open	34	**29**	64	17	14	14	58

Table 2.5 Results of pair-wise statistical tests for habitat preference data presented in Figs. 2.3 and 2.4

	Full data set		Data set with blue duiker removed	
	χ^2 value	p value	χ^2 value	p value
HP (PGS) vs. HP (GS)	5.37	0.0682	2.36	0.3073
HP (GS) vs. HP (GR)	35.45	**<0.0001**	24.93	**<0.0001**
HP (GR) vs. pHP2 low	1.12	0.5712	4.59	0.1008
pHP2 low vs. pHP2 upp	6.08	**0.0478**	9.86	**0.0072**
pHP2 upp vs. pHP1 low	n/a[a]	**0.0001**	n/a[a]	**0.0177**
pHP1 low vs. pHP1 upp	n/a[a]	0.3381	n/a[a]	0.2678

Bolded p-values are significant at $\alpha = 0.05$

[a] Sample sizes too small for appropriate use of chi-squared test; Fisher's Exact Test applied instead

(Table 2.6; each pair-wise comparison showed a statistically significant shift in the proportional data, but the highest chi-squared value is associated with the transition from the lower to the upper post-HP MSA 1.

A glance at the numbers reveals that much of this patterning is driven by the smallest ungulates falling out of the sequence; since blue duiker are included in that group, I again removed them from the analysis, but as was the case with the habitat preference data, the general trend remained (Fig. 2.6; Table 2.6).

Discussion

In thinking about the implications of these results, one cannot overlook two primary caveats: (1) they are based primarily on faunal data, and the incorporation of finer-grained botanical data and isotopic data (analyses in progress) will be critical for further demonstrating the validity of the patterns presented here, and (2) data from other sites preserving the HP to post-HP MSA transition (such as Diepkloof) will be necessary for demonstrating the broader applicability of these findings. Notwithstanding these issues, I argue that these data have at least two important implications.

First, they suggest that the changes in material culture and technology at the end of the HP are perhaps *not* best modeled as an adaptive response to environmental change. This implies that we must seek alternative models for explaining the disappearance of the HP (as suggested by Jacobs and Roberts 2008). While recognizing that demographic and/or social models for behavioral change during the MSA will be much more difficult to test than ecological ones, it seems that the time is ripe for us to focus on fleshing out alternative models, particularly when it comes to developing testable implications for these models. While this is not the place for such an endeavor, it is worth noting that faunal data may also be useful for addressing the models which imply that the HP to post-HP MSA transition

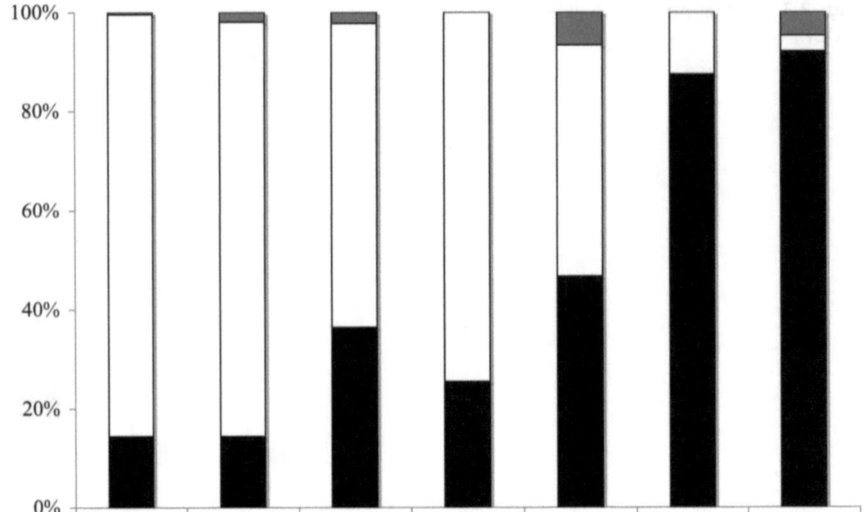

	HP (PGS)	HP (GS)	HP (GR)	pHP2 low	pHP2 upp	pHP1 low	pHP1 upp
Either	1	4	4	0	2	0	3
Closed/Semi-closed	202	169	108	50	14	2	2
Open	34	29	64	17	14	14	58

Fig. 2.4 Habitat preference data (without blue duiker; NISP)

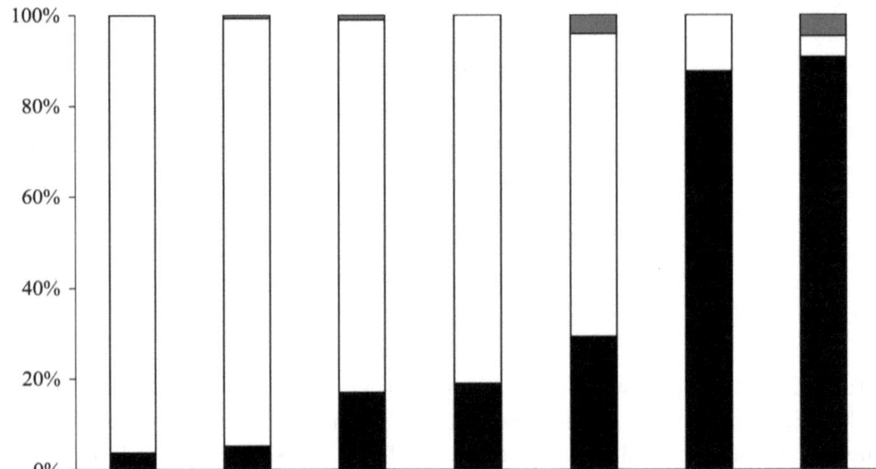

	HP (PGS)	HP (GS)	HP (GR)	pHP2 low	pHP2 upp	pHP1 low	pHP1 upp
Either	1	4	4	0	2	0	3
Closed/Semi-closed	864	527	310	73	32	2	3
Open	34	29	64	17	14	14	58

Fig. 2.5 Ungulate size class data (NISP)

Table 2.6 Results of pair-wise statistical tests for ungulate size class presented in Figs. 2.5 and 2.6

	Full data set		Data set with blue duiker removed	
	χ^2 value	p value	χ^2 value	p value
HP (PGS) vs. HP (GS)	64.39	**<0.0001**	31.41	**<0.0001**
HP (GS) vs. HP (GR)	38.69	**<0.0001**	25.89	**<0.0001**
HP (GR) vs. pHP2 low	64.77	**<0.0001**	13.89	**0.001**
pHP2 low vs. pHP2 upp	6.68	**0.0354**	8.26	**0.0161**
pHP2 upp vs. pHP1 low	29.19	**<0.0001**	19.98	**<0.0001**
pHP1 low vs. pHP1 upp	68.73	**<0.0001**	69.42	**<0.0001**

Bolded p-values are significant at $\alpha = 0.05$

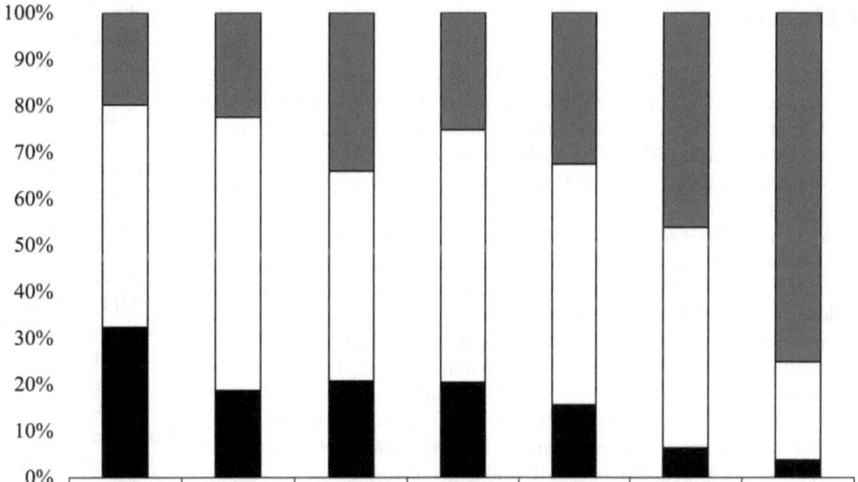

	HP (PGS)	HP (GS)	HP (GR)	pHP2 (low)	pHP2 (upp)	pHP1 (low)	pHP1 (upp)
■ Bov III + (+equids)	135	135	209	167	130	124	442
□ Bov II (+ suids)	325	353	277	360	207	127	124
■ Bov I	220	112	127	135	62	17	22

Fig. 2.6 Ungulate size class data (without blue duiker; NISP)

reflects population replacement (per Bar-Yosef 2002; Mellars 2006), as one might expect a different population using different technologies to also employ different animal exploitation strategies, not only in terms of what prey was captured, but also in terms of transport and processing strategies. And yet, as demonstrated by Adler and colleagues (Adler et al. 2006; Adler and Bar-Oz 2009), the exploitation strategies of late Middle Paleolithic and Early Upper Paleolithic populations (presumably representing Neanderthals and anatomically modern humans) were often times remarkably similar. There is no doubt that this is a thorny issue, but it is clearly important to our broader understanding of the adaptive capabilities and strategies of Late Pleistocene human populations.

The second implication also relates to the development and use of alternative models for technological change. Despite the seemingly major changes in hunting weaponry between the HP and the post-HP MSA—perhaps as dramatic as use of the bow and arrow and remote capture technology during the HP versus the use of hand-thrust or cast stone-tipped spears during the post-HP MSA—there do not appear to be any major changes in basic subsistence behavior across the transition between the phases. If the changes in hunting technology are not a function of different subsistence practices or needs, this implies that we may also need to focus more attention on social models for explaining technological change during the MSA.

Conclusions

Environmental change often serves as a default explanation for marked changes in the archaeological record during the Later Pleistocene. Given the apparent chronological association between the HP and a glacial period (MIS 4), the innovations evidenced during the HP have often been modeled as an adaptive response to a particular set of environmental conditions. However, high-resolution faunal data from HP and post-HP MSA deposits at Sibudu Cave indicate a marked disconnect in the timing and nature of changes in the local environment relative to the major changes in technology and material culture evidenced at the end of the HP. While not denying the important role that climate change can play in shaping human behavior, I argue that these data imply that we need to focus more attention on developing demographic and/or social explanations for behavioral change during the period—and particularly on deriving testable implications for these models.

Acknowledgments First and foremost, I would like to thank Lyn Wadley for her continued support—one could not ask for a better project director/colleague/friend. I also thank Grant McCall and the two anonymous reviewers for their useful commentary. Any mistakes that remain, are, of course, my own. Funding for this work was provided by the Leakey Foundation, the Palaeontological Scientific Trust, the National Science Foundation (DDIG # 0612606), and various units at the University of Michigan.

References

Adler, D. S., & Bar-Oz, G. (2009). Seasonal patterns of prey acquisition and inter-group competition during the Middle and Upper Paleolithic of the Southern Caucasus. In J.-J. Hublin & M. P. Richards (Eds.), *The evolution of hominin diets: Integrating approaches to the study of Palaeolithic subsistence* (pp. 127–140). Berlin: Springer.

Adler, D. S., Bar-Oz, G., Belfer-Cohen, A., & Bar-Yosef, O. (2006). Ahead of the game: Middle and Upper Paleolithic hunting behaviors in the Southern Caucasus. *Current Anthropology, 47*(1), 89–118.

Allott, L. F. (2006). Archaeological charcoal as a window on palaeovegetation and wood-use during the Middle Stone Age at Sibudu Cave. *Southern African Humanities, 18*(1), 173–201.

Ambrose, S. H., & Lorenz, K. G. (1990). Social and ecological models for the Middle Stone Age in Southern Africa. In P. Mellars (Ed.), *The emergence of modern humans: An archaeological perspective* (pp. 3–33). Ithica, NY: Cornell University Press.

Backwell, L., d'Errico, F., & Wadley, L. (2008). Middle Stone Age bone tools from the Howiesons Poort layers, Sibudu Cave, South Africa. *Journal of Archaeological Science, 35*(6), 1566–1580.

Bar-Yosef, O. (2002). The Upper Paleolithic revolution. *Annual Review of Anthropology, 31*, 363–393.

Chase, B. M. (2010). South African paleoenvironments during marine oxygen isotope stage 4: A context for the Howiesons Poort and Still Bay industries. *Journal of Archaeological Science, 37*, 1359–1366.

Clark, J. L. (2009). *Testing models on the emergence and nature of modern human behavior: Middle Stone Age Fauna from Sibudu Cave (South Africa)*. Unpublished Ph.D. thesis, University of Michigan, Ann Arbor.

Clark, J. L. (2011). The evolution of human culture during the later Pleistocene: Using fauna to test models on the emergence and nature of "modern" human behavior. *Journal of Anthropological Archaeology, 30*, 273–291.

Clark, J. L., & Ligouis, B. (2010). Burned bone in the Howieson's Poort and post-Howieson's Poort Middle Stone Age deposits at Sibudu (South Africa): Behavioral and taphonomic implications. *Journal of Archaeological Science, 37*, 2650–2661.

Clark, J. L., & Plug, I. (2008). Animal exploitation strategies during the South African Middle Stone Age: Howiesons Poort and post-Howiesons Poort fauna from Sibudu Cave. *Journal of Human Evolution, 54*(6), 886–898.

Cochrane, G. W. G. (2006). An analysis of lithic artefacts from the ~60 ka layers of Sibudu Cave. *Southern African Humanities, 18*(1), 69–88.

Glenny, W. (2006). Report on the micromammal assemblage analysis from Sibudu Cave. *Southern African Humanities, 18*(1), 279–288.

Goldberg, P., Miller, C. E., Schiegl, S., Ligouis, B., Conard, N. J., & Wadley, L. (2009). Bedding, hearths and site maintenance in the Middle Stone Age at Sibudu, KwaZulu-Natal, South Africa. *Archaeological and Anthropological Sciences, 1*, 95–122.

Grayson, D. K. (1984). *Quantitative zooarchaeology: Topics in the analysis of archaeological faunas*. Orlando: Academic Press.

Grayson, D. K., & Frey, C. J. (2004). Measuring skeletal part representation in archaeological faunas. *Journal of Taphonomy, 2*(1), 27–42.

Hall, G., Woodborne, S., & Scholes, M. (2008). Stable carbon isotope ratios from archaeological charcoal as palaeoenvironmental indicators. *Chemical Geology, 247*, 384–400.

Herries, A. I. R. (2006). Archaeomagnetic evdience for climate change at Sibudu Cave. *Southern African Humanities, 18*(1), 131–147.

Jacobs, Z., & Roberts, R. G. (2008). Testing times: Old and new chronologies for the Howieson's Poort and Still Bay Industries in environmental context. *South African Archaeological Society Goodwin Series, 10*, 9–34.

Jacobs, Z., Roberts, R. G., Galbraith, R. F., Deacon, H. J., Grun, R., Mackay, A., et al. (2008). Ages for the Middle Stone Age of Southern Africa: Implications for human behavior and dispersal. *Science, 322*, 733–735.

Lombard, M. (2007). The gripping nature of ochre: The association of ochre with Howiesons Poort adhesives and Later Stone Age mastics from South Africa. *Journal of Human Evolution, 53*(4), 406–419.

Lombard, M. (2008). Finding resolution for the Howiesons Poort through the microscope: Micro-residue analysis of segments from Sibudu Cave, South Africa. *Journal of Archaeological Science, 35*(1), 26–41.

Lombard, M., & Pargeter, J. (2008). Hunting with Howiesons Poort segments: pilot experimental study and the functional interpretation of archaeological tools. *Journal of Archaeological Science, 35*, 2523–2531.

Lombard, M., & Phillipson, L. (2010). Indicators of bow and stone-tipped arrow use 64,000 years ago in KwaZulu-Natal, South Africa. *Antiquity, 84*(325), 635–648.

Lyman, R. L. (2008). *Quantitative paleozoology*. New York: Cambridge University Press.

Marshall, F., & Pilgram, T. (1993). NISP vs. MNI in quantification of body-part representation. *American Antiquity, 58*(2), 261–269.

McCall, G. S. (2007). Behavioral ecological models of lithic technological change during the later Middle Stone Age of South Africa. *Journal of Archaeological Science, 34*(10), 1738–1751.

Mellars, P. (2006). Why did modern human populations disperse from Africa ca. 60,000 years ago? A new model. *Proceedings of the National Academy of the Sciences of the United States of America, 103*(25), 9381–9386.

Parkington, J. E., Poggenpoel, C., Rigaud, J.-P., & Texier, P.-J. (2005). From tool to symbol: The behavioural context of intentionally marked ostrich eggshell from Diepkloof, Western Cape. In F. d'Errico & L. Backwell (Eds.), *From tools to symbols: From early hominids to modern humans* (pp. 475–492). Johannesburg: Witwatersrand University Press.

Pickering, R. (2006). Regional geology, setting and sedimentology of Sibudu Cave. *Southern African Humanities, 18*(1), 123–129.

Schiegl, S., & Conard, N. J. (2006). The Middle Stone Age sediments at Sibudu: Results from FTIR spectroscopy and microscopic analyses. *Southern African Humanities, 18*(1), 149–172.

Skinner, J. D., & Chimimba, C. T. (2005). *The mammals of the Southern African sub-region*. Singapore: Cambridge University Press.

Texier, P.-J., Porraz, G., Parkington, J., Rigaud, J.-P., Poggenpoel, C., Miller, C. E., et al. (2010). A Howiesons Poort tradition of engraving ostrich eggshell containers dated to 60,000 years ago at Diepkloof Rock Shelter, South Africa. *Proceedings of the National Academy of Sciences of the United States of America, 107*, 6180–6185.

Wadley, L. (2006). Partners in grime: Results of multi-disciplinary archaeology at Sibudu Cave. *Southern African Humanities, 18*(1), 315–341.

Wadley, L. (2010). Were snares and traps used in the Middle Stone Age and does it matter? A review and case study from Sibudu, South Africa. *Journal of Human Evolution, 58*(2), 179–192.

Wadley, L., & Jacobs, Z. (2006). Sibudu Cave: Background to the excavations, stratigraphy and dating. *Southern African Humanities, 18*(1), 1–26.

Wadley, L., & Mohapi, M. (2008). A segment is not a monolith: Evidence from the Howiesons Poort of Sibudu, South Africa. *Journal of Archaeological Science, 35*(9), 2594–2605.

Wurz, S., & Lombard, M. (2007). 70,000-year-old geometric backed tools from the Howiesons Poort at Klasies River, South Africa: Were they used for hunting? *Southern African Humanities, 19*, 1–16.

Chapter 3
Middle Paleolithic Large-Mammal Hunting in the Southern Levant

John D. Speth

Introduction

A lot has been written about the "Broad Spectrum Revolution" (BSR) or, as it is sometimes called, the "Broadening Food Spectrum," since the idea was first introduced over four decades ago by Kent Flannery (1969). As originally envisioned, the BSR denotes a fairly rapid shift in resource exploitation strategies toward the end of the Pleistocene stemming from an imbalance between available resources and the number of mouths to be fed. This imbalance was the result of an ongoing influx into marginal habitats of "daughter" groups that had budded off from growing "donor" populations in richer "core" areas. The immigration of surplus population into the less productive habitats gave rise to resource stresses to which the "recipient" populations responded by intensifying their reliance on lower-ranked plant and animal foods, and on more labor-intensive or costly methods of procuring and processing these foods. In the Near East these changes in foodways are generally seen as critical initial steps that ultimately led to the domestication of plants and animals and the emergence of village-based farming economies (e.g., Stiner et al. 2000; Stiner 2001; Piperno et al. 2004; Weiss et al. 2004a, b; Munro 2003, 2004, 2009a, b; Stutz et al. 2009; Speth 2010a).

Archaeologists often see the BSR, a stress-related response by foragers to an imbalance between available resources and population, as roughly synonymous with increasing diet breadth—that is, an expansion of the diet under increasing levels of stress to include a wider range of food resources with lower net return rates, such as small mammals, birds, marine mollusks, terrestrial snails, and a host of comparatively labor-intensive wild plant foods, including the wild progenitors of wheat and barley (Flannery 1969; Winterhalder 1981). Unlike most traditional diet-breadth models in behavioral ecology, however, archaeologists also recognize that the BSR entailed an increase in the intensity of food-processing, incorporating techniques like parching, grinding, winnowing, roasting, baking, boiling, and undoubtedly a host of other methods as well, many currently invisible to archaeology, designed to enhance the nutritional yield of these lower-order, sometimes poorly digestible or even toxic food resources.

In his original paper, Flannery (1969: 74, 77) suggested that the BSR got underway roughly midway through the Upper Paleolithic, sometime prior to about 22 kyr. Many subsequent studies tended to shift the focus of attention upward in time, placing the onset of the BSR closer to the end of the Pleistocene, which, in the Southern Levant, would be during the Epipaleolithic, and especially toward the end of this slice of the Near Eastern archaeological record, the Natufian, now dated between ~17 and ~12 cal kBP (Hayden 1981; Blockley and Pinhasi 2011: 99).

During the past decade or so archaeologists have presented new evidence, both faunal and botanical, which suggests that the BSR may already have been underway before the Natufian, probably beginning during the early stages of the Epipaleolithic, or even further back into the preceding UP (e.g., Stiner 2009a; Stutz et al. 2009)—in other words, more or less in the timeframe Flannery had originally suggested (but see Bar-Oz et al. 1999, and Bar-Oz and Dayan 2003 for recent taphonomic studies of faunal remains from Israeli Epipaleolithic sites which do not seem to show the expected dietary trend). Perhaps the most compelling evidence in the Southern Levant for significant Late Pleistocene broadening of the subsistence base comes from the ~23 kyr Israeli site of Ohalo II, with its abundant remains of wild cereals, fish, birds, and other low-ranking or "second-order" resources (Richards et al. 2001; Piperno et al. 2004; Weiss et al. 2004a, b; see also Stiner 2003 and Aranguren et al. 2007 for comparably early evidence from Europe).

J. D. Speth (✉)
Museum of Anthropology, University of Michigan, Ann Arbor, MI 48109-1079, USA
e-mail: jdspeth@umich.edu

Jamie L. Clark and John D. Speth (eds.), *Zooarchaeology and Modern Human Origins: Human Hunting Behavior during the Later Pleistocene*, Vertebrate Paleobiology and Paleoanthropology, DOI: 10.1007/978-94-007-6766-9_3,

Much of the faunal work that has been conducted since the concept of the BSR was first introduced has been concerned primarily with changes in the proportion of large versus small taxa. Zooarchaeologists have also looked at the frequency of juvenile versus adult individuals (e.g., Davis 1983; Bar-Oz et al. 1999; Bar-Oz 2004; Atici 2009; Munro 2009a, b). More recently, Mary Stiner, Natalie Munro, and colleagues have pointed out that the BSR involved more than simply incorporating more small-game animals into the diet, it involved a switch toward more difficult-to-catch small game, such as hares and birds, and in some areas toward greater use of freshwater and marine resources (Richards et al. 2001; Stiner et al. 2000; Stiner 2001, 2009a; Munro 2003, 2004, 2009a, b; see also Stiner and Munro 2011 for an interesting recent study of this same process in Greece).

There is now limited but intriguing evidence that the BSR in the Southern Levant may actually have gotten underway even earlier than the early- to mid-UP date that nowadays is commonly accepted. In 2004 and 2006 Jamie L. Clark and I (Speth 2004a, 2010a; Speth and Clark 2006) suggested, largely on the basis of data from Kebara Cave (Israel), that in the Southern Levant the beginning of this process of resource intensification, reflected in particular by the overhunting of the largest available ungulates and the targeting of adults and juveniles of smaller ungulate taxa, may have gotten underway as early as 50–55 kyr during the latter part of the MP—in other words, the BSR, in this region at least, may have been initiated by Neanderthals—and by their quasi-contemporary Anatomically Modern Human neighbors who appear to have been hunting the same basic suite of ungulates—well in advance of the timeframe usually envisioned by archaeologists, a possibility already anticipated nearly a quarter of a century ago by Simon Davis and colleagues (1988). The suggestion that MP hominins in the Near East might have overhunted their big-game resources certainly seemed quite far out on the proverbial limb when Davis first suggested the possibility in 1988, and remained so as recently as 2006 when Jamie Clark and I published a more detailed look at the same issue using data from the same site. In an academic environment that saw Neanderthals as a different species from our own, and a decidedly inferior one at that—an "archaic" hominin who opportunistically scavenged already dead carcasses rather than a strategizing hunter of big game—the idea that the BSR, a process which ultimately led to the origins of agriculture, could have been initiated by a less-than-human ancestor seemed unlikely, if not downright preposterous. But in the last few years the intellectual environment has undergone a radical transformation. Almost no one today questions whether Neanderthals and their contemporaries across Asia could hunt. We now see them as formidable hunters, capable of killing the biggest and most dangerous animals on the Pleistocene landscape (Bratlund 2000; Gaudzinski 1998, 2000, 2006; Speth and Tchernov 2001, 2007; Adler et al. 2006; Adler and Bar-Oz 2009; Gaudzinski-Windheuser and Niven 2009; Stiner 2009b; Zhang et al. 2009). In fact, largely on the basis of nitrogen isotope data, it has become fashionable nowadays to see Neanderthals as "top predators," right up there with cave lions and hyenas (Bocherens 2009, 2011). Even more startling is the new genetic evidence that challenges the very foundations of the view that placed Neanderthals into a separate species of hominin. What may have gone extinct were *populations* of humans, not an entire species, and their disappearance left more than caves filled with artifacts and animal bones; Eurasians carry their genes as well (Garrigan and Kingan 2007; Hawks et al. 2008; Wall et al. 2009; Green et al. 2010; Yotova et al. 2011). So the idea that these "top predators" might have set the stage for the BSR by overexploiting their large-mammal resources seems less farfetched today than it did only a few short years ago.

The major obstacle now to entertaining this sort of scenario is the common assumption that Neanderthal populations, as well as quasi-contemporary populations of Anatomically Modern Humans who likely occupied the same region throughout much of the 200–250 kyr span of the MP, were too small, too mobile, and too widely scattered to have had any detectable impact on their prey populations (Kuhn and Stiner 2006; Shea 2008). But how small is small? Smaller than Clovis (early North American Paleoindian) populations that many have argued were directly or indirectly responsible for the extinction of mammoths and other megafauna in late Pleistocene North America (e.g., Alroy 2001; Haynes 2002; Johnson 2002; Barnosky et al. 2004; Brook and Bowman 2004; Lyons et al. 2004; Martin 2005; Surovell et al. 2005)? Smaller than Aboriginal populations in the arid central and western deserts of Australia during the Late Pleistocene who, not long after their entry into Sahul, may have driven their own megafauna to extinction armed with only wooden-tipped spears, and aided perhaps by fire (White 1977: 26; Roberts et al. 2001; Turney et al. 2001; Kershaw et al. 2002)? While we lack solid estimates of just how big or small Neanderthal populations might have been, and their numbers undoubtedly varied considerably across both space and over time, there is evidence, albeit limited and indirect, that in Europe, at least, Middle Paleolithic populations did grow toward the end of the MP and that this growth did not simply track the paleoclimatic record (e.g., Richter 2000; Lahr and Foley 2003; Zilhão 2007; Bocquet-Appel and Tuffreau 2009).

After about 90–100 kyr, Levantine Neanderthal populations (and perhaps those of Anatomically Modern Humans) may have grown as well. Thus, not only does the number of late MP sites increase, but many cave sites show a marked increase in the depth of their culture-bearing deposits, reflecting recurrent visits by their human occupants to the same locality. Many of these sites also display significantly greater densities of artifacts and faunal remains, as well as hearths stacked one upon another, many showing clear evidence of rebuilding and reuse, raked-out ash lenses and ash dumps, and actual trash middens along the perimeter of the living areas (e.g., Hovers 2006; Meignen et al. 2006; Speth 2006; Shea 2008: 2264; Speth et al. 2012). And finally, burials of both Neanderthals and Anatomically Modern Humans, including multiple interments, are unusually common in the Levant by comparison to other regions inhabited by MP hominins, and may well point to larger populations, greater residential stability, emerging corporateness, and increasing delineation of territorial space:

> Aquitaine and the Levant contain relatively large numbers of burials as well as places of multiple burial, which might suggest that burial was practised more widely in these areas, and that, by contrast, Neanderthals in other regions either did not bury their dead, or did not practise it frequently. These are regions where Mousterian archaeology suggests that Neanderthals were particularly numerous, and it is tempting to suggest that the practice of burial may have been connected to population size, and perhaps to a sense of territoriality (Pettitt 2010: 130).

It is intriguing to see how our preconceived notions color the way we interpret data of this sort. In eastern North America, archaeologists who deal with the Early and Middle Archaic, periods which fall squarely within the Holocene and are therefore unquestionably the product of fully modern humans like ourselves, almost universally see this sort of trend as compelling evidence for increasing populations, tighter packing of territories within a region, and declining overall levels of residential mobility, reflected most tellingly by limited use of non-local lithic raw materials, but also, at least initially, by the scarcity of other material evidence for intergroup interaction and exchange (Ford 1974; Speth 2004b). An analogous pattern is nicely documented in Australia by Pardoe (1988, 1994, 1995) using skeletal morphology and other data. So why is a scenario of this nature not even on the table for discussion when we think about late MP hominins? And if Neanderthals, as so many would argue, were "top predators," killing aurochs, bison, mammoth, reindeer, horses, rhinos, cave bears, wild boar, and other large, very dangerous animals, and at times using communal methods to do so (Gaudzinski-Windheuser and Niven 2009; Rendu et al. 2012), what is so implausible about their having been sufficiently effective at what they did to be capable of overexploiting them, especially the largest ones with very slow reproductive rates?

And what if MP hominins used fire to manipulate or manage their landscape and resources, much as virtually all historically and ethnographically documented hunters and gatherers did, and in some places still do (e.g., Lewis 1982)? Based on both palynological data and the frequency of microscopic charcoal particles in ocean cores, *as early as* 50 kyr landscape burning as a form of resource management may have been employed by small hunting and gathering populations as they colonized Southeast Asia, New Guinea, and Australia (Fairbairn et al. 2006; Barker et al. 2007: 256, 257), and anthropogenic fire may even have played a role in resource management and megafaunal extinctions during the Paleoindian colonization of the Americas (Pinter et al. 2011). If Neanderthals did, in fact, use fire in this manner, their impact on wildlife populations could have been substantial even if their own populations were very small. All of this is quite speculative, of course, but not beyond the realm of plausibility. Daniau et al. (2010: 7), in fact, explored this issue in Western Europe using long-term charcoal records. While they found no compelling evidence that either Neanderthals or their modern human successors made intensive use of fire to manage their resources, they do not rule it out either:

> At a macro level at least, the colonisation of Western Europe by Anatomically Modern Humans did not have a detectable impact on fire regimes. This, however, does not mean that Neanderthals and/or Modern Humans did not use fire for ecosystem management but rather that, if this were indeed the case, the impact on the environment of fire use is not detectable in our records, and was certainly not as pronounced as it was in the biomass burning history of Southeast Asia.

In the Levant there is fairly convincing evidence of humanly manipulated landscape burning extending back at least into the Epipaleolithic (Emery-Barbier and Thiébault 2005; Turner et al. 2010), but whether humans were using fire as a tool to manage their local resources prior to this remains an open question, one well worth exploring further.

Of course, even if it can be shown with some degree of certainty that the large-game resources exploited by Kebara's Neanderthal inhabitants declined over time, this does not demonstrate that the change was the result of human overexploitation relative to what was available to them on the landscape. Rather than overhunting, the decline could reflect climate-driven environmental changes that reduced local game populations. However, as I will attempt to show momentarily, paleoclimatic data, particularly isotopic studies of Israeli speleothems, reveal no obvious correlation between major climatic fluctuations and the proportional representation of the larger taxa in the Kebara assemblage.

A more difficult problem to resolve is whether the late MP decline in the larger taxa is strictly a local phenomenon to be found only at Kebara and its immediate environs, or one that affected the region more generally, and further work elsewhere in the Levant could, in fact, show that it did not occur farther afield. Fortunately, this is a potentially tractable matter that can be addressed through more regionally focused comparative studies of MP faunas. This will not be an easy or straightforward task, however, as only a few other MP assemblages have been published in detail as yet, and these are far from sufficient to cover the entire region or the more than 200 kyr-span of the MP.

Moreover, there are many other factors that can intervene to complicate matters. For example, even if overhunting were, in fact, widespread in the region toward the close of the MP, the impact of such intensified hunting practices may not become evident everywhere at the same time, or at the same rate, or to the same extent (some of the more important reasons for variability of this sort are discussed at the end of the paper). In addition, depletion of large-game resources at the regional scale may be difficult to see if the sites that are being compared were occupied at different times of the year, and especially if the occupations represent different functional poses within their respective settlement systems (e.g., basecamps vs. short-term hunting stations or transitory camps). Differences in site function may become particularly problematic if comparisons are made between open-air hunting locations situated close to fixed watering points where large game could be ambushed, such as Biq'at Quneitra (Davis et al. 1988; Goren-Inbar 1990) or Far'ah II (Gilead and Grigson 1984), and basecamp occupations in caves such as Kebara or Amud (Hovers et al. 1991; Rabinovich and Hovers 2004) to which game had to be transported. Ethnographic studies among the Hadza (Tanzania) and the Kalahari San or Bushmen (Botswana/Namibia) show that remains of large game are likely to be over-represented in open-air ambush localities, while easily transportable small game will be more evident in cave sites (O'Connell et al. 1988, 1990; Brooks 1996). The contrast between caves and open-air kill localities is likely to be even more pronounced in sites predating the UP, the earliest time period for which we have reasonably convincing evidence in the form of fire-cracked rock that hunters began to transport large-mammal vertebrae and other bones of low meat or marrow utility back to camp expressly for the purpose of grease-rendering (Stiner 2003; Speth 2010a).

So my goal in this paper is quite modest. What I hope to show is that Kebara provides reasonably clear evidence for a steady decline after about 55–60 kyr in the proportional representation of large game, particularly red deer (*Cervus elaphus*) and aurochs (*Bos primigenius*), and a concomitant increase in the importance of gazelles (*Gazella gazella*) and fallow deer (*Dama mesopotamica*), including juveniles of these taxa; and that these changes are very likely unrelated or at best only loosely driven by climatic fluctuations during the Late Pleistocene. We must await future zooarchaeological work elsewhere in the Southern Levant and beyond to decide whether Kebara's sequence is an isolated occurrence, or instead part of a broader trend brought about either by (1) growing human populations (via increases in fertility and/or survivorship of autochthonous residents, or the influx of MP hominins from more northerly regions of Eurasia where MIS 4 climatic conditions were deteriorating, as suggested many years ago by Bar-Yosef 1995: 516); or (2) by a significant change in the way Levantine hominins hunted large animals (e.g., the introduction of new hunting weapons, or a shift from reliance on solitary to communal methods of procurement, or possibly even the increased use of fire to procure or manage their plant and animal resources); or (3) some combination of these.

Kebara and Its Fauna

Kebara is a large cave on the western face of Mt. Carmel (Israel), about 30 km south of Haifa and 2.5 km east of the present-day Mediterranean shoreline (Fig. 3.1). Two major excavations at the site, the first conducted by Moshe Stekelis between 1951 and 1965 (Schick and Stekelis 1977), the second by a French-Israeli team co-directed by Ofer Bar-Yosef and Bernard Vandermeersch between 1982 and 1990 (Bar-Yosef 1991; Bar-Yosef et al. 1992), yielded many thousands of animal bones and stone tools from a 4-m deep

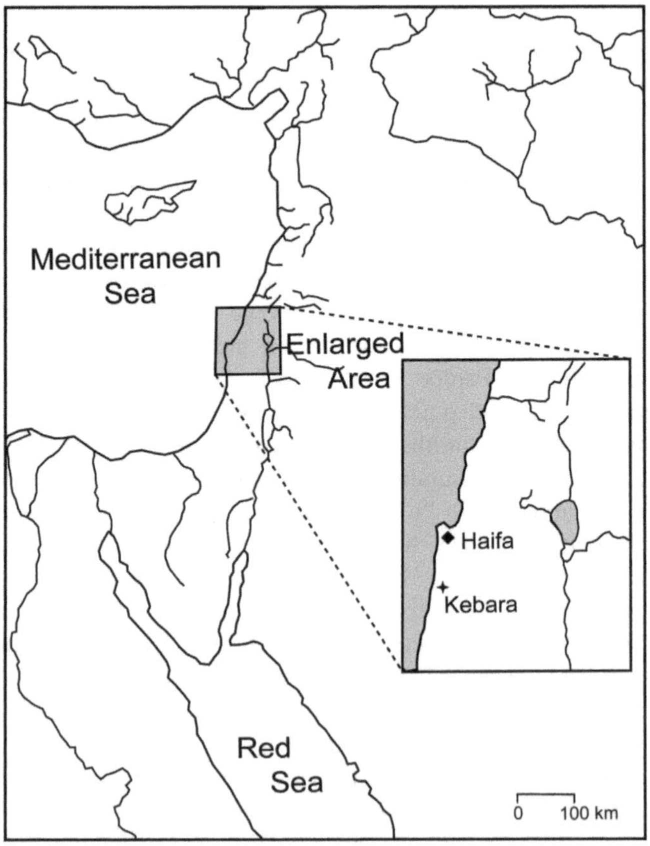

Fig. 3.1 Map of the Levant showing the location of Kebara Cave

sequence of MP deposits dating between approximately 60 and 48 kyr (Valladas et al. 1987).

Stekelis's excavations were conducted within 2 × 2-m grid squares using arbitrary horizontal levels or "spits," typically 10 or 20 cm in thickness. Almost all of the excavated deposits were screened and all faunal material, including unidentifiable bone fragments, were kept. Depths for levels were recorded in cm below a fixed datum. In the more recent work at the site, directed by Bar-Yosef and Vandermeersch, the excavators employed 1-m grid units (often divided into smaller subunits), many items (including fauna) were piece-plotted, and wherever possible they followed the natural stratigraphy of the deposits, using levels that seldom exceeded 5 cm in thickness. Depths were again recorded in cm below datum, using the same reference point that Stekelis had used. The newer excavations recognized nine natural stratigraphic levels (levels or *couches*) within the Mousterian sequence: level XIII (bottom) to level V (top). The Early Upper Paleolithic (EUP) sequence begins with level IV.

One of the problems that has plagued the analysis of the Kebara fauna from the outset is how to make effective use of all of the faunal materials recovered by both Stekelis and Bar-Yosef-Vandermeersch. The problem stems from their differing strategies for handling the site's complex stratigraphy. As already noted, Stekelis, like many of his contemporaries in the 1950s, excavated the site using relatively thick, arbitrary, horizontal spits; whereas, the more recent work used much thinner levels that for the most part tracked the natural, sometimes irregular or discontinuous, sometimes sloping, stratigraphy of the deposits. It is clear therefore that Stekelis's recovery methods pooled bones that in reality derived from stratigraphically different layers within the deposits. In an ideal world, we would be best off ignoring these materials altogether and working exclusively with the collections from the more recent excavations. Unfortunately, were we to do so, many of our analyses would be impossible because the Bar-Yosef-Vandermeersch assemblage, by itself, is simply too small.

Fortunately, both excavations used the same datum point to record the depth or z-coordinate of the artifacts and bones. So, for a number of our analyses we simply divide the depths into arbitrary half-meter increments. While admittedly crude, these arbitrarily pooled levels should suffice to reveal temporal patterns that are robust, of large magnitude, and unfold over long periods of time.

But for certain questions, especially those that seek to track change over time in the procurement and handling of the largest ungulates, or just the juveniles, even half-meter increments at times are precluded by small sample sizes. For these questions, we need to split the total MP assemblage into just two stratigraphic subsets or groups, a "lower" one and an "upper" one. In order to do this, however, we need a basis for deciding at what depth below datum to place the boundary between the two groups. As documented in previous publications (Speth 2006; Speth and Tchernov 2007), the Kebara MP sequence can be divided into a "midden period" (levels XI-IX), with evidence of intensive hunting that was most probably concentrated in the cooler months of the year, and the latest MP levels (levels VII-V), with little or no evidence of midden accumulation and less evidence of hunting activity (Speth and Clark 2006). This leaves us with the problem of where to place level VIII, which produced a small faunal assemblage that in many respects is transitional between the midden period and the later occupations. Although the decision is somewhat arbitrary, and I have vacillated over the years on this issue, my inclination is to group the bones from this level with the preceding midden period because it is not until level VII that midden accumulation dwindled to the point that it is no longer easily detectable in the composition or spatial patterning of the site's faunal remains. Thus, in temporal comparisons where I need to maximize the size of the samples, I dichotomize the MP material into two stratigraphic groups, the "lower" one consisting of levels XIII-VIII, the "upper" one consisting of levels VII-V.

That still leaves the question of how to link the Stekelis materials with the faunal remains recovered by the Bar-Yosef/Vandermeersch excavations. In other words, what should we use as an approximate depth below datum for the boundary between the "lower" and "upper" MP groups, knowing that the MP deposits are not entirely horizontal throughout the site? My solution to this problem is again somewhat crude, though relatively straight-forward (see Speth and Clark 2006). I generated two histograms, one showing the frequency distribution of depths below datum for the MP bones in Bar-Yosef and Vandermeersch's "lower" levels (XIII-VIII), the other showing the spread of depths for the material in their "upper" levels (VII-V). Fortunately, there is remarkably little overlap between the two histograms, with the boundary (*at least in those parts of the site that Bar-Yosef and Vandermeersch were able to sample*) lying at a depth of about 550 cm below datum. Thus, crude as it might be, the 550-cm figure will serve as the dividing point between the "lower" and "upper" MP assemblages in those temporal analyses where it is necessary to combine the Stekelis and Bar-Yosef/Vandermeersch assemblages.

I evaluate statistical significance using three methods: for the difference between two percentages, I use the arcsine transformation (t_s), as defined by Sokal and Rohlf (1969: 607–610); to evaluate differences between means I use standard unpaired t-tests (t); and to assess correlation I use the non-parametric Spearman's rho (r_s).

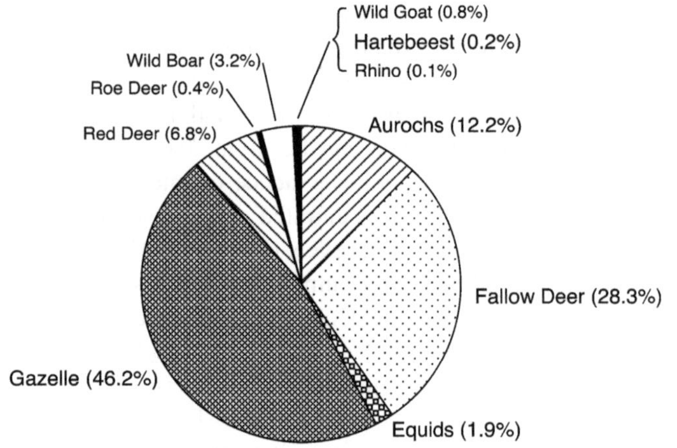

Fig. 3.2 Taxonomic composition of Kebara's MP ungulate fauna (% of total ungulate NISP)

Most of the larger mammal remains in Kebara's MP deposits, expressed as percentages of total ungulate NISP values, derive from just two taxa—mountain gazelle (*G. gazella*, 46.2 %) and Persian fallow deer (*D. mesopotamica*, 28.3 %). Other animals, represented by smaller numbers of specimens (also using NISP), include aurochs (*Bos primigenius*, 12.2 %), red deer (*Cervus elaphus*, 6.8 %), wild boar (*Sus scrofa*, 3.2 %), small numbers of equid remains (1.9 %), very likely from more than one species, wild goat (*Capra* cf. *aegagrus*, 0.8 %), roe deer (*Capreolus capreolus*, 0.4 %), hartebeest (*Alcelaphus bucephalus*, 0.2 %), and rhinoceros (*Dicerorhinus hemitoechus*) (<0.1 %) (see Fig. 3.2; Davis 1977; Eisenmann 1992; Speth and Tchernov 1998, 2001; Tchernov 1998).

During the MP, nearly half of Kebara's ungulate remains derive from an extremely dense concentration of bones which accumulated within a roughly 2 to 4-m wide zone close to the cave's north wall (the "north-wall midden" or just "midden"), and particularly from levels IX-XI (the "midden period"). In the central floor area of the cave, separated by a gap of several meters from the north-wall midden, and most clearly evident in level X (the *décapage*), bones were encountered in small, discrete concentrations or patches, separated from each other by zones with few or no bones (Meignen et al. 1998: 230, 231; Speth et al. 2012). Studies of the sediments on the cave floor, using on-site Fourier transform infrared spectrometry, indicate that these localized bone concentrations reflect the original burial distribution, not the end-product of selective dissolution following burial (Weiner et al. 1993). While the origin of these curious circular bone concentrations still eludes us, their form and content are not unlike the ubiquitous trash-filled pits characteristic of the Holocene archaeological record in both the Old and New World (Speth et al. 2012).

While there is clear evidence throughout the cave's MP sequence for the intermittent presence of carnivores, most notably spotted hyenas (*Crocuta crocuta*), the modest numbers of gnawed and punctured bones, the scarcity of gnaw-marks on midshaft fragments (Marean and Kim 1998: S84, S85), and the hundreds of cutmarked and burned bones, as well as hearths, ash lenses, and large numbers of lithic artifacts, clearly testify to the central role played by humans in the formation of the bone accumulations (see Speth and Tchernov 1998, 2001, 2007 for more detailed discussions of the taphonomy).

Evidence for Overhunting

Since several of the more obvious lines of evidence that are suggestive of overhunting at Kebara have been presented elsewhere (Speth 2004a; Speth and Clark 2006), here I will just briefly summarize them and introduce other data that I think further support this idea. I should hasten to point out, however, that the specter of equifinality is close at hand and will follow my arguments from beginning to end (Munro and Bar-Oz 2004). Since many of the changes that I view as likely evidence of overhunting, particularly the rapid fall off in the proportion of aurochs and red deer, occur at more or less the same time that midden accumulation declined or ceased in the cave, it is entirely conceivable that we are seeing a change in Kebara's function as a settlement, and/or the time of year when Neanderthals occupied the cave, and/or a shift in favored hunting areas, and/or the use of new or different hunting technologies or strategies, such as more communal hunting of gazelles, and perhaps fallow deer, rather than solitary ambush hunting of aurochs and red deer. And of course the decline could reflect a reduction in the availability of large game within range of the site brought about by paleoenvironmental fluctuations. Some of these possibilities are easier to deal with than others. Thus, as I will show momentarily, the fall-off in the proportions of large taxa in Kebara's ungulate assemblage is unlikely to be the result of major fluctuations in paleoclimate, since the decline in the frequency of these taxa continues unabated across major up and down oscillations in regional temperature and precipitation. Some of the other potential sources of equifinality, however, are not so easy to address and, as a consequence, cannot be convincingly dismissed.

One of the most striking features of Kebara's faunal record, and perhaps the most obvious sign that local populations some 55–60 kyr may have begun to overhunt their preferred large-game resources, is the sharp decline over the 4-m-long late MP sequence in the frequency of the two principal large-bodied animals—aurochs (*Bos primigenius*) and red deer (*Cervus elaphus*) (Fig. 3.3). In this figure, I use

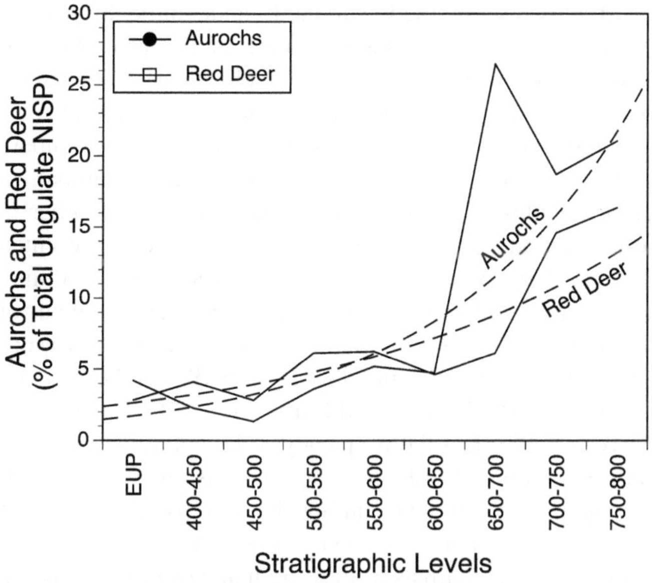

Fig. 3.3 Decline in proportion of red deer and aurochs over the course of the late MP and EUP by 0.5-m arbitrary horizontal levels (% of total ungulate NISP)

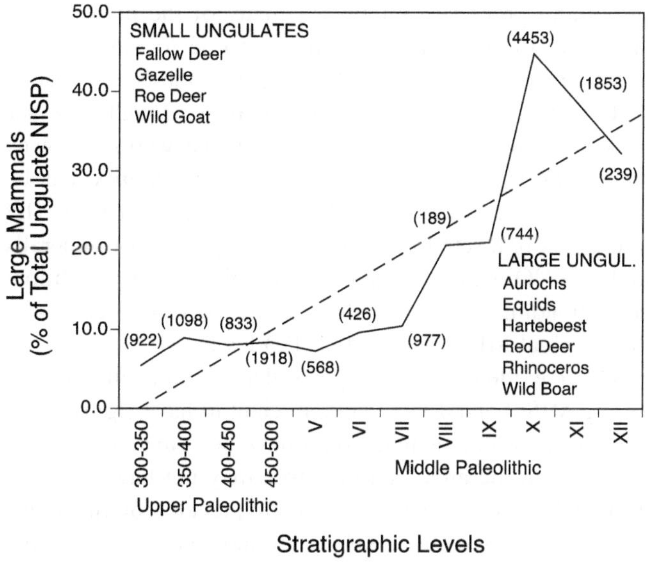

Fig. 3.4 Decline in proportion of large mammals over the course of the late MP and EUP by natural stratigraphic levels (MP) and 0.5-m arbitrary horizontal levels (EUP) (% of total ungulate NISP)

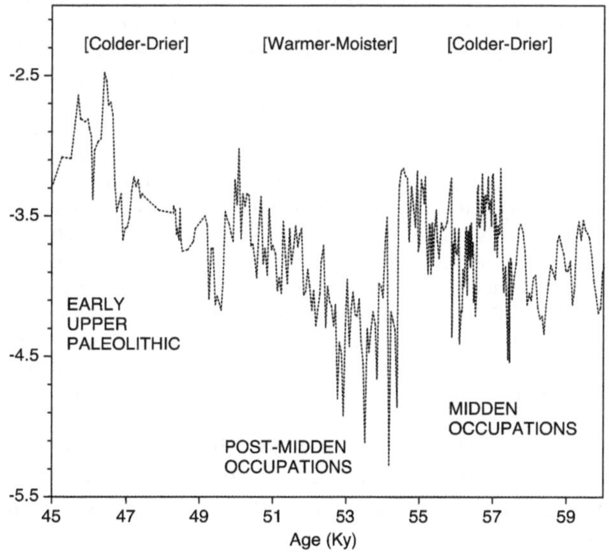

Fig. 3.5 Smoothed oxygen-isotope record ($\partial^{18}O$ ‰ PDB), derived from speleothems in Soreq Cave (Israel), for the period 45–60 kyr. Original data provided by M. Bar-Matthews (see Bar-Matthews et al. 1999: 88, their Fig. 1A for unsmoothed record)

the combined ungulate assemblages from the Stekelis and Bar-Yosef/Vandermeersch excavations in order to achieve the largest possible samples. As already discussed, this lumping procedure obviously comes with considerable loss of stratigraphic precision. Nevertheless, the pattern is very striking. Aurochs and red deer fall off sharply above about 600–650 cm below datum, which roughly corresponds to the end of midden accumulation in Kebara. The next figure (Fig. 3.4), which uses only the stratigraphically controlled materials for the MP fauna from the more recent Bar-Yosef/ Vandermeersch excavations, but which as a consequence has to lump all of the large-bodied taxa into a single group in order to achieve adequate sample sizes, shows this pattern more clearly. Large prey decline sharply above level VIII, or after the period of midden formation.

When the MP assemblage is divided, as discussed earlier, into just two stratigraphic subsets or groups, "lower" (bones below 550 cm) and "upper" (bones above 550 cm), the decline in the proportion of large ungulates is substantial and statistically significant ("lower" MP, 30.8 %, "upper" MP, 11.9 %, $t_s = 20.37$, $p = 0.0000$). Though handicapped by very small sample sizes, the proportion continues to drop slightly in the EUP (10.5 %), the difference nearly attaining statistical significance ($t_s = 1.71$, $p = 0.09$).

As already noted, large-scale paleoenvironmental changes are unlikely to be the cause of this decline in large ungulates; the trend continues unabated across several major swings in regional paleoclimate that are clearly evident in the speleothem-based oxygen-isotope record from Soreq Cave in central Israel (Fig. 3.5; data from Bar-Matthews et al. 1998, 1999). Very similar patterns are seen in the speleothem records from two other caves—Peqiin Cave in northern Israel and Ma'ale Efrayim Cave in the rain shadow to the east of the central mountain ridge of Israel (Bar-Matthews et al. 2003; Vaks et al. 2003). The isotope data shown in this figure have been "smoothed" using a cubic spline statistical interpolation procedure. This technique estimates a value for *y* using four values of *x* at a time (i.e., fixed window width). The program uses "…a series of cubic (third-order) polynomials to fit a moving window of data, four points at a time" (SAS Institute 1998: 227). This smoothing procedure eliminates many of the minor oscillations in the data, thereby allowing one to more readily perceive the major trends. According to the Soreq record, $\partial^{18}O$ values were generally lower between roughly 48 and 54 kyr, denoting a shift toward somewhat warmer-moister conditions. Chronologically, this correlates (approximately) to Kebara's levels VI and VII (Valladas et al. 1987). This interval is bracketed on either side by periods of generally higher $\partial^{18}O$ values indicative of colder-drier conditions corresponding, at least approximately, to levels XI-IX and the terminal MP (V) and EUP (IV) levels.

Other evidence also suggests that climatic fluctuations during MIS 3 and 4 were not sufficient to have significantly altered the resources that would have been available to MP hominins in the area, and therefore may not have been the principal cause of the declining frequencies of the larger ungulates that we see at Kebara, Amud, and elsewhere. For example, the rodent faunas from Amud show little evidence of dramatic change in the nature and composition of the habitats surrounding the cave during the late MP.

> The results of this study suggest that changes in relative abundance of micromammal species throughout the Amud Cave sequence are likely the result of taphonomic biases. Once such biases are addressed, there is no shift in the presence-absence; rank abundance and diversity measures of these communities in the time span 70-55 ka. The persistence of the micromammal community is consistent with low amplitude climate change. There is no indication for a decrease in productivity and aridification throughout the sequence of the cave, specifically toward the end of the sequence at 55 ka. The species present are suggestive of a mesic humid Oak woodland environment in Amud Cave and most of the contemporaneous Middle Paleolithic sites in northern Israel. Consequently, climate change may not have had a cause-and-effect relationship with the disappearance of the local Neanderthal populations from the southern Levant (Belmaker and Hovers 2011: 3207).

Carbon and oxygen isotope studies of goat (*Capra aegagrus*) and gazelle (*G. gazella*) tooth enamel from Amud Cave by Hallin et al. (2012: 71) reach very similar conclusions:

> …the disappearance of Neandertals after MIS 3 does not appear to be due to climate forcing. Our isotope data from Amud Cave and the species composition of its micromammals (Belmaker and Hovers 2011) provide no support for major climate change during MIS 3….

Interestingly, in discussions of the BSR only a handful of studies consider increases in the hunting of immature animals as part and parcel of the intensification process (see, for example, Davis 1983; Koike and Ohtaishi 1985, 1987; Broughton 1994, 1997; Munro 2004; Stiner 2006; Lupo 2007; Wolverton 2008). In one of the most explicit recent looks at the frequency of juvenile animals as evidence for overhunting during the Levantine Epipaleolithic, Stutz et al. (2009: 300) report a major change in the proportion of unfused versus fused gazelle first phalanges from values below 10 % in Kebaran and Geometric Kebaran assemblages to values in excess of 35 % in assemblages from the Natufian.

Usually, however, the frequency of immature animals in an archaeological faunal assemblage is discussed either in terms of taphonomy (immature bones are under-represented because they do not preserve as well as the bones of adults), or in terms of seasonality (lots of juveniles means a focus of hunting activity during the fawning or calving season) (e.g., Monks 1981; Klein 1982; Lyman 1994; Pike-Tay and Cosgrove 2002; Munson and Garniewicz 2003; Munson and Marean 2003). But the frequency of immature animals can reflect factors other than these two customary ones. For example, in communal hunting the scarcity or absence of juveniles can also be an inadvertent consequence of the behavior of the animals as they are maneuvered toward and into a kill (Speth 1997). American Bison (*Bison bison*) provide a case in point. When they are "gathered" and moved (but not yet stampeded) from a collecting area toward a trap or cliff, the animals often become strung out into a line or column, somewhat akin to the manner in which dairy cows move when they return to the barn at the end of the day. Spatially, the column is not a random mix of all ages and both sexes. The leaders are generally adults, typically females, with the calves lagging behind, and some of the adult bulls guarding the rear. Once the animals reach the trap or cliff, the hunters stampede the line from the rear, which effectively blocks the animals at the front of the column from turning and escaping, and pushes them instead into the trap or over the precipice. The animals farther back in the line or column, which may include many of the calves, are far more likely to escape. As a result, juveniles may end up under-represented or missing altogether in a communal kill for reasons unrelated to either season or taphonomy.

As far as gazelles are concerned, from modern wildlife management studies we do know that they can be gathered into modest-sized herds and that they will move along low walls and even white plastic strips lying on the ground rather than stepping or jumping over them (Speth and Clark 2006; Holzer et al. 2010), a behavior very similar to what has been observed in caribou, reindeer, and many other ungulates (e.g., Stefansson 1921: 400–402; Wolfe et al. 2000; Benedict 2005; Brink 2005). Unfortunately, we know regrettably little (at least judging by the published literature) about how gazelles distribute themselves spatially by age and sex as they are gathered and maneuvered toward and into a corral or trap. Hence, we do not know whether juveniles would end up being under-represented in the take from a communal hunt, and of course we have no idea whether MP hunters in the Southern Levant exploited gazelles in this manner. We do know that Eurasian Neanderthals used communal tactics to hunt reindeer, bison, and other large animals, so there is no justification for assuming *a priori* that Levantine Neanderthals or Anatomically Modern Humans lacked the wherewithal to communally hunt gazelles (Gaudzinski 2006; Rendu et al. 2012).

Under most circumstances, however, I suspect that the targeting of juveniles is a deliberate choice made by hunters. Viewed from the perspective of diet breadth models, the hunters are faced with a decision: should they invest time and effort to pursue, kill, transport, and process a small-bodied juvenile with limited subcutaneous and marrow fat deposits, or forego that opportunity and go after a prime adult that provides larger quantities of meat and is much more likely to be endowed with substantial deposits of fat? Seen in this way, elevated numbers of immature animals in an assemblage provide a pretty good indication of stress-related resource intensification (see also Stutz et al. 2009).

> If broad spectrum resources world-wide are generally expensive to collect and process relative to other foods, then, contrary to common argument, changes in their own abundance, however great, are unlikely to account for their adoption, not only in arid Australia, but anywhere (Edwards and O'Connell 1995: 775).

In a very dimorphic species, immature males grow much faster, and put on body fat sooner, than immature females, and hence hunters may treat them more like adults. This is testable in principle, but, unfortunately, in most cases archaeologists are hard put to sex the bones of juveniles, especially when the bones are in fragmentary condition (for an example of a bison kill where hunters did in fact selectively hunt juvenile males, see Speth and Clark 2006: 16, their Figs. 11, 12).

Viewed from a rather different vantage point, if one sees big-game hunting as an enterprise driven to any significant extent by social or political goals, such as prestige or costly signaling, one would come to more or less the same conclusion—hunters generally would be expected to focus their efforts on prime adults, not juveniles (e.g., Speth 2010b).

Of course there are some circumstances in which hunters will deliberately go after very young animals, even fetuses, primarily as delicacies because their meat is more tender than that of older adults. Soft skins may also be an important attraction (Binford 1978: 85, 86). It is doubtful, however, that hunting for either of these purposes in the MP or UP would have occurred often enough to produce recognizable signals in the archaeological record.

In sum, where preservation can be shown not to be a major factor, and setting aside the possibility that MP hominins hunted gazelles communally, some combination of seasonality and the degree of stress-related intensification will likely account for much if not most of the variability in the frequency of juvenile remains.

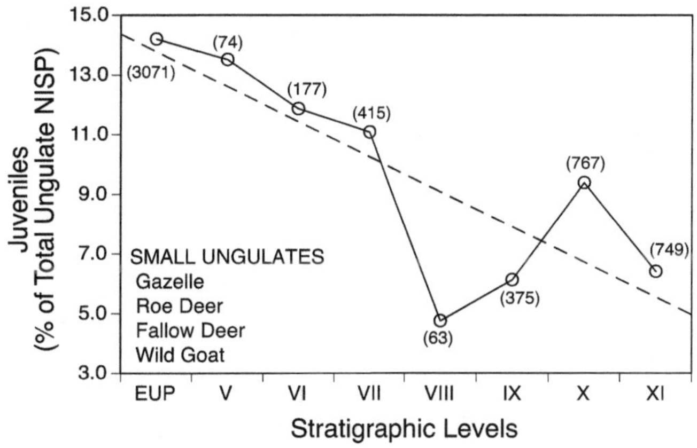

Fig. 3.6 Increase in proportion of small ungulate juveniles over the course of the late MP and EUP by natural stratigraphic levels (% of total ungulate NISP)

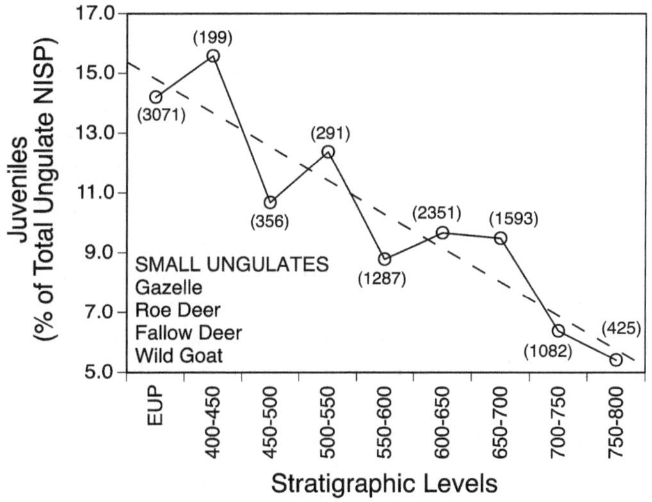

Fig. 3.7 Increase in proportion of small ungulate juveniles over the course of the late MP and EUP by 0.5-m arbitrary horizontal levels (% of total ungulate NISP)

The frequency of juveniles at Kebara may well present just such a case. The proportion of immature remains among the smaller ungulates (gazelle, roe deer, fallow deer, and wild goat)—age in this case based upon the fusion state of postcranial elements and teeth of immature animals—increases over the course of the MP sequence. This trend can be seen in Figs. 3.6 and 3.7. In the first of these figures, I use just the bones recovered in the Bar-Yosef/Vandermeersch excavations. These materials, because they were excavated following the natural stratigraphy of the deposits, provide the most reliable data set. But in so doing I am forced to exclude the wealth of material from the Stekelis excavations. When the materials from both excavations are combined, as shown in the second figure (Fig. 3.7), the substantially enlarged assemblage reveals the increase in

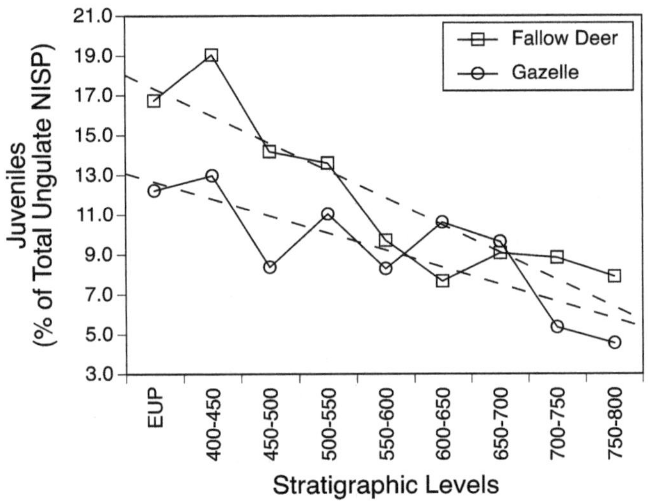

Fig. 3.8 Increase in proportion of juvenile gazelles and fallow deer over the course of the late MP and EUP by 0.5-m arbitrary horizontal levels (% of total ungulate NISP)

juveniles far more clearly. Comparing the aggregated "lower" and "upper" MP assemblages, as well as the EUP, the increase is significant, or nearly so, across all three samples ("lower" MP 8.7 % vs. "upper" MP 11.9 %, $t_s = 2.92$, $p < 0.01$; "upper" MP vs. EUP 14.2 %, $t_s = 1.73$, $p = 0.08$). Finally, Fig. 3.8 shows the increasing proportion of juveniles in the younger levels of Kebara's MP sequence, but in this case gazelles and fallow deer are plotted separately. Both species show similar trends, although the increase is more pronounced in the latter.

Despite very small sample sizes, a similar pattern of increasing juvenile representation, based again on the fusion state of postcranial bones, also seems to occur in the larger ungulates ("lower" MP 4.3 %, "upper" MP 7.4 %, EUP 8.8 %), although the only comparison that attains statistical significance is between the "lower" MP and the EUP ($t_s = 2.56$, $p = 0.01$). This result is very weakly echoed by the proportion of aurochs and red deer teeth (both taxa combined) from immature animals, using both eruption and crown height as described in Stiner (1994: 289–291; see also Speth and Tchernov 2007) to assign specimens to the immature and adult age classes. The proportion of juveniles is 14.9 % in the "lower" MP group, 27.3 % in the "upper" MP group, and 25.6 % in the EUP. None of these differences is statistically significant, however, given the miniscule sample sizes of these taxa in the younger levels, although if we combine the "upper" MP and EUP samples, the result approaches significance ($t_s = 1.84$, $p = 0.07$).

The targeting of juveniles changes, at least in gazelle, in another interesting and admittedly unexpected way as well. To see this, I compare the average crown height of gazelle lower deciduous fourth premolars (dP_4) in the "lower" and "upper" MP groups and the EUP. The average crown height provides a crude index of the age of immature animals, a smaller value indicating more heavily worn teeth or older animals, a larger value denoting less worn teeth or younger animals. In the Kebara assemblage, there is no significant difference in average crown height between the "lower" (5.0 mm) and "upper" (5.1 mm) MP stratigraphic groups ($t = 0.31$, $p = 0.76$), but both groups differ significantly, or nearly so, from the larger crown height value seen in the EUP (5.8 mm; "lower" MP vs. EUP, $t = 2.72$, $p = 0.008$; "upper" MP vs. EUP, $t = 1.79$, $p = 0.08$). These data suggest that, by the end of the MP, Kebara's hunters had not only begun to target more juvenile gazelle but perhaps also younger individuals *within* the juvenile age classes. This assumes, of course, that increasing numbers of younger juveniles is not the result of a shift in the seasonal timing of gazelle hunting activities, one of those ever-present problems of equifinality. No comparable trend is evident in the dP_4s of fallow deer.

Unfortunately, yet another issue of equifinality enters the picture here. The greater average crown height value seen in the EUP dP_4s could also reflect climatic or environmental shifts that, following the MP, reduced the abrasiveness of the forage available to the gazelles. We see no obvious way at present to examine this possibility, but it is a factor that should be kept in mind.

Finally, the larger crown height value might also indicate that the gazelles themselves had become larger in the EUP. Measurements on over 400 of Kebara's gazelle astragali (tali) suggest the opposite, however. The greatest lateral length (GLl in von den Driesch's 1976 terminology), for example, declines from a mean of 2.95 cm ($N = 293$) in the MP (all levels combined) to 2.86 cm ($N = 115$) in the EUP, a change that is highly significant ($t = 5.95$, $p < 0.0001$). Similarly, the maximum distal breadth (Bd in von den Driesch's terminology) declines from 1.76 cm ($N = 329$) in the MP to 1.72 cm ($N = 124$) in the EUP, again a significant change ($t = 3.79$, $p = 0.0002$). Whether the decline in the size of the astragali is due to an overall reduction in the average body size of the gazelles, or a higher proportion of females among the kills, or the increased representation of juveniles, or some combination of these, it seems clear that the increase in average crown height of the juvenile gazelles in the EUP is not likely to be due to an overall increase in the body size of the animals.

The gazelles reveal another interesting temporal pattern. Not only does the proportion of juveniles increase toward the end of the MP, as does the proportion of younger individuals among the juveniles, but the average age of the *adult* gazelles appears to decline as well. In Fig. 3.9, mean crown heights for the lower or mandibular third molar (M_3) of adult gazelles are plotted by arbitrary 1-m thick levels. Such coarse stratigraphic lumping was necessary to get minimally adequate sample sizes. Comparable data for

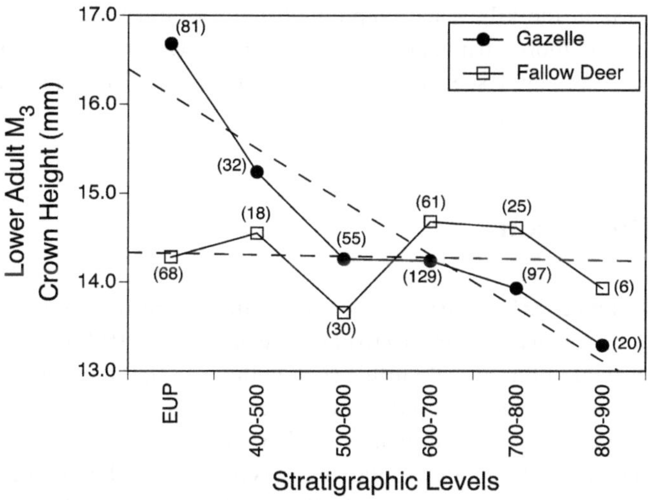

Fig. 3.9 Mean crown height values (mm) of *adult* gazelle and fallow deer lower third molars (M_3) over the course of the late MP and EUP by 1.0-m arbitrary horizontal levels

fallow deer are also included. This figure suggests that mean crown heights in adult gazelles increase from the beginning of the sequence into the EUP. While the graph shows what appears to be a reasonably clear trend, the sample sizes are small and only one of the pairwise comparisons—between the 400–500 cm MP level and the EUP—is significant, or nearly so ($t = 1.75$, $p = 0.08$). By pooling the data further and using only the two MP stratigraphic groups, the mean crown height values obtained for the M_3 of adult gazelles in the "lower" and "upper" MP groups do not differ significantly from each other ($t = 0.48$, $p = 0.63$), but both differ from the EUP value ("lower" MP vs. EUP, $t = 4.76$, $p < 0.0001$; "upper" MP vs. EUP, $t = 3.64$, $p < 0.001$). Finally, when the individual crown height values for the M_3 of adult gazelles are correlated with their actual depths below datum, the resulting coefficient, while not particularly strong, is negative, as expected, and statistically significant ($r_s = -0.23$, $p < 0.0001$).

Thus, although these data are not as clear-cut as one might like, the implication of Fig. 3.9 would seem to be that toward the close of the MP Kebara's Neanderthal hunters appear to have focused increasingly, not only on juvenile gazelles, but on younger adult gazelles as well, yet another likely sign of subsistence intensification. Interestingly, although the hunters also increased their reliance on juvenile fallow deer, as already shown, there is no evidence in the crown height data that would suggest their use of young adult deer increased toward the end of the sequence. Unfortunately, the sample of measurable red deer and aurochs M_3s is much too small to see if similar targeting of younger adults might have occurred in these taxa as well.

The Kebara faunal record shows another interesting trend that also appears to cross-cut the speleothem paleoclimate record, and could therefore be another sign of resource intensification. Among the ungulates being hunted by Neanderthals at Kebara, gazelles are by far the smallest (the only other comparably small ungulate, roe deer, is exceptionally rare throughout the sequence). Modern male gazelles, on average, weigh only about 25.2 kg; females weigh about 18.3 kg (Baharav 1974; Mendelssohn and Yom-Tov 1999: 250; Martin 2000). Persian fallow deer, though larger than gazelles, are still relatively small animals. Unfortunately, while body-weight data for European fallow deer (*Dama dama*) are widely available for both farmed and wild animals, reliable information for the Persian form (*D. mesopotamica*) is virtually non-existent. Nevertheless, it is clear that the Near Eastern cervid is larger than its European cousin, with adult males often exceeding 100 kg and females falling somewhere around 60 or 70 kg (e.g., Haltenorth 1959; Chapman and Chapman 1975; Mendelssohn and Yom-Tov 1987; Nugent et al. 2001). There is fairly abundant evidence in the ethnographic literature that hunters commonly transport the entire carcass back to camp when they are dealing with prey in the size range of gazelles and fallow deer, but become increasingly selective in what body parts of the bigger, heavier animals they jettison and what they transport home. In prey the size of red deer and aurochs, heads are one of the parts most often discarded at the kill (O'Connell et al. 1988, 1992: 332; Domínguez-Rodrigo 1999: 21). Kebara's data echo the ethnographic cases quite closely—there are proportionately many fewer heads of aurochs and red deer than there are of gazelle and fallow deer (gazelle, 17.1 %; fallow deer, 11.0 %; wild boar, 12.2 %; red deer, 8.4 %; aurochs, 9.4 %). Isolated teeth have been excluded in this comparison because they introduce their own bias—the proportion of teeth that are isolated from their bony sockets increases dramatically in the largest taxa, thereby greatly inflating the NISP counts for the heads of these animals (see Speth and Clark 2006).

Distance also enters into the hunters' calculus of what to carry home and what to abandon at the kill. Thus, within a given body-size class, the farther the carcass has to be transported, the more likely the heads, which are bulky low-utility elements, will be jettisoned, quite likely after on-site processing to remove the tongue, brain, and other edible tissues. Figure 3.10 is interesting in this regard; it shows that, upward in the sequence, the proportional representation of gazelle heads (both mandibles and crania) falls off almost monotonically, suggesting that Kebara's hunters may have had to travel increasingly greater distances to acquire these diminutive ungulates, increasing numbers of which were juveniles. Figure 3.11 shows that fallow deer heads also declined, although the pattern is much "noisier" than the trend for gazelles. Again, the implication is that the hunters toward the close of the MP had to travel farther to acquire deer, and, as in gazelles, despite the increasing

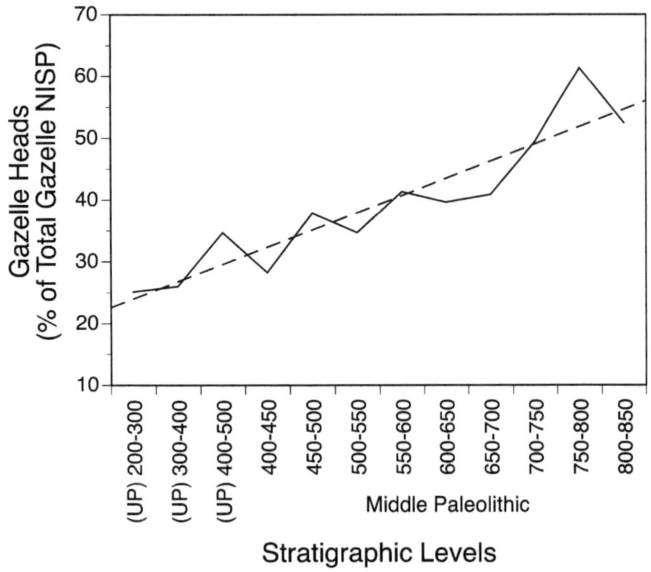

Fig. 3.10 Decrease in proportion of gazelle heads (mandibles, maxillae, and isolated teeth) by 0.5-m arbitrary horizontal levels (MP) and 1.0-m arbitrary horizontal levels (EUP) (% of total gazelle NISP)

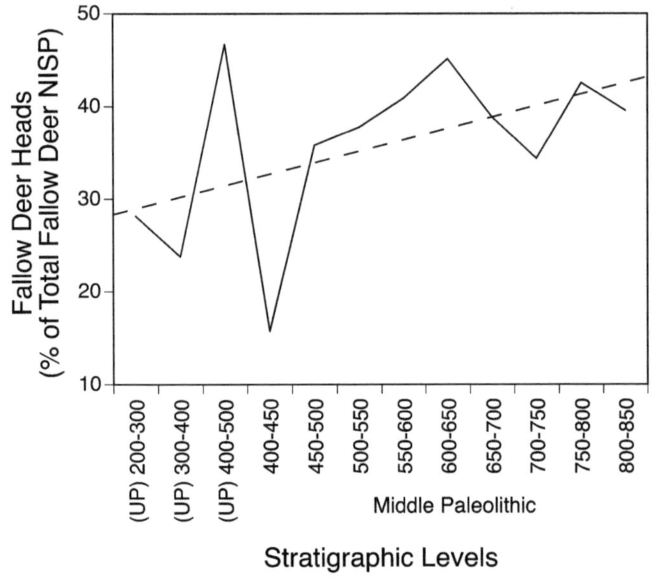

Fig. 3.11 Decrease in proportion of fallow deer heads (mandibles, maxillae, and isolated teeth) by 0.5-m arbitrary horizontal levels (MP) and 1.0-m arbitrary horizontal levels (EUP) (% of total fallow deer NISP)

transport costs they nonetheless brought back greater numbers of immature animals.

As usual, of course, issues of equifinality are never far away. If Neanderthal hunters toward the close of the MP began capturing larger numbers of gazelles, and possibly fallow deer, in communal drives, one might expect them to have abandoned more heads and other bulky, low-utility parts at the kill, since communal hunts would have necessitated more stringent culling decisions. Unfortunately, at the moment I see no reliable way to distinguish between these alternatives.

Discussion

As noted early on in this paper, until fairly recently it has been common in the Near East to place the onset of the BSR toward the end of the Epipaleolithic (i.e., within the Natufian), just a few millennia prior to the emergence of village-based farming economies. Exciting new work by both zooarchaeologists and paleoethnobotanists, however, is pushing the beginnings of resource intensification further back in time, back to at least 20–25 kyr, at the onset of the Kebaran, the period that marks the beginning of the Epipaleolithic (e.g., Piperno et al. 2004; Weiss et al. 2004a, b; Stiner 2009a; Stutz et al. 2009). The faunal data from Kebara Cave may push the beginnings of the process even further back in time, providing the first hint that significant resource intensification, at least in the Southern Levant, may in fact have already begun as much as 50–55 kyr, and reflect the impact of the region's growing MP populations on their large-game resources.

Stutz et al. (2009) assembled faunal data from a number of Levantine sites dating to the Epipaleolithic. They expressed these data using a series of indices which show the abundance of small "big game" (gazelle, roe deer, wild goat), medium "big game" (fallow deer, wild boar, hartebeest), and large "big game" (aurochs, red deer, equids, rhino) relative to each other and to "small game" (hares, tortoises, birds). Using these indices, the decline in the abundance of both large and medium "big game" across the span of the Epipaleolithic is readily apparent. Applying two of these same indices to the Kebara data—the large "big-game" index (LbgI) and the medium "big-game" index (MbgI)—one can readily see that the fall-off in the largest taxa (aurochs and red deer) very likely began well before the Epipaleolithic, perhaps as much as 50–55 kyr during the final stages of the MP (see Fig. 3.12).

Incidentally, while I calculated the indices in the same way that Stutz and colleagues did, I have changed their labels in order to make them a little easier to recognize and remember (see Stutz et al. 2009, their Tables 3 and 4). I should also point out that in calculating the indices I included fragments that were identifiable only to skeletal element and approximate body-size class (e.g., gazelle-sized, fallow deer-sized, red deer-sized, aurochs-sized, etc.). One reason for doing this, aside from increasing overall sample sizes, was to reduce the bias against the largest taxa. At Kebara, very few bones of the biggest animals, most notably aurochs, could be identified to species with any

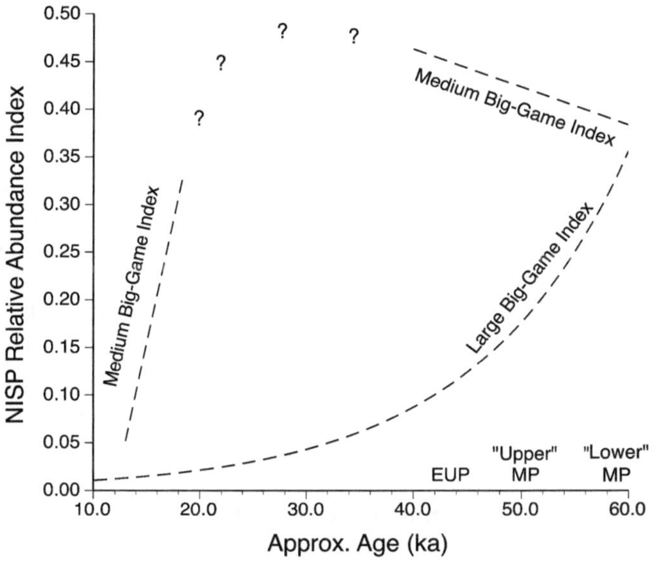

Fig. 3.12 Decline in large "big game" index (LbgI) and medium "big-game" index (MbgI) at Kebara, and at a series of Levantine Epipaleolithic sites, using procedures and index values from Stutz et al. (2009). Chronological placement of the three Kebara assemblages is approximate

degree of certainty. These bovids are represented mostly by large, unidentifiable pieces of cortical bone from limb shafts that, judging by the thickness of the shaft and its radius of curvature, clearly derived from an aurochs-sized animal. Had these fragments been excluded, red deer and especially aurochs would be severely under-represented in these analyses. Nor would reliance on teeth have eliminated the bias, because many fewer crania of the largest animals were transported back to the cave, a pattern widely documented among modern hunters and gatherers (Speth and Clark 2006: 19).

Returning now to our discussion of Fig. 3.12, the rapid decline at Kebara in the largest taxa is to some extent compensated for by a gradual *increase* in the medium-sized species, especially fallow deer, which, judging from Stutz et al.'s (2009) data, do not begin their own precipitous decline until later, during the Epipaleolithic, probably sometime after about 20 kyr. Gazelles, of course, are the most abundant ungulate species throughout the sequence at Kebara, varying between about 35 and 55 % of the total ungulate assemblage (based on NISP). Over this same period fallow deer make up some 25–35 % of the assemblage, while all of the other ungulate taxa together add up to only about a quarter of the total. The ascendancy of gazelles in Levantine faunal assemblages, in fact, seems to begin much earlier in the MP, perhaps as early as 200 kyr or more (e.g., ~38 % at Misliya Cave, Yeshurun et al. 2007; see also Stiner et al. 2011: 218 and Yeshurun 2013), and persists across major region-wide fluctuations in paleoclimate, suggesting that this animal's abundance in archaeological sites may be more a reflection of human subsistence choices than paleoenvironmental factors (Rowland 2006; Marder et al. 2011). Given the small size of the gazelles, and given the widely accepted view that MP hominins were "top predators," living at the apex of the food chain (Bocherens 2009, 2011), the fact that MP hunters in the Southern Levant already by ~200 kyr heavily focused their efforts on this diminutive ungulate could be an indication that the roots of the BSR predate the late MP, perhaps by a sizeable margin.

Despite their small size, gazelles may, of course, have become a favored target of MP hunters for largely pragmatic reasons. Although the behavior of *G. gazella* is not well understood, these animals may have been easier than other taxa to hunt successfully, perhaps because they could be stalked, snared, or trapped more readily than the larger ungulates, or because they were more easily taken in groups by parties of cooperating hunters (see also Yeshurun's discussion, 2013, concerning the possible use of long-distance projectile weapons for hunting gazelles).

But obviously people hunt, and at times overhunt, for many reasons other than, or in addition to, the procurement of food. Some of the more obvious of these non-food motivations for hunting include: (1) procuring hides or other animal parts for clothing, footwear, shelter, containers, weapons, tools, shields, hunting disguises, ceremonial costumes, rattles, and glues (e.g., Wissler 1910; Ewers 1958: 14, 15; Gramly 1977; Klokkernes 2007); (2) gaining prestige or attaining other social or political goals (Sackett 1979; Wiessner 2002; Bliege Bird et al. 2009; Speth 2010b); (3) underwriting periodic communal aggregations, fulfilling needs and requirements of male initiation rites, vision quests, and various other ritual and ceremonial performances and observances (Ewers 1958; Sackett 1979; Potter 2000; Zedeño 2008; Bliege Bird et al. 2009); (4) controlling or eliminating dangerous predators or pests (Headland and Greene 2011); and (5) procuring meat, hides, hair, antlers, ivory, oils, scent glands, hooves, and other commodities for use as gifts or items of exchange (Lourandos 1997). Some of these motivations are unlikely to favor a small ungulate like the gazelle, and others may not be relevant to the remote time period we are considering here. However, hides for clothing, as well as footwear, shelter, and containers, may have been important to Near Eastern MP hominins and, historically at least, gazelle skins are noted as having been particularly valued for such purposes (see Bar-Oz et al. 2011). The non-pastoral nomadic Solubba provide an interesting case in point:

> Several sources mention the distinctive Solubba dress...a garment made from 15 to 20 gazelle skins, the hair to the outside, open at the neck, with a hood, and long sleeves gathered at the wrist, extending down to cover the hands up to the fingers.... The men used to wear such garments either as their only

> clothing, or as shirts underneath the traditional bedouin [sic] costume. They reputedly were the only group to wear clothing made of skins (Betts 1989: 63).

I am not suggesting that MP hominins went around bedecked in fancy tailored garments fashioned from exquisitely tanned gazelle skins, each garment requiring in excess of a dozen animals for its manufacture! But if MP hunters and their families wore any clothing at all, and that seems increasingly plausible based on recent studies of the evolution of head and body ("clothing") lice in humans (Kittler et al. 2003; Reed et al. 2004; Leo and Barker 2005; Toups et al. 2011), gazelles may have been important for precisely this reason, their diminutive size notwithstanding. Estimates vary considerably as to when human head and clothing lice diverged, with dates ranging from as recently as ca. 70–80 kyr to in excess of 1 Myr. These studies seem to be in somewhat greater agreement, however, when it comes to identifying where clothing first came into use, pointing not toward the arctic or subarctic, but toward sub-Saharan Africa as the most probable source. Thus, the presence of clothing during the MP in the Near East seems quite likely. If this is true, then the extent to which MP hominins might have placed pressure on gazelles (and other large game) for reasons above and beyond their use as a food resource would become a question of the number of hides needed to make the clothing, the number of individuals requiring such clothing, and the length of time the items lasted before they became unwearable and had to be replaced. And of course these hominins may have used gazelle skins for many other purposes as well, including footwear, bedding, carrying containers, shelter, and so forth. Gramly (1977) provides an intriguing study of deer hunting by the historic Huron, a North American "First Nation" (Indian) tribe living in Ontario, Canada—he shows that their need for deer hides to make clothing and moccasins often outstripped their need for meat.

Returning our focus once again to the Near Eastern MP faunal record, ungulate data from other sites in the region yield medium and large "big game" indices which are roughly compatible with the overhunting scenario suggested for Kebara. For example, the MP faunal remains from layers B1 and B2 of Amud (Rabinovich and Hovers 2004), a cave located close to the Sea of Galilee and dated by thermoluminescence to about 55 kyr (Valladas et al. 1999), yield very low LbgI values (0.01 and 0.02, respectively). The MbgI values are also low (0.14 and 0.13, respectively), and well below the values for medium big-game at Kebara. Moving back in time, Qafzeh Cave, dated to between 90 and 100 kyr (Valladas and Valladas 1991), produced an LbgI value (0.62) that is higher than Kebara's (Rabinovich and Tchernov 1995). Qafzeh's MbgI is also high (0.45) but more or less in line with Kebara's values. Moving back further still to about ~200 kyr, Hayonim Cave's MP levels yielded an LbgI of 0.28, a value not all that different from the LbgI value for the earlier part of Kebara's MP sequence (Mercier et al. 2007; Stiner 2005: her Appendix 11). Hayonim's MbgI of 0.51 is somewhat higher than Kebara's. Taken together, the large and medium "big-game" indices from these sites support the view that the largest prey species declined in the Southern Levant in the latter part of the MP, sometime between about 60 and 50 kyr, with hunting increasingly focused on the medium- and small-sized ungulates, particularly fallow deer and gazelles, with fallow deer not declining significantly until sometime after about 30 kyr. Gazelles then became the principal "big-game" target of hunters in the region until the Prepottery Neolithic when their importance was rapidly eclipsed by domestic livestock (Davis 1982, 1983, 1989; Munro 2009a, b; Sapir-Hen et al. 2009).

Summary and Conclusions

The various patterns and relationships that over the years we have been able to tease out of the Kebara faunal data provide interesting insights into the hunting behavior of Southern Levantine Neanderthals during the last ten or so millennia prior to the UP (e.g., Speth and Tchernov 2001, 2007; Speth 2006; Speth and Clark 2006; Speth et al. 2012). The results of these studies shed light on how these rather enigmatic "pre-modern" humans went about earning their living, an interesting issue in its own right, and they have a bearing on our understanding of the MP-UP transition, the period when supposedly "archaic" foraging lifeways gave way to more or less "modern" hunters and gatherers. Here I briefly summarize what I feel are the most interesting findings of the present study, which has looked specifically at the possibility that Neanderthals (and, by implication, perhaps other quasi-contemporary MP hominins in the Near East), already some 50–60 kyr, if not before, began to overhunt their large-game resources, for food, very likely for hides, and possibly even for prestige or other social and political reasons, thereby initiating or augmenting a process of subsistence intensification—the BSR—that continued through the UP and into the early Holocene, ultimately setting the stage for the origins of agriculture.

(1) The two most common large taxa at Kebara—red deer and aurochs—decline steadily in numbers over the course of the sequence such that by the EUP they constitute a very small percentage of the total assemblage. This decline seems to be largely independent of the broad climatic swings that have been documented for the region in the speleothem isotope record. Overhunting, at least on a local scale, is strongly implicated by this pattern.

(2) The proportions of immature gazelles and fallow deer increase steadily over the course of the sequence. Because of their small size and limited body fat deposits, juveniles of both taxa were probably low-ranked resources by comparison to their adult counterparts and, as a consequence, MP hunters may often have excluded them from their "optimal diet." Thus, while the *presence* of juvenile individuals may be informative about the approximate time of year when hunting took place, *fluctuations in their numbers* probably say more about shifts in encounter rates with more highly-ranked adult prey than about seasonality (it could instead reflect a shift from intercept or ambush hunting to a greater reliance on communal methods of procurement). However, in a relatively dimorphic species such as fallow deer, sub-adult males put on muscle mass and body fat much faster than their female counterparts, and therefore may have been targeted more often by hunters than sub-adult males in a less dimorphic species like gazelle. Thus, all other things being equal, sub-adult males of the more dimorphic species should be better represented in the faunal assemblage than juveniles of either sex of the less dimorphic taxon. While I lack sufficient data at present to sex the juvenile fallow deer remains at Kebara, it is interesting that there are more juvenile deer than juvenile gazelles, and their number increases hand-in-hand with the number of adult deer being taken in the upper part of the MP sequence. In other words, as red deer and aurochs declined, Kebara's hunters increasingly focused their attention on fallow deer, taking both juveniles and adults. Someday, if reliable methods become available to sex the immature fallow deer remains, I would expect there to be a distinct bias toward older juvenile males.

(3) If one accepts the view that juveniles are low-ranked resources, regardless of their abundance on the landscape, then their increase in the younger MP horizons at Kebara points to a decline in encounter rates for higher-ranked adults.

(4) Based on average crown heights of gazelle lower deciduous fourth premolars (dP_4), the Kebara hunters not only targeted increasing numbers of juveniles, but also younger ones. Similarly, based on average crown heights of permanent lower third molars (M_3), the hunters also took greater numbers of young adult gazelles.

(5) In the younger levels of the MP and continuing into the EUP, Kebara's hunters brought back progressively fewer heads of both fallow deer and gazelles. Given their comparatively small body size, I had expected that the hunters would generally have transported complete or nearly complete carcasses of these animals back to the cave. The decline in heads toward the very end of the sequence, therefore, may suggest that the hunters had to travel greater distances to acquire these animals, forcing them to become more selective in which body parts they abandoned and which they transported home. Alternatively, of course, one could argue that, during the latter part of the MP sequence, many more of these animals were taken in mass drives, which might also necessitate greater selectivity on the part of the hunters. The implication might be similar, however, since communal hunting is commonly a response to resource depletion (Speth and Scott 1989).

> It seems that communal hunting is most practical when obtaining a given amount of meat per day is more crucial than the increased work effort associated with communal hunting (i.e., under resource-poor or commercial hunting conditions) (Hayden 1981: 371).

(6) While published faunal assemblages from other MP sites in the region, particularly cave sites, are few and far between, what data there are fit reasonably well with the pattern seen at Kebara. Earlier assemblages, for example Hayonim's at about 200 kyr and Qafzeh's at somewhere between 90 and 100 kyr, both have fairly high Large "Big-Game" Indices (LbgI), values that are more or less in line with Kebara's "lower" MP value. Amud Cave, whose faunal assemblages from Layers B_1 and B_2 are roughly the same age as Kebara's "upper" MP group (i.e., around 55 kyr), have exceedingly low LbgI values, implying that Amud's MP occupants were already taking the largest mammals at a rate little different from what Stutz et al. (2009) observed during the Epipaleolithic.

Suggestive as these results may be, none of this is as yet sufficient to prove that the BSR had its roots in the MP or that Neanderthals played a significant role in the early stages of the process. But the idea is intriguing, and even more so now that we are finally admitting Neanderthals into the human family (e.g., Green et al. 2010). To see if this view has real substance, however, we obviously need a lot more faunal data from many more sites over a much broader region of the eastern Mediterranean. We also need to explore other lines of evidence in much greater detail. Prime among these is information on the plant-food component of Neanderthal and Anatomically Modern Human diet. This is obviously a much more difficult domain of research than the fauna, since plants are notoriously vulnerable to decay and, even where they are preserved, it is often difficult to tell food plants from background "noise" (Abbo et al. 2008). Kebara is again interesting in this regard. The MP deposits, especially in close proximity to the many hearths and associated ash lenses, produced thousands of charred seeds (>3,300), the vast majority of which were from legumes (family Papilionaceae). Was this normal Neanderthal fare, or do these plant foods also reflect late MP intensification, much as Ohalo II's wild cereal grains and grinding slab with starch granules likely does for the early Epipaleolithic (Piperno et al. 2004)?

> Our findings indicate that broad spectrum foraging was thus a long-established human behavior pattern and included wild legumes as well as wild grasses and other fruits and seeds. This concept calls for a reevaluation or a more detailed definition of the notion of 'broad spectrum revolution' as a precursor phase in human subsistence strategies prior to agricultural origins (Lev et al. 2005: 482).

Phytoliths from the mature spikelets of grasses have been recovered from late MP deposits at Amud Cave and provide surprisingly early evidence that Neanderthals harvested the ripened seeds of these plants:

> Coupled with the presence of legumes in Kebara Cave, the Amud phytolith data constitute evidence that two families of plants, which would subsequently provide some of the earliest domesticates…, were already exploited in the Levant in the Late Middle Palaeolithic. The faunal and botanical records thus concur that exploitation of a broad spectrum of food resources was part of the Palaeolithic lifeways long before it became the foundation of, and a pre-requisite for, an economic revolution (Madella et al. 2002: 715).

Miller Rosen (2003), on the basis of phytoliths, starch grains, and spores, suggests that the late MP occupants of Tor Faraj in Jordan may have utilized wild dates (*Phoenix dactylifera*), horse-tail rush (*Equisetum* sp.), possibly pistachio nuts (*Pistacia* sp.), and probably other food plants as well, including roots and tubers.

Starch grains retrieved directly from Neanderthal dental calculus likewise point to widespread use of plant foods in MP hominin diet:

> The timing of two major hominin dietary adaptations, cooking of plant foods and an expansion in dietary breadth or 'broad spectrum revolution,' which led to the incorporation of a diversity of plant foods such as grass and other seeds that are nutritionally rich but relatively costly to exploit, has been of central interest in anthropology…. Our evidence indicates that both adaptations had already taken place by the Late Middle Paleolithic, and thus the exploitation of this range of plant species was not a new strategy developed by early modern humans during the Upper Paleolithic or by later modern human groups that subsequently became the first farmers (Henry et al. 2011: 489, 490).

Likewise, recent macro- and microwear studies of the teeth of Near Eastern MP hominins point to eclectic diets that very likely included a substantial plant food component (e.g., El Zaatari 2007; Fiorenza et al. 2011). Even the nitrogen isotope data extracted from the collagen of European Neanderthals, the data that archaeologists and paleoanthropologists routinely cite nowadays as proof that these "archaic" humans were "top predators," could mask a substantial dietary contribution of plant foods.

> A small percent of meat already increases very significantly the Δ^{15}N value, and contributions of plant food as high as 50% do not yield Δ^{15}N values lower than 1 standard-deviation of the average hyena collagen Δ^{15}N value…. This example clearly illustrates that the collagen isotopic values of Neanderthal collagen provide data on the relative contribution of different protein resources, but it does not preclude a significant amount of plant food with low nitrogen content, *as high as half the dry weight dietary intake* (Bocherens 2009: 244; emphasis added).

Future Directions

Tantalizing as these data are, we still have a long ways to go to fully grasp the nature, timing, and cause(s) of the "broad spectrum revolution." Most archaeologists today seem to agree that the BSR was a largely stress-driven response of foragers, late in the Pleistocene, to an increasing imbalance between available food resources and the number of mouths to be fed. More specifically, what most scholars see as being "broadened" was the number of different food types (i.e., taxa) that were added to the larder, usually by incorporating plant and animal resources with lower return rates, such as grass seeds, tubers, molluscs, snails, reptiles, birds, and small mammals, all resources that had been available before but largely or entirely ignored.

Addressing the animal component of the BSR, Mary Stiner and colleagues (e.g., Stiner et al. 1999; Stiner 2001: 6995; Munro 2003) sharpened our perspective by pointing out that a shift toward greater reliance on smaller, lower-ranked animal resources becomes evident "…only when small animals [are] classified according to development rates and predator escape strategies, rather than by counting species or genera or organizing prey taxa along a body-size gradient." Similar arguments have been applied to the plant food component as well (Weiss et al. 2004a, b).

Refining our understanding of the nature, timing, and causes of the BSR is obviously very much handicapped by what is visible to us in the Paleolithic and Epipaleolithic archaeological record. And since animal bones are usually better preserved and easier to recover and study than plant-food remains, the lion's share of our efforts, not surprisingly, have been directed at the fauna. But such a focus has an inherent risk—we tend to operate with the assumption, usually implicit, that population-resource imbalances will inevitably show up in some form or fashion in the faunal remains if we just look closely enough. But, as I will outline below, this assumption at times may be off the mark. People can respond to resource stresses in a myriad of ways that may or may not lead to an increase in diet breadth—plant or animal. In fact, at times these responses may not detectably alter the diversity of foods that are eaten, nor the way they are processed and cooked, nor even involve significant changes in technology—*they may be largely social.* One likely consequence of this is that foragers may experience and respond to subsistence-related stresses long before there are noticeable changes in either diet breadth or technology. Unfortunately, as will become clearer shortly, our theoretical understanding of how foragers, past or present, select among the wide array of potential responses when faced with subsistence-stresses that differ in frequency, magnitude, duration, and predictability is woefully underdeveloped, and "reality" is likely to be far more complex than we currently assume.

Let us look briefly at some of the diverse ways that foragers can respond to population-resource imbalances, beginning first with some of the more obvious technological options. This discussion is anything but exhaustive, its purpose being simply to point out the complexity that is subsumed under the catchy rubric of the "Broad Spectrum Revolution." One can enhance the value of many starchy plant foods by grinding, pounding, or soaking them, by baking them in pits or earth ovens, by leaching them to remove toxic secondary compounds, or by adding lime, wood ash, clay, or other substances to make the foods more palatable or nutritionally worthwhile (Katz et al. 1974; Stahl 1989; Johns 1990; Wandsnider 1997). For animal resources, this may mean exploiting marrow bones on a more systematic or regular basis, or making greater use of bones that are marginal sources of marrow, such as phalanges and mandibles (Binford 1978; James 1990). It can also mean hunting greater numbers of immature or elderly individuals, even though the fat reserves of animals in these age classes are generally much smaller than what one could expect to get from prime adults (Speth 2004a). And it can mean beginning to render grease from bones by boiling, perhaps first without heated stones in skin or gut containers as described, for example, by Herodotus in the fifth-century BC in his monumental "Histories" (Rawlinson 1859: 52, 53; see also Ryder 1966, 1969), then with heated stones, and finally in pots.

> As Scythia, however, is utterly barren of firewood, a plan has had to be contrived for boiling the flesh, which is the following. After flaying the beasts, they take out all the bones, and (if they possess such gear) put the flesh into boilers made in the country, which are very like the cauldrons of the Lesbians, except that they are of a much larger size; then, placing the bones of the animals beneath the cauldron, they set them alight, and so boil the meat. If they do not happen to possess a cauldron, they make the animal's paunch hold the flesh, and pouring in at the same time a little water, lay the bones under and light them. The bones burn beautifully, and the paunch easily contains all the flesh when it is stript [sic] from the bones, so that by this plan your ox is made to boil himself, and other victims also to do the like. When the meat is all cooked, the sacrificer offers a portion of the flesh and of the entrails, by casting it on the ground before him. They sacrifice all sorts of cattle, but most commonly horses.

Technological intensification—squeezing more calories and nutrients out of the same foods; and adding new foods to the diet that were previously inedible, or of only limited food value, or which required considerably more time and effort to harvest or prepare—did not suddenly materialize out of thin air in the Epipaleolithic or Upper Paleolithic; quite the contrary, it has been transforming forager foodways since the first appearance of the genus *Homo*. It began in earnest in East Africa some 2.6 Myr with the development of the first stone tool technologies, the so-called "Oldowan," making possible a wide range of cutting, slicing, and chopping activities that would have been much more difficult, if not impossible, to do without these technical aids (Semaw 2000). Perhaps the most significant addition to the repertoire of human culinary technology, appearing (at least according to some) by at least 1.0–1.5 Myr, if not earlier, was the control of fire (Wrangham and Conklin-Brittain 2003; Carmody and Wrangham 2009; Wrangham 2009). With fire, many plant foods which might otherwise have been toxic or indigestible, could become regular contributors to the foragers' larder (Ames 1983; Johns 1990; Nabhan 2004; Dominy et al. 2008). Then, beginning by at least 30 kyr (my own hunch is that in the Southern Levant it began even earlier), came an inflection point in the rate of food-related intensification, with the addition in rapid succession of many new food types and many new practices and technologies for processing these foods, such as baking, grinding, pounding, steaming, parching, smoking, leaching, fermenting, and technologies that permitted boiling and stewing (Stahl 1989; Wandsnider 1997; Benison 1999; Manne et al. 2005; Holt and Formicola 2008; Wollstonecroft et al. 2008; Jones 2009: 177, 178; Nakazawa et al. 2009; Thoms 2009). The impact of these innovations in food-processing technologies is indirectly reflected in the dramatic changes that have occurred over the Late Pleistocene and Holocene in tooth size and craniofacial structure (e.g., Brace et al. 1987; Lieberman et al. 2004). While these skeletal alterations very likely stem in large part from reduction of the biomechanical stresses and strains associated with mastication, they almost certainly are also linked to major changes in the nutritional properties and digestibility of our foods (Lucas 2006; Lucas et al. 2009; Carmody and Wrangham 2009; Wrangham 2009). There are many other important technologies, each of which added significantly to the array of foods humans could exploit and the way they were processed for consumption. Unfortunately, for most of these we have little or no direct archaeological evidence that would help us pin down the timing of their introduction. Among these—and this is by no means an exhaustive list—are spears, atlatls, bows and arrows, pit-traps, digging sticks, snares, traps, starch presses, leaching baskets, geophagy or clay-eating (Johns and Duquette 1991; Rowland 2002), nets, decoys, weirs, fish hooks, atlatls, boomerangs, bolas stones, seed-beaters, and many other hunting and gathering technologies. Each of these altered what foods could be procured and in what quantity, as well as their toxicity, digestibility, and nutritional properties.

There are other ways to respond to food stress instead of, or in addition to, the largely technological ones I have just mentioned; there are many social mechanisms that can be brought into play to cope with subsistence-related stresses. This is the most understudied part of the BSR, and the part that really deserves a much larger share of our attention. While these social mechanisms are going to be very difficult to "see" in the archaeological record, I think we could make considerable progress by devoting more of our time and effort to developing a suitable theoretical framework—in other words, what should we expect in the way of responses under different sets of circumstances? Such a framework would then give us a better idea of what to look for in the archaeological record. A few of the more likely social responses that we need to consider include: working harder, "tightening one's belt," traveling farther to exploit resources, relocating to more productive areas, driving competitors out of one's foraging range, borrowing or begging from close kin, tapping the resources of more distant kin and non-kin, storing (hoarding) surpluses, exchanging with others or raiding their resources or food stores instead, and many others. We owe much of the pioneering thought along these lines to the work of Elizabeth Colson (1979), and since then many others have pursued facets of this interesting issue (e.g., Minnis 1985; Minc 1986; Minc and Smith 1989; Halstead and O'Shea 1989; Lupo and Schmitt 2002). Unfortunately, we still are far from understanding the specific constellation of conditions that might favor one particular response, or set of responses, over others. What does seem probable, however, is that there is likely to be a hierarchy in which these responses are adopted, such that those coping strategies which are the least costly, or which entail the least long-term or permanent disruption or alteration of the social fabric, are the ones most likely to be employed first and, only under conditions of persisting, worsening, or rapidly recurring stress, would one shift to higher-level responses (Slobodkin 1968; Slobodkin and Rapoport 1974; Ware 1983; Minnis 1985). In other words, one might expect the adoption of such strategies as "belt-tightening," expanding one's diet to incorporate lower-ranked food types, or grease-rendering to precede the adoption of food storage practices, since the latter would necessitate changes in mobility patterns, encourage raiding, increase the chances of food losses to decay and pests, and is more difficult to accommodate within an egalitarian ideology.

Developing reliable means for retrieving, preserving, and transmitting critical knowledge and information might also be considered a component of subsistence-related intensification (Whallon 2011). As Whallon points out, even in small-scale foraging societies not everyone possess all of the knowledge and skills needed for the group to survive and persist. There may be only one shaman who carries the group's sacred knowledge in his or her head, only one or two good spear-makers, only a few skilled hunters, only a few good weavers or basket-makers, and only a handful of individuals who have experienced a particular type of life-threatening crisis in the past, such as a devastating crash of the reindeer population, and who therefore know what to do should an analogous crises recur (Minc 1986; Minc and Smith 1989; Funk 2011). Random accidents, disease, or other disasters can instantly eliminate a critical component of the group's knowledge base. Thus, as hominin technology and society become increasingly complex, and the knowledge base more specialized and diversified, mechanisms must be developed and fine-tuned for preserving and transmitting vital information, obviously including those aspects of knowledge that relate directly to subsistence.

When viewed from this enlarged perspective, the BSR becomes a very complex, multi-dimensional phenomenon. In fact, I prefer to think of the BSR as a subset of issues within a much broader domain encompassing social as well as economic "intensification." Much more is involved than simply adding a few less-desirable or difficult-to-catch foods to the daily fare, or figuring out new ways to process them. Thus, Kebara's faunal record, while interesting, provides only the tiniest of hints that some form of resource intensification might have been occurring along Israel's coastal plain during the latter part of the MP. It remains to be determined whether this was a spatially restricted, low-level response to stressful conditions that persisted for a few millennia and then abated, or instead reflects a more widespread and persistent population-resource imbalance that could have begun long before the late MP, and may have continued to worsen following the late MP, engendering an ever-broadening and increasingly costly and irreversible chain of subsistence-related technological *and social responses* over the course of the ensuing Upper Paleolithic and Epipaleolithic that culminated in the dramatic transformations that mark the beginnings of agriculture and settled village life in the Near East. We still have much to do…

Acknowledgments I thank S. Bar-David, G. Bar-Oz, O. Bar-Yosef, J. L. Clark, A. Belfer-Cohen, S. J. M. Davis, T. Dayan, P. Goldberg, N. Goren-Inbar, L. K. Horwitz, E. Hovers, S. Kuhn, L. Meignen, H. Monchot, E. Morin, N. Munro, J. F. O'Connell, R. Rabinovich, B. Shalmon, M. C. Stiner, A. Stutz, and the late E. Tchernov, as well as my anonymous reviewers, for their many very helpful suggestions in the development of the ideas presented here. Partial support for this research was received from the U.S.-Israel Binational Science Foundation, L. S. B. Leakey Foundation, National Endowment for the Humanities Fellowship for University Teachers, Hebrew University (Jerusalem), and the University of Michigan's Museum of Anthropology, Office of the Vice-President for Research, and Horace H. Rackham School of Graduate Studies.

References

Abbo, S., Zezak, I., Schwartz, E., Lev-Yadun, S., Kerem, Z., & Gopher, A. (2008). Wild lentil and chickpea harvest in Israel: Bearing on the origins of Near Eastern farming. *Journal of Archaeological Science, 35*, 3172–3177.

Adler, D. S., & Bar-Oz, G. (2009). Seasonal patterns of prey acquisition and inter-group competition during the Middle and Upper Palaeolithic of the southern Caucasus. In J.-J. Hublin & M. P. Richards (Eds.), *The evolution of hominin diets: Integrating approaches to the study of Palaeolithic subsistence* (pp. 127–140). New York: Springer.

Adler, D. S., Bar-Oz, G., Belfer-Cohen, A., & Bar-Yosef, O. (2006). Ahead of the game: Middle and Upper Palaeolithic hunting behaviors in the southern Caucasus. *Current Anthropology, 47*, 89–118.

Alroy, J. (2001). A multispecies overkill simulation of the end-Pleistocene megafaunal mass extinction. *Science, 292*, 1893–1896.

Ames, B. N. (1983). Dietary carcinogens and anticarcinogens: Oxygen radicals and degenerative diseases. *Science, 221*, 1256–1264.

Aranguren, B., Becattini, R., Lippi, M. M., & Revedin, A. (2007). Grinding flour in Upper Palaeolithic Europe (25 000 years bp). *Antiquity, 81*, 845–855.

Atici, A. L. (2009). Implications of age structures for Epipaleolithic hunting strategies in the western Taurus Mountains, southwest Turkey. *Anthropozoologica, 44*, 13–39.

Baharav, D. (1974). Notes on the population structure and biomass of the mountain gazelle, *Gazella gazella gazella*. *Israel Journal of Zoology, 23*, 39–44.

Barker, G., Barton, H., Bird, M., Daly, P., Datan, I., Dykes, A., et al. (2007). The 'human revolution' in lowland tropical Southeast Asia: The antiquity and behavior of Anatomically Modern Humans at Niah Cave (Sarawak, Borneo). *Journal of Human Evolution, 52*, 243–261.

Bar-Matthews, M., Ayalon, A., & Kaufman, A. (1998). Palaeoclimate evolution in the eastern Mediterranean region during the last 58,000 years as derived from stable isotopes of speleothems (Soreq Cave, Israel). In *Isotope Techniques in the Study of Environmental Change. Proceedings of an International Symposium on Isotope Techniques in the Study of Past and Current Environmental Changes in the Hydrosphere and the Atmosphere* (pp. 673–682). Vienna: International Atomic Energy Agency.

Bar-Matthews, M., Ayalon, A., Kaufman, A., & Wasserburg, G. J. (1999). The eastern Mediterranean paleoclimate as a reflection of regional events: Soreq Cave, Israel. *Earth and Planetary Science Letters, 166*, 85–95.

Bar-Matthews, M., Ayalon, A., Gilmour, M., Matthews, A., & Hawkesworth, C. J. (2003). Sea-land oxygen isotopic relationships from planktonic foraminifera and speleothems in the eastern Mediterranean region and their implication for paleorainfall during interglacial intervals. *Geochimica et Cosmochimica Acta, 67*, 3181–3199.

Barnosky, A. D., Bell, C. J., Emslie, S. D., Goodwin, H. T., Mead, J. I., Repenning, C. A., et al. (2004). Exceptional record of mid-Pleistocene vertebrates helps differentiate climatic from anthropogenic ecosystem perturbations. *Proceedings of the National Academy of Sciences of the United States of America, 101*, 9297–9302.

Bar-Oz, G. (2004). *Epipalaeolithic subsistence strategies in the Levant: A zooarchaeological perspective*. Boston: Brill Academic Publishers.

Bar-Oz, G., & Dayan, T. (2003). Testing the use of multivariate inter-site taphonomic comparisons: The faunal analysis of Hefzibah in its Epipalaeolithic cultural context. *Journal of Archaeological Science, 30*, 885–900.

Bar-Oz, G., Dayan, T., & Kaufman, D. (1999). The Epipalaeolithic faunal sequence in Israel: A view from Neve David. *Journal of Archaeological Science, 26*, 67–82.

Bar-Oz, G., Zeder, M., & Hole, F. (2011). Role of mass-kill hunting strategies in the extirpation of Persian gazelle (*Gazella subgutturosa*) in the northern Levant. *Proceedings of the National Academy of Sciences of the United States of America, 108*, 7345–7350.

Bar-Yosef, O. (1991). The history of excavations at Kebara Cave. In O. Bar-Yosef & B. Vandermeersch (Eds.), *Le Squelette Moustérien de Kébara 2* (pp. 17–27). Paris: Editions du Centre National de la Recherche Scientifique.

Bar-Yosef, O. (1995). The role of climate in the interpretation of human movements and cultural transformations in western Asia. In E. S. Vrba, G. H. Denton, T. C. Partridge, & L. H. Burckle (Eds.), *Paleoclimate and evolution, with emphasis on human origins* (pp. 507–523). New Haven: Yale University Press.

Bar-Yosef, O., Vandermeersch, B., Arensburg, B., Belfer-Cohen, A., Goldberg, P., Laville, H., et al. (1992). The excavations in Kebara Cave, Mt. Carmel. *Current Anthropology, 33*, 497–550.

Belmaker, M., & Hovers, E. (2011). Ecological change and the extinction of the Levantine Neanderthals: Implications from a diachronic study of micromammals from Amud Cave, Israel. *Quaternary Science Reviews, 30*, 3196–3209.

Benedict, J. B. (2005). Tundra game drives: An arctic-alpine comparison. *Arctic, Antarctic, and Alpine Research, 37*, 425–434.

Benison, C. J. (1999). Burned rock complexes, baked clay objects, steatite, and ceramics: Evolutionary implications for Plains/Eastern Woodlands cooking technologies. *North American Archaeologist, 20*, 287–317.

Betts, A. (1989). The Solubba: Nonpastoral nomads in Arabia. *Bulletin of the American Schools of Oriental Research, 274*, 61–69.

Binford, L. R. (1978). *Nunamiut Ethnoarchaeology*. New York: Academic Press.

Bliege Bird, R. L., Codding, B. F., & Bird, D. W. (2009). What explains differences in men's and women's production? Determinants of gendered foraging inequalities among Martu. *Human Nature, 20*, 105–129.

Blockley, S. P. E., & Pinhasi, R. (2011). A revised chronology for the adoption of agriculture in the Southern Levant and the role of Lateglacial climatic change. *Quaternary Science Reviews, 30*, 98–108.

Bocherens, H. (2009). Neanderthal dietary habits: Review of the isotopic evidence. In J.-J. Hublin & M. P. Richards (Eds.), *The evolution of hominin diets: Integrating approaches to the study of Palaeolithic subsistence* (pp. 241–250). Dordrecht: Springer.

Bocherens, H. (2011). Diet and ecology of Neanderthals: Implications from C and N isotopes—insights from bone and tooth biogeochemistry. In N. J. Conard & J. Richter (Eds.), *Neanderthal lifeways, subsistence and technology: One hundred fifty years of Neanderthal study* (pp. 73–85). New York: Springer.

Bocquet-Appel, J.-P., & Tuffreau, A. (2009). Technological responses of Neanderthals to macroclimatic variations (240,000–40,000 BP). *Human Biology, 81*, 287–307.

Brace, C. L., Rosenberg, K. R., & Hunt, K. D. (1987). Gradual change in human tooth size in the late Pleistocene and post-Pleistocene. *Evolution, 41*, 705–720.

Bratlund, B. (2000). Taubach revisited. *Jahrbuch des Römisch-Germanischen Zentralmuseums Mainz, 46*, 61–174.

Brink, J. W. (2005). *Inukshuk*: caribou drive lanes on southern Victoria Island, Nunavut, Canada. *Arctic Anthropology, 42*, 1–28.

Brook, B. W., & Bowman, D. M. J. S. (2004). The uncertain blitzkrieg of Pleistocene megafauna. *Journal of Biogeography, 31*, 517–523.

Brooks, A. S. (1996). Open air sites in the Middle Stone Age of Africa. In N. J. Conard & F. Wendorf (Eds.), *Middle Palaeolithic and Middle Stone Age settlement system, International Union of Prehistoric and Protohistoric Sciences. Proceedings of the XIII International Congress of Prehistoric and Protohistoric Sciences, Forli, Italy, 8–14 September 1996, Vol. 6, Workshops, Tome 1* (pp. 249–253). Forli: ABACO Edizioni.

Broughton, J. M. (1994). Declines in mammalian foraging efficiency during the late Holocene, San Francisco Bay, California. *Journal of Anthropological Archaeology, 13*, 371–401.

Broughton, J. M. (1997). Widening diet breadth, declining foraging efficiency, and prehistoric harvest pressure: Ichthyofaunal evidence from the Emeryville Shellmound, California. *Antiquity, 71*, 845–862.

Carmody, R. N., & Wrangham, R. W. (2009). The energetic significance of cooking. *Journal of Human Evolution, 57*, 379–391.

Chapman, D. I., & Chapman, N. G. (1975). *Fallow deer: Their history, distribution and biology*. Lavenham, Suffolk: Terence Dalton.

Colson, E. (1979). In good years and in bad: Food strategies of self-reliant societies. *Journal of Anthropological Research, 35*, 18–29.

Daniau, A.-L., d'Errico, F., & Sánchez Goñi, M. F. (2010). Testing the hypothesis of fire use for ecosystem management by Neanderthal and Upper Palaeolithic modern human populations. *PLoS One, 5*. doi:10.1371/journal.pone.0009157

Davis, S. J. M. (1977). The ungulate remains from Kebara Cave. In B. Arensburg & O. Bar-Yosef (Eds.), *Moshé Stekelis Memorial Volume* (pp. 150–163). Jerusalem: Israel Exploration Society.

Davis, S. J. M. (1982). Climatic change and the advent of domestication: The succession of ruminant artiodactyls in the late Pleistocene-Holocene in the Israeli region. *Paléorient, 8*, 5–15.

Davis, S. J. M. (1983). The age profiles of gazelle predated by ancient man in Israel: Possible evidence for a shift from seasonality to sedentism in the Natufian. *Paléorient, 9*, 55–62.

Davis, S. J. M. (1989). Why did prehistoric people domesticate food animals? The bones from Hatoula 1980-86. In O. Bar-Yosef & B. Vandermeersch (Eds.), *Investigations in South Levantine Prehistory (Préhistoire du Sud-Levant)* (pp. 43–59). Oxford: British Archaeological Reports.

Davis, S. J. M., Rabinovich, R., & Goren-Inbar, N. (1988). Quaternary extinctions and population increase in western Asia: The animal remains from Biq'at Quneitra. *Paléorient, 14*, 95–105.

Domínguez-Rodrigo, M. (1999). The study of skeletal part profiles: An ambiguous taphonomic tool for zooarchaeology. *Complutum, 10*, 15–24.

Dominy, N. J., Vogel, E. R., Yeakel, J. D., Constantino, P., & Lucas, P. W. (2008). Mechanical properties of plant underground storage organs and implications for dietary models of early hominins. *Evolutionary Biology, 35*, 159–175.

Edwards, D. A., & O'Connell, J. F. (1995). Broad spectrum diets in arid Australia. *Antiquity, 69*, 769–783.

Eisenmann, V. (1992). Systematic and biostratigraphical interpretation of the equids from Qafzeh, Tabun, Skhul and Kebara (Acheulo-Yabrudian to Upper Paleolithic of Israel). *Archaeozoologia, 5*, 43–62.

El Zaatari, S. (2007). *Ecogeographic variation in Neandertal dietary habits: Evidence from microwear texture analysis*. Ph.D. dissertation, State University of New York, Stony Brook.

Emery-Barbier, A., & Thiébault, S. (2005). Preliminary conclusions on the Late Glacial vegetation in south-west Anatolia (Turkey): The complementary nature of palynological and anthracological approaches. *Journal of Archaeological Science, 32*, 1232–1251.

Ewers, J. C. (1958). *The Blackfeet: Raiders on the Northwestern Plains*. Norman: University of Oklahoma Press.

Fairbairn, A. S., Hope, G. S., & Summerhayes, G. R. (2006). Pleistocene occupation of New Guinea's highland and subalpine environments. *World Archaeology, 38*, 371–386.

Fiorenza, L., Benazzi, S., Tausch, J., Kullmer, O., Bromage, T. G., & Schrenk, F. (2011). Molar Macrowear reveals Neanderthal eco-geographic dietary variation. *PLoS One, 6*. doi: 10.1371/journal.pone.0014769

Flannery, K. V. (1969). Origins and ecological effects of early domestication in Iran and the Near East. In P. J. Ucko & G. W. Dimbleby (Eds.), *The domestication and exploitation of plants and animals* (pp. 73–100). Chicago: Aldine.

Ford, R. I. (1974). Northeastern archaeology: Past and future directions. *Annual Review of Anthropology, 3*, 385–413.

Funk, C. L. (2011). Yup'ik Eskimo gendered information storage patterns. In R. E. Whallon, W. A. Lovis, & R. K. Hitchcock (Eds.), *Information and its role in Hunter-Gatherer Bands* (pp. 29–58). Los Angeles: University of California, Cotsen Institute of Archaeology Press.

Garrigan, D., & Kingan, S. B. (2007). Archaic human admixture: A view from the genome. *Current Anthropology, 48*, 895–902.

Gaudzinski, S. (1998). Large mammal hunting strategies in the Palaeolithic of Europe: A taphonomic approach. In J. Bayley (Ed.), *Science in archaeology: An agenda for the future* (pp. 47–62). London: English Heritage.

Gaudzinski, S. (2000). On the variability of Middle Palaeolithic procurement tactics: The case of Salzgitter Lebenstedt, northern Germany. *International Journal of Osteoarchaeology, 10*, 396–406.

Gaudzinski, S. (2006). Monospecific or species-dominated faunal assemblages during the Middle Paleolithic in Europe. In E. Hovers & S. L. Kuhn (Eds.), *Transitions before the transition: Evolution and stability in the Middle Paleolithic and Middle Stone Age* (pp. 137–147). New York: Springer.

Gaudzinski-Windheuser, S., & Niven, L. (2009). Hominin subsistence patterns during the Middle and Late Paleolithic in northwestern Europe. In J.-J. Hublin & M. P. Richards (Eds.), *The evolution of hominin diets: Integrating approaches to the study of Palaeolithic subsistence* (pp. 99–111). New York: Springer.

Gilead, I., & Grigson, C. (1984). Far'ah II: A Middle Palaeolithic open-air site in the northern Negev, Israel. *Proceedings of the Prehistoric Society, 50*, 71–97.

Goren-Inbar, N. (Ed.). (1990). *Quneitra: A Mousterian site on the Golan heights*. Qedem, Monograph 31. Jerusalem: Hebrew University of Jerusalem, Institute of Archaeology.

Gramly, R. M. (1977). Deerskins and hunting territories: Competition for a scarce resource of the northeastern woodlands. *American Antiquity, 42*, 601–605.

Green, R. E., Krause, J., Briggs, A. W., Maricic, T., Stenzel, U., Kircher, M., et al. (2010). A draft sequence of the Neandertal genome. *Science, 328*, 710–722.

Hallin, K. A., Schoeninger, M. J., & Schwarcz, H. P. (2012). Paleoclimate during Neandertal and anatomically modern human occupation at Amud and Qafzeh, Israel: The stable isotope data. *Journal of Human Evolution, 62*, 59–73.

Halstead, P. L. J., & O'Shea, J. M. (1989). Introduction: Cultural responses to risk and uncertainty. In P. L. J. Halstead & J. M. O'Shea (Eds.), *Bad year economics: Cultural responses to risk and uncertainty* (pp. 1–7). Cambridge: Cambridge University Press.

Haltenorth, T. (1959). Beitrag zur kenntnis des Mesopotamischen damhirsches—*Cervus (Dama) mesopotamicus* Brooke, 1875—und zur stammes- und verbreitungsgeschichte der damhirsche allgemein. *Säugetierkundliche Mitteilungen, 7*, 1–89.

Hawks, J. D., Cochran, G., Harpending, H. C., & Lahn, B. T. (2008). A genetic legacy from archaic *Homo*. *Trends in Genetics, 24*, 19–23.

Hayden, B. (1981). Subsistence and ecological adaptations of modern hunter-gatherers. In R. S. O. Harding & G. Teleki (Eds.), *Omnivorous primates: Gathering and hunting in human evolution* (pp. 344–421). New York: Columbia University Press.

Haynes, G. (2002). The catastrophic extinction of North American mammoths and mastodonts. *World Archaeology, 33*, 391–416.

Headland, T. N., & Greene, H. W. (2011). Hunter-gatherers and other primates as prey, predators, and competitors of snakes. *Proceedings of the National Academy of Sciences of the United States of America, 108*, E1470–E1474.

Henry, A. G., Brooks, A. S., & Piperno, D. R. (2011). Microfossils in calculus demonstrate consumption of plants and cooked foods in Neanderthal diets (Shanidar III, Iraq; Spy I and II, Belgium). *Proceedings of the National Academy of Sciences of the United States of America, 108*, 486–491.

Holt, B. M., & Formicola, V. (2008). Hunters of the ice age: The biology of Upper Paleolithic people. *American Journal of Physical Anthropology, 137*, 70–99.

Holzer, A., Avner, U., Porat, N., & Horwitz, L. K. (2010). Desert kites in the Negev Desert and northeast Sinai: Their function, chronology and ecology. *Journal of Arid Environments, 74*, 806–817.

Hovers, E. (2006). Neandertals and modern humans in the Middle Paleolithic of the Levant: What kind of interaction? In N. J. Conard (Ed.), *When Neanderthals and modern humans met* (pp. 65–85). Tübingen: Kerns Verlag.

Hovers, E., Rak, Y., & Kimbel, W. H. (1991). Amud Cave—1991 season. *Journal of the Israel Prehistoric Society, 24*, 152–157.

James, S. R. (1990). Monitoring archaeofaunal changes during the transition to agriculture in the American Southwest. *Kiva, 56*, 25–43.

Johns, T. A. (1990). *With bitter herbs they shall eat it: Chemical ecology and the origins of human diet and medicine*. Tucson: University of Arizona Press.

Johns, T. A., & Duquette, M. (1991). Detoxification and mineral supplementation as functions of geophagy. *American Journal of Clinical Nutrition, 53*, 448–456.

Johnson, C. N. (2002). Determinants of loss of mammal species during the late Quaternary 'megafauna' extinctions: Life history and ecology, but not body size. *Proceedings of the Royal Society of London, Series B. Biological Sciences, 269B*, 2221–2227.

Jones, M. (2009). Moving north: Archaeobotanical evidence for plant diet in Middle and Upper Paleolithic Europe. In J.-J. Hublin & M. P. Richards (Eds.), *The evolution of hominin diets: Integrating approaches to the study of Palaeolithic subsistence* (pp. 171–180). Dordrecht: Springer.

Katz, S. H., Hediger, M. L., & Valleroy, L. A. (1974). Traditional maize processing techniques in the New World: Anthropological and nutritional significance. *Science, 184*, 765–773.

Kershaw, A. P., Clark, J. S., Gill, A. M., & D'Costa, D. M. (2002). A history of fire in Australia. In R. A. Bradstock, J. E. Williams, & A. M. Gill (Eds.), *Flammable Australia: The fire regimes and biodiversity of a continent* (pp. 3–25). Cambridge: Cambridge University Press.

Kittler, R., Kayser, M., & Stoneking, M. (2003). Molecular evolution of *Pediculus humanus* and the origin of clothing. *Current Biology, 13*, 1414–1417.

Klein, R. G. (1982). Age (mortality) profiles as a means of distinguishing hunted species from scavenged ones in Stone Age archaeological sites. *Paleobiology, 8*, 151–158.

Klokkernes, T. (2007). *Skin processing technology in Eurasian reindeer cultures: A comparative study in material science of Sàmi and Evenk methods—Perspectives on deterioration and preservation of museum artefacts*. Rudkøbing: Langelands Museum, LMR Press.

Koike, H., & Ohtaishi, N. (1985). Prehistoric hunting pressure estimated by the age composition of excavated Sika Deer (*Cervus nippon*) using the annual layer of tooth cement. *Journal of Archaeological Science, 12*, 434–456.

Koike, H., & Ohtaishi, N. (1987). Estimation of prehistoric hunting rates based on the age composition of Sika Deer (*Cervus nippon*). *Journal of Archaeological Science, 14*, 251–269.

Kuhn, S. L., & Stiner, M. C. (2006). What's a mother to do? The division of labor among Neandertals and modern humans in Eurasia. *Current Anthropology, 47*, 953–980.

Lahr, M. M., & Foley, R. A. (2003). Demography, dispersal and human evolution in the Last Glacial period. In T. H. van Andel & W. Davies (Eds.), *Neanderthals and modern humans in the European landscape during the last glaciation: Archaeological results of the Stage 3 Project* (pp. 241–256). Cambridge: McDonald Institute for Archaeological Research.

Leo, N. P., & Barker, S. C. (2005). Unravelling the evolution of the head lice and body lice of humans. *Parasitology Research, 98*, 44–47.

Lev, E., Kislev, M. E., & Bar-Yosef, O. (2005). Mousterian vegetal food in Kebara Cave, Mt. Carmel. *Journal of Archaeological Science, 32*, 475–484.

Lewis, H. T. (1982). Fire technology and resource management in Aboriginal North America and Australia. In N. M. Williams & E. S. Hunn (Eds.), *Resource managers: North American and Australian hunter-gatherers* (pp. 45–68). Boulder: Westview Press.

Lieberman, D. E., Krovitz, G. E., Yates, F. W., Devlin, M., & St. Claire, M. (2004). Effects of food processing on masticatory strain and craniofacial growth in a retrognathic face. *Journal of Human Evolution, 46*, 655–677.

Lourandos, H. (1997). *Continent of hunter-gatherers: New perspectives in Australian Prehistory*. Cambridge: Cambridge University Press.

Lucas, P. W. (2006). Facial dwarfing and dental crowding in relation to diet. In T. G. Bromage, A. Vidal, F. Ramón Areces, E. Aguirre, & A. Perez-Ochoa (Eds.), *Integrative Approaches to Human Health and Evolution. Proceedings of the International Symposium "Integrative Approaches to Human and Evolution," Madrid, 18–20 April, 2005* (pp. 74–82). International Congress Series 1296. Amsterdam: Elsevier.

Lucas, P. W., Sui, Z., Ang, K. Y., Tan, H. T. W., King, S. H., Sadler, B., et al. (2009). Meals versus snacks and the human dentition and diet during the Paleolithic. In J.-J. Hublin & M. P. Richards (Eds.), *The evolution of hominin diets: Integrating approaches to the study of Palaeolithic subsistence* (pp. 31–41). Dordrecht: Springer.

Lupo, K. D. (2007). Evolutionary foraging models in zooarchaeological analysis: Recent applications and future challenges. *Journal of Archaeological Research, 15*, 143–189.

Lupo, K. D., & Schmitt, D. N. (2002). Upper Paleolithic net-hunting, small prey exploitation, and women's work effort: A view from the ethnographic and ethnoarchaeological record of the Congo Basin. *Journal of Archaeological Method and Theory, 9*, 147–179.

Lyman, R. L. (1994). *Vertebrate taphonomy*. Cambridge: Cambridge University Press.

Lyons, S. K., Smith, F. A., & Brown, J. H. (2004). Of mice, mastodons and men: Human-mediated extinctions on four continents. *Evolutionary Ecology Research, 6*, 339–358.

Madella, M., Jones, M. K., Goldberg, P., Goren, Y., & Hovers, E. (2002). The exploitation of plant resources by Neanderthals in Amud Cave (Israel): The evidence from phytolith studies. *Journal of Archaeological Science, 29*, 703–719.

Manne, T. H., Stiner, M. C., & Bicho, N. F. (2005). Evidence for bone grease rendering during the Upper Paleolithic at Vale Boi (Algarve,

Portugal). In N. F. Bicho (Ed.), *Animais na Pré-história e Arqueologia da Península Ibérica. Actas do IV Congresso de Arqueologia Peninsular. Congresso de Arqueologia Peninsular, Faro, 2004* (pp. 145–158). Promontoria Monográfica 3. Faro: Centro de Estudos de Património, Universidade do Algarve.

Marder, O., Yeshurun, R., Lupu, R., Bar-Oz, G., Belmaker, M., Porat, N., et al. (2011). Mammal remains at Rantis Cave, Israel, and Middle-Late Pleistocene human subsistence and ecology in the Southern Levant. *Journal of Quaternary Science, 26*, 769–780.

Marean, C. W., & Kim, S. Y. (1998). Mousterian large-mammal remains from Kobeh Cave: Behavioral implications for Neanderthals and early modern humans. *Current Anthropology, 39*, S79–S113.

Martin, L. (2000). Gazelle (*Gazella* spp.) behavioral ecology: Predicting animal behaviour for prehistoric environments in South-west Asia. *Journal of Zoology (London), 250*, 13–30.

Martin, P. S. (2005). *Twilight of the mammoths: Ice age extinctions and rewilding of America*. Berkeley: University of California Press.

Meignen, L., Beyries, S., Speth, J. D., & Bar-Yosef, O. (1998). Acquisition, traitement des matières animales et fonction du site au Paléolithique Moyen dans la Grotte de Kébara (Israël): approche interdisciplinaire. In J.-P. Brugal, L. Meignen, & M. Patou-Mathis (Eds.), *Économie Préhistorique: Les Comportements de Subsistance au Paléolithique* (pp. 227–241). Sophia Antipolis: APDCA.

Meignen, L., Bar-Yosef, O., Speth, J. D., & Stiner, M. C. (2006). Middle Paleolithic settlement patterns in the Levant. In E. Hovers & S. L. Kuhn (Eds.), *Transitions before the transition: Evolution and stability in the Middle Paleolithic and Middle Stone Age* (pp. 149–170). New York: Springer.

Mendelssohn, H., & Yom-Tov, Y. (1987). *Plants and animals of the land of Israel, Vol. 7, Mammals*. Tel Aviv: Israel Defense Ministry Publishing House [in Hebrew].

Mendelssohn, H., & Yom-Tov, Y. (1999). *Mammalia of Israel. Fauna Palaestina*. Jerusalem: Israel Academy of Sciences and Humanities.

Mercier, N., Valladas, H., Froget, L., Joron, J.-L., Reyss, J.-L., Weiner, S., et al. (2007). Hayonim Cave: A TL-based chronology for this Levantine Mousterian sequence. *Journal of Archaeological Science, 34*, 1064–1077.

Miller Rosen, A. (2003). The microbotanical evidence. In D. O. Henry (Ed.), *Neanderthals in the Levant: Behavioral organization and the beginnings of human modernity* (pp. 156–171). London: Continuum.

Minc, L. D. (1986). Scarcity and survival: The role of oral tradition in mediating subsistence crises. *Journal of Anthropological Archaeology, 5*, 39–113.

Minc, L. D., & Smith, K. (1989). The spirit of survival: Cultural responses to resource variability in north Alaska. In P. L. J. Halstead & J. M. O'Shea (Eds.), *Bad year economics: Cultural responses to risk and uncertainty* (pp. 8–39). Cambridge: Cambridge University Press.

Minnis, P. E. (1985). *Social adaptation to food stress: A prehistoric southwestern example*. Chicago: University of Chicago Press.

Monks, G. G. (1981). Seasonality studies. *Advances in Archaeological Method and Theory, 4*, 177–240.

Munro, N. D. (2003). Small game, the Younger Dryas, and the transition to agriculture in the Southern Levant. *Mitteilungen der Gesellschaft für Urgeschichte, 12*, 47–72.

Munro, N. D. (2004). Zooarchaeological measures of hunting pressure and occupation intensity in the Natufian: Implications for agricultural origins. *Current Anthropology, 45*, S5–S33.

Munro, N. D. (2009a). Epipaleolithic subsistence intensification in the Southern Levant: The faunal evidence. In J.-J. Hublin & M. P. Richards (Eds.), *The evolution of hominid diets: Integrating approaches to the study of Palaeolithic subsistence* (pp. 141–155). Dordrecht: Springer.

Munro, N. D. (2009b). Integrating inter- and intra-site analyses of Epipalaeolithic faunal assemblages from Israel. *Before Farming, 1*, 1–18.

Munro, N. D., & Bar-Oz, G. (2004). Debating issues of equifinality in ungulate skeletal part studies. *Journal of Taphonomy, 2*, 1–13.

Munson, P. J., & Garniewicz, R. C. (2003). Age-mediated survivorship of ungulate mandibles and teeth in canid-ravaged faunal assemblages. *Journal of Archaeological Science, 30*, 405–416.

Munson, P. J., & Marean, C. W. (2003). Adults only? A reconsideration of Middle Paleolithic 'prime-dominated' reindeer hunting at Salzgitter Lebenstedt. *Journal of Human Evolution, 44*, 263–273.

Nabhan, G. P. (2004). *Why some like it hot: Food, genes, and cultural diversity*. Washington, DC: Island Press.

Nakazawa, Y., Straus, L. G., González-Morales, M. R., Solana, D. C., & Saiz, J. C. (2009). On stoneboiling technology in the Upper Paleolithic: Behavioral implications from an early Magdalenian hearth in El Mirón Cave, Cantabria, Spain. *Journal of Archaeological Science, 36*, 684–693.

Nugent, G., Fraser, K. W., Asher, G. W., & Tustin, K. G. (2001). Advances in New Zealand mammalogy 1990–2000: Deer. *Journal of the Royal Society of New Zealand, 31*, 263–298.

O'Connell, J. F., Hawkes, K., & Blurton Jones, N. G. (1988). Hadza hunting, butchering, and bone transport and their archaeological implications. *Journal of Anthropological Research, 44*, 113–162.

O'Connell, J. F., Hawkes, K., & Blurton Jones, N. G. (1990). Reanalysis of large mammal body part transport among the Hadza. *Journal of Archaeological Science, 17*, 301–316.

O'Connell, J. F., Hawkes, K., & Blurton Jones, N. G. (1992). Patterns in the distribution, site structure and assemblage composition of Hadza kill-butchering sites. *Journal of Archaeological Science, 19*, 319–345.

Pardoe, C. (1988). The cemetery as symbol: The distribution of prehistoric Aboriginal burial grounds in southeastern Australia. *Archaeology in Oceania, 23*, 1–16.

Pardoe, C. (1994). Bioscapes: The evolutionary landscape of Australia. *Archaeology in Oceania, 29*, 182–190.

Pardoe, C. (1995). Riverine, biological and cultural evolution in southeastern Australia. *Antiquity, 69*, 696–713.

Pettitt, P. B. (2010). *The Palaeolithic origins of human burial*. London: Routledge.

Pike-Tay, A., & Cosgrove, R. (2002). From reindeer to wallaby: Recovering patterns of seasonality, mobility, and prey selection in the Palaeolithic Old World. *Journal of Archaeological Method and Theory, 9*, 101–146.

Pinter, N., Fiedel, S., & Keeley, J. E. (2011). Fire and vegetation shifts in the Americas at the vanguard of Paleoindian migration. *Quaternary Science Reviews, 30*, 269–272.

Piperno, D. R., Weiss, E., Holst, I., & Nadel, D. (2004). Processing of wild cereal grains in the Upper Palaeolithic revealed by starch grain analysis. *Nature, 430*, 670–673.

Potter, J. M. (2000). Pots, parties, and politics: Communal feasting in the American Southwest. *American Antiquity, 65*, 471–492.

Rabinovich, R., & Hovers, E. (2004). Faunal analysis from Amud Cave: Preliminary results and interpretations. *International Journal of Osteoarchaeology, 14*, 287–306.

Rabinovich, R., & Tchernov, E. (1995). Chronological, paleoecological and taphonomical aspects of the Middle Paleolithic site of Qafzeh, Israel. In H. Buitenhuis & H.-P. Uerpmann (Eds.), *Archaeozoology of the Near East, II, Proceedings of the Second*

International Symposium on the Archaeozoology of Southwestern Asia and Adjacent Areas (pp. 5–44). Leiden: Backhuys.

Rawlinson, G. (1859). *The history of Herodotus, Vol. III, Book IV—Melpomene*. London: John Murray.

Reed, D. L., Smith, V. S., Hammond, S. L., Rogers, A. R., & Clayton, D. H. (2004). Genetic analysis of lice supports direct contact between modern and archaic humans. *PLoS Biology, 2*, e340. doi: 10.1371/journal.pbio.0020340

Rendu, W., Costamagno, S., Meignen, L., & Soulier, M.-C. (2012). Monospecific faunal spectra in Mousterian contexts: Implications for social behavior. *Quaternary International, 247*, 50–58.

Richards, M. P., Pettitt, P. B., Stiner, M. C., & Trinkaus, E. (2001). Stable isotope evidence for increasing dietary breadth in the European mid-Upper Paleolithic. *Proceedings of the National Academy of Sciences of the United States of America, 98*, 6528–6532.

Richter, J. (2000). Social memory among late Neanderthals. In J. Orschiedt & G.-C. Weniger (Eds.), *Neanderthals and modern humans—Discussing the transition: Central and Eastern Europe from 50.000-30.000 B.P.* (pp. 123–132). Wissenschaftliche Schriften des Neanderthal Museums 2. Mettmann: Neanderthal Museums.

Roberts, R. G., Flannery, T. F., Ayliffe, L. K., Yoshida, H., Olley, J. M., Prideaux, G. J., et al. (2001). New ages for the last Australian megafauna: Continent-wide extinction about 46,000 years ago. *Science, 292*, 1888–1892.

Rowland, M. J. (2002). Geophagy: An assessment of implications for the development of Australian indigenous plant processing technologies. *Australian Aboriginal Studies, 2002*, 51–66.

Rowland, J. C. (2006). *Inter- and intra-tooth isotopic variation in mammalian tooth enamel from Western Israel: Implications for paleoenvironmental and paleoclimate change over the past 350 kyr*. M.S. Thesis, University of Arizona, Tucson.

Ryder, M. L. (1966). Can one cook in a skin? *Antiquity, 40*, 225–227.

Ryder, M. L. (1969). Paunch cooking. *Antiquity, 43*, 218–220.

Sackett, L. (1979). The pursuit of prominence: Hunting in an Australian Aboriginal community. *Anthropologica, 21*, 223–246.

Sapir-Hen, L., Bar-Oz, G., Khalaily, H., & Dayan, T. (2009). Gazelle exploitation in the early Neolithic site of Motza, Israel: The last of the gazelle hunters in the Southern Levant. *Journal of Archaeological Science, 36*, 1538–1546.

SAS Institute. (1998). *StatView reference* (2nd ed.). Cary, NC: SAS Institute.

Schick, T., & Stekelis, M. (1977). Mousterian assemblages in Kebara Cave, Mount Carmel. In B. Arensburg & O. Bar-Yosef (Eds.), *Moshé Stekelis Memorial Volume* (pp. 97–149). Jerusalem: Israel Exploration Society.

Semaw, S. (2000). The world's oldest stone artefacts from Gona, Ethiopia: Their implications for understanding stone technology and patterns of human evolution between 2.6-1.5 million years ago. *Journal of Archaeological Science, 27*, 1197–1214.

Shea, J. J. (2008). Transitions or turnovers? Climatically-forced extinctions of *Homo sapiens* and Neanderthals in the east Mediterranean Levant. *Quaternary Science Reviews, 27*, 2253–2270.

Slobodkin, L. B. (1968). Toward a predictive theory of evolution. In R. C. Lewontin (Ed.), *Population biology and evolution* (pp. 187–205). Syracuse: University of Syracuse Press.

Slobodkin, L. B., & Rapoport, A. (1974). An optimal strategy of evolution. *Quarterly Review of Biology, 49*, 181–200.

Sokal, R. R., & Rohlf, F. J. (1969). *Biometry: The principles and practice of statistics in biological research.* San Francisco: W. H. Freeman.

Speth, J. D. (1997). Communal bison hunting in western North America: Background for the study of Paleolithic bison hunting in Europe. In M. Patou-Mathis & M. Otte (Eds.), *L'Alimentation des Hommes du Paléolithique: Approche Pluridisciplinaire* (pp. 23–57). Etudes et Recherches Archéologiques de l'Université de Liège (ERAUL) 83. Liège: Université de Liège.

Speth, J. D. (2004a). Hunting pressure, subsistence intensification, and demographic change in the Levantine late Middle Paleolithic. In N. Goren-Inbar & J. D. Speth (Eds.), *Human Paleoecology in the Levantine Corridor* (pp. 149–166). Oxford: Oxbow Books.

Speth, J. D. (2004b). News flash: Negative evidence convicts Neanderthals of gross mental incompetence. *World Archaeology, 36*, 519–526.

Speth, J. D. (2006). Housekeeping, Neandertal-style: Hearth placement and midden formation in Kebara Cave (Israel). In E. Hovers & S. L. Kuhn (Eds.), *Transitions before the transition: Evolution and stability in the Middle Paleolithic and Middle Stone Age* (pp. 171–188). New York: Springer.

Speth, J. D. (2010a). Boiling vs. roasting in the Paleolithic: Broadening the "broadening food spectrum". *Journal of the Israel Prehistoric Society (Mitekufat Ha'even), 40*, 63–83.

Speth, J. D. (2010b). *The Paleoanthropology and archaeology of big-game hunting: Protein, fat or politics?* New York: Springer.

Speth, J. D., & Clark, J. L. (2006). Hunting and overhunting in the Levantine late Middle Palaeolithic. *Before Farming, 3*, 1–42.

Speth, J. D., & Scott, S. L. (1989). Horticulture and large-mammal hunting: The role of resource depletion and the constraints of time and labor. In S. Kent (Ed.), *Farmers as hunters* (pp. 71–79). New York: Cambridge University Press.

Speth, J. D., & Tchernov, E. (1998). The role of hunting and scavenging in Neandertal procurement strategies: New evidence from Kebara Cave (Israel). In T. Akazawa, K. Aoki, & O. Bar-Yosef (Eds.), *Neandertals and modern humans in Western Asia* (pp. 223–239). New York: Plenum.

Speth, J. D., & Tchernov, E. (2001). Neandertal hunting and meat-processing in the Near East: Evidence from Kebara Cave (Israel). In C. B. Stanford & H. T. Bunn (Eds.), *Meat-eating and human evolution* (pp. 52–72). Oxford: Oxford University Press.

Speth, J. D., & Tchernov, E. (2007). The Middle Paleolithic occupations at Kebara Cave: A faunal perspective. In O. Bar-Yosef & L. Meignen (Eds.), *Kebara Cave, Mt. Carmel, Israel: The Middle and Upper Paleolithic Archaeology, Part I* (pp. 165–260). American School of Prehistoric Research Bulletin 49. Cambridge: Harvard University, Peabody Museum of Archaeology and Ethnology.

Speth, J. D., Meignen, L., Bar-Yosef, O., & Goldberg, P. (2012). The spatial organization of Middle Paleolithic occupation X in Kebara Cave (Israel): The concentrations of animal bones. *Quaternary International, 247*, 85–102.

Stahl, A. B. (1989). Plant-food processing: Implications for dietary quality. In D. R. Harris & G. C. Hillman (Eds.), *Foraging and farming: The evolution of plant exploitation* (pp. 171–194). London: Unwin Hyman.

Stefansson, V. (1921). *The friendly arctic: The story of five years in polar regions*. New York: Macmillan.

Stiner, M. C. (1994). *Honor among thieves: A zooarchaeological study of Neandertal Ecology*. Princeton: Princeton University Press.

Stiner, M. C. (2001). Thirty years on the "broad spectrum revolution" and Paleolithic demography. *Proceedings of the National Academy of Sciences of the United States of America, 98*, 6993–6996.

Stiner, M. C. (2003). Zooarchaeological evidence for resource intensification in Algarve, southern Portugal. *Promontoria, Revista do Departamento de Historia, Arqueologia e Patrimonio da Universidade do Algarve, 1*, 27–61.

Stiner, M. C. (2005). *The Faunas of Hayonim Cave, Israel: A 200,000-Year record of Paleolithic diet, demography, and society*. American School of Prehistoric Research Bulletin 48. Cambridge:

Harvard University, Peabody Museum of Archaeology and Ethnology.

Stiner, M. C. (2006). Middle Paleolithic subsistence ecology in the Mediterranean region. In E. Hovers & S. L. Kuhn (Eds.), *Transitions before the transition: Evolution and stability in the Middle Paleolithic and Middle Stone Age* (pp. 213–231). New York: Springer.

Stiner, M. C. (2009a). Prey choice, site occupation intensity and economic diversity in the Middle-early Upper Palaeolithic at the Üçağızlı Caves, Turkey. *Before Farming, 3*, 1–20.

Stiner, M. C. (2009b). The antiquity of large game hunting in the Mediterranean Paleolithic: Evidence from mortality patterns. In J. J. Shea & D. E. Lieberman (Eds.), *Transitions in prehistory: Papers in honor of Ofer Bar-Yosef* (pp. 105–125). Harvard University, American School of Prehistoric Research. Cambridge: Peabody Museum Press/Brill.

Stiner, M. C., & Munro, N. D. (2011). On the evolution of diet and landscape during the Upper Paleolithic through Mesolithic at Franchthi Cave (Peloponnese, Greece). *Journal of Human Evolution, 60*, 618–636.

Stiner, M. C., Munro, N. D., Surovell, T. A., Tchernov, E., & Bar-Yosef, O. (1999). Paleolithic population growth pulses evidenced by small animal exploitation. *Science, 283*, 190–194.

Stiner, M. C., Munro, N. D., & Surovell, T. A. (2000). The tortoise and the hare: Small-game use, the broad-spectrum revolution, and Paleolithic demography. *Current Anthropology, 41*, 39–73.

Stiner, M. C., Gopher, A., & Barkai, R. (2011). Hearth-side socioeconomics, hunting and paleoecology during the late Lower Paleolithic at Qesem Cave, Israel. *Journal of Human Evolution, 60*, 213–233.

Stutz, A. J., Munro, N. D., & Bar-Oz, G. (2009). Increasing the resolution of the broad spectrum revolution in the Southern Levantine Epipaleolithic (19-12 ka). *Journal of Human Evolution, 56*, 294–306.

Surovell, T. A., Waguespack, N., & Brantingham, P. J. (2005). Global archaeological evidence for proboscidean overkill. *Proceedings of the National Academy of Sciences of the United States of America, 102*, 6231–6236.

Tchernov, E. (1998). The faunal sequence of the Southwest Asian Middle Paleolithic in relation to hominid dispersal events. In T. Akazawa, K. Aoki, & O. Bar-Yosef (Eds.), *Neandertals and modern humans in Western Asia* (pp. 77–90). New York: Plenum.

Thoms, A. V. (2009). Rocks of ages: Propagation of hot-rock cookery in western North America. *Journal of Archaeological Science, 36*, 573–591.

Toups, M. A., Kitchen, A., Light, J. E., & Reed, D. L. (2011). Origin of clothing lice indicates early clothing use by Anatomically Modern Humans in Africa. *Molecular Biology and Evolution, 28*, 29–32.

Turner, R., Roberts, N., Eastwood, W. J., Jenkins, E., & Rosen, A. (2010). Fire, climate and the origins of agriculture: Micro-charcoal records of biomass burning during the Last Glacial-Interglacial transition in Southwest Asia. *Journal of Quaternary Science, 25*, 371–386.

Turney, C. S. M., Kershaw, A. P., Moss, P., Bird, M. I., Fifield, L. K., Cresswell, R. G., et al. (2001). Redating the onset of burning at Lynch's Crater (North Queensland): Implications for human settlement in Australia. *Journal of Quaternary Science, 16*, 767–771.

Vaks, A., Bar-Matthews, M., Ayalon, A., Schilman, B., Gilmour, M., Hawkesworth, C. J., et al. (2003). Paleoclimate reconstruction based on the timing of speleothem growth and oxygen and carbon isotope composition in a cave located in the rain shadow in Israel. *Quaternary Research, 59*, 182–193.

Valladas, H., & Valladas, G. (1991). Datation par la thermoluminescence de silex chauffés des grottes de Kébara et de Qafzeh. In O. Bar-Yosef & B. Vandermeersch (Eds.), *Le Squelette Moustérien de Kébara 2* (pp. 43–47). Cahiers de Paléoanthropologie. Paris: CNRS.

Valladas, H., Joron, J.-L., Valladas, G., Arensburg, B., Bar-Yosef, O., Belfer-Cohen, A., et al. (1987). Thermoluminescence dates for the Neanderthal burial site at Kebara in Israel. *Nature, 330*, 159–160.

Valladas, H., Mercier, N., Froget, L., Hovers, E., Joron, J.-L., Kimbel, W. H., et al. (1999). TL dates for the Neanderthal site of the Amud Cave, Israel. *Journal of Archaeological Science, 26*, 259–268.

von den Driesch, A. (1976). *A guide to the measurement of animal bones from archaeological sites*. Peabody Museum Bulletin 1. Cambridge: Harvard University, Peabody Museum of Archaeology and Ethnology.

Wall, J. D., Lohmueller, K. E., & Plagnol, V. (2009). Detecting ancient admixture and estimating demographic parameters in multiple human populations. *Molecular Biology and Evolution, 26*, 1823–1827.

Wandsnider, L. (1997). The roasted and the boiled: Food composition and heat treatment with special emphasis on pit-hearth cooking. *Journal of Anthropological Archaeology, 16*, 1–48.

Ware, J. A. (1983). Environmental change and sociocultural causality. In J. E. Smith (Ed.), *Proceedings of the Anasazi Symposium 1981* (pp. 125–132). Mesa Verde National Park, Colorado: Mesa Verde Museum Association.

Weiner, S., Goldberg, P., & Bar-Yosef, O. (1993). Bone preservation in Kebara Cave, Israel using on-site Fourier transform infrared spectrometry. *Journal of Archaeological Science, 20*, 613–627.

Weiss, E., Kislev, M. E., Simchoni, O., & Nadel, D. (2004a). Small-grained wild grasses as staple food at the 23,000-year-old site of Ohalo II, Israel. *Economic Botany, 58*, S125–S134.

Weiss, E., Wetterstrom, W., Nadel, D., & Bar-Yosef, O. (2004b). The broad spectrum revisited: Evidence from plant remains. *Proceedings of the National Academy of Sciences of the United States of America, 101*, 9551–9555.

Whallon, R. E. (2011). An introduction to information and its role in hunter-gatherer bands. In R. E. Whallon, W. A. Lovis, & R. K. Hitchcock (Eds.), *Information and its role in hunter-gatherer bands* (pp. 1–27). Los Angeles: University of California, Cotsen Institute of Archaeology Press.

White, J. P. (1977). Crude, colourless and unenterprising? Prehistorians and their views on the stone age of Sunda and Sahul. In J. Allen, J. Golson, & R. Jones (Eds.), *Sunda and Sahul* (pp. 13–30). New York: Academic Press.

Wiessner, P. (2002). Hunting, healing, and hxaro exchange: A long-term perspective on !Kung (Ju/'hoansi) large-game hunting. *Evolution and Human Behavior, 23*, 407–436.

Winterhalder, B. (1981). Optimal foraging strategies and hunter-gatherer research in anthropology: Theory and models. In B. Winterhalder & E. A. Smith (Eds.), *Hunter-gatherer foraging strategies: Ethnographic and archaeological analyses* (pp. 13–35). Chicago: University of Chicago Press.

Wissler, C. (1910). *Material culture of the Blackfoot Indians*. Anthropological Paper (Vol. 5, Issue 1). New York: American Museum of Natural History.

Wolfe, S. A., Griffith, B., & Wolfe, C. A. G. (2000). Response of reindeer and caribou to human activities. *Polar Research, 19*, 63–73.

Wollstonecroft, M. M., Ellis, P. R., Hillman, G. C., & Fuller, D. Q. (2008). Advances in plant food processing in the Near Eastern Epipalaeolithic and implications for improved edibility and nutrient bioaccessibility: An experimental assessment of *Bolboschoenus maritimus* (L.) Palla (sea club-rush). *Vegetation History and Archaeobotany, 17*, S19–S27.

Wolverton, S. (2008). Harvest pressure and environmental carrying capacity: An ordinal-scale model of effects on ungulate prey. *American Antiquity, 73*, 179–199.

Wrangham, R. W. (2009). *Catching fire: How cooking made us human*. New York: Basic Books.

Wrangham, R. W., & Conklin-Brittain, N. L. (2003). Cooking as a biological trait. *Comparative Biochemistry and Physiology. Part A. Molecular and Integrative Physiology, 136*, 35–46.

Yeshurun, R., Bar-Oz, G., & Weinstein-Evron, M. (2007). Modern hunting behavior in the early Middle Paleolithic: Faunal remains from Misliya Cave, Mount Carmel, Israel. *Journal of Human Evolution, 53*, 656–677.

Yotova, V., Lefebvre, J.-F., Moreau, C., Gbeha, E., Hovhannesyan, K., Bourgeois, S., et al. (2011). An X-linked haplotype of Neandertal origin is present among all non-African populations. *Molecular Biology and Evolution, 28*, 1957–1962.

Zedeño, M. N. (2008). Bundled worlds: The roles and interactions of complex objects from the North American plains. *Journal of Archaeological Method and Theory, 15*, 362–378.

Zhang, S., Li, Z., Zhang, Y., & Gao, X. (2009). Mortality profiles of the large herbivores from the Lingjing Xuchang Man Site, Henan Province and the early emergence of the modern human behaviors in East Asia. *Chinese Science Bulletin, 54*, 3857–3863.

Zilhão, J. (2007). The emergence of ornaments and art: An archaeological perspective on the origins of "behavioral modernity". *Journal of Archaeological Research, 15*, 1–54.

Chapter 4
Middle Paleolithic Prey Choice Inferred from a Natural Pitfall Trap: Rantis Cave, Israel

Reuven Yeshurun

> ...as archaic humans increasingly come to be seen as intelligent and capable hunters, it becomes imperative that we find new ways to explore the complex interplay between environmental or ecological factors that determined what was available on the landscape and cultural factors that determined what was selected and in what proportions (Speth and Tchernov 2007: 199).

Introduction

One of the most desired goals in paleoanthropology is the interpretation of human behavior in relation to the natural environment. What did Paleolithic humans choose to exploit out of the spectrum of resources naturally available to them, and why? The agreement of the archaeological remains (e.g., procured game and plants, lithic raw materials and minerals) with their (often implicitly assumed) natural availability, or the divergence of human preferences from the natural affordances, is potentially informative about human behavioral ecology, cognitive capabilities and social structure (e.g., Klein 1989; Stiner 2002; Grayson and Delpech 2008; Adler and Bar-Oz 2009). It may reveal expansion of the spectrum of utilized resources stemming from demographic pulses, cognitive shifts, socio-cultural factors or technological innovation/intensification. Ultimately, it serves to answer whether the course of human evolution and Paleolithic human behavior were by and large determined by environmental factors, or whether the social networks and cognitive or innovative capacities of humans had an equally great, or even greater influence (e.g., Gamble 1999; Kaufman 2002; Bar-Yosef 2004; Klein 2009; Speth 2010).

Specifically in the field of zooarchaeology, one of the most basic questions is human prey selection versus natural availability of game. Did Paleolithic humans largely base their hunting choices on the natural availability of game in their environment, or diverge from this availability to acquire rarer animals for whatever reason, such as greater fit to their technology, ease of hunting, greater prestige or better taste? The prey-choice patterns of past human populations convey a great deal of information regarding hunting capabilities, hunting technology, ecological adaptation and social factors, sometimes intertwined. However, isolating prey-choice patterns requires one to take into consideration the pool of resources locally and temporally available to the hunters (for recent discussions on Late Pleistocene prey choice patterns see Faith 2008; Dusseldorp 2010; Weaver et al. 2011). If the prey choice of hunters, as attested to by archaeofaunal remains in their camps, closely resembles the 'natural' living community, this could be taken as evidence of non-selective or environmentally-determined hunting. On the other hand, if the hunted spectrum markedly diverges from the natural presence and relative abundance of game species, then specific prey-choice patterns can be determined and explanations for these patterns may be sought. Thus, in theory, the presence and relative abundance of game species in an archaeofaunal assemblage should be compared to the presence and relative abundance of a non-anthropogenic assemblage, reflecting the 'live' community through different mechanisms.

However, this kind of comparison is rarely made, because finding a natural control against which to compare the anthropogenic faunas is challenging. Many regions of the world do not possess natural collections of Pleistocene faunas that are unbiased by hominin predation and thus can provide a proxy for the availability of animal resources. In fact, the natural availability of game in a spatio-temporal setting suitable for comparison to the archaeological record is hardly ever known, but is instead inferred from the archaeological remains themselves (thus creating circular reasoning), or estimated from independent paleoenvironmental proxies that correlate tentatively with the archaeological record.

R. Yeshurun (✉)
Zinman Institute of Archaeology, University of Haifa,
Mount Carmel, 31905 Haifa, Israel
e-mail: ryeshuru@research.haifa.ac.il

Jamie L. Clark and John D. Speth (eds.), *Zooarchaeology and Modern Human Origins: Human Hunting Behavior during the Later Pleistocene*, Vertebrate Paleobiology and Paleoanthropology, DOI: 10.1007/978-94-007-6766-9_4,

This study aims to focus on a subset of this problem, namely the prey-choice patterns of human hunters in the Middle Paleolithic (MP) period of the Levant (ca. 250–47 ka). My approach here is to use the finds from the newly discovered site of Rantis Cave, interpreted as a natural pitfall trap whose contents reflect the structure of a faunal 'live community' in MP times, as a natural control against which to compare the relative taxonomic frequencies in the anthropogenic faunas. The outcome of this exercise is the isolation of a hunting pattern that diverges from the natural availability of game. I will argue that this pattern could be interpreted as a case for specific human prey-choice during the Levantine MP and discuss its possible implications.

The Middle Paleolithic Archaeofaunas of the Southern Levant

The Middle Paleolithic of the Southern Levant figures prominently in debates concerning human evolution, as it has yielded important Neanderthal and anatomically modern human (AMH) fossils, numerous cave and open-air sites displaying rich cultural layers, and has been the focus of a great deal of groundbreaking research since the 1930s (e.g., Garrod and Bate 1937; Bar-Yosef and Vandermeersch 1993; Kaufman 1999; Shea 2003 and references therein; Weinstein-Evron et al. 2003; Bar-Yosef and Meignen 2007; Hovers 2009). This small, yet archaeologically-rich and intensively-explored region, displays marked ecological variability, making it an ideal area to explore the nature of game procurement relative to what was available in the environment. Faunal studies concerning the MP period of the Levant effectively commenced with Bate's (1937) now-classic quantitative account of mountain gazelle (*Gazella gazella*; hence referred to as simply 'gazelle') and Mesopotamian fallow deer (*Dama mesopotamica*; hence 'fallow deer') frequencies in the Mount Carmel caves (Tabun and el-Wad). She attributed shifts from fallow deer to gazelle dominance in the Pleistocene faunal assemblages to changes in regional vegetation and paleoclimate. An abundance of woodland-dwelling fallow deer was taken to indicate a humid phase and contracting 'open' biomes, while the steppe-adapted gazelles signaled desiccation and contracting woodlands. Bate's conclusions were later widely debated, because of unsystematic methods of collection and analysis that could alter the relative frequencies, inappropriate tallying methods and the lack of taphonomic and contextual data clarifying how these animals reached the cave in the first place (Jelinek et al. 1973; Davis 1982; Speth and Tchernov 2007; Bar-Oz et al. in press). As human hunters were responsible for accumulating most of the faunal remains in these caves (Marín-Arroyo in press), this debate cannot be truly resolved unless the issue of human prey-choice, and its adherence to the relative paleoenvironmental abundance of game species, is addressed.

Detailed zooarchaeological and taphonomic studies have shown in recent years that the MP archaeofaunas of the Southern Levant were largely the result of systematic human hunting, carcass transport, butchery and discard throughout this ca. 200 ka period (e.g., Speth and Tchernov 1998, 2007; Rabinovich and Hovers 2004; Rabinovich et al. 2004; Stiner 2005; Speth and Clark 2006; Yeshurun et al. 2007a). The ungulate species present in these assemblages are similar (chiefly *G. gazella, D. mesopotamica, Bos primigenius, Cervus elaphus, Capra aegagrus* and *Sus scrofa*), but some variations have been noted in their relative frequencies. These variations are hard to interpret, given the still-small number of published archaeofaunal assemblages for this region and period and the variety of taphonomic processes that may have acted on the assemblages. Recovery and analyst biases in old excavations, differential transport patterns aimed at reducing carcasses of larger ungulates (e.g., Rabinovich and Hovers 2004; Speth and Clark 2006; Yeshurun et al. 2007a) or reflecting different site types such as open-air localities versus cave habitations (e.g., Gilead and Grigson 1984; Rabinovich 1990; Hovers et al. 2008; Sharon et al. 2010), and differential post-depositional fragmentation (Yeshurun et al. 2007b) could all alter the relative taxonomic abundance of ungulates in these sites. Another, more elusive factor is, of course, human hunting preferences, which are the subject of this paper.

Holding the paleoenvironment roughly constant by considering only faunas from the Mediterranean climatic region of the Southern Levant (Danin 1988), which most probably covered a similar territory in this region during the MP (Enzel et al. 2008), may enable the consideration of taphonomic biases and subsequently anthropogenic prey preferences. But while pre- and post-depositional destruction processes of vertebrate faunas are relatively well-understood and their study is enabled by the actualistic and experimental literature, the study of human prey choices suffers from a lack of a 'natural control' in this region.

Importantly, in recent years several Levantine MP large-mammal faunal collections that were accumulated mainly by natural agents (e.g., carnivore dens and pitfall traps) have been recognized, thereby providing the necessary comparative framework for the presence and abundance of game on the paleolandscape that was available for human exploitation. These are Geula Cave (Monchot 2005), Rantis Cave (Marder et al. 2011) and perhaps the area beneath the 'chimney' of Tabun B (Garrard 1982; Marín-Arroyo in press). The case of Rantis Cave is especially illuminating. In the next section I present this newly discovered site and argue that the cave acted as a natural pitfall trap for ungulates, reflecting the ungulate composition through different, non-anthropogenic accumulation processes. I then compare the relative frequency of animal species selected for procurement in archaeological sites (usually by systematic hunting) to the relative frequency of species at Rantis Cave to elucidate human subsistence, in particular the choice of certain animals over others by MP hunters.

Rantis Cave

Rantis Cave is a unique site for the Pleistocene Southern Levant, containing a rich faunal assemblage along with meager evidence for human occupation. This recently discovered filled doline in west-central Israel yielded micro- and macromammalian remains, as well as a few lithic artifacts (Marder et al. 2011). The cave, a filled chamber with a roof opening, is located on the western slopes of the Samaria Hills, within the western flanks of the Ramallah anticline, at an elevation of 220 m above sea level (Fig. 4.1). The site lies in the Mediterranean phytogeographic zone of the Southern Levant (Danin 1988), in an area of low limestone hills forming the transition from the coastal plain to the highlands. Today the region experiences a Mediterranean climate with rainy winters and dry summers. Mean annual precipitation is 600 mm and mean annual temperature is 19 °C. The cave is an isolated karstic chamber, truncated by an abrasion-denudation terrace, and subsequently filled by sediments.

The site was discovered in 2004 during construction activities. A cave chamber, filled with brown *terra rosa* soil with numerous animal bones and sporadic flint artifacts, was noticed in the section (Fig. 4.2a, b). A salvage excavation was conducted at Rantis Cave on behalf of the Israel

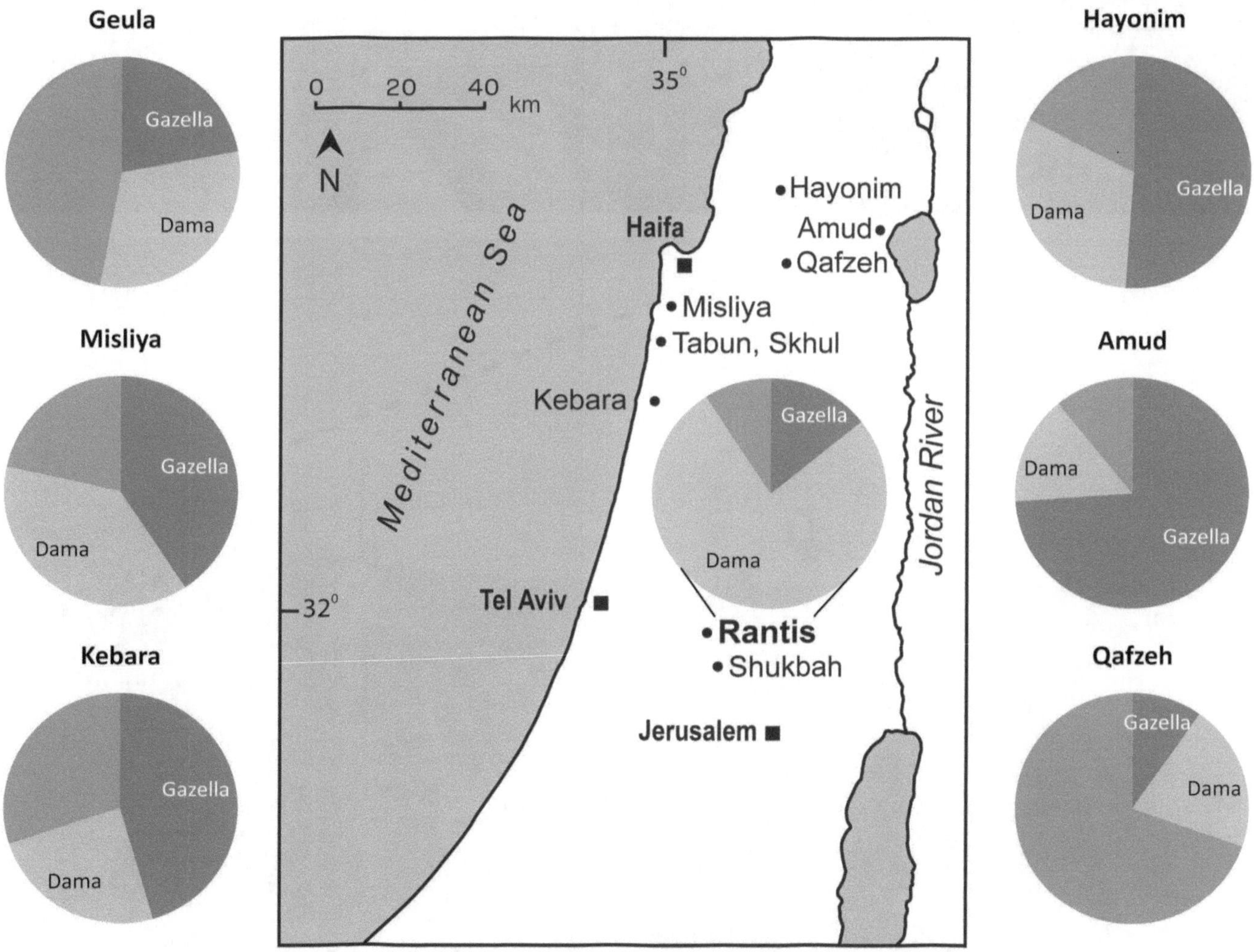

Fig. 4.1 Map of Israel showing the Middle Paleolithic cave sites mentioned in the text. All are located within the present-day Mediterranean climatic zone. The *pie charts* present the relative taxonomic composition in each site, simplified to three groups: *Gazella*, *Dama* and all other ungulates

Fig. 4.2 Rantis Cave during excavation. **a** *Top view* showing the round outline of the doline, interpreted as an ancient pitfall trap; **b** *Side view* of the cave, exposed during road construction. Note the clear borders of the filled cave; **c** Fallow deer metapodial, showing its high degree of completeness and severe in situ attrition; **d** Fallow deer mandible in situ

Antiquities Authority in 2005, directed by O. Marder and R. Lupu, in an area of 12 m^2. The sediments were systematically sieved. Faunal remains were the most ubiquitous find (Fig. 4.2c, d); only a few lithic artifacts were found scattered among various levels, and no hearths, living levels or other human-made features were encountered, in stark contrast to most other MP cave sites in the region (compare Albert et al. 2003; Bar-Yosef and Meignen 2007; Shahack-Gross et al. 2008). Micromammal remains were numerous and dominated by the social vole *Microtus guenetheri*. Preliminary taphonomic observations suggest that the micromammals were probably deposited by raptors (Marder et al. 2011).

Attempts were made to date the Rantis Cave sediments using paleomagnetic, OSL and Th/U techniques. The soil sequence of the cave is of normal magnetization and the age is therefore younger than 780 ka. A flowstone close to the bottom of the section was dated by U-Th to yield an age of ~140 ka, indicating that most of the cave fill (from 0.5 m above the cave bottom upwards), as well as human and animal activities, postdate ~140 ka. Unfortunately, due to the complex depositional processes in the cave, including recent pedogenesis at the upper layer, OSL dating efforts have been unable to provide finer chronological resolution, even though the age of a significant portion of the sediment grains corresponded to the Th-U age. The small lithic assemblage (n = 39, including chips) was assigned to the Levantine MP period, in agreement with the radiometric dates. Thus, the Rantis Cave fill can be assigned to the later half of the Levantine MP (Marder et al. 2011: Table 1).

The Large Mammal Assemblage of Rantis Cave

A detailed description of the large mammal assemblage and the methods used in the analysis are given by Marder et al. (2011). A summary of the most relevant characteristics is given here. The bone assemblage of Rantis Cave is quite fragmented, although complete bone elements do occur. The identified assemblage (Number of Identified Specimens; NISP) consists of 241 bone and tooth fragments, representing a minimum number (MNI) of 22 individual animals. Ungulate species (NISP = 228, including 130 fragments identified to size-class) dominate the assemblage, followed by carnivores (NISP = 11) and two hyrax (*Procavia* sp.) specimens. The most common species (Fig. 4.3a) is Mesopotamian fallow deer (*D. mesopotamica*, 68 % of NISP). Other ungulates represented are mountain gazelle (*G. gazella*, 13 %) and to a lesser extent, aurochs (*B. primigenius*), goat (*Capra* sp.), and wild boar (*S. scrofa*). A similar representation was also found for specimens that were identified to size-classes corresponding to the three major ungulate species (i.e., gazelle-size, fallow deer-size and aurochs-size). Species representation is similar among stratigraphic phases. The carnivore remains include five species: leopard (*Panthera pardus*, NISP = 4), wolf (*Canis lupus*, NISP = 2), brown bear (*Ursus arctos*, NISP = 2), a single undetermined species of hyena (*Hyaena/Crocuta*) and a small canid (Fig. 4.3a; see also Marder et al. 2011: Appendices 1, 2).

The age structure of the ungulate species was analyzed on the basis of tooth eruption and wear, following Stiner's (1990, 2005) three-cohort age system (juvenile, prime adult and old adult). The fallow deer dental sample was the largest and the only one with a satisfactory, albeit small, sample size (n = 13). The fallow deer age structure may resemble a 'catastrophic' age profile, where juveniles form the largest age class (n = 6), followed by prime-age adults (n = 4) and to a lesser extent by old adults (n = 3; Fig. 4.3c). This mortality profile is considered to conform to a hypothetical age profile of a living herd (Klein and Cruz-Uribe 1984; Stiner 1990). However, since little data are currently available on the natural age structure of Mesopotamian fallow deer, the interpretation of the Rantis age profile cannot by itself provide strong evidence for the depositional origin; suffice it to say that such an abundance of juveniles is unusual in the context of Levantine MP cave sites (e.g., Speth and Tchernov 1998; Yeshurun et al. 2007a).

The bone cortical surfaces generally are reasonably well preserved, enabling the search for various types of surface modifications. Human-induced butchery and consumption marks are nearly absent in the assemblage. Only a single burned specimen (a gazelle petrous bone), and one cut-marked specimen (a fallow deer mandible) were noted. None of the bones bear evidence of hammerstone percussion marks. Indications of animal modifications were also extremely rare in the assemblage. No unambiguous carnivore gnaw marks were found, despite systematic microscopic analysis, and only a single rodent gnaw mark was detected. Most bone surface modifications in the assemblage are the result of abiotic post-depositional agents, such as subaerial weathering, root marks, trampling striations and a few bone edges that underwent abrasion and rounding (Fig. 4.3b). The bone surface modification data suggest that neither humans nor carnivores played an important role in the formation and modification of the Rantis Cave assemblage and that the bones were exposed to the elements and buried slowly. Similarly, processes such as fluvial transport do not seem to have modified the bone assemblage further. It should be noted that little intra-site variation was found (Marder et al. 2011).

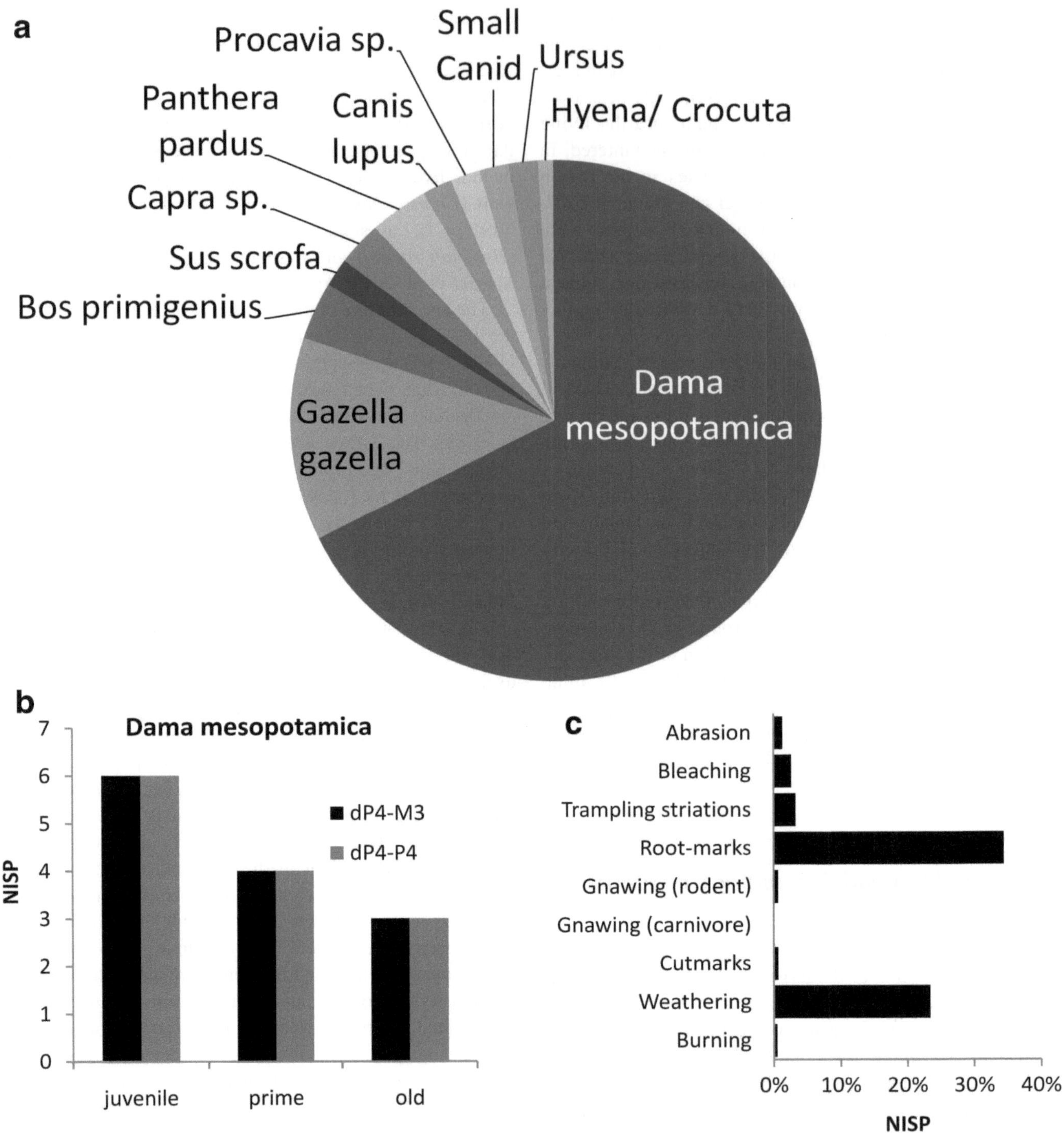

Fig. 4.3 The Rantis Cave large-mammal assemblage (see also Table 4.1). **a** Taxonomic composition based on specimens identified to the species level (n = 130); **b** mortality profile of Mesopotamian fallow deer according to two possible dentition sequences (dP_4-M_3 and dP_4-P_4); **c** taphonomic variables (all species and size-classes combined)

Interpretation

Taking into consideration the very limited anthropogenic signature at the site and on the faunal remains themselves, as well as the fact that carnivore remains are few and scattered, that signs for carnivore consumption are lacking, and that the age profile of the fallow deer is unusual, I interpret the large-mammal remains of Rantis Cave as an accumulation resulting from natural deaths either in the cave or in its immediate vicinity (Marder et al. 2011). Such 'natural death' sites are known globally, created either because of the existence of a pitfall trap or by a catastrophic event (e.g., Shield Trap Cave, Oliver 1989; Untermassfeld, Kahlke and Gaudzinski 2005), but they are very rare in the Pleistocene of the Southern Levant (a notable exception is Bear's Cave in the Upper Galilee of Israel, Tchernov and Tsoukala 1997, and possibly also the area beneath the 'chimney' of Tabun B, Garrard 1982; Marín-Arroyo in press). Humans and carnivores were intermittently attracted to Rantis Cave, possibly exploiting the dead ungulates (though positive evidence for this is nearly absent), but they never lived in it.

Several processes can explain the natural deaths of large mammals in Rantis Cave. Some of the animal remains may have been washed into the cave, yet its interfluve location suggests that it has not been significantly affected by surface streams. This is reinforced by the fresh, unabraded state of most of the artifacts and bones. However, some water probably accumulated within the doline. The animals most likely were attracted to the water, and then were trapped in the sinkhole, with the vertical and overhanging walls offering no escape route. The probable standing water and the geometry of the sinkhole likely ruled out occupation by non-flying animals and by humans (Fig. 4.2). Thus, Rantis Cave probably acted as a natural pitfall that trapped the animals inhabiting the landscape (Marder et al. 2011: 777).

Comparison of Rantis Cave to Anthropogenic Assemblages

Several important cave and open-air sites are known from the MP period in the Southern Levant, yielding predominantly anthropogenic faunas. Since Rantis Cave reflects the 'living' ungulate composition through non-anthropogenic accumulation, the relative frequencies of the animal species selected for procurement in the archaeological sites (usually by systematic hunting) may be compared to the relative frequency of species at Rantis Cave in order to elucidate hominin subsistence, in particular the choice of certain animals over others by MP hunters.

Here, only MP sites that belong to broadly the same topographical, ecological and cultural *milieu* are compared, that is, cave sites from the Mediterranean phytogeographic zone of the Southern Levant. I selected only assemblages that were retrieved and analyzed using modern, relatively unbiased methods (Marean et al. 2004), such as full retention of bone elements, and published in enough detail to allow comparison. Other faunal assemblages, such as those of the key site of Tabun Cave, Mount Carmel, were not used in the comparison because of the highly biased nature of the faunal collections that were excavated in the 1930s. I also refrained from using open-air sites from the Southern Levant, some of which have yielded important faunal collections, such as Quneitra (Davis et al. 1988; Rabinovich 1990), Far'ah II (Gilead and Grigson 1984), and more recently 'Ein Qashish (Hovers et al. 2008), Nahal Mahanayeem Outlet (Sharon et al. 2010) and Nesher-Ramle Quarry (Zaidner et al. 2010). These sites, with the possible exception of the recently discovered and particularly rich occupation of Nesher-Ramle Quarry, probably represent shorter visits and/or fewer repeated occupations of human groups that left behind their animal food refuse, sometimes following less intensive processing than at the repeatedly occupied cave sites; the role of open-air sites in the game-provisioning systems of the Levantine MP is yet to be explored and is outside of the scope of this paper. It is nonetheless clear that such sites cannot be directly compared to caves for the purpose of quantifying relative taxonomic abundance, because of the differences in transport and processing considerations associated with caves and open-air sites. Faunas found in cave sites represent MP meat procurement as the averaging of numerous episodes of hunting, transport, butchery and consumption. The abundance of aurochs, in particular, as well as other large ungulate taxa, is elevated in open-air sites in comparison to caves.

An additional factor is keeping the environmental location of sites as similar as possible, so that habitations surrounded by broadly similar settings, and supposedly similar animal communities, are compared. For this purpose I focused on the small, topographically mild and well-studied region of the Mediterranean Southern Levant, excluding faunal assemblages from the caves of the more humid and more wooded Mediterranean Northern Levant and from the deserts or semi-deserts to the east and south (e.g., Griggo 2004). The selected sites are all located in a small and defined region, exhibiting considerable similarity in their mammal communities along a north-south humidity gradient (Danin 1988; see Belmaker and Hovers 2011). Rantis Cave, the southernmost of the considered sites (Fig. 4.1), is thus expected to yield a somewhat more xeric Mediterranean fauna. This expectation was met by the micromammal data (Marder et al. 2011), requiring an explanation for the abundance of woodland-adapted fallow deer at the site. Preliminary tooth mesowear analysis indicated that at Rantis Cave fallow deer, animals that are normally browsers, included a relatively high proportion of graze in their diet (Marder et al. 2011), in accordance with the southern location of the site. This highlights the adaptive flexibility of this species and attests to the fact that the immediate environment of Rantis was not more wooded than that surrounding the northern sites, and thus would not be expected to capture more woodland animals.

As the selected sites are widely spaced in time, the issue of climate change through the MP requires a note here. The extent and nature of climate fluctuations and their implications for human ecology and the natural availability of game in the Southern Levant have long been the focus of debate (e.g., Bate 1937; Jelinek et al. 1973; Davis 1982; Garrard 1982; Tchernov 1992; Bar-Matthews et al. 2003; Vaks et al. 2006; Shea 2008). Much data from various proxies (e.g., pollen, isotopes, paleohydrology, archaeofaunal and archaeobotanical remains) exist for the reconstruction of paleoenvironments in the Levantine MP. The proxies that reflect human ecology in the most direct manner, i.e., animals and plants exploited or otherwise deposited at the sites, generally point at remarkable taxonomic persistence throughout the MP period in this region, with changes in the relative frequency of species seemingly unrelated to global climatic events. Some moderate climatic fluctuations notwithstanding, it thus seems that the heart of the present-day Mediterranean ecosystem of the Levant displays remarkable persistence through glacial and interglacial periods, even if adjacent regions such as the eastern deserts and North Africa experienced more marked climate changes (e.g., Bar-Matthews et al. 2003; Lev et al. 2005; Vaks et al. 2006; Belmaker 2008; Enzel et al. 2008; Frumkin et al. 2011; Belmaker and Hovers 2011; Yeshurun et al. 2011; Marder et al. 2011; but see Tchernov 1992; Shea 2008 for a different view emphasizing climatic fluctuations and linking them to human ecology). At this stage I proceed with the assumption that, in this coarse-grained level of analysis, the MP archaeofaunas of the Mediterranean Southern Levant may be used in inter-site comparisons, with the natural presence and abundance of large-mammal species not remarkably transformed over time by environmental fluctuations.

Having concentrated on analytically reasonably unbiased faunal assemblages from the MP caves of the Mediterranean Southern Levant, six sites provide a good comparison to the natural accumulation at Rantis: Misliya, Kebara and Geula caves in Mount Carmel, Hayonim and Qafzeh caves in the Galilee and Amud Cave at the margin of the Jordan Rift Valley (Fig. 4.1; Table 4.1). All of these sites except Geula are primarily composed of repeated human occupations full of faunal and lithic remains, sometimes with hearths and sometimes with AMH or Neanderthal burials. Taxonomic composition, bone-surface modification data, bone fragmentation data and ungulate age profiles indicate that the ungulate collections accumulated as a result of systematic hunting of mostly prime-age individuals and their subsequent transport to the cave (sometimes preceded by initial field butchery), where they were processed, consumed and usually discarded on the spot (Speth and Tchernov 1998, 2001; Rabinovich et al. 2004; Rabinovich and Hovers 2004; Stiner 2005; Speth and Clark 2006; Yeshurun et al. 2007a; and see a summary in Table 4.1). The assemblages in these sites therefore largely represent the hunters' choices. Geula Cave, which recently underwent a taphonomic analysis by Monchot (2005), is different in its unique carnivore- and porcupine-rich species composition, the ubiquity of carnivore and porcupine gnawing marks on ungulate bones, the absence of human tool marks and scarcity of lithic artifacts and anthropogenic features. All of these point to Geula being sporadically visited by humans, but primarily serving as a hyena den and porcupine lair (Monchot 2005). The ungulate assemblage at Geula thus represents mostly the kill or scavenging activities of hyenas and may be an additional important reference for discerning MP human prey-choices, one that was formed by different processes than we see in the natural pitfall at Rantis on the one hand and in the anthropogenic caves on the other hand.

The two most abundant ungulates in all of the assemblages except Qafzeh are *Gazella* and *Dama*, which are the most ubiquitous big-game procured during the entire MP period in the Mediterranean Southern Levant and which continues until the terminal Pleistocene (Bate 1937; Davis 1982; Bar-Oz et al. in press). The relative frequencies of gazelle and fallow deer specimens out of all ungulate specimens that were identified to the species level were compared and plotted on a scattergram (Fig. 4.4). This inter-site comparison reveals that Rantis Cave is especially rich in fallow deer, significantly more than any anthropogenic cave site examined (*Dama* vs. other ungulates at Rantis vs. Misliya: $\chi^2 = 41.91$, $p < 0.001$; Rantis vs. Hayonim: $\chi^2 = 88.3$, $p < 0.001$; Rantis vs. Qafzeh: $\chi^2 = 144.65$, $p < 0.001$; Rantis vs. Amud: $\chi^2 = 148.13$, $p < 0.001$; Rantis vs. Kebara: $\chi^2 = 141.57$, $p < 0.001$) and also significantly more than at Geula Cave, interpreted as a hyena den ($\chi^2 = 81.87$, $p < 0.001$).

Figure 4.4 reveals that at Rantis alone fallow deer greatly outnumber gazelle, while at four out of five anthropogenic sites (Misliya, Hayonim, Amud and Kebara) the two species are either equally represented or gazelle clearly outnumber fallow deer. At Qafzeh Cave *Dama* outnumbers *Gazella*, but the two species together comprise just about one-third of NISP, as the unique faunal composition of Qafzeh is dominated by *Cervus*, *Dama* and *Bos* in roughly equal proportions—the only MP site in the Southern Levant to yield such a faunal composition. The natural (hyena-porcupine) site of Geula is also different than the four abovementioned caves, in that *Dama* somewhat outnumber *Gazella* there and in the relatively high abundance of *Bos*.

To conclude the intersite comparisons, it seems that fallow deer were available in the landscape in great numbers, and in any case were not rarer than gazelle, as their abundance at Rantis points out. The fact that gazelles were transported in roughly equal or greater numbers to four out of five cave habitations and usually comprise over 40 % of identified ungulate specimens could point to the deliberate human decision to hunt gazelle rather than the abundant fallow deer during the MP of the Southern Levant (Fig. 4.4).

Table 4.1 Zooarchaeological and taphonomic data used in this study

Reference	Misliya Yeshurun et al. (2007a)		Hayonim Stiner (2005)		Qafzeh Rabinovich et al. (2004)		Rantis Marder et al. (2011)		Amud Rabinovich and Hovers (2004)		Kebara[a] Speth and Clark (2006)		Geula Monchot (2005)	
Ungulate NISP														
Gazella	98	40.7 %	1,912	51.1 %	91	9.7 %	14	14.3 %	291	73.9 %	8,810	45.4 %	221	22.1 %
Dama	91	37.8 %	1,178	31.5 %	193	20.6 %	75	76.5 %	60	15.2 %	4,752	24.5 %	309	30.9 %
Capra	8	3.3 %	8	0.2 %	97	10.4 %	3	3.1 %	18	4.6 %	182	0.9 %	13	1.3 %
Cervus	6	2.5 %	155	4.1 %	198	21.1 %	0	0.0 %	5	1.3 %	1,557	8.0 %	44	4.4 %
Sus	11	4.6 %	176	4.7 %	62	6.6 %	2	2.0 %	17	4.3 %	759	3.9 %	118	11.8 %
Bos	26	10.8 %	281	7.5 %	171	18.2 %	4	4.1 %	1	0.3 %	2,902	15.0 %	209	20.9 %
Capreolus	1	0.4 %	11	0.3 %	5	0.5 %	0	0.0 %	1	0.3 %	N/A		38	3.8 %
Equus[b]	0	0.0 %	22	0.6 %	78	8.3 %	0	0.0 %	0	0.0 %	424	2.2 %	21	2.1 %
Dicerorhinus	0	0.0 %	2	0.1 %	35	3.7 %	0	0.0 %	1	0.3 %	N/A		22	2.2 %
Alcelaphus	0	0.0 %	0	0.0 %	7	0.7 %	0	0.0 %	0	0.0 %	N/A		5	0.5 %
Hippopotamus	0	0.0 %	0	0.0 %	0	0.0 %	0	0.0 %	0	0.0 %	N/A		1	0.1 %
Total	241	100.0 %	3,745	100.0 %	937	100.0 %	98	100.0 %	394	100.0 %	19,386	100.0 %	1,001	100.0 %
Taphonomy														
Burning	**Frequent**		**Frequent**		Present		*Absent*		**Frequent**		Present		*Absent*	
Cutmarks	Present		Present		Present		*Very rare*		Present		Present		*Absent*	
Carnivore gnawing	*Scarce*		*Scarce*		Present		*Absent*		*Scarce*		Present		**Frequent**	
Age profile of dominant ungulates	Prime-dominated		Prime-dominated		Mainly adults		'Catastrophic'		Prime-dominated		Prime-dominated		Prime-dominated	
Inferred primary agent of accumulation	Human hunting, butchery and consumption		Human hunting, butchery and consumption		Human hunting, butchery and consumption		Natural pitfall trap		Human hunting, butchery and consumption		Human hunting, butchery and consumption		Hyena den and porcupine lair	

[a] For Kebara NISP counts are available only for the main prey taxa (*Gazella, Dama, Capra, Cervus, Sus and Bos*)
[b] Two or more equid species were probably present in the Levantine MP. They are lumped here, as in some of the reports I used

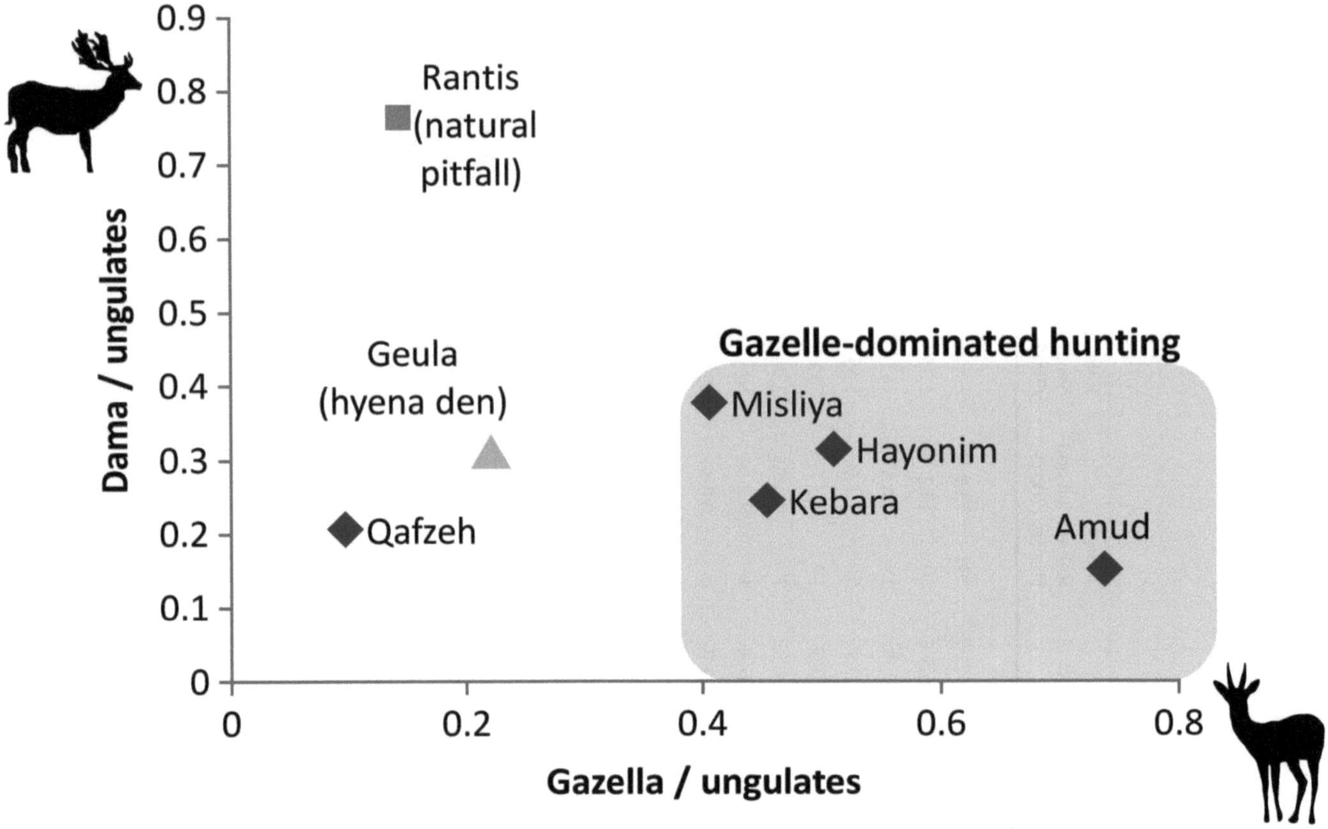

Fig. 4.4 Comparison of *Gazella* and *Dama* frequencies (expressed as their percent out of the total ungulates in the assemblage) of five assemblages that represent the hunter's choice, as well as a natural pitfall trap and a hyena den. Data are from Table 4.1

Discussion

I have shown that MP hunters who occupied caves in the Southern Levant generally procured gazelle rather than fallow deer. In light of the inverse pattern of greater fallow deer abundance in the natural control case of Rantis Cave, I interpreted this as a reflection of the human preference for hunting gazelle. Importantly, fallow deer were found to be the dominant ungulate species in two other suggested cases of non-anthropogenic accumulation in the region, i.e., Geula Cave (Monchot 2005) and the 'chimney' of Tabun Cave (Marín-Arroyo in press) and hence the pattern of fallow deer abundance in 'natural' accumulations versus gazelle abundance in anthropogenic accumulations is quite consistent. Given the rarity, or sometimes absence, of gazelle in late Lower Paleolithic cave assemblages (e.g., Bate 1937; Stiner et al. 2011), this pattern seems to commence in the early MP and continues well into the Upper Paleolithic and Epipaleolithic of the region (Davis 1982; Rabinovich 2003; Bar-Oz 2004; Munro 2009). Before I discuss possible causes and consequences of this suggestion, I should discuss some of the pitfalls of inferring human behavioral patterns from Number of Identified Specimens (NISP) values.

Zooarchaeological species abundance data, even in its raw form of NISP, may not always constitute a straightforward measure of the proportions of game that were procured. As mentioned above, aside from the hunter's choice, other factors may distort NISP values to some degree, including selective transport of body parts to the cave, intensity of processing, differential pre-depositional and post-depositional fragmentation, and analytical techniques. The early MP faunal assemblage of Misliya Cave in Mount Carmel may serve here to demonstrate these issues. At Misliya, a multivariate taphonomic analysis (Yeshurun et al. 2007a) showed that the assemblage was formed and primarily modified by human hunting, transport, butchery, consumption and discard, thus in principle allowing behavioral inferences to be drawn. However, the Misliya hunters treated the two main game species—fallow deer and gazelle—differently. The small gazelles were transported intact to the cave, while the larger fallow deer underwent some field butchery and low-bulk and meat-rich body parts were primarily transported. This pattern was demonstrated by the differential skeletal element profiles of the two animals, taking into account the density-mediated attrition of skeletal elements. Thus, counting by NISP would favor gazelles in this case, because more gazelle elements were transported and discarded in the cave than fallow deer elements. To complicate matters, the Misliya gazelles show somewhat reduced fragmentation (higher completeness rates of most skeletal elements) than fallow deer, a pattern interpreted as stemming from stronger post-depositional actions acting on larger bones (Yeshurun et al. 2007b). Since the relative frequency of game is based only on skeletal elements identified to the species level and therefore on complete or near-complete elements (mainly teeth and long bone articular ends), gazelles may be overrepresented because of this process, too. Additionally, the most resistant elements, and those that are most easily assigned to the species level—teeth—may not suffer greater or lesser fragmentation, but are underrepresented for fallow deer as skulls were transported less often than mandibles for this species (Yeshurun et al. 2007a). A possible solution could be to base the comparative analysis presented here on Minimum Number of Individuals (MNI) counts, which theoretically may be less affected by differential transport and fragmentation. At Misliya MNI counts do not alter the relative frequency to a great extent (MNI counts are seven for *Dama* and six for *Gazella*). This, however, raises additional questions about the conceptual and statistical validity of using MNI counts for comparisons of what are essentially fragmented and time-averaged assemblages formed over the course of numerous human occupations (e.g., Grayson and Delpech 2008; Lyman 2008), a discussion which is beyond the scope of this paper. The problems demonstrated here for the Misliya game counts may exist to some extent in all other assemblages used in this study and these issues were indeed addressed in their publications (see references in Table 4.1); it is imperative, therefore, that one uses well-collected assemblages with published and detailed zooarchaeological and taphonomic analyses, allowing the recognition of biasing factors when comparing and attempting a behavioral reconstruction.

Moving back to Rantis Cave, differential human transport is probably not an issue here, at least as far as the primary deposition of the material is concerned, and no marked patterns stand out when comparing the fragmentation and preservation of gazelle versus fallow deer, albeit the sample size for the former is small (Marder et al. 2011: Table S1). Still, the ecological context of this pitfall trap could affect species abundance. Had this pitfall been located in a primarily mesic environment, more woodland-adapted fallow deer would be expected to be 'sampled' by this natural hole (compare Jelinek et al. 1973). However, as evident from preliminary analysis of micromammal remains and fallow deer tooth mesowear data, the Rantis landscape was somewhat more xeric than the more northern sites located in the Carmel and Galilee (Marder et al. 2011) and fallow deer, therefore, should not be preferentially deposited there. The great natural abundance of fallow deer in even the more xeric MP landscapes of the Southern Levant further emphasizes the peculiarly inverse prey-choice pattern seen in the anthropogenic caves.

If my suggestion of deliberate human choice for hunting gazelle rather than fallow deer is accepted, what could be the cause of such behavior? Optimal foraging models can hardly explain it. The Mesopotamian fallow deer is much larger on average than gazelle (ca. 50–100 kg vs. ca. 20 kg, respectively; Mendelssohn and Yom-Tov 1999), containing significantly more meat and fat at presumably the same cost of processing, and therefore would yield greater nutritional benefits than gazelle. Perhaps the inferred preference for gazelles is related to novel hunting technologies that are more suitable to hunting in open (gazelle-favored) terrain, such as the development of *long-distance* projectiles—the bow and arrow or spear-thrower (atlatl) and dart. The timing and place in which human populations developed the capacity for 'killing at a distance' is debated. Lithic evidence suggests that this technology came to regular use as late as the late African Middle Stone Age (MSA) or the Eurasian Upper Paleolithic (Shea 2006; Churchill and Rhodes 2009; Shea and Sisk 2010). However, gazelles—which were hunted regularly during the Levantine MP—may be more suitable for capturing with just this complex projectile technology than by spearing from close quarters, as traditionally thought for MP hunting techniques. Indeed, it is hard to envision how a small, agile and steppe-dwelling antelope such as gazelle, which tends to flee at the slightest sign of danger, could be approached closely enough to kill with a thrusting or short-range throwing spear. Close-range hunting with spears is always associated in the ethnographic literature with hunting of larger ungulates, sometimes in wooded settings where it is feasible to approach the animal without being exposed, or to ambush it (e.g., Oswalt 1973; Churchill 1993; Ellis 1997). In spite of the contradictory lithic evidence, it is not inconceivable that long-distance projectiles, perhaps made of perishable materials (see Thieme 1997), were widely used by Levantine MP populations to procure gazelles. Direct evidence for using MP stone points for hunting exists only in conjunction with hunting of larger ungulates (e.g., a wild ass vertebra with an embedded Levallois point, Boëda et al. 1999), not gazelles.

If complex projectile technology was already in use in MP times, the use of such weapons may have been better suited to capturing game that live in open landscapes, where they could be more easily encountered and captured. Such a long-distance hunting technology may have been both safer for the hunter (no need to approach the game closely) and would have provided more predictable results, because of the higher encounter rates in open (gazelle-favored) biomes as opposed to closed biomes, where fallow deer prefer to forage most of the time. Once the ability of hunting at a distance became available, hunters might have preferred to obtain smaller ungulates (gazelles) at lower search costs than larger ungulates (fallow deer) that are harder to locate. The ecotonal settings of all the sites presented in this study would have allowed access to several types of biomes, enabling the hunters to choose their favorite hunting grounds and consequently their preferred type of weapons and prey.

Long-distance projectile technology is commonly assigned to the modern humans of the late MSA and especially the Upper Paleolithic periods (Shea and Sisk 2010 and references therein). Interestingly, the only site in this study that can be linked to AMH—Qafzeh Cave—is the one most diverging from the general Southern Levantine hunting pattern, displaying relatively low frequency of gazelles and high abundance of larger ungulates (Rabinovich et al. 2004). On the other hand, two Neanderthal sites (Amud and Kebara) conform to the common hunting pattern of preferring open-terrain gazelles (Rabinovich and Hovers 2004; Speth and Tchernov 2007) which may have necessitated some kind of long-distance projectile technology for their capture. Intriguing as these patterns may be, more human fossil data and better chronological resolution are needed to tie human types to prey preferences in the MP of the Levant.

Conclusion

Humans in the MP period of the Southern Levant habitually obtained animal meat and fat by hunting, but their choice of prey taxa—reflecting on the nature of their interactions with the surrounding paleoenvironment—is obscure because of the general lack of non-anthropogenic faunal collections from this region and period that could serve as a natural reference. The recently discovered Rantis Cave, being unbiased by human predation, may serve as such a natural reference, providing an independent measure of the presence and abundance of ungulates in the human landscape. The subsequent comparison suggests that hunters preferred to procure mountain gazelles over fallow deer, despite the ubiquity of the latter and their greater nutritional gains. This pattern seems to commence in the early MP and continues well into the Upper Paleolithic and Epipaleolithic of the region. A possible explanation (but one that relies on circumstantial evidence) may be related to the development of complex projectile technology already in the early MP period, leading to exploitation of more open habitats.

Regardless of whether these suggestions are accepted or not, Rantis Cave demonstrates the value of the newly recognized 'natural' accumulations for the purpose of reconstructing human prey-choice in the Levantine MP. Were hunting preferences based upon opportunistic encounters with game, dictated by the natural availability of game in the region (e.g., Bar-Yosef 2004; Stiner 2002); or were the hunter's preferences strongly guided instead by traditions, cognitive capacities, ease of capture, technological means or socio-political reasons (e.g., Klein 1989; Speth 2010)? With the flow of new data and the discovery of sites of various types in recent years, we may be able to progress towards deciphering this fundamental problem in a significant period of human evolution.

Acknowledgments I thank Jamie Clark and John Speth for organizing the ICAZ session (Paris, 2010) where this paper was presented, and for inviting me to contribute to this volume. I owe gratitude to the research team working on Rantis Cave, whose members contributed to the ideas expressed here and made this study possible: Ofer Marder and Ronit Lupu (direction of excavation), Guy Bar-Oz and Miriam Belmaker (collaboration on studying the faunal remains) and Amos Frumkin, Naomi Porat and Hagai Ron (geology and dating). Additionally I thank Guy Bar-Oz, Erella Hovers, Daniel Kaufman, Ofer Marder, Ron Shimelmitz, Mina Weinstein-Evron and two anonymous referees for improving previous drafts of this paper. The Rantis Cave excavation and analysis of finds were undertaken on behalf of the Israel Antiquities Authority and financed by the Israel Ministry of Defense.

References

Adler, D. S., & Bar-Oz, G. (2009). Seasonal patterns of prey acquisition and inter-group competition during the Middle and Upper Palaeolithic of the Southern Caucasus. In J.-J. Hublin & M. P. Richards (Eds.), *The evolution of hominin diets: Integrating approaches to the study of Palaeolithic subsistence* (pp. 127–140). Dordrecht: Springer.

Albert, R. M., Bar-Yosef, O., Meignen, L., & Weiner, S. (2003). Quantitative phytolith study of hearths from the Natufian and Middle Palaeolithic levels of Hayonim Cave (Galilee, Israel). *Journal of Archaeological Science, 30*, 461–480.

Bar-Matthews, M., Ayalon, A., Gilmour, M., Matthews, A., & Hakesworth, C. J. (2003). Sea-land oxygen isotopic relationships from planktonic foraminifera and speleothems in the Eastern Mediterranean region and their implications for palaeorainfall during interglacial intervals. *Geochimica et Cosmochimica Acta, 67*, 3181–3199.

Bar-Oz, G. (2004). *Epipaleolithic subsistence strategies in the Levant: A zooarchaeological perspective*. Boston: Brill.

Bar-Oz, G., Yeshurun, R., & Weinstein-Evron, M. (in press). Specialized hunting of gazelle in the Natufian: Cultural cause or climatic effect? In O. Bar-Yosef & F. R. Valla (Eds.), *Natufian Foragers in the Levant: Terminal Pleistocene Social Changes in Western Asia*. Ann Arbor: International Monographs in Prehistory.

Bar-Yosef, O. (2004). Eat what is there: Hunting and gathering in the world of Neanderthals and their neighbors. *International Journal of Osteoarchaeology, 14*, 333–342.

Bar-Yosef, O., & Meignen, L. (2007). *Kebara Cave, Mt. Carmel, Israel. The Middle and Upper Palaeolithic Archaeology*, Part 1. Cambridge, MA: Peabody Museum of Archaeology and Ethnology.

Bar-Yosef, O., & Vandermeersch, B. (1993). Modern humans in the Levant. *Scientific American, 4*, 94–100.

Bate, D. M. A. (1937). Palaeontology: The fossil fauna of the Wadi el-Mughara caves. In D. A. E. Garrod & D. M. A. Bate (Eds.), *The Stone Age of Mount Carmel, Vol. 1. Excavations at the Wadi Mughara* (pp. 135–240). Oxford: Clarendon Press.

Belmaker, M. (2008). Analysis of ungulate diet during the last glacial (MIS 5-2) in the Levant: Evidence for long-term stability in a

Mediterranean ecosystem. *Journal of Vertebrate Paleontology, 28*, 50a.

Belmaker, M., & Hovers, E. (2011). Ecological change and the extinction of the Levantine Neanderthals: Implications from a diachronic study of micromammals from Amud Cave, Israel. *Quaternary Science Reviews, 30*, 3196–3209.

Boëda, E., Geneste, J. M., Griggo, C., Mercier, N., Muhesen, S., Reyss, J. L., et al. (1999). A Levallois point embedded in the vertebra of a wild ass (*Equus africanus*): Hafting, projectiles and Mousterian hunting weapons. *Antiquity, 73*, 394–402.

Churchill, S. E. (1993). Weapon technology, prey size selection and hunting methods in modern hunter-gatherers: Implications for hunting in the Palaeolithic and Mesolithic. In G. L. Peterkin, H. M. Bricker, & P. Mellars (Eds.), *Hunting and animal exploitation in the Later Palaeolithic and Mesolithic of Eurasia* (pp. 11–24). Archaeological Papers of the American Anthropological Association 4.

Churchill, S. E., & Rhodes, J. A. (2009). The evolution of the human capacity for "killing at a distance": The human fossil evidence for the evolution of projectile weaponry. In J.-J. Hublin & M. P. Richards (Eds.), *The evolution of hominin diets: Integrating approaches to the study of Palaeolithic subsistence* (pp. 201–210). Dordrecht: Springer.

Danin, A. (1988). Flora and vegetation of Israel and adjacent areas. In Y. Yom Tov & E. Tchernov (Eds.), *The zoogeography of Israel* (pp. 129–159). Dordrecht: Dr. W. Junk Publishers.

Davis, S. J. M. (1982). Climatic change and the advent of domestication: The succession of ruminant artiodactyls in the Late Pleistocene-Holocene in the Israel region. *Paléorient, 8*, 5–15.

Davis, S. J. M., Rabinovich, R., & Goren-Inbar, N. (1988). Quaternary extinctions and population increase in western Asia: The animal remains from Biq'at Quneitra. *Paléorient, 14*, 95–105.

Dusseldorp, G. L. (2010). Prey choice during the South African Middle Stone Age: Avoiding dangerous prey or maximising returns? *African Archaeology Review, 27*, 107–133.

Ellis, C. J. (1997). Factors influencing the use of stone projectile tips: An ethnographic perspective. In H. Knecht (Ed.), *Projectile technology* (pp. 37–78). New York: Plenum Press.

Enzel, Y., Amit, R., Dayan, U., Crouvi, O., Kahana, R., Ziv, B., et al. (2008). The climatic and physiographic controls of the Eastern Mediterranean over the late Pleistocene climates in the Southern Levant and its neighboring deserts. *Global and Planetary Change, 60*, 165–192.

Faith, J. T. (2008). Eland, buffalo, and wild pigs: Were Middle Stone Age humans ineffective hunters? *Journal of Human Evolution, 55*, 24–36.

Frumkin, A., Bar-Yosef, O., & Schwarcz, H. P. (2011). Possible paleohydrologic and paleoclimatic effects on hominin migration and occupation of the Levantine Middle Paleolithic. *Journal of Human Evolution, 60*, 437–451.

Gamble, C. (1999). *The Palaeolithic societies of Europe*. Cambridge: Cambridge University Press.

Garrard, A. N. (1982). The environmental implications of the re-analysis of the large mammal fauna from the Wadi el-Mughara Caves, Palestine. In J. L. Bintliff & W. Van Zeist (Eds.), *Paleoclimates, paleoenvironments and human communities in the Eastern Mediterranean Region in Later Prehistory* (pp. 165–187). Oxford: BAR International Series S133.

Garrod, D. A. E., & Bate, D. M. A. (1937). *The Stone Age of Mount Carmel, Vol. 1. Excavations at the Wadi Mughara*. Oxford: Clarendon Press.

Gilead, I., & Grigson, C. (1984). Far'ah II: A Middle Paleolithic open-air site in the Northern Negev, Israel. *Proceedings of the Prehistoric Society, 50*, 71–97.

Grayson, D. K., & Delpech, F. (2008). The large mammals of Roc de Combe (Lot, France): The Châtelperronian and Aurignacian assemblages. *Journal of Anthropological Archaeology, 27*, 338–362.

Griggo, C. (2004). Mousterian fauna from Dederiyeh Cave and comparisons with fauna from Umm el Tlel and Douara Cave. *Paléorient, 30*, 149–162.

Hovers, E. (2009). The Middle-to-Upper Paleolithic transition: What news? In M. Camps & P. Chauhan (Eds.), *Sourcebook of Paleolithic transitions: Methods, theories, and interpretations* (pp. 455–462). New York: Springer.

Hovers, E., Buller, A., Ekshtain, R., Oron, M., & Yeshurun, R. (2008). Ein Qashish—a new Middle Paleolithic open-air site in northern Israel. *Journal of the Israel Prehistoric Society, 38*, 7–40.

Jelinek, A. J., Farrand, W. R., Hass, G., Horowitz, A., & Goldberg, P. (1973). New excavations at the Tabun Cave, Mount Carmel, Israel: A preliminary report. *Paléorient, 1*, 151–183.

Kahlke, R.-D., & Gaudzinski, S. (2005). The blessing of a great flood: Differentiation of mortality patterns in the large mammal record of the Lower Pleistocene fluvial site of Untermassfeld (Germany) and its relevance for the interpretation of faunal assemblages from archaeological sites. *Journal of Archaeological Science, 32*, 1202–1222.

Kaufman, D. (1999). *Archaeological perspectives on the origins of modern humans: A view from the Levant*. Westport: Bergin and Garvey.

Kaufman, D. (2002). Re-evaluating subsistence skills of Levantine Middle and Upper Palaeolithic hunters: A comparison of the faunal assemblages. *Oxford Journal of Archaeology, 21*, 217–229.

Klein, R. G. (1989). Biological and behavioural perspectives on modern human origins in Southern Africa. In P. Mellars & C. Stringer (Eds.), *The human revolution: Behavioural and biological perspectives on the origins of modern humans* (pp. 529–546). Edinburgh: Edinburgh University Press.

Klein, R. G. (2009). *The human career*. Chicago: University of Chicago Press.

Klein, R. G., & Cruz-Uribe, K. (1984). *The analysis of animal bones from archaeological sites*. Chicago: University of Chicago Press.

Lev, E., Kislev, M. E., & Bar-Yosef, O. (2005). Mousterian vegetal food in Kebara Cave, Mt. Carmel. *Journal of Archaeological Science, 32*, 475–484.

Lyman, R. L. (2008). *Quantitative paleozoology*. Cambridge: Cambridge University Press.

Marder, O., Yeshurun, R., Lupu, R., Bar-Oz, G., Belmaker, M., Porat, N., et al. (2011). Mammal remains at Rantis Cave, Israel, and Middle-Late Pleistocene human subsistence and ecology in the Southern Levant. *Journal of Quaternary Science, 26*, 769–780.

Marean, C. W., Dominguez-Rodrigo, M., & Pickering, T. R. (2004). Skeletal element equifinality in zooarchaeology begins with method: The evolution and status of the "shaft critique". *Journal of Taphonomy, 2*, 69–98.

Marín-Arroyo, A. B. (in press). Palaeolithic human subsistence in Mount Carmel (Israel): A taphonomic assessment of Middle and Early Upper Palaeolithic faunal remains from Tabun, Skhul and el-Wad. *International Journal of Osteoarchaeology*. doi: 10.1002/oa.1241

Mendelssohn, H., & Yom-Tov, Y. (1999). *Fauna Palaestina: Mammalia of Israel*. Jerusalem: The Israel Academy of Sciences and Humanities.

Monchot, H. (2005). Un assemblage original au Paléolithique moyen: le repaire à hyènes, porcs-épics et hominidés de la grotte Geula (Mont Carmel, Israël). *Paléorient, 31*, 27–42.

Munro, N. D. (2009). Epipalaeolithic subsistence intensification in the Southern Levant: The faunal evidence. In J.-J. Hublin & M. P. Richards (Eds.), *The evolution of hominin diets: Integrating approaches to the study of Palaeolithic subsistence* (pp. 141–155). Dordrecht: Springer.

Oliver, J. S. (1989). Analogues and site context: Bone damages from Shield Trap Cave (24CB91), Carbon County, Montana, U.S.A. In R. Bonnichsen & M. H. Sorg (Eds.), *Bone modification* (pp. 73–98). Orono: University of Maine, Center for the Study of the First Americans.

Oswalt, W. H. (1973). *Habitat and technology: The evolution of hunting*. New York: Holt, Rinehart and Winston.

Rabinovich, R. (1990). Taphonomic research on the faunal assemblage from the Quneitra site. In N. Goren-Inbar (Ed.), *Quneitra: A Mousterian Site on the Golan Heights* (Vol. 31, pp. 189–219). Jerusalem: Hebrew University of Jerusalem, Monographs of the Institute of Archaeology, Qedem.

Rabinovich, R. (2003). The Levantine Upper Palaeolithic faunal record. In A. N. Goring-Morris & A. Belfer-Cohen (Eds.), *More than meets the eye: Studies on Upper Palaeolithic diversity in the Near East* (pp. 33–48). Oxford: Oxbow.

Rabinovich, R., & Hovers, E. (2004). Faunal analysis from Amud Cave: Preliminary results and interpretations. *International Journal of Osteoarchaeology, 14*, 287–306.

Rabinovich, R., Bar-Yosef, O., Vandermeersch, B., & Horwitz, L. K. (2004). Hominid-carnivore interactions in the Palaeolithic site of Qafzeh Cave, Israel. *Revue de Paléobiologie, Genève, 23*, 627–637.

Shahack-Gross, R., Ayalon, A., Goldberg, P., Goren, Y., Ofek, B., Rabinovich, R., et al. (2008). Formation processes of cemented features in karstic cave sites revealed using stable oxygen and carbon isotopic analysis: A case study at Middle Paleolithic Amud Cave, Israel. *Geoarchaeology, 23*, 43–62.

Sharon, G., Grosman, L., Fluck, H., Melamed, Y., Rak, Y., Rabinovich, R., et al. (2010). The first two excavation seasons at NMO: A Mousterian site at the bank of the Jordan River. *Eurasian Prehistory, 7*, 135–157.

Shea, J. J. (2003). The Middle Paleolithic of the East Mediterranean Levant. *Journal of World Prehistory, 17*, 313–394.

Shea, J. J. (2006). The origins of lithic projectile point technology: Evidence from Africa, the Levant, and Europe. *Journal of Archaeological Science, 33*, 823–846.

Shea, J. J. (2008). Transitions or turnovers? Climatically-forced extinctions of *Homo sapiens* and Neanderthals in the East Mediterranean Levant. *Quaternary Science Reviews, 27*, 2253–2270.

Shea, J. J., & Sisk, M. L. (2010). Complex projectile technology and *Homo sapiens* dispersal into Western Eurasia. *PaleoAnthropology, 2010*, 100–122.

Speth, J. D., & Clark, J. L. (2006). Hunting and overhunting in the Levantine Late Middle Palaeolithic. *Before Farming, 3*, 1–42.

Speth, J. D., & Tchernov, E. (1998). The role of hunting and scavenging in Neandertal procurement strategies: New evidence from Kebara Cave, Israel. In T. Akazawa, K. Aoki, & O. Bar-Yosef (Eds.), *Neandertals and modern humans in Western Asia* (pp. 223–239). New York: Plenum Press.

Speth, J. D., & Tchernov, E. (2001). Neandertal hunting and meat processing in the Near East: Evidence from Kebara Cave (Israel). In C. B. Stanford & H. T. Bunn (Eds.), *Meat-eating and human evolution* (pp. 52–72). New York: Oxford University Press.

Speth, J. D., & Tchernov E. (2007). The Middle Palaeolithic occupations at Kebara Cave: A faunal perspective. In O. Bar-Yosef & L. Meignen (Eds.), *Kebara Cave, Mt. Carmel, Israel. The Middle and Upper Palaeolithic Archaeology*, Part 1 (pp. 165–260). Cambridge, MA: Peabody Museum of Archaeology and Ethnology.

Speth, J. D. (2010). *The paleoanthropology and archaeology of big-game hunting: Protein, fat or politics?* New York: Springer.

Stiner, M. C. (1990). The use of mortality patterns in archaeological studies of hominid predatory adaptations. *Journal of Anthropological Archaeology, 9*, 305–351.

Stiner, M. C. (2002). Carnivory, coevolution, and the geographic spread of the genus *Homo*. *Journal of Archaeological Research, 10*, 1–63.

Stiner, M. C. (2005). *The Faunas of Hayonim Cave (Israel): A 200,000-year record of Paleolithic diet, demography and society*. Cambridge, MA: Peabody Museum of Archaeology and Ethnology.

Stiner, M. C., Gopher, A., & Barkai, R. (2011). Hearth-side socioeconomics, hunting and paleoecology during the late Lower Paleolithic at Qesem Cave, Israel. *Journal of Human Evolution, 60*, 213–233.

Tchernov, E. (1992). Biochronology, paleoecology, and dispersal events of hominids in the Southern Levant. In T. Akazawa, K. Aoki, & T. Kimura (Eds.), *The evolution and dispersal of modern humans in Asia* (pp. 149–188). Tokyo: Hokusen-Sha.

Tchernov, E., & Tsoukala, E. (1997). Middle Pleistocene (Early Toringian) carnivore remains from Northern Israel. *Quaternary Research, 42*, 122–136.

Thieme, H. (1997). Lower Palaeolithic hunting spears from Germany. *Nature, 385*, 807–810.

Vaks, A., Bar-Matthews, M., Ayalon, A., Matthews, A., Frumkin, A., Dayan, U., et al. (2006). Paleoclimate and location of the border between Mediterranean climate region and the Saharo-Arabian desert as revealed by speleothems from the northern Negev Desert, Israel. *Earth and Planetary Science Letters, 249*, 384–399.

Weaver, T. D., Steele, T., & Klein, R. G. (2011). The abundance of eland, buffalo, and wild pigs in Middle and Later Stone Age sites. *Journal of Human Evolution, 60*, 309–314.

Weinstein-Evron, M., Bar-Oz, G., Zaidner, Y., Tsatskin, A., Druck, D., Porat, N., et al. (2003). Introducing Misliya Cave: A new continuous Lower/Middle Palaeolithic sequence in the Levant. *Eurasian Prehistory, 1*, 31–55.

Yeshurun, R., Bar-Oz, G., & Weinstein-Evron, M. (2007a). Modern hunting behavior in the early Middle Paleolithic: Faunal remains from Misliya Cave, Mount Carmel, Israel. *Journal of Human Evolution, 53*, 656–677.

Yeshurun, R., Marom, N., & Bar-Oz, G. (2007b). Differential fragmentation of different ungulate body-size—comparison of gazelle and fallow deer bone fragmentation in Levantine prehistoric assemblages. *Journal of Taphonomy, 5*, 137–148.

Yeshurun, R., Zaidner, Y., Eisenmann, V., Martinez-Navarro, B., & Bar-Oz, G. (2011). Lower Paleolithic hominin ecology at the fringe of the desert: Faunal remains from Bizat Ruhama and Nahal Hesi, northern Negev, Israel. *Journal of Human Evolution, 60*, 492–507.

Zaidner, Y., Tsatskin, A., Yeshurun, R., Frumkin, A., Weissbrod, L., Weinstein-Evron, M., et al. (2010). The new Mousterian sequence at Nesher-Ramle Quarry: Preliminary results from the 2010 season. Paper presented at the 40th Annual Meeting of the Israel Prehistoric Society, Tel-Hai.

Chapter 5
New Opportunities for Previously Excavated Sites: Paleoeconomy as a Human Evolutionary Indicator at Tabun Cave (Israel)

Ana B. Marín-Arroyo

Introduction

In terms of human evolution, the Near East is an exceptional region. It was the only area where Neanderthals and Anatomically Modern Humans (AMH) were known to have inhabited the same geographical territory during the Middle Paleolithic ($\sim$250–40 ka), sharing also the same lithic technology, known as the "Levantine Mousterian." This interesting scenario took place during Marine Isotope Stages (MIS) 7-3, a period involving several abrupt climatic changes that threatened the survival of both human species, thus provoking two subsequent replacements, AMH by Neanderthals and these later in turn by *Homo sapiens* (Bar-Yosef 2000; Shea 2003b) although probably with no direct confrontation. This makes the Levant a unique place to assess the adaptive skills of Neanderthals and AMH independently, without the disturbing effect of distinctive culture, as is the case during the Middle to Upper Paleolithic transition. In this sense, the paleoeconomic differences found between them can reasonably be related both to biological or cognitive dissimilarities or environmental conditions, the latter being known by means of paleoclimatic reconstructions.

Thanks to the large number of Mousterian assemblages found in the region, the archaeozoological record of the period is well known. However, many of the existing sites were excavated in the first half of the twentieth century (Turville-Petre 1932; Garrod and Bate 1937; McCown and Keith 1939; Neuville 1951) with old-fashioned analytical techniques, the aim of which was principally to recover human fossils, to document the succession of lithic industries and to reconstruct the paleoenvironment as a way to understand the available food resources. However, little attention was paid to characterizing human subsistence strategies or to understanding the accumulative agents of the bone assemblages (de la Torre and Domínguez-Rodrigo 2000). Nonetheless, this limitation was progressively overcome with new projects starting in the 1970–1980s (including the sites of Tabun, Kebara, Hayonim, Qafzeh and Amud), where not only excavation procedures and dating techniques were radically improved, but also where a wide range of multidisciplinary approaches were applied in order to better understand the way each site was formed and preserved and to accurately reconstruct the behavior of the human groups who lived there (see for example Rabinovich and Tchernov 1995; Speth and Tchernov 2001 or Stiner 2005).

Based on this new information, there is now a wide consensus on the importance that large-game hunting had during the Levantine Middle Paleolithic (Shea 2003a), the residual significance of scavenging as represented in preferences for certain anatomical parts and taphonomic alterations (Speth and Tchernov 1998) and the influence of climate change on human dispersal and replacement in the region (Frumkin et al. 2011; Shea 2008). However, data regarding comparative hunting skills between AMH and archaic populations are still scarce, probably due to the fact that both human species inhabited similar topographic areas but very different ecological scenarios (Bar-Yosef 2000) although the lack of well preserved faunal assemblages associated with anatomically modern humans in the region, limited only to Qafzeh and Skhul, is also a constraint. The resulting variations in the type and abundance of available ungulate species due to climatic oscillations (Tchernov 1998) might have then concealed any particular pattern in human behavior.

Given this background, detailed new studies are clearly needed to boost our understanding of human economic evolution, as relates to both environmental and biological changes, leading thus to a better evaluation of the reasons behind the final success of *Homo sapiens* during the Middle to Upper Paleolithic transition. Until new sites are discovered

A. B. Marín-Arroyo (✉)
Instituto Internacional de Investigaciones Prehistóricas de Cantabria, Universidad de Cantabria, Santander, Spain
e-mail: anabelen.marin@unican.es

Jamie L. Clark and John D. Speth (eds.), *Zooarchaeology and Modern Human Origins: Human Hunting Behavior during the Later Pleistocene*, Vertebrate Paleobiology and Paleoanthropology, DOI: 10.1007/978-94-007-6766-9_5,

and analyzed, an interesting option to expand our paleoeconomic knowledge is to pay new attention to classic sites, but with a modern methodological approach that can overcome the existing biases of old excavation procedures. One of these classic sites is Tabun Cave, located in the foothills of Mount Carmel (Israel), in a cliff-line overlooking the nearby coastal plain. The importance of Tabun in the cultural sequencing of the Near East, through the characterisation of its industrial phases, is well-established (Bar-Yosef 1998; Copeland 1975; Jelinek 1982b) as is its key place in paleoanthropological studies due to the fact that it holds human remains of both AMH (Quam and Smith 1998; Rak 1998) and Neanderthals (Coppa et al. 2005; McCown and Keith 1939) dated to the Middle Paleolithic. In addition, Tabun Level D is one of the few known Early Middle Paleolithic sites of the region, thus providing a unique chronological and evolutionary sequence throughout MIS 7-MIS 5 that can reveal relevant information about economic differences between AMH and archaic populations.

However, the long-available faunal assemblage has only been studied from a paleontological point of view (Bate 1937; Garrard 1981, 1982), usually without any taphonomic considerations. Not surprisingly, the lack of adequate excavation procedures during the 1930s, with a marked preference for the more diagnostic elements (i.e. epiphyses and teeth) and the almost total neglect of the shafts of long bones and axial remains, has discouraged a deeper analysis. Despite this, with the necessary precautions, the deposit might still provide new insights into subsistence strategies, if exhaustive taphonomic techniques to maximize the information that can be extracted are used. In this sense, the differential collection of faunal remains that occurred during the excavations may make the study more difficult, but does not prevent it. On one hand, it is possible that the biases were not as significant as previously thought, as the results of the skeletal profile study will show later. On the other hand, being aware of this problem, there are still useful comparative techniques that can be applied.

Consequently, the objective of the study presented here is to reappraise the faunal material of Garrod's excavation at Tabun Cave from a new paleoeconomic perspective, and to contextualize the obtained data with other well-known Levantine Middle Paleolithic sites. This has allowed the identification of behavioral differences between the two human species that occupied the site. As a result, a new hypothesis of human mobility and hunting preferences has been formulated, which is in agreement with biological and cognitive changes. Bearing this information in mind, the reasons behind the replacement of AMH at the end of MIS 5 by Neanderthals have been evaluated.

Materials and Methods

Tabun Cave is located on Mount Carmel (northern Israel) 45 m above present sea level in the Wadi el-Mughara valley (see Fig. 5.1). The entrance faces northwest, overlooking the Mediterranean coastal plain. The site contained a very long stratigraphic sequence from Lower Paleolithic to historical times. Dorothy Garrod first excavated Tabun between 1929 and 1934 (Garrod and Bate 1937), with rather expedited procedures, although this did not prevent her from recovering one of the richest Middle Paleolithic assemblages in the Levant. Garrod designated the Levantine Mousterian levels as B, C and D based mainly on differences in lithic industries. Between 1967 and 1972, Jelinek carried out a new excavation of the site, but unfortunately neither human nor faunal remains were recovered (Jelinek et al. 1973; Jelinek 1982a). There have been different

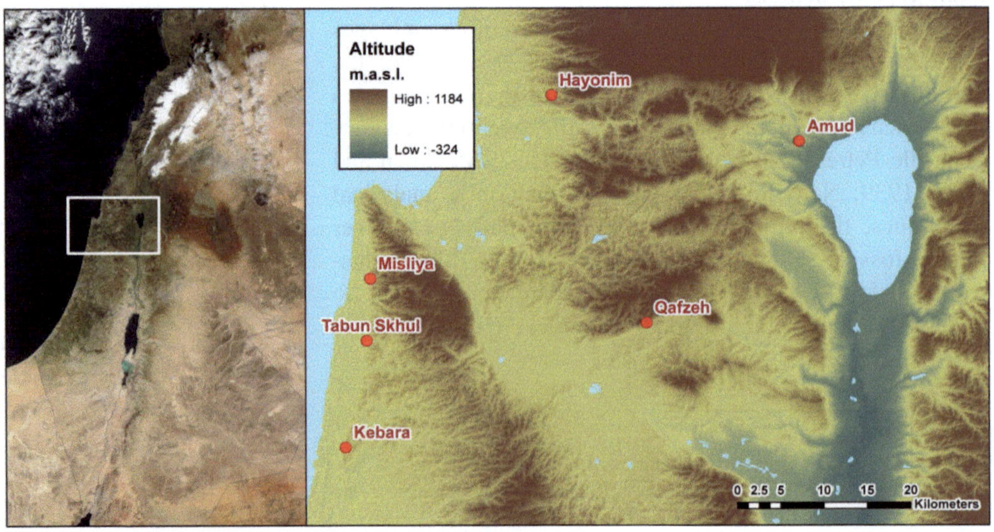

Fig. 5.1 Location of Tabun Cave and other sites cited in the text. Sea level at its present position

estimates of the age of the various levels at Tabun over time. At present, thermoluminescence, electron spin resonance and U-series absolute dates are available for Tabun (Grün and Stringer 2000: 602; Mercier and Valladas 2003), placing Level D, attributed to archaic human species, at the end of MIS 8 or second half of MIS 7, Level C, with AMH fossils, in the first part of MIS 5, and Level B at the end of MIS 5, with Neanderthal remains present (Shea 2003a, b, 2008). I have followed Bar-Yosef and Callander's (1999) assumption that the Neanderthal skeleton C1 is better placed within Level B and Rak's (1998) study when he attributes specimen C2 to an AMH jaw.

The current analysis was conducted in the Department of Paleontology at the Natural History Museum in London (NHM), where the material is currently curated. Only macromammalian remains were considered. Starting with the taxonomic and anatomical identification made by Garrard (1981), newly found specimens that were previously ignored, together with remains a priori unidentified but now successfully recognized using the large comparative collection of the NHM, were added to the study. The following indices were applied in terms of quantification: Number of Identified Specimens (NISP) (Payne 1975), Minimum Number of Individuals (MNI; Klein and Cruz-Uribe 1984), Minimum Number of Skeletal Elements (MNE; Marín-Arroyo 2009a) and Minimum Animal Units (MAU; Binford 1978). Biomass calculation was made by multiplying values of useable meat given by Garrard (1998: 63) by the MNI. This might be representative of human diet regardless of the differential transport that each prey might have undergone, because most edible parts would have been consumed should primary access to carcasses have occurred. In order to favor comparative analysis, different taxa were grouped together according to their size into large (*Rhinoceros/Hippopotamus/Bos/Equus/Camelids*), medium (*Cervus/Dama/Capra/Sus*) and small (*Gazella/Capreolus*) ungulate groups. Mountain gazelle and ibex are considered mountain species whereas the remaining ones are better suited to plains environments. The latter classification will allow the identification of trends in hunting preferences as a function of topographic relief, as this variable is not altered by climate change, unlike vegetation cover, which, on the other hand, can not be utilized due to a lack of adequate paleoenvironmental reconstructions in space and time.

Every element was examined under a 40× binocular microscope in order to detect any anthropogenic bone surface modifications, such as cut marks [identified as evisceration, skinning, dismembering and defleshing marks following Binford (1981) and Pérez Ripoll (1992)], hammerstone percussion marks [including conchoidal notches (Bunn 1981; Capaldo and Blumenschine 1994; Pickering and Egeland 2006), type and angle of fracturing (fresh-green versus old-dry), thermal alterations and trampling (Behrensmeyer et al. 1986). Carnivore and rodent gnawing marks and traces of digestion were also identified (Binford 1981; Stiner 1994) as well as other biological and physico-chemical alterations, such as weathering (Behrensmeyer 1978), root etching, carbonate deposits (Shipman 1981; Fisher 1995), dissolution or mineral coatings (mainly manganese, see Marín-Arroyo et al. 2008).

The ungulate mortality patterns (i.e. juvenile, prime or old adult) was assessed by both dental eruption and wear stage of dp4, P4 and M3, following Stiner (1991, 2005), and fusion of the bone epiphyses. For gazelle, the work of Munro et al. (2009) was used. The diet breadth and the degree of anthropogenic use of the environment has been evaluated with the Inverse of Simpson's Index (1949), whose maximum value equals the number of consumed species only when they are in the same proportion.

Differential transport of carcasses and the related functionality of the site was investigated using bivariable correlations of %MAU with utility (in terms of %MGUI following Binford 1978) and bone survivorship (represented by bone density following Lam et al. 1999). Pielou's evenness index (1966) was used to assess skeletal completeness for each size-class.

Finally, the relationship between the consumed fauna and the environment was investigated by estimating the catchment areas associated with several contemporaneous sites (Tabun, Misliya, Hayonim, Amud, Kebara, Skhul and Qafzeh) and characterizing the local relief, following the methodology described by Marín-Arroyo (2009b). This approach is based on the hypothesis that more carcass processing at the kill-site would arise when animals are hunted far from the base camp, implying thus non-desirable larger transport costs of low-yield anatomical parts. The Central Place Prey Foraging Model (Cannon 2003) predicts that small animals will likely be transported complete but that there is also a threshold of distance from the base camp beyond which the energetic yield of hunting this size of animals will diminish, favoring the exploitation of large game. Encounter rates will also affect these decisions, so that the shortage of a type of prey should prevent their usual consumption, above all when longer travel times are needed. Biotope abundance has been used as a proxy for encounter rates and so, the higher the biotope surface at a certain distance from the base camp, the larger the likelihood of hunting animals related to that kind of environment compared with other animals located at the same distance. From a practical point of view, a digital model of the terrain around the sites has been produced including the continental shelf (ASTER GLOBAL DEM and ETOPO1 data). An average sea level height of −80 and −40 m has been estimated for glacial and interglacial periods respectively (Siddall et al. 2006). Traveling times across the territory have been estimated with empirical formulae depending on distance, slope angles, movement direction (uphill or downhill) and the existence of insurmountable barriers (Marín-Arroyo 2008). In addition, to define the preferred biotopes or habitats for plains and mountain species within the boundaries determined by the catchment area, a threshold value of 10 % slope has been fixed to differentiate areas related to one or the other group of taxa. Beyond that value, grazing suitability is depleted (Holechek et al. 1998).

Results

Quantification of the Faunal Assemblage

The faunal assemblage is mainly dominated by complete bones. As previously mentioned, shaft fragments are rare and usually associated with epiphyses. A total of 3,402 specimens were accounted for, with roughly two-thirds belonging to Tabun B. 46 new bone remains were added to Garrard's total by means of a thorough evaluation of previously unidentified fragments. As a result of the recovery bias during Garrod's excavation, almost 99 % of the assemblage was identifiable taxonomically and anatomically. Table 5.1 shows the final figures for NISP, MNE and MNI together with biomass estimations. As can be seen, the NISP/MNE ratio (1.4 in Tabun D, 1.1 in Tabun C and 1.07 in Tabun B) is extremely low due to the same recovery problem.

Species Representation

A total of 25 mammal species were identified through the Tabun sequence; the carnivore frequency is low and ungulates clearly dominate the assemblage. In terms of NISP, *Dama mesopotamica* and *Gazella gazella* stand out, followed by a relatively large number of *Bos primigenius* (see Table 5.2). Taxonomic representation based on MNI values do not differ significantly. However, there are variations between levels in NISP. Thus, Tabun B shows a high presence of fallow deer (78.5 % NISP) in comparison with Tabun C (8.9 %) and Tabun D (31.3 %) while the percentage of gazelle decreases through time between 41.5 % in Tabun D to 15.1 % in Tabun B.

The significant difference in faunal composition of Tabun B was explained by Bate (1937) as being a result of climatic change, assuming then that the animal remains were introduced into the cave by humans, although she also considered that some of the Layer B fauna had been washed into the cave (1931: 150). However, as will be explained later, the origin of the accumulation of Tabun B was in fact mainly natural, and therefore the data from this level must not be interpreted economically. Other ungulates, such as *Cervus elaphus*, *Capreolus capreolus*, *Sus scrofa*, *Capra* sp. and *Equus* sp., are also represented in the deposit, although in small quantities, while large mammals such as *Dicerorhinus mercki* and *Hippopotamus amphibius* appear only in Levels D and C.

The percentage of large game, represented by rhinoceros, hippopotamus, horse, aurochs and camel, is highest in Tabun C (see Fig. 5.2). This is also the case with other Middle Paleolithic sites with known AMH accumulations, such as Skhul B (Marín-Arroyo 2011) and Qafzeh (Rabinovich and Tchernov 1995). In contrast, occupations by archaic hominins during the Early and Late Middle Paleolithic, such as Misliya (Yeshurun et al. 2007), Hayonim E (Stiner 2005), Amud B (Rabinovich and Hovers 2004) and Kebara VI-XIII (Speth and Tchernov 1998; Speth and Clark 2006), show a higher preference for the consumption of small ungulates, in these cases mountain gazelle. A similar comparison regarding the percentage of bone remains (NISP) belonging to taxa related to plains or mountain environments shows evidence of a greater use of rough terrain by archaic populations (see Fig. 5.2), which is not clearly related to the topographic relief around each settlement (see the heterogeneity of relative abundance of plains against mountains in Table 5.6). The diet breadth represented by the inverse of Simpson's Index notably reflects a more diverse use of the available resources during AMH occupations, with the inclusion of a larger proportion of high-ranked prey species typical of plains environments.

When biomass is calculated, the role of large fauna in all levels is more emphasized, as it comprises the most important caloric input in the diet, leaving gazelle as of secondary importance. This clearly shows the need to reconsider the role of each taxon in the subsistence strategies, based on the quantification of NISP (Marín-Arroyo and González Morales 2009), and opens new interpretative possibilities, as it would imply a more selective hunting strategy based on high-ranked species during the Middle Paleolithic, especially with regards to AMH. It is important to point out, however, that biomass calculations will establish the upper contribution of a species in the diet while NISP will define the lower. The real contribution will depend on both transport strategies and the role of scavenging in the paleoeconomy.

Bone Surface Modifications

Table 5.3 shows the abundance of different types of bone surface alterations in the assemblage. Due to the almost complete absence of bone shafts, cut marks were found mostly on epiphyses and near-epiphyses. In Tabun D (3.7 %) and especially Tabun C (9.2 %), the frequency of

Table 5.1 Assemblage quantification of Tabun Cave

	Tabun D				Tabun C				Tabun B			
	NISP	MNE	MNI	Biomass	NISP	MNE	MNI	Biomass	NISP	MNE	MNI	Biomass
Dicerorhinus mercki	12	11	2	2,084	26	26	2	2,500				
Hippopotamus amphibius	7	5	1	1,250	35	31	3	2,918				
Bos primigenius	31	31	3	934	100	97	3	1,200	49	45	3	1,200
Equus hydruntinus									39	39	3	600
Equus hemionus	3	3	1	150	4	4	2	300	4	4	1	150
Equus ferus					11	11	2	300				
Camellus sp.					1	1	1	150				
Cervus elaphus	46	37	3	200	21	17	2	150	17	17	2	150
Dama mesopotamica	141	115	11	696	43	38	3	174	1,729	1,599	78	4,022
Capreolus capreolus	5	5	1	8					3	3	1	8
Capra sp.	6	6	1	29	15	14	2	86	21	21	5	215
Gazella gazella	187	162	11	106	187	174	8	77	332	318	23	242
Sus scrofa	13	13	2	88	38	36	3	141	9	9	1	35
Lepus capensis					2	2	1					
Procavia capensis					3	2	2		1	1	1	
Hystrix indica	25	25	1						13	13	1	
Hyaena hyaena	1	1	1						9	9	2	
Crocuta crocuta	5	5	2		2	2	1					
Panthera pardus					2	2	1		48	46	4	
Felis cf. silvestris	1	1	1						2	2	1	
Canis aureus	2	2	1						4	4	1	
Canis lupus	27	27	2						2	2	1	
Vulpes vulpes					6	6	1		42	42	6	
Ursus arctos	2	2	1		15	15	1					
Martes foina									3	3	1	
Megafauna size					17	6						
Large mammal size					13	10						
Medium mammal size					3	2			3	3		
Small mammal size					10	8						
Indeterminate					1	1			3	3		
Total Ungulates	432	372	33		420	392	26		2,203	2,055	117	
Total Carnivores	38	38	8		25	25	4		110	108	16	
% Ungulates	84	82	73		76	78	68		94	94	87	
Total biomass				5,545				7,996				6,622
Total	**514**	**451**	**45**	**5,545**	**555**	**505**	**38**	**7,996**	**2,333**	**2,183**	**135**	**6,622**

tool-marked bones is high enough to confirm a human origin of the faunal deposit, whereas in Tabun B (1.1 %) the accumulating agent is unclear. The abundance of burnt bones and anthropogenic fractures confirms this assumption, with values of 20.7 and 10.6 % in Tabun C respectively. This is undoubtedly the level where more carcass processing occurred, perhaps due to a more permanent occupation (see some examples in Fig. 5.3). This is coherent with the abundant thin layers (red, white and black) resulting from repeated fires found at this level (Jelinek et al. 1973: 158; Albert et al. 1999). In addition, out of the 29 cut marks identified on the main ungulate remains from Level C, one corresponds to skinning, 24 to dismembering and five to defleshing (Marín-Arroyo 2011). They are located on meat-bearing skeletal elements such as the pelvis, ribs, radius and vertebrae, suggesting primary access to carcasses.

Carnivore activity was very scarce. Only 0.4 % of the assemblage shows gnawing marks, indicating that, at least in Tabun D and C, humans were the primary accumulating

Table 5.2 Taxonomic representation of Tabun Cave

	% NISP											
	Tabun D				Tabun C				Tabun B			
	NISP	MNE	MNI	Biomass	NISP	MNE	MNI	Biomass	NISP	MNE	MNI	Biomass
Dicerorhinus mercki	2.7	2.8	5.6	37.6	5.4	5.8	6.5	31.3	–	–	–	–
Hippopotamus amphibius	1.6	1.3	2.8	22.5	7.3	6.9	9.7	36.5	–	–	–	–
Bos primigenius	6.9	8.0	8.3	16.8	20.8	21.6	9.7	15.0	2.2	2.2	2.6	18.1
Equus hydruntinus	–	–	–	–	–	–	–	–	1.8	1.9	2.6	9.1
Equus hemionus	0.7	0.8	2.8	2.7	0.8	0.9	6.5	3.8	0.2	0.2	0.9	2.3
Equus ferus	–	–	–	–	2.3	2.4	6.5	3.8	–	–	–	–
Camellus sp.	–	–	–	–	0.2	0.2	3.2	1.9	–	–	–	–
Cervus elaphus	10.2	9.5	8.3	3.6	4.4	3.8	6.5	1.9	0.8	0.8	1.7	2.3
Dama mesopotamica	31.3	29.6	30.6	12.6	8.9	8.5	9.7	2.2	78.5	77.8	66.7	60.7
Capreolus capreolus	1.1	1.3	2.8	0.1	–	–	–	–	0.1	0.1	0.9	0.1
Capra sp.	1.3	1.5	2.8	0.5	3.1	3.1	6.5	1.1	1.0	1.0	4.3	3.2
Gazella gazella	41.5	41.8	30.6	1.9	38.9	38.8	25.8	1.0	15.1	15.5	19.7	3.7
Sus scrofa	2.9	3.4	5.6	1.6	7.9	8.0	9.7	1.8	0.4	0.4	0.9	0.5

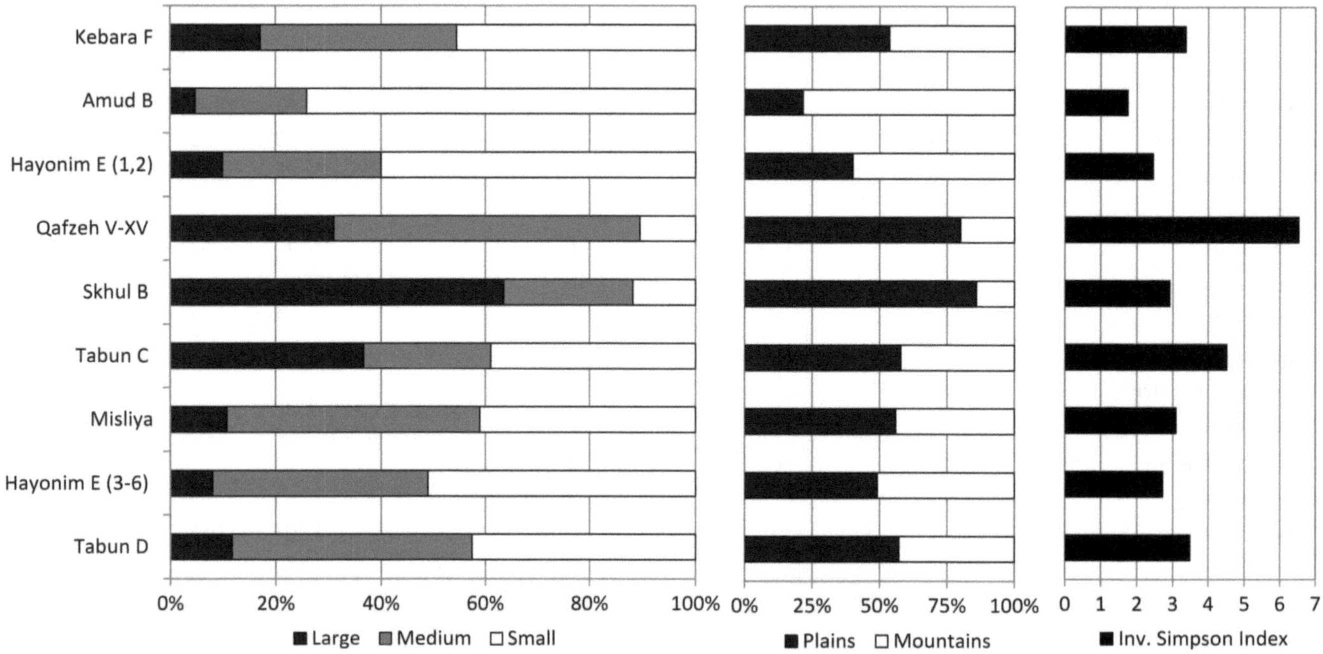

Fig. 5.2 Diet comparison of Levantine Middle Paleolithic in terms of NISP

agent with carnivores having a secondary and less important role (see results from actualistic studies in Marean 1998: 129; Marean and Spencer 1991; Marean et al. 1992).

The case with Tabun B is rather different. In addition to a low, although not negligible, level of human activity, the overall appearance of the bone assemblage is different throughout. On the one hand, bones show a white colouring with some manganese dendrites, related to a likely period of ponding in the cave, contrasting clearly with the general brown colouring of the bones in Tabun C and D (high presence of organic material in the sediment as a result of intense human occupation, see Marín-Arroyo et al. 2008). The action of water is also evidenced by the larger amount of dissolution and subaerial weathering. On the other hand, some bones appear in anatomical articulation and predominantly complete. The breakage pattern is typical of mechanical destruction produced by pressure applied by stones/bones within the sediment (Fisher 1995). Bearing this in mind, a natural accumulation of carcasses caused by animals falling through the hole in the cave roof can be

Table 5.3 Assemblage taphonomy of Tabun Cave

Bone modifications	%			
	Tabun B	Tabun C	Tabun D	Total
Cut marks	1.1	9.2	3.7	2.7
Impact scar	0.1	0.4	0.4	0.2
Green-bone fracture	2.3	10.6	2.9	3.6
Burnt	1.6	20.7	3.3	4.7
Gnaw-marks	0.2	1.6	0.4	0.4
Rodent-marks	0.0	0.4	0.2	0.1
Weathering	4.9	0.8	1.6	3.7
Trampling	3.5	0.6	0.4	2.5
Insect/fungus	0.0	1.4	0.0	0.2
Root etching	0.7	1.8	0.2	0.8
Carbonate	10.5	15.5	16.7	12.0
Coating (mineral)	12.6	10.6	11.9	12.0
Dissolution	9.7	2.2	2.3	7.3
Polishing	0.2	0.2	0.0	0.1

Skeletal Element Profiles

The MNE values for each skeletal element and class-size are presented in Table 5.4. Tabun B, which was not quantified anatomically by Garrard (1981), is also shown. The vast majority of skeletal elements in the assemblage are teeth (61 %), followed by feet (metapodial + phalanx) (24 %). The representation of long bones is only 6 % and it is mainly characterized by epiphyses of small-medium sized taxa, mostly with shafts attached. The axial skeleton is only represented by a proportion of 1 %.

The study of skeletal profiles at Tabun is problematic due to the excavation bias. There is now a wide consensus on the importance of recovering and quantifying long bone shafts in order not to hinder the economic interpretation (Marean and Frey 1997). However, taking into account the limitations of the available information, there is still room for drawing relevant conclusions if comparative analyses are applied, hence playing down individual absolute results. In this way, the effect of the bias in the collection can be side-stepped to a certain extent, as its influence would have been similar in the whole assemblage.

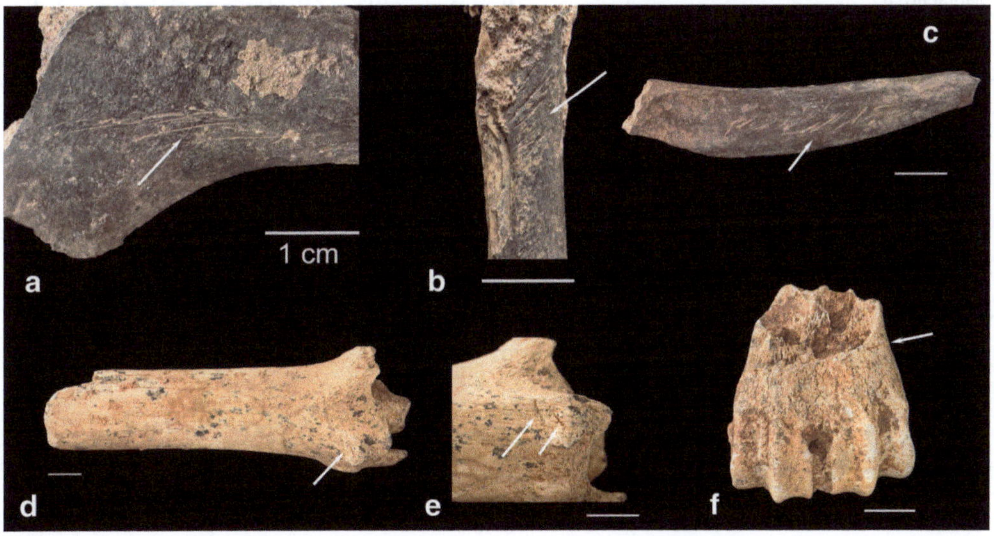

Fig. 5.3 Examples of tool-marked bones identified in Tabun Cave. **a** Lumbar vertebra of aurochs from Tabun C. **b** Metapodial of fallow deer from Tabun C with cut marks in the caudal proximal meta-diaphysis to cut tendons. **c** Shaft of gazelle femur from Tabun C with defleshing marks. **d** Tibia of fallow deer from Tabun B with dismembering marks. **e** Close-up of D. **f** Metacarpal of fallow deer from Tabun B

confirmed (Bate 1937: 150). However, this does not imply that the Tabun B assemblage is completely paleontological. The presence of cut marks on some fallow deer remains, as well as thermally alterated bones (identified on 1.6 % of the remains), seems to suggest an occasional human consumption of ungulates, possibly those accumulated naturally and found later by humans who knew the potential of the cave as a natural trap.

Apart from the cut marks found on ungulate remains, there are also some tool-marked carnivore bones, such as a metatarsal of *Panthera pardus* in Tabun B, probably related to skinning, and a radius of *Ursus* sp. in Tabun C that shows fresh breakage, burn traces and several dismembering marks, which could mean meat and marrow exploitation as well as skin use.

Figure 5.4 shows the skeletal profiles in terms of MAU. In general, small ungulates are more evenly represented than larger game, whereas long bones are less abundant in medium-size game. These results are consistent with a mostly whole transport of mountain gazelle and roe deer carcasses, whose smaller weight does not require the processing of the prey at the kill site (Cannon 2003), and with an intense defleshing of large fauna before transport, in order to maximize the energetic contribution to the base camp, particularly if the animal was hunted a long distance away (O'Connell et al. 1988, 1989). Taking into account

Table 5.4 Skeletal representation of large, medium and small ungulates in Tabun Cave

MNE w isolated teeth/tarsals	Tabun B			Tabun C			Tabun D		
	LU	MU	SU	LU	MU	SU	LU	MU	SU
Antler	–	2	–	–	2	4	–	1	11
Maxilar	22	61	7	1	1	4	–	10	4
Mandible	–	82	10	3	16	10	2	11	14
Teeth	50	976	234	124	29	82	31	114	91
Atlas	–	–	–	–	–	–	–	–	–
Axis	–	2	–	–	1	–	–	1	–
Cervical vert	–	2	–	–	1	–	–	–	–
Thoracic vertebrae	–	3	1	–	–	–	–	–	2
Lumbar vertebrae	–	4	–	1	–	1	–	–	–
Ribs	–	2	–	1	–	2	–	–	–
Sternum	–	10	–	–	–	–	–	–	–
Scapula	–	2	–	–	–	6	1	1	1
Humerus	–	11	4	–	2	4	1	1	5
Radius	–	14	4	1	2	2	3	1	4
Carpals	–	112	1	3	–	–	1	–	–
Metacarpal	1	26	10	1	7	4	3	5	7
Pelvis	–	1	–	–	1	3	–	–	–
Sacrum	–	1	–	–	–	–	–	–	–
Femur	–	13	6	–	2	3	3	–	–
Tibia	–	10	5	1	1	2	1	2	3
Astragalus	1	30	6	3	8	12	–	4	2
Calcaneus	–	3	2	1	1	5	–	–	1
Metatarsal	1	13	4	–	5	–	–	4	–
Tarsals	3	31	1	–	–	2	–	–	–
Proximal phalanx	4	107	14	6	6	9	2	10	11
Medial phalanx	6	70	6	2	7	11	–	2	7
Distal phalanx	–	50	3	1	8	7	1	4	3
Totals	**88**	**1,638**	**318**	**149**	**100**	**173**	**49**	**171**	**167**

LU Large Ungulates (rhinoceros, hippopotamus, horse, aurochs and camel), *MU* Medium Ungulates (red deer, capra, fallow deer and wild boar), *SU* Small Ungulates (gazella, roe deer)

that there are 24 different anatomical elements, Pileou's evenness index has been also calculated for each body-size class and level using MAU values. The higher values correspond to small ungulates in each level (average value of 0.76) followed by medium and large prey (average values of 0.74 and 0.53 respectively). Spearman's correlations between large, medium and small ungulates have been calculated for each level in order to better evaluate dissimilarities. As a result, significant positive correlations have been found between species in Tabun B ($\rho_s = 0.80$, $p = 0$ between medium and small ungulates, and $\rho_s = 0.62$, $p = 0.002$ between large and medium ungulates), thus corroborating the probable natural origin of this assemblage. In Tabun D, there is also a significant positive correlation between medium and small ungulates ($\rho_s = 0.46$, $p = 0.05$) and higher similarities between body-size skeletal profiles than in Tabun C, which can be interpreted as a greater importance of prey weight in transport decisions within AMH groups, probably due to longer travel times.

To better assess this assumption, correlations between %MAU and %MGUI, as well as between %MAU and maximum bone density, have been calculated (see Table 5.5). Correlations with bone survivorship are always positive (in three cases significantly), which could be related to the existence of moderate attrition. Not surprisingly, correlations with utility are low or negative (even significantly in two cases) as a result of bone attrition and the known bias in the assemblage recovery methods against long bone shafts and axial elements, the most meat-bearing parts of the skeleton. However, as regards medium and small ungulates, Tabun D shows more negative values compared with Tabun C, which could indicate less

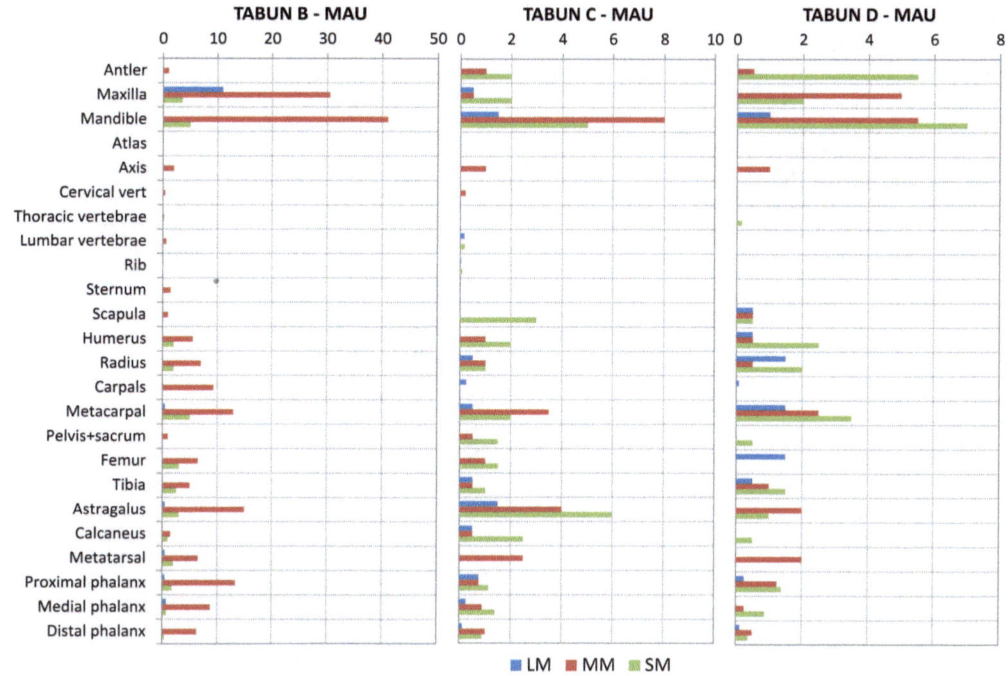

Fig. 5.4 Skeletal profiles of large, medium and small ungulates in terms of MAU

processing at the kill-site as a consequence of smaller catchment areas or a less residential use of the settlement.

In addition, to estimate how much the non-collection of shafts might have biased the economic interpretation of the assemblage, the MNE of the long bones (humerus, radio-ulna, femur, tibia and metapodials) was multiplied by two and by four, and the correlations with %MGUI and maximum bone density were repeated. The aim of this analysis was not to reconstruct the original assemblage, but rather to evaluate the possible loss of information that might have occurred. The results (see Table 5.5) point towards similar interpretations regarding the differential transport (the correlations are only a little less negative, but with no appreciable changes), which would support the use of the present information as representative of the subsistence behavior of the human groups. However, the presence of more long bones would have resulted in a more clearly-defined influence of attritional processes.

Seasonality and Ungulate Mortality Profiles

The mortality profiles of Tabun were assessed mainly by tooth eruption and wear stages and in some cases by bone fusion, for some species such as gazelle. Figure 5.5 shows the Tabun results together with those from other nearby

Table 5.5 Spearman's correlations between skeletal part representation and utility/survivorship of bones

MNE long bones	Correlation		Tabun B			Tabun C			Tabun D		
			Large	Medium	Small	Large	Medium	Small	Large	Medium	Small
x1	%MAU/%MGUI	ρs	**−0.42**	−0.30	0.18	−0.24	−0.28	−0.07	0.16	**−0.43**	−0.32
		p	**0.05**	0.17	0.40	0.27	0.20	0.76	0.47	**0.04**	0.14
	%MAU/Bone density	ρs	0.40	0.32	**0.63**	0.47	0.20	0.19	**0.72**	0.19	**0.74**
		p	0.08	0.18	**0.02**	0.14	0.49	0.52	**0.02**	0.56	**0.00**
x2	%MAU/%MGUI	ρs	**−0.42**	−0.20	−0.14	−0.21	−0.16	0.02	0.17	−0.37	−0.31
		p	**0.05**	0.37	0.53	0.33	0.47	0.94	0.45	0.08	0.15
	%MAU/Bone density	ρs	0.81	**0.57**	**0.84**	**0.62**	0.48	0.50	**0.81**	0.46	**0.75**
		p	0.10	**0.01**	**0.00**	**0.04**	0.07	0.07	**0.00**	0.13	**0.00**
x4	%MAU/%MGUI	ρs	**−0.42**	−0.14	−0.15	−0.22	−0.14	0.02	0.17	−0.34	−0.28
		p	**0.05**	0.52	0.49	0.32	0.54	0.91	0.43	0.11	0.19
	%MAU/Bone density	ρs	0.81	**0.73**	**0.89**	**0.87**	**0.59**	**0.73**	**0.85**	**0.69**	**0.77**
		p	0.10	**0.00**	**0.00**	**0.00**	**0.02**	**0.00**	**0.00**	**0.01**	**0.00**

Values in bold imply statistically significant correlations

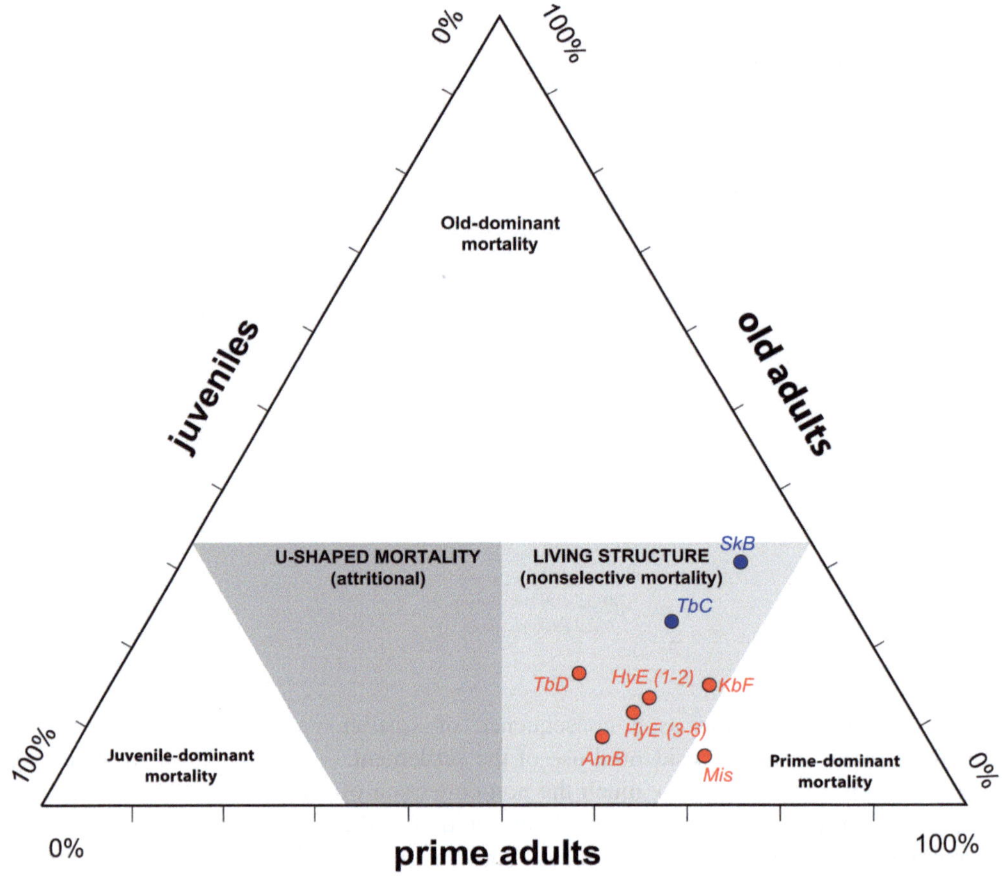

Fig. 5.5 Mortality profiles of ungulates at Tabun C and D, Skhul B (Marín-Arroyo 2011), Misliya (Yeshurun et al. 2007), Hayonim E (Stiner 2005), Amud B (Rabinovich and Hovers 2004) and Kebara VI-XIII (Speth and Tchernov 1998; Speth and Clark 2006). AMH occupation (*open circles*) and archaic ones (*filled circles*)

sites; Qafzeh was not included due to the fact that only bone fusion was used to determine age, which neglects the possibility of identifying old individuals. The mortality profiles reflect a general predominance of prime-age individuals, with a percentage in agreement with their typical abundance in natural populations. However, within this general trend a decrease can also be seen in the number of juvenile individuals in favor of elderly ones in the levels associated with AMH occupations (Tabun C and Skhul) which might imply a greater preference for more productive animals. Although not shown in the graphic, the age profile of Tabun B lies right in the centre of the living structure area, agreeing thus with the proposed natural origin of the deposit.

Concerning seasonality, apart from the winter occupations estimated by Lieberman (1993) for these levels, the identification of three mandibles of juvenile individuals with teeth in the first stage of eruption also points towards a late spring/early summer use of the cave in Levels C and D.

Human Mobility

Figure 5.6 shows the catchment areas corresponding to 1, 1.5 and 2 h of travel time around Tabun and other Middle Paleolithic sites. The surface area of plains and mountain biotopes, as identified with a 10 % slope threshold, is presented in Table 5.6, together with the so-called topographic factor (TF), obtained as a quotient between plain and mountain extensions (in km^2).

As can be seen, the higher proportion of plains in the surroundings of AMH occupations corresponds to a travel time of 1.5 h, whilst the lower proportion in the surroundings of archaic occupations is established in 1 h travel time. Given the comparative preference for plains ungulates among AMH (see Fig. 5.2), this would mean that they usually travelled longer distances to hunt their prey. Moreover, taking into account only Tabun levels C and D, the slightly higher dependence on plains ungulates seen in Tabun C (AMH occupation) is at odds with the reduction in plains biotopes in the surroundings of the cave due to the rise in sea level during MIS 5. A probable explanation for this incongruence would again be that AMH relied more heavily on logistic mobility than their predecessors, leading to greater travel distances during their expeditions which, in turn, would mean less residential mobility. A greater degree of carcass processing should then be expected then in Tabun C, which is consistent with the taphonomy of the analyzed assemblage.

To objectively test this hypothesis, bivariate correlations have been made between the topographic factor and the faunal preference for plains ungulates, represented by the

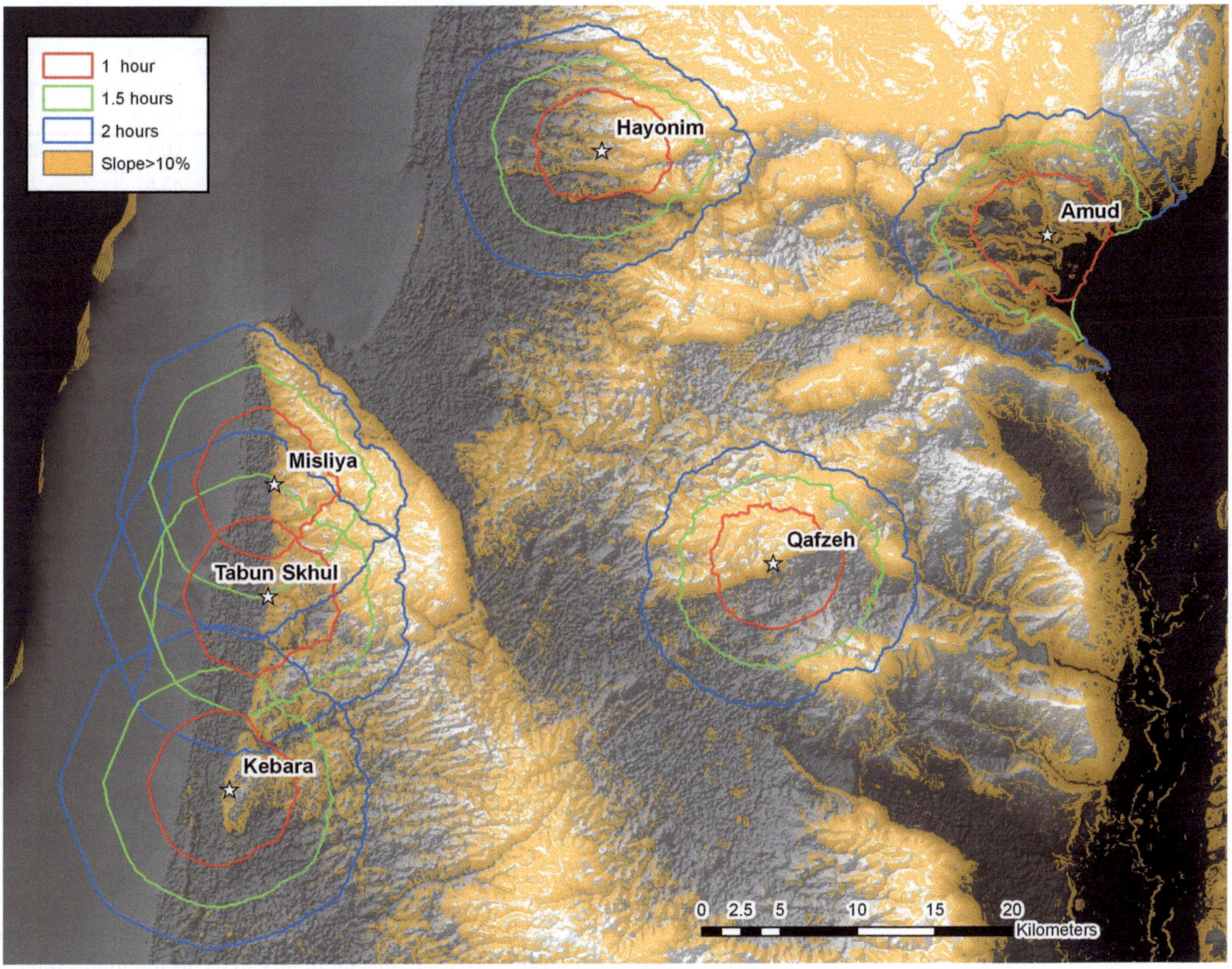

Fig. 5.6 Catchment areas for 1, 1.5 and 2 h of travel time over Digital Elevation Model showing zones considered less suitable for plains ungulates

quotient between plains and mountain NISP values at each site. Nine possible combinations of the three different travel times proposed for AMH and archaic populations (1, 1.5 or 2 h) were examined. The most probable combination in terms of Spearman's test ($\rho_s = 0.58$, $p = 0.10$) points towards travel times for AMH and archaic populations of 1.5 and 1 h respectively (see Fig. 5.7) thus reinforcing the idea of longer logistic movements amongst AMH.

Discussion

Subsistence Patterns in the Levantine Middle Paleolithic

The re-analysis of the faunal assemblage from Tabun Cave presented here, and its proper contextualisation with other Levantine Middle Paleolithic sites, allows the drawing of relevant conclusions on the subsistence patterns adopted by both AMH and archaic hominin populations in this key region for human evolutionary studies.

First, the origin of the assemblage has been clarified. Thus, Tabun B seems to have acted as a natural trap for ungulates (but also some carnivores) that would have fallen through the chimney (Bate 1937: 150; but see Kaufman 2002). Nontheless, there is evidence of human exploitation of some carcasses, probably scavenging the animals that perished there. On the other hand, the anthropogenic origin of Tabun C and D is clear, due to the presence of tool-marked bones and traces of burning, above all in level C, and the almost complete absence of carnivore marks. The location and abundance of butchering marks, including filleting, also corroborates the assumption that the human groups that inhabited the cave were capable of regular hunting episodes. In fact, the frequency of anthropogenic

Table 5.6 Topographic characterization of catchment areas in the Levantine Middle Paleolithic

Site	Human species	1 h catchment area			1.5 h catchment area			2 h catchment area		
		Plain (km^2)	Mountain (km^2)	TF	Plain (km^2)	Mountain (km^2)	TF	Plain (km^2)	Mountain (km^2)	TF
Tabun D	Archaic/ Neand.	51.5	20.3	2.54	120.7	44.7	2.70	217.2	80.1	2.71
Hayonim E (3–6)		20.6	26.4	0.78	69.1	50.0	1.38	142.1	85.6	1.66
Misliya		41.3	19.3	2.14	98.1	46.9	2.09	175.3	81.2	2.16
Hayonim E (1,2)		20.6	26.4	0.78	69.1	50.0	1.38	142.1	85.6	1.66
Amud B		24.2	27.3	0.88	42.9	63.0	0.68	73.2	110.2	0.66
Kebara F		57.6	11.8	4.87	143.1	24.1	5.93	264.7	43.8	6.04
Tabun C	AMH	51.5	20.3	2.54	117.2	44.7	2.62	180.6	80.1	2.25
Skhul B		51.5	20.3	2.54	117.2	44.7	2.62	180.6	80.1	2.25
Qafzeh V-XV		29.2	22.2	1.31	71.7	52.5	1.37	128.1	97.8	1.31
Average Archaic/Neand.				2.00			2.36			2.5
Average AMH				2.13			2.20			1.9

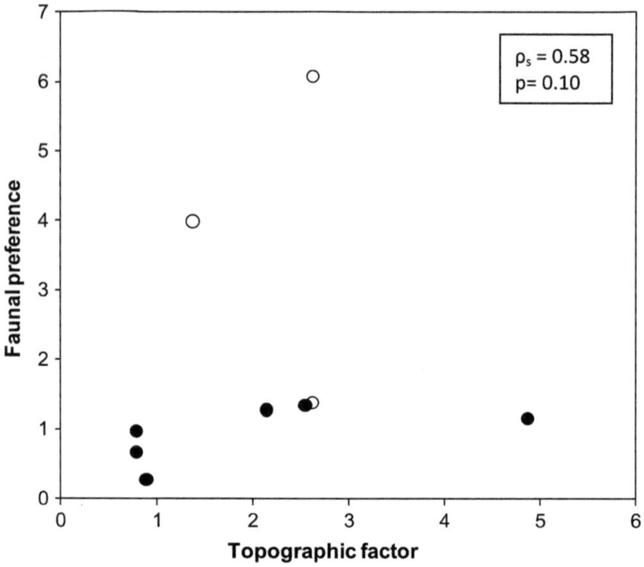

Fig. 5.7 Correlation between topographic factor in the surroundings of each settlement and faunal preference for the most probable combination of typical travel times: 1.5 h for AMH (*open circles*) and 1 h for archaic populations (*filled circles*)

evidence in the assemblage is higher than that from other well known human settlements, such as Qafzeh, where 2.3 % of bone remains show cut-marks or burn traces (Rabinovich and Hovers 2004) or Misliya, where 3–4 % of bone specimens bear cut-marks and 4–6 % impact notches (Yeshurun et al. 2007). The anthropogenic alteration of Tabun C is even more intense than in Kebara, which is seen as a long-term seasonal basecamp (Bar-Yosef et al. 1992). Scavenging as a usual subsistence strategy can then be ruled out in Tabun C and D, even more so taking into account the presence of large and relatively dangerous taxa in the assemblage, such as aurochs, rhinoceros, red deer or wild boar (Bar-Yosef et al. 1992; Speth and Clark 2006).

Secondly, there is a certain distinction in hunting strategies between AMH and archaic populations. The former would have depended more on high-ranked prey during MIS 5, dismissing small mountain gazelle to some extent. This trend could be seen as an incipient specialisation strategy (Rabinovich and Tchernov 1995; Rabinovich 1990) above all taking into account that it would mean longer expeditions to more distant plains areas, as ungulates related to plains habitats were bigger in the area of study. This trend towards more productive strategies is also slightly reflected in the mortality profiles, where juvenile individuals were partially substituted by prime-age or old ones (Stiner 2005; Stiner and Tchernov 1998). In terms of diet-breadth, the emphasis on large game typical of grassy plains or woodlands resulted in a significant broadening that could reflect the even exploitation of several different environments. The recognition of more efficient strategies and a higher territorial control is usually related to optimal foraging behavior, implying more advanced cognitive capacities. This would imply that the cognitive changes that were common during the Upper Paleolithic might have already been taking place (Klein 2009).

And thirdly, the more specialized economy recognized before is usually related to lower residential mobility. In the case of the Levantine Middle Paleolithic, this fact has also been tested by means of comparing the topographic relief in the surroundings of each site with the faunal preference. As a result, a greater reliance on logistic mobility has been suggested among AMH (with travel times usually 0.5 h longer), a fact that was also proposed by Wallace and Shea (2006) by analyzing differences in the production of lithic tools, and by Lieberman (1998) based on the age of death of gazelles. In addition, body size of carcasses seems to have played a greater role in transport decisions, probably involving more intense butchering at the kill-site (with the subsequent contribution of less-useful parts to the base camp) as prey-weight increases (Marín-Arroyo 2009c). Rabinovich and Tchernov (1995) also identified in Qafzeh a selective transport strategy of hunted prey for rhinoceros and red deer. In contrast, Stiner (2005) suggests an almost complete transport of carcasses during the archaic occupation of MIS 6 at Hayonim.

This pattern of mobility was probably favored among AMH by their particular anatomy; hence, their longer lower limbs and smaller body mass (Weaver and Steudel-Numbers 2005; MacDonald et al. 2009) would have reduced their energetic expenditure while travelling (from/to base camp) in their hunting expeditions, allowing them efficient exploitation of certain distant taxa. It is still unknown whether this more mobile pattern of AMH behavior and the subsequent depletion of the resources located close to the base camp led them to focus more on larger game in order to maintain the necessary energetic income, or perhaps instead it was changes in cognitive skills permitted them to better discern between different productive choices, allowing them to maintain more permanent occupations.

Implications for AMH Replacement in the Region

Interestingly enough, the economic behavior just noted amongst AMH did not provide them with a clear evolutionary advantage in terms of survivorship or adaptive skills, as they ceased to occupy the region at the end of MIS 5, being replaced by Neanderthals coming from northern latitudes (Tchernov 1998; Shea and Bar-Yosef 2005). Two possible explanations are currently proposed for this failure. The first involves climatic deterioration that would have significantly decreased terrestrial productivity during the MIS 5/4 transition ca. 75 ka (Shea 2008). This episode would have affected the number of available taxa (Davies et al. 2011; Dynesius and Jansson 2000) which, together with the great dependency on large ungulates seen in the archaeozoological record, would have reduced their numbers and even provoked their extinction due to overhunting (Haynes 2002; Miller et al. 2005; Marín-Arroyo 2009c; Speth 2004). In this scenario, AMH would have had to cope with dietary stress that, in other conditions, would have led to an expansion in diet-breadth, including previously considered low-ranked prey (Stiner et al. 1999). In a similar way, droughts and rapid shifts to colder conditions would have seriously limited *Homo sapiens*' ability to disperse during the Upper Paleolithic (Finlayson and Carrión 2007). However, either small or difficult-to-hunt animals such as lagomorphs, birds, etc. were not sufficient to maintain the existing human population due to their low calorific yield/low abundance, or the available technology of AMH was not sufficient to effectively exploit these kind of resources. For example, Shea and Sisk (2010) argue that it was not until projectile weapons of the kind used by ethnographic hunter-gatherers, such as the bow and arrow, and spear-thrower and dart, were developed in Africa between 50 and 100 ka that ecological niche broadening strategies were enabled (Brooks et al. 2006; d'Errico and Henshilwood 2007), which then allowed *Homo sapiens* to successfully colonize Eurasia at the Middle to Upper Paleolithic transition.

An alternative hypothesis would dismiss the effect that climatic worsening would have had on biomass production and consequently on resource availability (Frumkin et al. 2011), due to the existence of reliable and spatially dispersed seasonal food resources (Bar-Yosef et al. 1992; Hovers 2006). If that were the case, AMH would have not abandoned the region when Neanderthal populations arrived from northern latitudes, and thus, some kind of competition would have occurred between both human species (Shea and Bar-Yosef 2005) in which Neanderthals might possibly have prevailed, due to a better physical adaptation to the increasing glacial conditions of MIS 4.

There are several climate studies in the Levant that can shed light on this dilemma, above all isotope and growth-rate studies of speleothems, such as those from Soreq and Peqi'in Caves (Bar-Matthews et al. 1997, 2003a), Negev Caves (Vaks et al. 2006) and Jerusalem West Caves (Frumkin et al. 1999). Thus, according to Vaks et al. (2007) the arrival of AMH in the Levant was favored by a humid period in the Negev desert region between 140 and 110 ka, coinciding with an intensification of monsoonal effects in the Arabian Peninsula (Fleitmann et al. 2003) and the Sahara (Osmond and Dabous 2004), which would have provided a suitable land bridge between sub-Saharan latitudes and the Levant. This bridge seems to have disappeared afterwards, which would have isolated AMH populations during MIS 5, a period that became drier as seen both in the growth-rate of calcite (Vaks et al. 2006; Lisker et al. 2010) and in water level reconstruction of Lake Samra, the precursor of the Holocene Dead Sea (Waldmann et al. 2009). AMH populations probably dealt with this situation by migrating to the more humid central and northern regions to the West of the central mountain ridge of Israel (in caves such as Tabun, Skhul and Qafzeh). This isolated situation might have increased survival risks at times of rapid environmental change, as it constrained migration possibilities as well as reduced genetic flow. Why AMH did not expand farther north through Taurus-Zagros Mountains is still unknown. As a possible explanation, Frumkin et al. (2011) argued that the heterogeneity of the Levant ecosystems together with a colder climate in the mountain ridges, quite different from the African one, kept AMH within the region.

Based on the characterization of spelothem fluid inclusions of Soreq Cave, McGarry et al. (2004) determined a drop of around 4 °C in temperature between 80–86 and 73–74 ka. In addition, based on: (1) present measures of $\delta^{18}O$ abundance of rain and drip-water at the Soreq Cave site; (2) the similar isotopic depletion between the Eastern Mediterranean Sea and cave locations observed during the last 250 ka as evidenced in the $\delta^{18}O$ record of planktonic foraminifera species *Globigerinoides ruber* of a maritime core (Fontugne and Calvert 1992); (3) the similar climate of interglacial periods; and (4) the same $\delta^{18}O$ composition of the sea in Sapropel S3 event in comparison with present times (Kallel et al. 2000). Bar-Matthews et al. (2003b) were able to estimate an annual decrease in rainfall between 82 and 77 ka of 300 mm/year (from 500 to 200 mm/year). This trend has been confirmed by Almogi-Labin et al. (2009) when they eliminated the first-order "source effect" from the $\delta^{18}O$ record of Soreq Cave with the help of two eastern Mediterranean maritime cores. These authors found a significant drop in second-order $\delta^{18}O$ during the MIS 5/4 transition that was related to the so-called "amount effect" (Dansgaard 1964) and that also indicates a pronounced decrease in rainfall. This meteorological shift was also hydrologically recorded in the water level reconstruction of Lake Samra, which presents a 10 m drop between 80.2 and 71.5 ka (Waldmann et al. 2009) and in the $\delta^{13}C$ record throughout the region that reflects an increase in C4 vegetation.

The impact of this change in terrestrial productivity can be assessed by means of estimation of real evapotranspiration (ETr), i.e. the amount of water used by the vegetation cover, which has a direct relationship with existing biomass. There are two major forces that affect evapotranspiration: the availability of water (i.e. rainfall) and the availability of solar radiation and heat. The annual ETr can be roughly estimated as a function of mean annual temperature following Turc's equation (1954), later modified by Pike (1964). Therefore ETr at the onset of the MIS 5/4 transition would have been 450 mm/year whilst 10 ka later it would have decreased to 195 mm/year. According to Budyko (1974), this would have changed the environment from subhumid to subarid conditions, implying a major modification of resource availability.

Given this, the possibility of an AMH extinction before the arrival of Neanderthals is considered more plausible, above all as they did not possess the necessary technology to effectively broaden their diet (Shea 2009), which would have improved their survival chances (Betinger 1991), especially if they had intensively depended on large ungulates as has been argued before. There is, however, no direct proof for this assumption, as there is currently a lack of archaeozoological assemblages from this period that could show evidence of dietary stress, such as a progressive widening of diet-breadth or a greater consumption of old and juvenile individuals (Marín-Arroyo 2009c).

Acknowledgments I would like to especially thank John Speth and Jamie Clark for inviting me to participate in this fascinating ICAZ session. This work could not have been done without the kind help of NHM (London) staff, above all from Mark Lewis, Adrian Lister, Chris Stringer and Andy Currant. Mark Lewis also edited the English for this manuscript. The research was funded with a British Academy Postdoctoral Fellowship in LCHES (University of Cambridge), where Robert Foley and Marta Lahr-Mirazón gave me all their support and encouragement. Finally, my sincere thanks to David Ocio for all his constant support.

References

Albert, R., Lavi, O., Estroff, L., Weiner, S., Tsatskin, A., Ronen, A., et al. (1999). Mode of occupation of Tabun Cave, Mt Carmel, Israel during the Mousterian Period: A study of the sediments and phytoliths. *Journal of Archaeological Science, 26*, 1249–1260.

Almogi-Labin, A., Bar-Matthews, M., Shriki, D., Kolosovsky, E., Paterne, M., Schilman, B., et al. (2009). Climatic variability during the last ~90 ka of the southern and northern Levantine Basin as evident from marine records and speleothems. *Quaternary Science Reviews, 28*, 2882–2896.

Bar-Matthews, M., Ayalon, A., & Kaufman, A. (1997). Late Quaternary paleoclimate in the eastern Mediterranean region from stable isotope analysis of speleothems at Soreq Cave, Israel. *Quaternary Research, 47*, 155–168.

Bar-Matthews, M. A., Ayalon, M., Gilmour, A., Matthews, A., & Hawkesworth, C. J. (2003a). Soreq and Peqiin Caves, Israel Speleothem Stable Isotope Data, IGBP PAGES/World Data Center for Paleoclimatology Data Contribution Series #2003-061. NOAA/NGDC Paleoclimatology Program, Boulder, CO, USA.

Bar-Matthews, M., Ayalon, A., Gilmour, M., Matthews, A., & Hawkesworth, C. J. (2003b). Seaeland oxygen isotopic relationships from planktonic foraminifera and speleothems in the Eastern Mediterranean region and their implication for paleorainfall during interglacial intervals. *Geochimica et Cosmochimica Acta, 67*(17), 3181–3199.

Bar-Yosef, O. (1998). The chronology of the Middle Paleolithic of the Levant. In T. Akazawas, K. Aoki, & O. Bar-Yosef (Eds.), *Neanderthals and modern humans in Western Asia* (pp. 39–56). New York: Plenum.

Bar-Yosef, O. (2000). The Middle and early Upper Paleolithic in Southwest Asia and neighbouring regions. In O. Bar-Yosef & D. Pilbeam (Eds.), *The geography of Neanderthals and modern humans in Europe and the Greater Mediterranean* (pp. 107–156). Cambridge, MA: Peabody Museum of Archaeology and Ethnology, Bulletin No. 8.

Bar-Yosef, O., & Callander, J. (1999). The woman from Tabun: Garrod's doubts in historical perspective. *Journal of Human Evolution, 37*, 879–885.

Bar-Yosef, O., Vandermeersch, B., Arensburg, B., Belfer-Cohen, A., Goldberg, P., Laville, H., et al. (1992). The excavations in Kebara Cave, Mt. Carmel. *Current Anthropology, 33*, 497–550.

Bate, D. M. A. (1937). Paleontology: The fossil fauna of Wady el-Mughara caves. In D. A. E. Garrod & D. M. A. Bate (Eds.), *The Stone Age of Mount Carmel, Part 2* (pp. 137–240). Oxford, UK: Clarendon Press.

Behrensmeyer, A. K. (1978). Taphonomic and ecological information from bone weathering. *Paleobiology, 4*, 150–162.

Behrensmeyer, A. K., Gordon, K. D., & Yanagi, G. T. (1986). Trampling as a cause of bone surface damage and pseudo-cutmarks. *Nature, 319*, 768–771.

Betinger, R. L. (1991). *Hunter-gatherers: Archaeological and evolutionary theory*. New York: Plenum Press.

Binford, L. R. (1978). *Nunamiut Ethnoarchaeology*. New York: Academic Press.

Binford, L. R. (1981). *Bones: Ancient men and modern myths*. New York: Academic Press.

Brooks, A. S., Yellen, J. E., Nevell, L., & Hartman, G. (2006). Projectile technologies of the African MSA: Implications for Modern Human Origins. In E. Hovers & S. L. Kuhn (Eds.), *Transitions before the transition* (pp. 233–256). New York: Plenum/Kluwer.

Budyko, M. I. (1974). *Climate and life*. Orlando, FL: Academic Press.

Bunn, H. T. (1981). Archaeological evidence for meat-eating by Plio-Pleistocene hominids from Koobi Fora and Olduvai Gorge. *Nature, 291*, 574–577.

Cannon, M. D. (2003). A model of central place forager prey choice and an application to faunal remains from the Mimbres Valley, New Mexico. *Journal of Anthropological Archaeology, 22*, 1–25.

Capaldo, S. D., & Blumenschine, R. J. (1994). A quantitative diagnosis of notches made by hammerstone percussion and carnivore gnawing on bovid long bones. *American Antiquity, 59*, 724–748.

Copeland, L. (1975). The Middle and Upper Paleolithic of Lebanon and Syria in the light of recent research. In F. Wendorf & A. E. Marks (Eds.), *Problems in prehistory: North Africa and the Levant* (pp. 317–350). Dallas: SMU Press.

Coppa, A., Grün, R., Stringer, C., Eggins, S., & Vargiu, T. (2005). Newly recognized Pleistocene human teeth from Tabun Cave, Israel. *Journal of Human Evolution, 49*, 301–315.

d'Errico, F., & Henshilwood, C. S. (2007). Additional evidence for bone technology in the southern African Middle Stone Age. *Journal of Human Evolution, 52*, 142–163.

Dansgaard, W. (1964). Stable isotopes in precipitation. *Tellus, 16*, 436–468.

Davies, J., Buckley, L. B., Grenyer, R., & Gittleman, J. L. (2011). The influence of past and present climate on the biogeography of modern mammal diversity. *Philosophical Transactions of the Royal Society of Biological Sciences, 366*(1577), 2526–2535.

De la Torre, I., & Domínguez-Rodrigo, M. (2000). El Paleolítico Medio en el Próximo Oriente. Una síntesis regional. *Complutum, 11*, 9–27.

Dynesius, M., & Jansson, R. (2000). Evolutionary consequences of changes in species' geographical distributions driven by Milankovitch climate oscillations. *Proceedings of the National Academy of Sciences of the United States of America, 97*(16), 9115–9120.

Finlayson, C., & Carrión, J. S. (2007). Rapid ecological turnover and its impact on Neanderthal and other human populations. *Trends in Ecology & Evolution, 22*(4), 213–222.

Fisher, J. W. (1995). Bone surface modifications in zooarchaeology. *Journal of Archaeology of Method and Theory, 2*, 7–68.

Fleitmann, D., Burns, S. J., Neff, U., Mangini, A., & Matter, A. (2003). Changing moisture sources over the last 330000 years in northern Oman from fluid-inclusion evidence in speleothems. *Quaternary Research, 60*, 223–232.

Fontugne, M., & Calvert, S. E. (1992). Late Pleistocene variability of the carbon isotopic composition of organic matter in the Eastern Mediterranean: Monitor of changes in carbon sources and atmosphere CO_2 concentrations. *Paleoceanography, 7*, 1–20.

Frumkin, A., Ford, D. C., & Schwarcz, H. P. (1999). Continental oxygen isotopic record of the last 170,000 years in Jerusalem. *Quaternary Research, 51*, 317–327.

Frumkin, A., Bar-Yosef, O., & Schwarcz, H. P. (2011). Possible paleohydrologic and paleoclimatic effects on hominin migration and occupation of the Levantine Middle Paleolithic. *Journal of Human Evolution, 60*, 437–451.

Garrard, A. (1981). *Man-animal-plant relationship during the Upper Pleistocene and early Holocene of the Levant*. PhD dissertation, University of Cambridge, Cambridge.

Garrard, A. (1982). The environmental implications of the re-analysis of the large mammal fauna from the Wadi-el-Mughara caves, Palestine. In J. L. Bintliff & W. Van Zeist (Eds.), *Paleoclimates, paleoenvironments and human communities in the Eastern Mediterranean region in later prehistory* (pp. 165–198). Oxford: British Archaeological Reports.

Garrard, A. (1998). Food procurement by Middle Paleolithic hominids as Ras el- Kelb Cave in Lebanon. In L. Copeland & N. Moloney (Eds.), *The Mousterian Site of Ras el-Kelb, Lebanon* (pp. 45–65). Oxford: British Archaeological Reports.

Garrod, D. A. E., & Bate, D. M. A. (1937). *The Stone Age of Mount Carmel. Excavations at the Wadi Mughara* (Vol. I). Oxford: Clarendon Press.

Grün, R., & Stringer, C. (2000). Tabun revisited: Revised ESR chronology and new ESR and U-series analyses of dental material from Tabun C1. *Journal of Human Evolution, 39*, 601–612.

Haynes, G. (2002). The catastrophic extinction of North American mammoths and mastodonts. *World Archaeology, 33*(3), 391–416.

Holechek, J. L., Pieper, R. D., & Herbel, C. H. (1998). *Range management principles and practices* (3rd ed.). Upper Saddle River, NJ: Prentice-Hall, Inc.

Hovers, E. (2006). Neanderthals and modern humans in the Middle Paleolithic of the Levant: What kind of interaction? In N. J. Conard (Ed.), *When Neanderthals and moderns met* (pp. 65–85). Tübingen: Kerns Verlag.

Jelinek, A. J. (1982a). The Tabun cave and Paleolithic man in the Levant. *Science, 216*, 1369–1375.

Jelinek, A. J. (1982b). The Middle Paleolithic in the Southern Levant with comments on the appearance of modern *Homo sapiens*. In A. Ronen (Ed.), *The transition from Lower to Middle Paleolithic and the origins of modern man* (pp. 57–104). Oxford: British Archaeological Reports.

Jelinek, A., Farrard, W. R., Haas, G., Horowitz, A., & Golberg, P. (1973). New excavations at the Tabun cave, Mount Carmel, Israel, 1967–1972: A preliminary report. *Paléorient, 1*(2), 151–183.

Kallel, N., Duplessy, J.-C., Labeyrie, L., Fontugne, M., Paterne, M., & Montacer, M. (2000). Mediterranean pluvial periods and sapropel formation during the last 200,000 years. *Paleogeography, Paleoclimatology, Paleoecology, 157*, 45–58.

Kaufman, D. (2002). Re-evaluating subsistence skills of Levantine Middle and Upper Paleolithic hunters: A comparison of the faunal assemblages. *Oxford Journal of Archaeology, 21*, 217–229.

Klein, R. (2009). *The human career*. Chicago: University of Chicago Press.

Klein, R. G., & Cruz-Uribe, K. (1984). *The analysis of animal bones from archaeological sites*. Chicago: The University of Chicago Press.

Lam, Y. M., Chen, X., & Pearson, O. M. (1999). Intertaxonomic variability in patterns of bone density and the differential representation of bovid, cervid and equid elements in the archaeological record. *American Antiquity, 64*, 343–362.

Lieberman, D. E. (1993). The rise and fall of seasonal mobility among hunter-gatherers. *Current Anthropology, 34*, 599–631.

Lieberman, D. E. (1998). Neanderthal and early modern human mobility patterns: Comparing archaeological and anatomical evidence. In T. Akazawa, K. Aoki, & O. Bar-Yosef (Eds.), *Neanderthals and modern humans in Western Asia* (pp. 263–276). New York: Plenum.

Lisker, S., Vaks, A., Bar-Matthews, M., Porat, R., & Frumkin, A. (2010). Late Pleistocene paleoclimatic and paleoenvironmental reconstruction of the Dead Sea area (Israel) based on speleothems and cave stromatolites. *Quaternary Science Reviews, 29*, 1201–1211.

MacDonald, K., Roebroeks, W., & Verpoorte, A. (2009). An energetic perspective on the Neanderthal Record. In J. J. Hublin & M. P. Richards (Eds.), *The evolution of hominin diets: Integrating approaches to the study of Paleolithic Subsistence* (pp. 211–220). Vertebrate Paleobiology and Paleoanthropology Series. Dordrecht: Springer.

Marean, C. W. (1998). A critique of the evidence for scavenging by Neanderthals and early modern humans: New data from Kobeh Cave (Zagros Mountains, Iran) and Die Kelders Cave 1 layer 10 (South Africa). *Journal of Human Evolution, 35*, 111–136.

Marean, C. W., & Frey, C. (1997). Animal bones from caves to cities: Reverse utility curves as methodological artefacts. *American Antiquity, 62*, 698–716.

Marean, C. W., & Spencer, L. M. (1991). Impact of carnivore ravaging on zooarchaeological measures of element abundance. *American Antiquity, 56*, 645–658.

Marean, C. W., Spencer, L. M., Blumenschine, R. J., & Capaldo, S. (1992). Captive hyena bone choice and destruction, the Schlepp Effect, and Olduvai archaeofaunas. *Journal of Archaeological Science, 19*, 101–121.

Marín-Arroyo, A. B., & González Morales, M. R. (2009). Comportamiento económico de los últimos cazadores-recolectores y primeras evidencias de domesticación en el occidente de Asturias. La Cueva de Mazaculos II. *Trabajos de Prehistoria, 66*(1), 47–74.

Marín-Arroyo, A. B. (2008). Patrones de movilidad y control del territorio en el Cantábrico Oriental durante el Tardiglacial. *Trabajos de Prehistoria, 65*(1), 29–45.

Marín-Arroyo, A. B. (2009a). A comparative study of analytic techniques for skeletal part profile interpretation at El Mirón Cave (Cantabria, Spain). *Archaeofauna, 18*, 79–98.

Marín-Arroyo, A. B. (2009b). The use of Optimal Foraging Theory to estimate Late Glacial site catchment areas from a central place. The case of eastern Cantabria, Spain. *Journal of Anthropological Archaeology, 28*, 27–36.

Marín-Arroyo, A. B. (2009c). Economic adaptations during the Late Glacial in northern Spain. A simulation approach. *Before Farming, 2*, 1–18.

Marín-Arroyo, A. B. (2011). Paleolithic human subsistence in Mount Carmel (Israel). A taphonomic assessment of Middle and Early Upper Paleolithic faunal remains from Tabun, Skhul and el-Wad. *International Journal of Osteoarcaheology*. doi:10.1002/oa.1241.

Marín-Arroyo, A. B., Landete, D., Vidal, G., Seva, R., González Morales, M., & Straus, L. G. (2008). Archaeological implications of human-derived manganese coatings: A study of blackened bones in El Mirón Cave, Cantabrian Spain. *Journal of Archaeological Science, 35*(3), 801–813.

McCown, T. D., & Keith, A. (1939). *The Stone Age of Mt. Carmel, Vol. 2: The fossil human remains from the Levallois-Mousterian*. Oxford: Clarendon Press.

McGarry, S., Bar-Matthews, M., Matthews, A., Vaks, A., Schilman, B., & Ayalon, A. (2004). Constraints on hydrological and paleotemperature variations in the Eastern Mediterranean region in the last 140 ka given by the δD values of speleothem fluid inclusions. *Quaternary Science Reviews, 23*, 919–934.

Mercier, N., & Valladas, H. (2003). Reassessment of TL age estimates of burnt flints from the Paleolithic site of Tabun Cave, Israel. *Journal of Human Evolution, 45*, 401–409.

Miller, G. H., Fogel, M. L., Magee, J. W., Gagan, M. K., Clarke, S. J., & Johnson, B. J. (2005). Ecosystem collapse in Pleistocene Australia and a human role in megafaunal extinction. *Science, 309*(5732), 287–290.

Munro, N. D., Bar-Oz, G., & Stutz, A. J. (2009). Aging mountain gazelle (*Gazella gazella*): Refining methods of tooth eruption and wear and bone fusion. *Journal of Archaeological Science, 36*, 752–763.

Neuville, R. (1951). *Le Paléolithique et le Mésolithique de Désert de Judée*. Paris: Masson et Cie Editeurs.

O'Connell, J. F., Hawkes, K., & Blurton Jones, N. (1988). Hadza hunting, butchering, and bone transport and their archaeological implications. *Journal of Anthropological Research, 44*, 113–161.

O'Connell, J. F., Hawkes, K., & Blurton-Jones, N. (1989). Reanalysis of large mammal body part transport among the Hadza. *Current Anthropology, 29*, 356–363.

Osmond, J. K., & Dabous, A. A. (2004). Timing and intensity of groundwater movement during Egyptian Sahara pluvial periods by U-series analysis of secondary U in ores and carbonates. *Quaternary Research, 61*, 85–94.

Payne, S. (1975). Partial recovery and sample bias. In A. Classon (Ed.), *Archaeozoological studies* (pp. 7–17). Ámsterdam: North Holland.

Pérez Ripoll, M. (1992). *Marcas de carnicería, fracturas intencionadas y mordeduras de carnívoros en huesos prehistóricos del Mediterráneo español*. Alicante: Instituto de Cultura Juan Gil-Albert.

Pickering, T. R., & Egeland, C. P. (2006). Experimental patterns of hammerstone percussion damage on bones: Implications for inferences of carcass processing by humans. *Journal of Archaeological Science, 33*, 459–469.

Pielou, E. C. (1966). The measurement of diversity in different types of biological collections. *Journal of Theoretical Biology, 13*, 131–144.

Pike, J. G. (1964). The estimation of annual runoff from meteorological data in a tropical climate. *Journal of Hydrology, 2*, 116–123.

Quam, R. M., & Smith, F. H. (1998). A reassessment of the Tabun C2 Mandible. In T. Akazawa, K. Aoki, & O. Bar-Yosef (Eds.), *Neanderthals and modern humans in Western Asia* (pp. 405–422). New York: Plenum.

Rabinovich, R. (1990). Taphonomic research on the faunal assemblage from the Quneitra site. In N. Goren-Inbar (Ed.), *Quneitra: A Mousterian site on the Golan heights* (pp. 189–219). Jerusalem: The Hebrew University.

Rabinovich, R., & Hovers, E. (2004). Faunal analysis from Amud Cave: Preliminary results and interpretations. *International Journal of Osteoarchaeology, 14*, 287–306.

Rabinovich, R., & Tchernov, E. (1995). Chronological, paleoecological and taphonomical aspects of the Middle Paleolithic site of Qafzeh, Israel. In H. Buitenhuis & H. P. Uerpmann (Eds.), *Archaeozoology of the Near East II* (pp. 5–44). Leiden: H. Backhuys.

Rak, Y. (1998). Does any Mousterian Cave present evidence of two species? In T. Akazawa, K. Aoki, & O. Bar-Yosef (Eds.), *Neanderthals and modern humans in Western Asia* (pp. 353–366). New York: Plenum.

Shea, J. J. (2003a). The Middle Paleolithic of the East Mediterranean Levant. *Journal of World Prehistory, 17*(4), 313–394.

Shea, J. J. (2003b). Neanderthals, competition, and the origin of modern human behavior in the Levant. *Evolutionary Anthropology, 12*, 173–187.

Shea, J. J. (2008). Transitions or turnovers? Climatically-forced extinctions of *Homo sapiens* and Neanderthals in the east Mediterranean Levant. *Quaternary Science Reviews, 27*, 2253–2270.

Shea, J. (2009). The impact of projectile weaponry on Late Pleistocene hominin evolution. In J.-J. Hublin & M. P. Richards. (Eds.), *The evolution of hominin diets: Integrating approaches to the study of Paleolithic subsistence* (pp. 189–199). Vertebrate Paleobiology and Paleoanthropology Series. Dordrecht: Springer.

Shea, J. J., & Bar-Yosef, O. (2005). Who were the Skhul/Qafzeh people? An archaeological perspective on Eurasia's earliest modern humans. *Journal of the Israel Prehistoric Society, 35*, 449–466.

Shea, J. J., & Sisk, M. L. (2010). Complex projectile technology and *Homo sapiens* dispersal into Western Eurasia. *PaleoAnthropology, 2010*, 100–122.

Shipman, P. (1981). *Life history of a fossil: An introduction to taphonomy and paleoecology*. Cambridge: Harvard University Press.

Siddall, M., et al. (2006). Red Sea sea Level Reconstruction. IGBP PAGES/World Data Center for Paleoclimatology. Data Contribution Series # 2006-063. NOAA/NCDC Paleoclimatology Program, Boulder, CO, USA.

Simpson, E. H. (1949). Measurement of diversity. *Nature, 163*, 688.

Speth, J. D. (2004). Hunting pressure, subsistence intensification and demographic change in the Levantine late Middle Paleolithic. In N. Goren-Inbar & J. D. Speth (Eds.), *Human paleoecology in the Levantine Corridor* (pp. 149–166). Oxford: Oxbow Press.

Speth, J. D., & Clark, J. (2006). Hunting and overhunting in the Levantine late Middle Paleolithic. *Before Farming, 3*, 1–42.

Speth, J. D., & Tchernov, E. (1998). The role of hunting and scavenging in Neanderthal procurement strategies: New evidence from Kebara Cave. In T. Akazawa, K. Aoki, & O. Bar-Yosef (Eds.), *Neanderthals and modern humans in Western Asia* (pp. 223–240). New York: Plenum.

Speth, J. D., & Tchernov, E. (2001). Neanderthal hunting and meat processing in the Near East: Evidence from Kebara Cave (Israel). In C. B. Stanford & H. T. Bunn (Eds.), *Meat eating and human evolution* (pp. 52–72). New York: Oxford University Press.

Stiner, M. C. (1991). Food procurement and transport by human and no-human predators. *Journal of Archaeological Science, 18*, 455–482.

Stiner, M. C. (1994). *Honour among thieves: A zooarchaeological study of Neanderthal ecology*. Princeton: Princeton University Press.

Stiner, M. C. (2005). *The Faunas of Hayonim Cave (Israel): A 200,000-year record of Paleolithic diet, demography and society*. Cambridge: Peabody Museum of Archaeology and Ethnology.

Stiner, M. C., & Tchernov, E. (1998). Pleistocene species trends at Hayonim Cave: Changes in climate versus human behavior. In T. Akazawa, K. Aoki, & O. Bar-Yosef (Eds.), *Neanderthals and modern humans in Western Asia* (pp. 241–262). New York: Plenum.

Stiner, M. C., Munro, N., Surovell, T. A., Tchernov, E., & Bar-Yosef, O. (1999). Paleolithic growth pulses evidenced by small animal exploitation. *Science, 283*, 190–194.

Tchernov, E. (1998). The faunal sequences of the southwest Asian Middle Paleolithic in relation to hominid dispersal events. In T. Akazawa, K. Aoki, & O. Bar-Yosef (Eds.), *Neanderthals and modern humans in Western Asia* (pp. 77–90). New York: Plenum.

Turc, L. (1954). Le bilan d'eau des sols. Relation entre la précipitation, l'évaporation et l'écoulement. *Annales agronomiques, 5*, 491–569.

Turville-Petre, F. (1932). Excavations in the Mugharet el-Kebarah. *Journal of the Royal Anthropology Institute, 62*, 271–276.

Vaks, A., Bar-Matthews, M., Ayalon, A., Matthews, A., Frumkin, A., Dayan, U., et al. (2006). Paleoclimate and location of the border between Mediterranean climate region and the Saharo-Arabian Desert as revealed by speleothems from the northern Negev Desert, Israel. *Earth Planetary Science Letters, 249*, 384–399.

Vaks, A., Bar-Matthews, M., Ayalon, A., Halicz, L., & Frumkin, A. (2007). Desert speleothems reveal climatic window for African exodus of anatomically modern humans. *Geology, 35*, 831–834.

Waldmann, N., Stein, M., Ariztegui, D., & Starinsky, A. (2009). Stratigraphy, depositional environments and level reconstruction of the last interglacial Lake Samra in the Dead Sea basin. *Quaternary Research, 72*, 1–15.

Wallace, I., & Shea, J. (2006). Mobility patterns and core technologies in the Middle Paleolithic of the Levant. *Journal of Archaeological Science, 33*, 1293–1309.

Weaver, T., & Steudel-Numbers, K. (2005). Does climate or mobility explain the differences in body proportions between Neanderthals and their Upper Paleolithic successors? *Evolutionary Anthropology, 14*, 218–223.

Yeshurun, R., Bar-Oz, G., & Weinstein-Evron, M. (2007). Modern hunting behavior in the early Middle Paleolithic: Faunal remains from Misliya Cave, Mount Carmel, Israel. *Journal of Human Evolution, 53*, 656–677.

Chapter 6
New Contributions on Subsistence Practices during the Middle-Upper Paleolithic in Northern Spain

José Yravedra Sainz de los Terreros

Introduction

Subsistence studies demand a cross-disciplinary analysis, incorporating spatial data about mobility, the management and acquisition strategies of raw materials, and a detailed examination of animal and plant resource exploitation. In this paper I analyze the different strategies chosen by hominids for animal acquisition in northern Spain during the Middle-Upper Paleolithic transition. The taphonomic studies presented here provide information from new sites, as well as an alternative interpretation for previously excavated ones, supplementing earlier research on the topic (Freeman 1973; Straus 1976, 1977, 1992; Bernardo de Quirós 1980, 1982; Altuna 1989, 1992; Yravedra 2002a, b).

Traditional studies of the Middle-Upper Paleolithic transition in northern Spain were based on a series of sites including Morín, Pendo, Castillo, Amalda, Lezetxiki and Axlor. The new sites discussed in this paper enlarge this list, incorporating Otero, Hornos de la Peña, Esquilleu and Covalejos. Furthermore, new taphonomic data from the caves of Morín, Cuco, Cofresnedo, Ruso and Amalda are considered.

Most previous work assumed that the faunal remains found in these sites were accumulated by human beings, suggesting that both Neanderthals and *Homo sapiens* practiced specific hunting behaviors (Altuna 1972, 1989, 1990, 1992; Freeman 1973; Straus 1976, 1977; Bernaldo de Quirós 1980), whereas scavenging was restricted to Neanderthals (Straus 1982, 1992; Klein and Cruz Uribe 1994). Nevertheless, these interpretations are problematic when taphonomic studies are incorporated, as exemplified by the case of Amalda.

At Amalda, chamois dominates the faunal record. Lately, Yravedra (2005, 2006a, 2007, 2009, 2010a, b) has questioned the classic interpretation of an anthropogenic origin for all faunal remains found at the site (Altuna 1990; Altuna and Mariezkurrena 2010). Based on a taphonomic analysis of the sample, in which an abundance of tooth and digestion marks—as well as an absence of cut and percussion marks—were identified on small-sized ungulates (the latter of which were otherwise frequent on larger ungulates), Yravedra concluded that although larger animals such as deer, horse and aurochs were consumed by humans, small animals such as chamois were not. However, there is little agreement on this topic. Some authors (e.g., Mallye et al. 2010) also question the degree of human involvement with the small-sized ungulates at Amalda and Noissier caves, while others (e.g., Altuna and Mariezcurrena 2010) support the idea of an anthropogenic origin for these animals.

Other Cantabrian sites present a varied situation. The fauna from some levels of Mirón has been interpreted as being accumulated by vultures (Marín-Arroyo et al. 2009), and in Lezetxiki cave and Esquilleu levels III-IV, carnivores seem to have had a great impact on the bone assemblage (Arrizabalaga et al. 2005; Yravedra 2005, 2006b). Pendo's record is not significant in this case due to serious stratigraphic problems, which resulted in a palimpsest of materials from many levels (Montes et al. 2005). On the other hand, new studies at Castillo, Morín 17 and Esquilleu VI-XXX confirm that humans played a significant role in animal resource accumulation (Martínez 1998; Dari 2003; Yravedra 2005, 2006b; Landry and Burke 2006).

Faunal remains from Paleolithic sites in northern Spain clearly require detailed taphonomic analyses in order to identify factors that may affect the interpretation of these assemblages. The taphonomic studies presented here aim at filling this gap, allowing for an analysis of subsistence behaviors in the Middle and Upper Palaeolithic in northern Spain, and a consideration of whether Neanderthal and *Homo sapiens* subsistence strategies were similar.

J. Yravedra Sainz de los Terreros (✉)
Departamento de Prehistoria, Universidad Complutense de Madrid, Madrid, Spain
e-mail: joyravedra@hotmail.com

Jamie L. Clark and John D. Speth (eds.), *Zooarchaeology and Modern Human Origins: Human Hunting Behavior during the Later Pleistocene*, Vertebrate Paleobiology and Paleoanthropology, DOI: 10.1007/978-94-007-6766-9_6,

I propose here another innovation in faunal studies for the area. So far, subsistence strategies in northern Spain have been considered at a general level, grouping a number of sites together (Altuna 1989, 1992; Straus 1992; Yravedra 2002a, b). In this paper, I analyze subsistence practices during the Middle and Upper Paleolithic from a particularistic perspective, selecting sites that cover the complete chronology and considering each site separately. After the individual analysis, however, the data is synthesized to describe general patterns for the whole period.

Thus, this paper enlarges on the small number of sites in northern Spain from which taphonomic data for the Middle and Upper Paleolithic are available—mainly Castillo (Dari 2003; Landry and Burke 2006) and Amalda caves (Yravedra 2005, 2007)—by presenting data from several sites with Mousterian and Aurignacian-Gravettian levels such as Otero, Morín, Ruso, Hornos de la Peña and Covalejos.

Methods

Understanding taphonomy as the discipline that evaluates the multiple processes operating on any archaeological site, this paper presents the taxonomic and anatomical identification of the assemblages, an estimation of age and seasonality patterns and, finally, the taphonomic analysis of the aforementioned sites.

Regarding methodology, taxonomic identification was based on reference material and the models described by Lavocat (1966), Pales and Lambert (1971), Fernandez (2001), and Barba and Domínguez-Rodrigo (2005). Where exact taxonomic determination was not feasible, epiphysis and shaft fragments were assigned to approximate animal weight/size classes, separating then into three categories: (1) small-sized carcasses: <150 kg (i.e. *Capra pyrenaica, Rupicapra rupicapra, Capreolus capreolus*); (2) medium- and large-sized carcasses: >150 kg (i.e. *Cervus elaphus, Equus caballus, Bos primigenius, Bison priscus*); and (3) very large-sized carcasses: >800 kg.

The quantification of remains was based on the estimation of NISP and MNI in order to determine which measures better describe the taxonomic representation of the fauna. Both variables were also compared to describe their differences at each site. NISP determination follows Lyman's synthesis (1994), whereas MNI is based on Brain's (1969) model, which includes bone laterality and animal age. Furthermore, skeletal profiles and MNI calculation consider shaft thickness, section shape and the properties of the medullar surface (Barba and Domínguez-Rodrigo 2005).

Bones were divided into four anatomical regions: cranial (including antlers, skull, mandible and dentition), axial (vertebrae, ribs, pelvis and scapula, sensu Yravedra and Domínguez-Rodrigo 2009), upper limbs (humerus, radius, ulna, femur, patella, fibula and tibia) and lower limbs (metapodials, carpals, tarsals, phalanges and sesamoids, sensu Blumenschine 1986). Additionally, a systematic examination of bone surface modifications such as cut, percussion and tooth marks was carried out with 10X–20X hand lenses and different types of lighting, as proposed by Blumenschine (1995). The diagnostic criteria defined by Bunn (1982) and Potts and Shipman (1981) guided the identification of cut marks. Tooth marks were recorded following Binford (1981) and Blumenschine (1988, 1995). Finally, percussion marks were identified according to Blumenschine and Selvaggio (1988), and Blumenschine's (1995) criteria.

For comparative purposes, surface modifications were recorded using the values for epiphysis and shaft areas defined by Blumenschine (1988, 1995). Modifications were also quantified by type of element and bone section (Domínguez-Rodrigo 1997; Domínguez-Rodrigo and Barba 2005), based on NISP values. The presence of tooth, percussion and cut marks was quantified for the total number of remains, whereas estimated percentages include only well-preserved bone surfaces.

Considering the current view that anthropogenic bone concentrations yield a higher degree of fragmentation than assemblages accumulated by carnivores, bone fragmentation was studied from different perspectives. Characterization of the type of shaft circumference followed Bunn (1982), and the length of the fragment was measured against the total surface of the original bone. This author proposes three categories, where (1) stands for shaft circumference <25 %; (2) covers the 25–75 % range; and (3) groups remains with >75 % of the shaft circumference. The total dimensions were considered for fragmentation indexes and later compared to the complete bone, with fragments classified into three categories: (1) for <3 cm elements; (2) for 3–5 cm; and (3) for >3 cm.

The dental wear on *Capra* specimens which were less than 3 years old was used for age-pattern and seasonality determination (Coutourier 1962; Pérez Ripoll 1988). Finally, deer were analyzed following Steele (2002).

The Sample

The Cantabrian coast in northern Spain extends from eastern Galicia to the Basque country, and from the Cantabrian Sea to a series of mountain ranges including the Picos de Europa. The sites studied here are concentrated in the Cantabrian province, in the middle of the region (Fig. 6.1; Table 6.1). The proximity of the mountains to the coast produces a peculiar geography where valleys, mountains

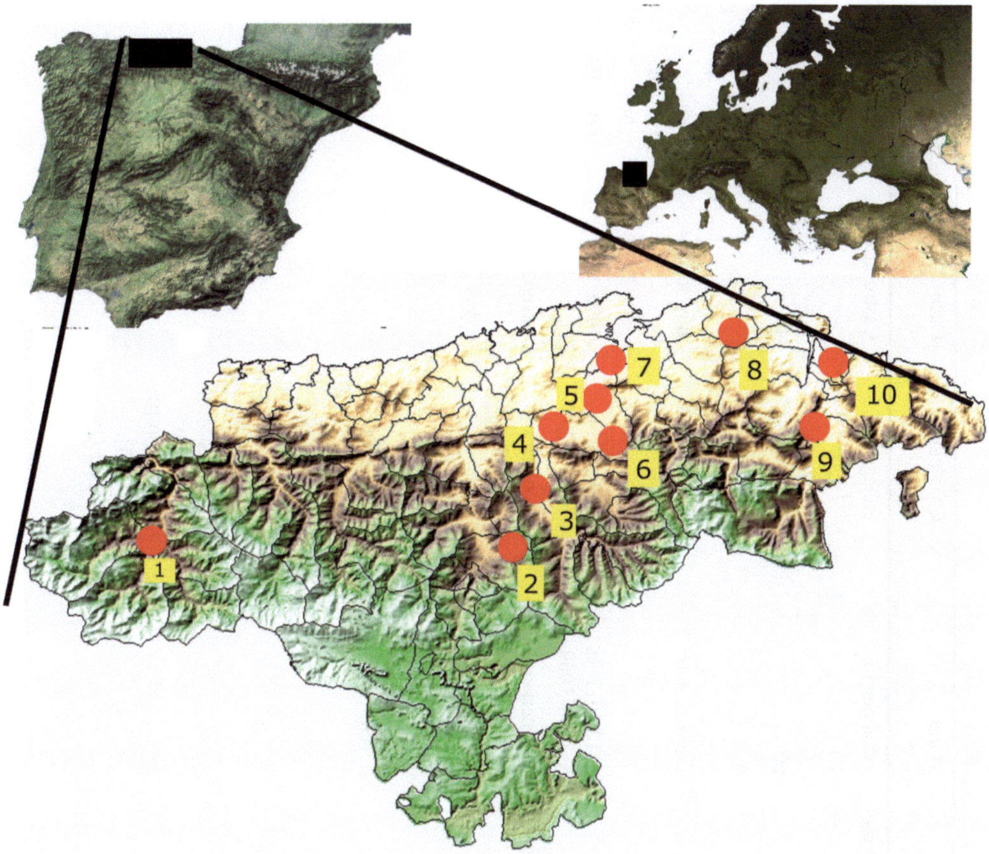

Fig. 6.1 Geographical location of the sites: *1* Esquilleu, *2* Hornos de la Peña, *3* Castillo, *4* Covalejos, *5* Pendo, *6* Morín, *7* Ruso, *8* Otero, *9* Cofresnedo, *10* Cuco

and plains are interspersed, resulting in a highly variable landscape with a large number of animal and plant resources which would have supported populations during the Middle and Upper Paleolithic.

Esquilleu Cave (Cillorigo de Liébana, X:371520, Y:4790700, page M. T. N: 1:50000 Carreña-Cabrales; 350 MASL) is found in the steep Hermida cliff limestone complex, some 40 km from the coast. It was excavated from 1997 to 2006 by J. Baena, revealing 35 Mousterian levels dated from 34,380 ± 670 ^{14}C BP to 53,491 ± 5,114 BP (TL) (Baena et al. 2005).

Hornos de la Peña (Tarriba, San Felices de Buelna; 280 MASL) is located 18 km from the coast in the Corrales de Buelna valley, characterized by a mountain landscape open to many different ecological niches. Excavated in 1906 by Alcalde del Río and in 1909–1912 by H. Breuil, H. Obermaier and J. Bouyssonie, the stratigraphy was first published by Breuil and Obermaier (1912) and later adapted by Bernaldo de Quirós (1982). Its five levels span from the Mousterian to the Neolithic. In this paper, only levels V (Mousterian) and IV and III (Aurignacian; 20,930 ± 370 ^{14}C BP; Soto Barreiro 2003) will be considered.

Ruso (Igollo, Camargo, X:427800, Y:4808670 UTM; 60 MASL) is an archaeological site situated in the coastal plain of Santander Bay, some 6 km from the coast. Excavations at this site, supervised by E. Muñoz Fernández, started in 1984. The site presents seven archaeological levels from the Mousterian (levels V and VI, and Evolved Aurignacian level IVb, 27,620 ± 180 ^{14}C BP) to the Bronze Age (Muñoz Fernández 1991; Muñoz Fernández and Serna 1994). Otero (Secadura, Voto, N:43°21′10″, E:0°09′30″, pages 18–35, plane 1/50000, I.G.C.; 60 MASL), is located some 12 km from the coast. Excavated in 1963 by J. González Echegaray, M. A. García Guinea and A. Begines Ramírez (González Echegaray et al. 1966), it yielded nine archaeological levels from the Mousterian to the Azilian. Only Mousterian (level 9) and Aurignacian (levels 8, 6, 5 and 4) levels are discussed here.

Morín (Villanueva de la Escusa, N:0°10′10″, E:43°21′43″, page 34 Torrelavega, plane 1/50000 I.G.C.; 60 MASL) is situated 6 km from the coast. Many fieldwork seasons (J. Carballo and Sierra in 1912; J. Carballo in 1917–1919; J. González Echegaray and L. Freeman in 1966–1969) revealed 22 archaeological levels that cover an important sequence with Azilian, Magdalenian, Solutrean, Aurignacian, Châtelperronian and Mousterian levels. We include the Aurignacian (levels 5–10; from 36,590 ± 770 ^{14}C BP to 27,336 ± 735 ^{14}C BP; Maillo et al. 2001), Châtelperronian (level 10; 35.905 ± 6,585 ^{14}C BP; González Echegaray and Freeman 1978) and Mousterian

Table 6.1 Main features of the sites mentioned

	Esquilleu	Hornos de la Peña	Ruso	Otero	Morín	Covalejos	Cuco	Cofresnedo
Location	Cillorigo de Liébana	Tarriba, San Felices Buelna	Igollo Camargo	Secadura Voto	Villanueva Escusa	Piélagos, Cantabria	Castro urdiales	Cibullas Matienzo
MASL (m)	350	280	60	60	60	110	43	160
Kilometer from coast	40	18	6	12	6	3	1.5	>20
Landscape	Steep Hermida Cliff Complex limestone	Variable ecologic niche	Bay seashore	Bay seashore	Bay seashore	Bay seashore	Bay seashore	Variable ecologic niche
Year of excavation	1997–2006	1906–1912	1984	1963	1912/1917–1919, 1966–1969	1879, 1914, 1997–2003	1996	2000–2001
Culture	Mousterian	Mousterian-Neolithic	Mousterian-Bronze Age	Mousterian-Azilian	Mousterian-Azilian	Mousterian-Aurignacian	Aurignacian-Gravettian	Mousterian-Roman
All levels	III–XXXV	I–V	I–VI	1–9	1–22	A–Q	I–XIV	–
Moust. level	III–XXXV	V	VI and V	9	22, 17–11	D–Q	–	–
Moust. chronol.	34,380 ± 670 ^{14}C BP (AA-37883; AMS; level VI), 36,500 ± 830 ^{14}C BP (AA-37883; level VIF), 36,500 ± 830 ^{14}C BP (AA-37882; AMS; level XIF), 39,000 ± 300 ^{14}C BP (Beta-149320; AMS; level XIII), 49,700 ± 1,600 ^{14}C BP (OXA-11414; AMS; level XVIII), 51,034 ± 5,114 BP (Mad 3299; TL; level XXId), 53,491 ± 5,114 BP (Mad 3300; TL; level XXIII)	–	–	–	39,770 ± 730 ^{14}C BP (GIF 96264; level 11)	40,650 + 2,300 – 1,800 ^{14}C BP (AMS-level D top), 41,640 + 650 – 530 ^{14}C BP (AMS-level D bottom), 101,000 BP (U/TH, level Q)	–	–

(continued)

Table 6.1 (continued)

	Esquilleu	Hornos de la Peña	Ruso	Otero	Morín	Covalejos	Cuco	Cofresnedo
Chat. level	–	–	–	–	10	–	–	–
Chat. chronol.	–	–	–	–	35,905 ± 6,585 ^{14}C BP (SI-951a)	–	–	–
Aurig. Level	–	IV and III	IV b	8, 6, 5 and 4	9-5b	B and C	VII–XIV	4
Aurig. chronol.	–	20,930 ± 370 ^{14}C BP (BM-1883R; level III)	27,620 ± 180 ^{14}C BP (BETA-12036; AMS)	–	36,590 ± 770 ^{14}C BP (GIF-96263; level 8), 27,685 ± 1,285 ^{14}C BP (SI-956; level 8a); 27,336 ± 735 ^{14}C BP (SI-952), 29,515 ± 840 ^{14}C BP (SI-955; level 7), 30,465 ± 901 ^{14}C BP (SI-954; level 6–7)	30,380 ± 250 ^{14}C BP (GRA 22443 – AMS; level B), 32,840 ± 280 ^{14}C BP (GRA-24200-AMS; level C)	30,020 + 160 – 150 ^{14}C BP (GrA 32436; AMS; level XIII)	31,360 ± 310 BP (GrA-20267, level 4)
Grav. level	–	–	–	–	5a-4	–	III–VI	–
Grav. chronol.	–	–	–	–	29,710 ± 340 ^{14}C BP (SI-953; level 5b)	–	23,400 ± 250 ^{14}C BP (GrA-32097; AMS)	–
Zooarch. ref.	Yravedra (2005)	Yravedra (2010b)	Yravedra et al. (2010)	Yravedra and Gómez-Castanedo (2010a)	Altuna (1971, 1973)	Castaños (2005)	Castaños and Castaños (2007)	Castaños (2003)
Taphon. ref.	Yravedra (2005)	Yravedra (2010b)	Yravedra et al. (2010)	Yravedra and Gómez-Castanedo (2010a)	Yravedra and Gómez-Castanedo (2010b)	Yravedra (submitted)	Yravedra Ms.	Yravedra Ms.
Bibliog. ref.	Baena et al. (2005)	Breuil and Obermaier (1912) Bernaldo de Quirós (1982)	Muñoz Fernández (1991) Muñoz Fernández and Serna (1994)	González Echegaray et al. (1966)	González Echegaray and Freeman (1978); Maillo et al. (2001)	Sanguino and Montes (2005)	Muñoz Fernández and Montes (2007)	Ruiz and Smith (2003)

levels (levels 11–17, 22; 39,770 ± 730 ^{14}C BP; Maillo et al. 2001).

Covalejos (Piélagos, Cantabria, UTM: X:424560, Y:4805500; 110 MASL) is 3 km from the coast in an undulating landscape. Pedraja first excavated this site in 1879 and later H. Obermaier, M. R. Sallcras and L. Rozas excavated in 1914. A new excavation in 1997–2003 by J. Sanguino and R. Montes (2005) recovered the bones analyzed here, from the two Aurignacian (30,380 ± 250 ^{14}C BP to 32,840 ± 280 ^{14}C BP) and 10 Mousterian (40,650 + 2,300 − 1,800 ^{14}C BP to 101,000 BP) levels.

Cuco rockshelter (Urdiales, Castro Urdiales UTM: X:481507, Y:4804428; 43 MASL) is found 1.5 km from the coast. The site was excavated in 2006 by E. Muñoz, P. Rasines, S. Santamaría and J. M. Morlote, who documented 14 Gravettian (23,400 ± 250 ^{14}C BP, level III) and Aurignacian (30,020 + 160 − 150 ^{14}C BP, level XIII) levels (Muñoz Fernández and Montes 2007).

The excavation at Cofresnedo (Cibullas, Matienzo; 160 MASL) in 2000–2001 was directed by J. Ruiz and yielded a series of discontinuous occupations by humans and carnivores alternatively dated in the Middle and Upper Paleolithic (31,360 ± 310 ^{14}C BP, level 4; Ruiz and Smith 2003).

All these sites cover a similar chrono-cultural range, spanning from the end of the Middle Paleolithic to the beginning of the Upper Paleolithic (40,000–25,000 years ago). Nevertheless, each of them presents certain peculiarities regarding environmental and climatic contexts. They were also excavated at different times, with a variety of field methods. Due to these differences, it is crucial to assess them individually in order to draw an accurate picture of the area in both periods.

Results

El Esquilleu

Capra pyrenaica dominates all levels, followed by chamois and deer (Table 6.2). *Bos/Bison* is only found in levels VIII, XI, XIF and XIII, which differ from the rest of the site due to the presence of animals from distant areas and the incorporation of new raw materials, suggesting longer-distance mobility patterns (Baena et al. 2005). Additionally, the determination of seasonality for these levels indicates extensive occupations spanning several seasons (Table 6.2), while in the remaining levels, human occupation was restricted to one specific season.

The taphonomic analysis also indicated some differences throughout the sequence. The upper levels (III-V) constitute a palimpsest with badly defined human occupations, where carnivores were the main accumulators of meat resources (Yravedra 2006a, b). In these units, bones with cut marks were not frequent (Fig. 6.2a, b), whereas tooth marks were present in more than half of the remains (Table 6.2). Moreover, the fragmentation index was lower in these two levels than in the rest of the sequence: only 65 % of the remains were <3 cm, whereas more than 80 % of the remains from other contexts fell into this category. The circumference of the diaphyses also differed in level III—28 % of the sample preserved more than 25 % of the circumference, whereas in the rest of the levels almost 99 % of the diaphyses yield values lower than 25 % (Table 6.2).

From level VI to XIII, cut and percussion marks increased, as did the percentage of bones showing thermal alteration (Fig. 6.2a, b). These levels also showed a higher degree of bone fragmentation. At the same time, evidence of carnivore action decreased (Table 6.2). Skeletal profiles indicate the transport of complete ibex to the site in all levels, together with deer in level XIF (Yravedra and Domínguez Rodrigo 2009). Finally, in levels XXI, XXIII, XXV, XXVIII and XXIX, the concentrations of burnt and intensively fragmented bones may be interpreted as fuel for hearths (Yravedra et al. 2005, 2010).

Hornos de la Peña

Given the field methodology in fashion at the beginning of the twentieth century, bone was rare at Hornos de la Peña, restricting the faunal analysis to a few observations. The Mousterian and the Aurignacian levels together yielded 655 remains. During the Mousterian, chamois is the best-represented animal both in NISP and MNI, followed by horse and deer (Table 6.3). The Aurignacian level 4 shows a predominance of horse over deer, chamois, ibex, roe deer and aurochs. In level 3, however, deer dominates over horse and other animals (Table 6.3).

In both the Middle and Upper Paleolithic, the only animals that preserve evidence for human activity are deer, horse and aurochs (Fig. 6.2b). Evidence for human involvement with *Rupicapra* is lacking; apparently at Hornos de la Peña, these animals were not consumed by humans (Fig. 6.3). This suggests that any differences in hunting strategies during the Middle and Upper Paleolithic should have been negligible. Taking into account both periods, horse dominates in levels 5 and 4, amounting to 56 and 43 % of the remains, as well as 58 and 55 % of MNI respectively. Deer follows with 43 % of NISP in both levels, and 36 and 33 % of MNI respectively, while *Bos/Bison* makes up less than 10 % of the MNI (Fig. 6.3; Table 6.3).

Table 6.2 Main zooarchaeological and taphonomic features in Esquileu

Esquilleu		*Capra*		*Rupicapra*		*Cervus*		*Bos/Bison*		Seasonality	Cut mark (%)		Tooth mark (%)		Burnt bone (%)	Bone <3 cm (%)	Shaft circumference <25 %
	NISP	NISP	MNI	NISP	MNI	NISP	MNI	NISP	MNI		Small	Large	Small	Large			
III	369	195	5	27	4	5	2	–	–	Spring-autumn	2.7	–	53	12,5	9	65	72
IV	136	57	2	14	2	6	2	–	–	–	7.6	–	77	50	5	82	94
V	513	106	2	27	4	11	2	–	–	Summer-winter	14	100	42	–	19	84	95
VI	1,994	306	9	13	4	34	2	–	–	Summer-winter	42	31	31	28	33	74	99
VII	1,459	111	4	19	1	34	2	–	–	Summer-autumn	30	52	17	14	46	80	99
VIII	1,197	192	2	4	3	23	2	2	1	Autumn	39	42	4	37	31	88	97
IX	1,203	49	2	12	1	14	1	–	–	¿?	35	–	10	14	34	89	99
X	66	9	1	–	–	–	–	–	–	¿?	–	–	–	–	–	94	–
XI	1,382	133	5	9	2	15	1	1	1	¿?	27	50	2	13	15	86	99
XIF	2,3402	2,257	16	126	9	404	9	16	1	All year round	27	24	3	3	16	83	98
XII	778	64	2	1	1	14	1			¿?	17	50	2	–	21	86	99
XIII	7,312	983	5	33	3	103	2	27	1	Summer-autumn	21	35	2	3.6	30	89	100
XIV	1,144	376	1	–	–	11	1	–	–	¿?	Bad preservation of bone surfaces				63	95	100
XV	1,014	188	1	–	–	–	–	–	–						80	99	100
XVI	202	76	1	–	–	–	–	–	–						77	100	100
XVII	263	68	1	–	–	1	1	–	–						79	100	100
XVIII	189	62	–	–	–	–	–	–	–						77	100	100
XIX	376	72	1	–	–	–	–	–	–		19	–	–	–	89	99	100
XX	3672	398	2	4	1	9	1	–	–		20	–	–	–	100	100	100
XXI	17,017	726	12	–	–	2	1	–	–		18	–	–	–	100	100	100
XXII	1,704	297	1	–	–	3	1	–	–		14	–	–	–	99	100	100
XXIII	2,374	95	1	–	–	4	1	–	–		Bad preservation of bone surfaces				99	99	100
XXIV	111	29	1	–	–	–	–	–	–						100	100	100
XXV	42	13	1	–	–	–	–	–	–						95	100	100
XXVI	243	57	1	–	–	–	–	–	–						78	100	100
XXVII	263	59	1	–	–	–	–	–	–						92	100	100
XXVIII	11	1	1	–	–	–	–	–	–						100	100	100
XXIX	48	1	1	–	–	–	–	–	–						96	100	100
XXX	9	9	1	–	–	–	–	–	–						100	100	100

Ibex seasonality is estimated from the emergence of the dental crown (Pérez Ripoll 1988). Mark frequencies do not include teeth and bones with scarce bone surface preservation. Small animals include *Capra, Rupicapra, Capreolus* and small sized ungulates. Large animals refer to *Cervus, Bos, Equus and* large and medium sized animals

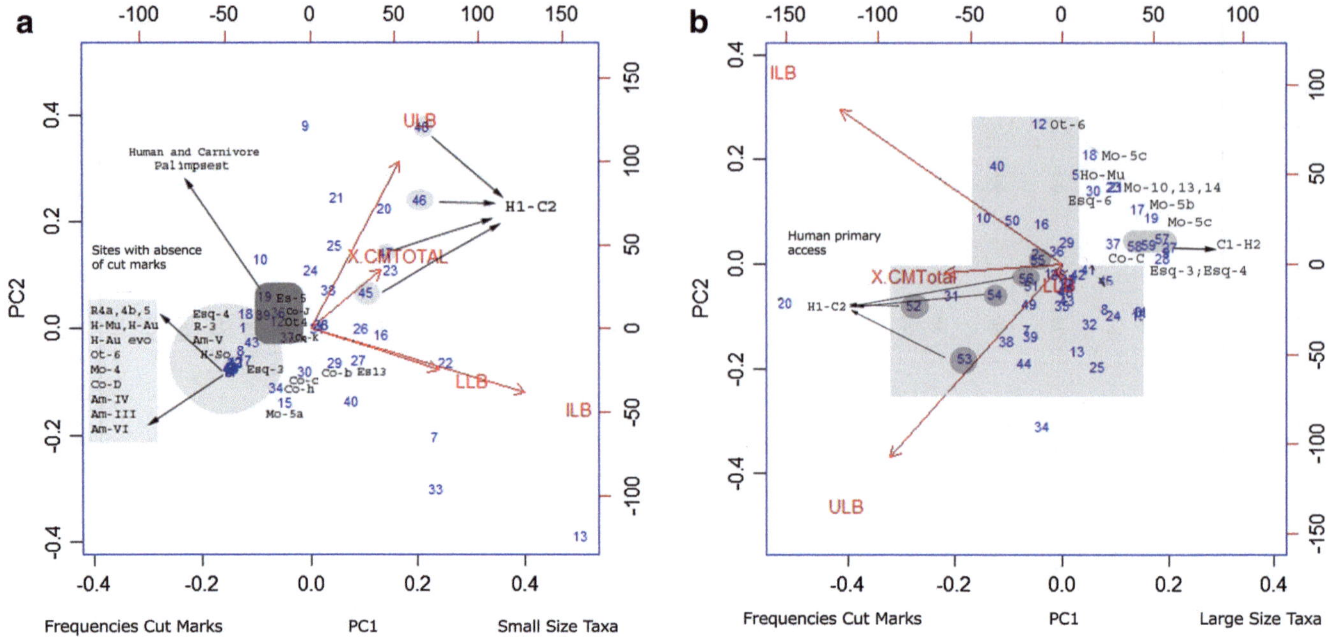

Fig. 6.2 **a** Plot graph showing the scores of the Principal Component Analysis of small animals (*Capra* + *Rupicapra* + *Capreolus* + smaller animals) from selected Mousterian and Upper Paleolithic sites, and the experimental samples described in Domínguez-Rodrigo and Barba (2005). Points 45–48 indicate the experiments where humans had a primary access to the prey. **b** Plot graph showing the scores of the Principal Component Analysis of large animals (*Bos* + *Equus* + *Cervus* + medium and large-sized indet.) from selected Mousterian and Upper Paleolithic sites. Points 52–55 corresponds to the experimental samples described in Domínguez-Rodrigo (1997). Lupo and O'Conell's (2002) point 56 shows values corresponding to a human intervention on the prey. Experiments regarding secondary access (Domínguez-Rodrigo, 1997) are exemplified in points 57–59. *Key* Cut marks frequencies on ULB (humerus and femur), ILB (tibia and radius), LLB (metapodial), CM total (total frequencies of cut marks in limbs, cranial and axial bones)

From level 3 upwards, some differences are evident. Deer dominates this unit, accounting for 60 % of MNI and 66 % of NISP. It is also the first time evidence of human processing is found on remains of chamois, ibex and roe deer (Yravedra 2010b), probably implying a change towards hunting strategies that incorporated a wider range of prey, including small-sized ungulates: from this unit, cut and percussion marks are found on small-sized animals with age profiles which show a large number of adult individuals (Yravedra 2010b).

Carnivore action was identified on deer, horse and aurochs bones, but there are also cut marks related to meat processing, suggesting human interaction with these animals as well (Fig. 6.2b). The presence of cut marks on ribs and upper appendicular (humerus and femur) or intermediate (tibia and radius) bones in levels 5–3 (Fig. 6.2b), together with fracture patterns defined by the absence of furrowing, and a predominance of adult individuals, are also characteristic of human action.

According to my observations, no significant differences exist between the taxa present in the Mousterian level 5 and in the Aurignacian level 4. In both levels, horse and deer are dominant. The absence of cut or percussion marks, together with a number of tooth marks, suggest chamois was not introduced by humans: in both Mousterian and Aurignacian times, chamois in Hornos de la Peña appear to have been introduced by carnivore activity. Differences, however, appear from level 3: deer constitute the main species in both MNI and NISP, and roe deer, ibex and chamois were incorporated into the range of human processed prey (Fig. 6.2a).

Ruso

Ruso is a small cave close to the Paleolithic sites of Pendo, Juyo and Santián. The Mousterian and Aurignacian materials from Pendo were described as similar, with a dominance of deer, followed by horse and large bovids (Fuentes 1980); however, this site is in fact a palimpsest of many different archaeological levels (Montes et al. 2005). Deer is also dominant in Santián, although the nature of the collection of this material raises questions about its value. In this context, the faunal assemblage excavated at Ruso is fundamental for defining a local sequence with contextualized materials spanning from the Mousterian to the Solutrean.

Table 6.3 Main zooarchaeological and taphonomic features in Hornos de la Peña

H. de la Peña		*Capra*		*Rupicapra*		*Cervus*		*Capreolus*		*Equus*		*Bos/Bison*		Carnivore		Cut mark (%)		Tooth mark (%)	
	NISP	NISP	MNI	NISP	MNI	NISP	MNI	NISP	MNI	NISP	MNI	NISP	MNI	NISP	MNI	Small	Large	Small	Large
V	168	4	2	77	10	27	4	8	2	29	7	7	1	13	2	0	13.3	28	13.3
IV	75	5	3	12	3	21	4	–	–	26	6	2	1	6	2	0	3.6	75	0
III	412	4	3	92	9	183	18	12	2	82	10	11	1	10	4	0	9.1	48	42

Frequencies of cut and tooth marks are determined from NISP, excluding teeth and badly preserved bones

The taxonomic analysis showed that deer was dominant both in Mousterian level V and Aurignacian level IVb, accounting for 71 % of NISP and 33–38 % of MNI. Horse was second in importance in both periods, with 31 % of MNI in level V, and 20 % in level IV. After them, large bovids and other animals were scarcely represented (Fig. 6.4). Age profiles closely coincided in both periods: *Bos/Bison* correspond to adults only, young equids represent 66 and 50 % of the sample in the Aurignacian and the Mousterian respectively, and young deer ranged from 33 to 50 % of the sample (Table 6.4).

Taphonomically, the situation is similar to Hornos de la Peña. Deer, horse and aurochs yielded evidence of human activity in the form of cut and percussion marks (Figs. 6.2b, 6.5). On the other hand, chamois, ibex and roe deer showed no indication of human activity before the Solutrean (level III; Fig. 6.2a).

Both the recorded fracture patterns and the location and number of cut and percussion marks on highly nutritional parts (upper bones and axial elements) indicate humans had early access to meat resources, be they deer, horse or large bovids (Yravedra et al. 2010). Conversely, the scarcity of human marks and the abundance of tooth marks on small ungulates suggest accumulation by carnivores rather than human beings (Table 6.4; Fig. 6.2a). Medium- and large-sized animals also preserved tooth marks, but they were located over cut marks typical of defleshing and disarticulation, suggesting that carnivores may have acted as scavengers of human consumed prey. Nevertheless, the presence of a small number of cut and percussion marks on small ungulates from the Mousterian and Aurignacian deposits may suggest humans had occasionally hunted these animals. In the case of ibex, the lower degree of fragmentation, as well as the low frequencies of anthropogenic marks, probably indicates that they were not frequently hunted (Fig. 6.5). Summing up, in both the Mousterian and the Aurignacian periods, humans focused on bovids, horses and deer, showing a persistence of basic subsistence strategies.

Otero

Due to the antiquity of the excavation, only 700 bones were available from this site. This sample was both biased and pre-selected, with unequal representation of NISP and MNI, few indeterminate bones, and a clear dominance of certain parts in the skeletal profiles, as well as a higher representation of epiphysis versus diaphysis fragments (Yravedra and Gómez-Castanedo 2010a). Any conclusions about this site should thus consider this bias.

Throughout the sequence, deer dominates the assemblage, followed by horse and other ungulates such as

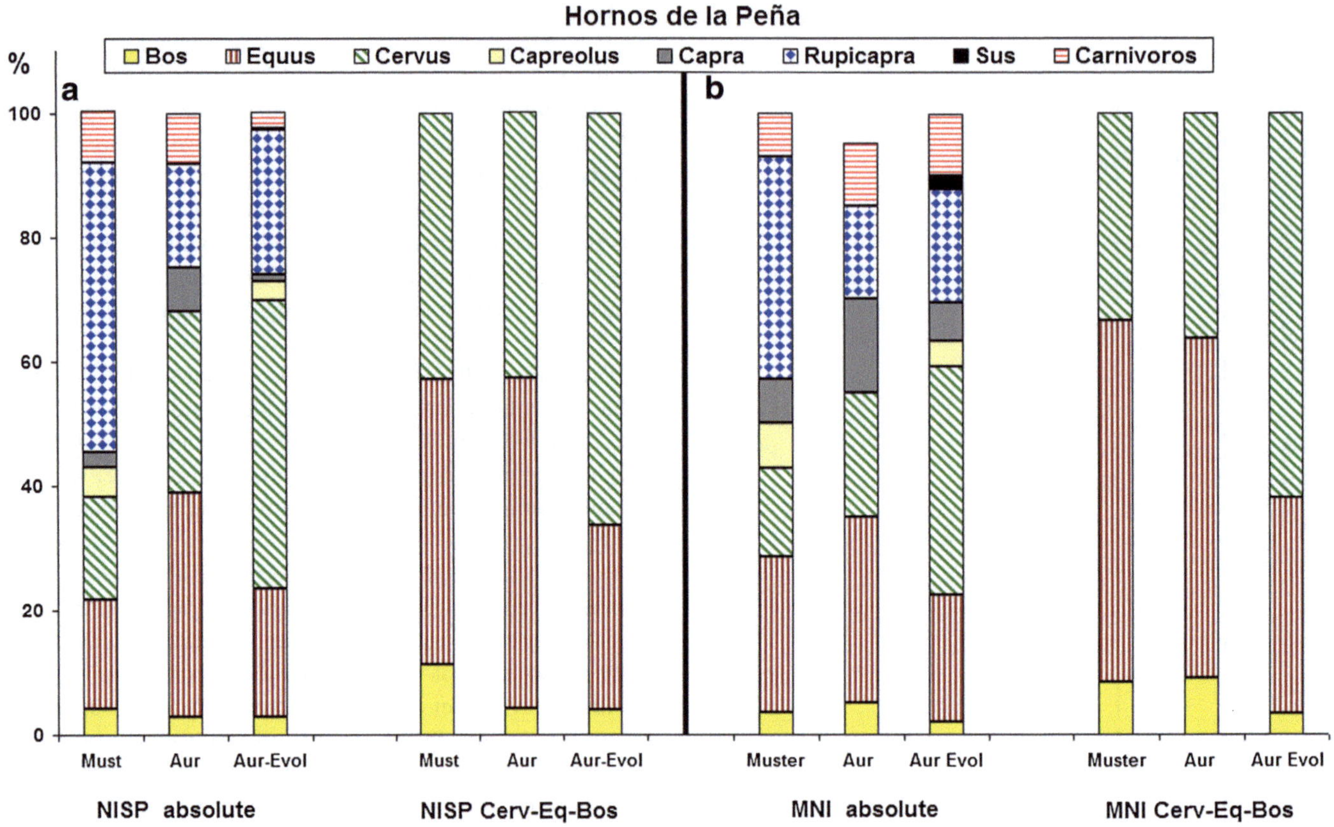

Fig. 6.3 Taxonomic profiles for Hornos de la Peña. **a** Complete sample in NISP on the *left* and taxa with anthropogenic marks on the *right*. **b** Idem for MNI

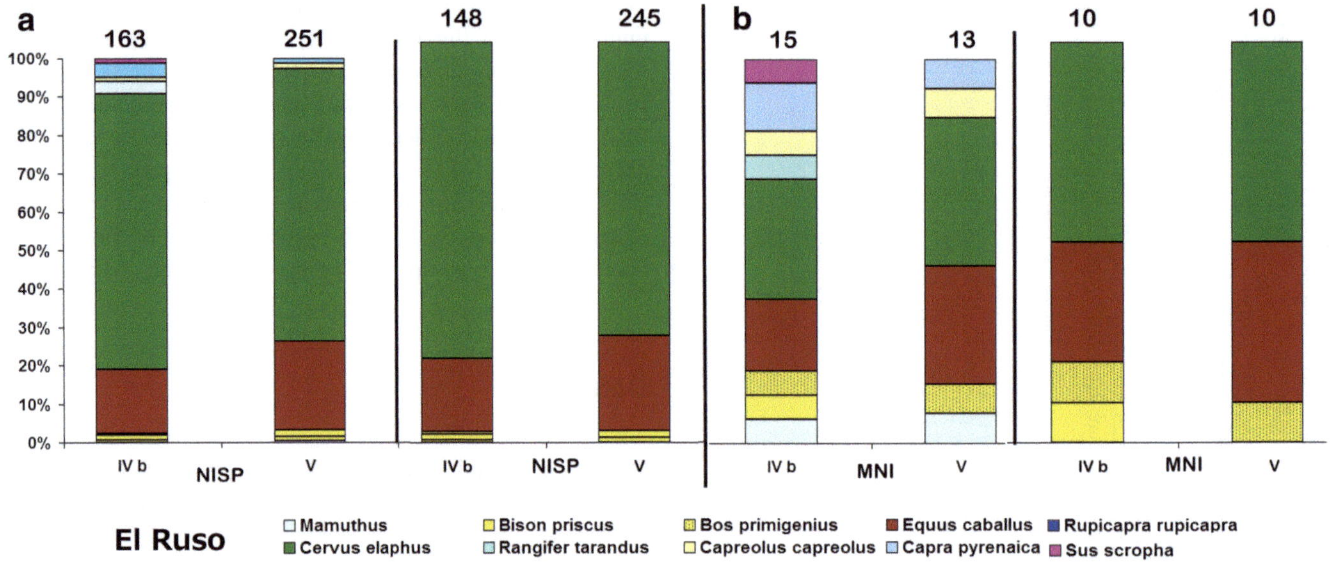

Fig. 6.4 Taxonomic representation in Ruso; NISP on the *left* and MNI on the *right*. Both variables are compared with the total number of animals present (on the *right*) and the animals introduced by humans to the site

chamois, reindeer, roe deer and ibex. When comparing Mousterian and Aurignacian assemblages, some differences become evident. Carnivore action is unquestionable in the Mousterian, but horse dominates over deer in both NISP and MNI, which is in contrast to the lower Aurignacian deposits, at least as far as the reduced sample of less than 60

Table 6.4 Main zooarchaeological and taphonomic features in Ruso

Ruso		*Capra*		*Cervus*		*Capreolus*		*Equus*		*Bos/Bison*		Carnivorous		Cut mark (%)		Tooth mark (%)		Percussion mark (%)	
	NISP	NISP	MNI	NISP	MNI	NISP	MNI	NISP	MNI	NISP	MNI	NISP	MNI	Small	Large	Small	Large	Small	Large
V Mous	1423	3	1	179	6	4	1	58	4	4	1	39	6	1.9	3.5	7.5	3.4	2.2	2.6
IV b Aur	667	6	2	117	5	2	1	27	3	7	2	29	6	2.5	8.4	35	8.4	0	6.4
	Frequencies of taxa associated to human behavior												Seasonality from *Cervus elaphus*						
V Mous	–	–	–	73 %	50 %	–	–	24 %	40 %	3 %	10 %	–	Winter-summer and beginning of autumn						
IV b Aur	–	–	–	79 %	50 %	–	–	18 %	30 %	3 %	20 %	–	No data						

Seasonality is established from the emergence of the dental crown (Steele 2002). Frequencies of cut and tooth marks are determined from NISP, excluding teeth and badly preserved bones. Both large (*Bos* + *Equus* + *Cervus* + medium and large sized indet.) and small sized animals (*Capra* + *Rupicapra* + *Capreolus* + small sized ungulates) are included

specimens from levels 5 and 6 can be considered (Table 6.5). However, the better-represented level 4 has dominance of deer, followed by horse. Regardless of the unit, adult individuals outnumber juveniles and infants. Seasonality could only be estimated for deer in level 4, indicating a year-round hunting strategy, especially at the beginning of autumn and the end of winter.

Due to the bias in bone collection, taphonomic analysis was not possible for all levels. The absence of postcranial bones in the Mousterian assemblage prevented the determination of the agent that collected the animals. Between levels 2–6, cut marks on upper appendicular and axial bones of deer suggest human consumption (Fig. 6.2b); however, in other species evidence for human involvement was unclear. For some taxa, such as aurochs or ibex, no cut marks were identified, and for others like horse, roe deer or chamois, they appeared only in some levels (Table 6.5). Small-sized animals preserved human-generated marks only in level 4 and in the Magdalenian Unit 2 (Fig. 6.2a; Table 6.5).

Morín Cave

From the 6,500 remains from Morín Cave, deer was the most numerous species, followed by *Bos/Bison* and horses, confirming Altuna's analysis (1971, 1973). The comparison between Middle and Upper Paleolithic levels suggested some differences. Throughout the sequence, the most common animals were deer and *Bos/Bison*, with a rate of 40–50 % of the MNI for the former, depending on the level. Nevertheless, from the Aurignacian level 6 upwards the number of small-sized ungulates (ibex, chamois and roe deer) increased, matching the progressive reduction of *Bos/Bison* and horse from the Gravettian.

Altuna (1971, 1973) noted no evidence of chronological changes in age profiles or seasonality, while Pike Tay et al. (1999) identified occupation from the end of autumn to springtime in both the Mousterian and the Aurignacian.

Furthermore, no differences were seen in the skeletal profiles. For both periods, the representation of anatomical sections for small- and medium-sized animals suggests that the entire animal was incorporated into the site. On the other hand, skeletal profiles for large-sized animals suggest a differential transport strategy that favored upper limbs elements over axial and lower appendicular bones (Altuna 1971, 1973).

The results of the taphonomic analysis suggest that the absence of certain anatomical sections may be conditioned by carnivore and water action (Yravedra and Gómez-Castanedo 2010b). Tooth marks on epiphyses and axial bones prove that carnivores participated in site formation.

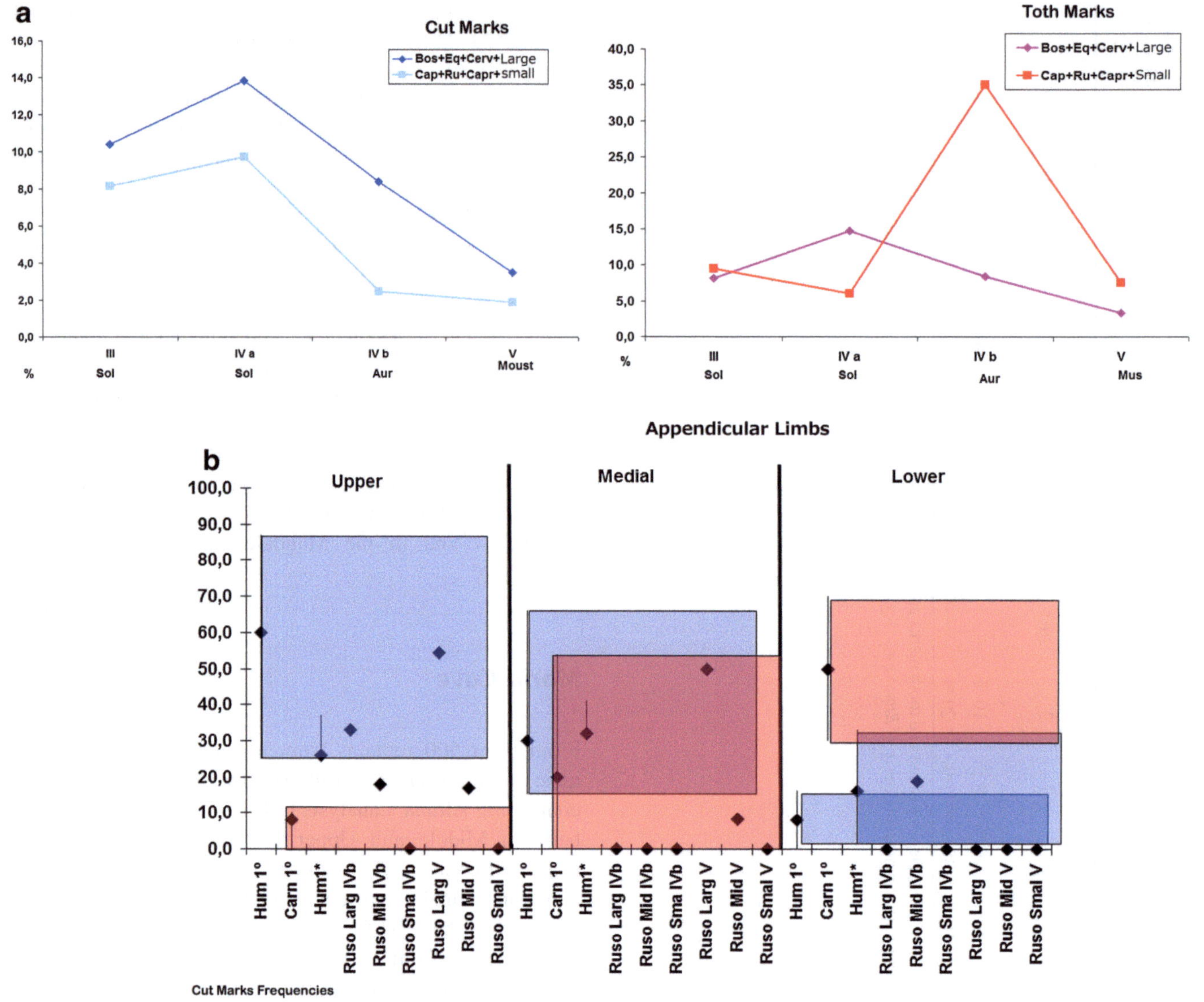

Fig. 6.5 **a** Absolute frequencies of cut marks and tooth marks by levels, discriminating small sized ungulates from large ones. The low frequencies of small sized ungulates with human-related marks are remarkable. On the opposite, bones with carnivore-generated marks are most numerous for smaller taxa. **b** Comparison with the referential framework for the identification of first consumption by humans. Appendicular elements were divided into upper (humerus, femur), intermedial (tibiae and radio) and lower bones (metapodials). Large sized animals data correspond to human consumption whereas in smaller ones human participation is scarce

Water was also important for sedimentation and water-action generated of a large number of rounded and polished bones (Yravedra and Gómez-Castanedo 2010b). And yet, when considered alongside the distribution of cut and percussion marks, it may be concluded that in all levels, humans had access to meat portions from all taxa (Table 6.6; Fig. 6.2a, b).

The frequency and distribution of cut marks suggests that human beings had primary access to meat from small-, medium- and large-sized animals in levels 17, 5a, 4, 3 and 2 (Yravedra and Gómez-Castanedo 2010b). The sample from the remaining levels was not representative, but some evidence of anthropogenic defleshing marks was identified on many taxa (Table 6.6).

Both the taxonomic and taphonomic data suggest some persistence in subsistence strategies between the Mousterian and the Aurignacian: in both periods, the same prey were hunted. In the case of small-sized ungulates such as ibex, chamois and roe deer, information on the agent of accumulation was scarce due to the low representation of bones at the site. In level 17, however, these species seem to have been accumulated by human hunting behavior; a few cut and percussion marks found on the scarce remains from the other levels (Fig. 6.2a, b) may be interpreted likewise.

Table 6.5 Main zooarchaeological and taphonomic features in Otero

Otero		*Capra*		*Rupic*		*Cervus*		*Capreolus*		*Equus*		*Bos/Bison*		Carnivorous		Cut mark (%)		Tooth mark (%)	
	NISP	NISP	MNI	NISP	MNI	NISP	MNI	NISP	MNI	NISP	MNI	NISP	MNI	NISP	MNI	Small	Large	Small	Large
9 Mous	28					6	1			12	3	2	2	7	2	0	0	0	0
6 Aur	86					14	2	3	1	6	1					0	20	100	40
5 Aur	37	4	1			14	2			4	1					0	0	0	0
4 Aur	408	13	2	20	4	177	10	4	2	71	10	12	4	7	2	6.3	7.7	19	18
	Frequencies of taxa associated to human behavior														Seasonality from *Cervus elaphus*				
9 Mous	¿?														¿?				
6 Aur	20					70 %	67 %			30 %	23 %				¿?				
5 Aur	18					75 %	67 %			25 %	23 %				¿?				
4 Aur	190			10 %	29 %	90 %	71 %								All year				

Seasonality is established from the emergence of the dental crown (Steele 2002). Frequencies of cut and tooth marks are determined from NISP, excluding teeth and badly preserved bones. Both large (*Bos* + *Equus* + *Cervus* + medium and large sized indet.) and small sized animals (*Capra* + *Rupicapra* + *Capreolus* + small sized ungulates) are included

Covalejos

The Mousterian and Aurignacian bone sample at Covalejos consists of 49,799 extremely fragmented remains, of which 64 % were indeterminate fragments (Yravedra 2011). As already described in Castaños (2005), the dominant animal in the whole sequence is deer and there are no significant differences between Mousterian and Aurignacian levels, except for an increase in the MNI of *Capreolus* and *Equus* in level B and the reduction of NISP and MNI percentages for deer in level B (Fig. 6.6). Furthermore, the Aurignacian unit C and the Mousterian unit D were similar in terms of taxonomic representation. The age profile shows a dominance of young adults (age 5–7) in all levels. Seasonality data indicate some variation in prey exploitation; during the Mousterian, hunting took place all year round, whereas in the Aurignacian, hunting was frequent between the end of summer and the beginning of winter (Yravedra 2011).

Taphonomically, the analyzed assemblage is well-preserved except for levels C, D and H, where 60 % of the bones yielded water-generated rolling, polishing and abrasion marks (Yravedra 2011). Other processes recorded in the site included spalling, breaching, calcification, and, above all, intense fragmentation that reduced most bones to less than 3 cm.

Despite this situation, taphonomic analysis of the well-preserved bones suggested different human uses of animal resources (Table 6.7). Cut marks were frequently identified on all species and showed the distribution typical of primary human consumption (Table 6.7; Figs. 6.2a, b, 6.6). Besides the anthropogenic marks, carnivore tooth marks were also found. However, their small size, the absence of digested bones, the high fragmentation, and the superposition of tooth marks over previous human traces suggests a secondary role for carnivore consumption. It is only in level H that an increase of tooth mark frequencies and a lower percentage of bones with human modifications indicate that small ungulates may have been consumed by carnivores (Fig. 6.2a).

Broadly speaking, from the taxonomic and taphonomic analyses, and the consideration of age profiles, a persistence of subsistence strategies may be postulated for the Mousterian and Aurignacian at Covalejos; the same species were hunted in similar frequencies, especially in units D and C.

Cuco

The sample from Cuco includes 12,000 extremely fragmented remains, from which less than 2 % are taxonomically diagnostic. According to Castaños and Castaños (2007), deer is the dominant species, followed by *Bos/Bison*

Table 6.6 Main zooarchaeological and taphonomic features in Morín

Morín	*Equus*		Small ungulates		*Cervus*		*Bos/Bison*		Cut mark (%)		Tooth mark (%)		Percussion mark (%)	
	NISP	MNI	NISP	MNI	NISP	MNI	NISP	MNI	Small	Large	Small	Large	Small	Large
2	12	3	36	8	260	9	8	1	3.7	15	1.8	4.9	0	4.2
3	3	2	10	5	51	5	4	1	32	24.6	0	0	0	1.8
4	39	2	63	8	271	11	19	2	3.7	7.1	11.1	6,6	0	2.6
5a	30	4	106	10	388	12	29	3	1.8	7.2	23.6	13.8	0	3.8
5b	12	2	42	7	147	8	17	3	0	5.8	13.3	13.2	0	0
5c	6	6	24	16	120	21	9	8	0	2.9	0	0	0	1.4
6	19	2	23	5	77	4	31	4	21.4	2.1	28.6	2.1	0	1.5
7	8	1	8	2	86	5	26	2	0	3.2	0	0	0	0
8	8	6	2	1	7	3	4	3	0	28.6	0	0	0	0
8a	2	1	1	1	10	2	3	1	0	0	0	0	0	0
8b	2	2	–	–	3	2	8	1	0	0	0	0	0	0
9	1	1	1	1	7	2	3	1	0	57.1	50	0	0	7.1
10	1	1	–	–	3	3	25	2	0	6.1	0	3	0	0
13	–	–	–	–	4	1	5	1	0	16.7	0	16.7	0	0
14	14	3	1	1	16	2	40	2	0	6.3	0	3.1	0	3.1
15	44	4	1	2	55	3	60	2	0	8.7	0	10.3	0	2.9
16	5	1	1	1	3	1	8	1	0	15.4	0	0	0	0
17	124	11	16	4	142	8	215	10	4.8	9	4.8	7.4	0	4.5

Frequencies of cut and tooth marks are determined from NISP, excluding teeth and badly preserved bones. Both large (*Bos* + *Equus* + *Cervus* + medium and large sized indet.) and small sized animals (*Capra* + *Rupicapra* + *Capreolus* + small sized ungulates) are included

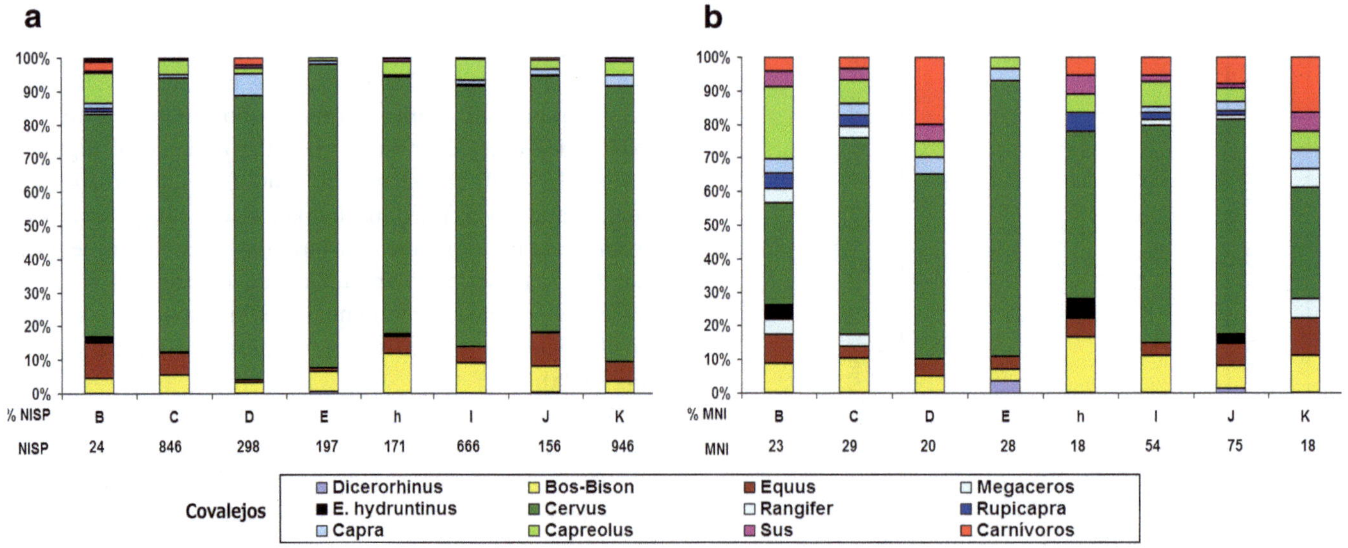

Fig. 6.6 Taxonomic representation of species introduced by humans in Covalejos. NISP on the *left*; MNI on the *right*

and *Equus* (Table 6.8). The high fragmentation rate in Cuco may have resulted from a number of agents. Nevertheless, the fracture patterns and presence of cut, percussion and tooth marks suggest intentional damage produced by both carnivores and humans.

Cut marks on bones with nutritious muscle packages such as the humerus, femur and axial bones suggest primary meat consumption. In the same way, assemblages with high fragmentation rates are typical of human accumulations. From the patterns found in Cuco, the bone assemblages

Table 6.7 Main zooarchaeological and taphonomic features in Covalejos

Covalejos	*Cervus*		*Capreolus*		*Cap* + *Rup* + *Capreo*		*Equus*		*Bos/Bison*		Carnivores		Cut mark (%)		Tooth mark (%)		Percussion mark	
	NISP	MNI	NISP	MNI	NISP	MNI	NISP	MNI	NISP	MNI	NISP	MNI	Small	Large	Small	Large	Small	Large
B	162	7	22	5	7	4	29	2	11	2	7	1	9	9	7.5	6.2	1.9	1.4
C	689	17	36	2	12	4	57	1	46	3	4	2	10.5	13.2	8.4	12.5	0.8	1.1
D	252	11	5	1	23	3	2	1	10	1	7	4	7.5	16.2	3.8	5.7	1.9	7.6
E	178	23	2	1	2	1	2	1	12	1	0	0	3.6	3.4	4.3	5.1	0.7	1
H	131	9	7	1	2	2	10	2	29	3	1	1	8.3	15.6	12.5	15.6	0	1
I	517	35	41	4	16	4	32	2	59	6	0	0	6.8	6.7	4.7	4.6	0.6	1.9
J	1198	48	43	3	31	5	159	7	120	5	13	6	7	3.9	8.3	8.7	1	1.9
K	776	6	37	1	40	5	54	2	33	2	5	3	6.5	6.8	8.5	10.9	1.8	3.7
Q	41	3	0	0	0	0	0	0	52	2	2	1	0	10.5	2.1	5.4	0	1.3

Frequencies of cut and tooth marks are determined from NISP, excluding teeth and badly preserved bones. Both large (*Bos* + *Equus* + *Cervus* + medium and large sized indet.) and small sized animals (*Capra* + *Rupicapra* + *Capreolus* + small sized ungulates indet.) are included

seem to have been first accumulated by humans and later consumed by carnivores. Further evidence for human activity is found in the circumference data, where 100 % of the sample presents values lower than 25 % of the circumference of the diaphysis.

Cofresnedo

The fauna from the Aurignacian levels studied by Castaños (2003) shows a dominance of medium- and large-sized animals such as horse, deer and large bovids (Table 6.8). Taphonomic analysis identified some bones as having concretions, patinas and rounding produced by water, together with tooth marks from carnivores (Table 6.8). Although the incidence of tooth marks is negligible, pit dimensions between 3.5 and 5.4 mm suggest the presence of animals such as wolf, hyena, bear or large felids.

However, the presence and frequency of cut and percussion marks (Table 6.8), as well as the intense bone fragmentation of appendicular elements, suggest an anthropogenic accumulation of bones. In the case of large- and medium-sized animals, cut marks on upper appendicular elements indicate early access to flesh-bearing parts (Fig. 6.2b). Small sized animals, on the other hand, were scarce and preserved few traces of human activity (Fig. 6.2a), except for some defleshing marks on diaphyses and axial bones, and some dismembering marks on limbs, which could indicate primary human consumption.

Other Paleolithic Sites

Besides the sites analyzed here, other sites, such as Castillo and Amalda, can provide useful information for reconstructing subsistence practices during the Middle and Upper Paleolithic. At Castillo, zooarchaeological and taphonomic studies carried out by Dari (2003) and Landry and Burke (2006) concluded that humans were the main contributors of animals to the site. Both studies estimated that the occupation in Castillo during Mousterian level 20 and Aurignacian level 18 took place from the end of summer to the beginning of winter. Age profiles indicated a dominance of young adults in both levels. Both papers reported the taxonomic dominance of deer, followed by chamois, *Bos/Bison* and horse (Dari 2003; Landry and Burke 2006).

According to the information available for the Middle and the Upper Paleolithic at this site, some persistence in human hunting behavior could be postulated. In both periods, animals from different environments were incorporated. A similar situation was reported for the nearby area,

Table 6.8 Main zooarchaeological and taphonomic features in Cuco and Cofresnedo

	Cervus NISP	*Cap* + *Rup* + *Capreo* NISP	*Equus* NISP	*Bos/Bison* NISP	Carnivores NISP	Cut mark (%)		Tooth mark (%)		Percussion mark (%)	
						Small	Large	Small	Large	Small	Large
Cuco											
Grav	27	9	3	34	2	50	3.1	0	2.9	0	2.1
Aur	115	14	1	14	18	14	21.2	9.5	10.1	3.6	7.3
Cofresnedo											
NISP	42	27	28	72	4	28.0	20.5	7	6.4	9.3	3.5

Frequencies of cut and tooth marks are determined from NISP, excluding teeth and badly preserved bones. Both large (*Bos* + *Equus* + *Cervus* + medium and large sized indet.) and small sized animals (*Capra* + *Rupicapra* + *Capreolus* + small sized ungulates indet.) are included

such as the Mousterian site of Flecha, with a dominance of deer supplemented by other taxa like horses, bovids, chamois and ibex.

Amalda is another cave with evidence for persistence in subsistence strategies between the Middle and Upper Palaeolithic. The record from the Gravettian and Mousterian levels suggested that humans focused on the exploitation of cervids, equids and large bovids, and used the site seasonally, between summer and autumn (Yravedra 2005, 2006a, 2007).

Lezetxiki lacks taphonomic information relating to bone assemblage formation. Taxonomic representation was similar in Mousterian level 5 and Aurignacian level 4; deer dominates, accounting for 63 % of the remains in level 5 and 54 % in level 4. Large bovids follow, comprising 25 % of the assemblage in unit 5 and 18 % in unit 4, supplemented by ibex and chamois (Altuna 1972).

Due to the absence of taphonomic data for the remaining Middle and Upper Paleolithic sites in the area that preserve faunal remains (i.e. Labekokoba, Axlor, Arrillor, Atxagorta, Bolinkoba, Arenillas, Rascaño, etc.), it is impossible to determine the role of human activity in the formation of these assemblages. My reluctance to incorporate sites lacking taphonomic analysis is soundly justified by examples like the Mousterian sites of Moros de Gabasa (Blasco 1995, 1997) or Abauntz (Altuna et al. 2002), where different species were accumulated by different agents, with human responsible for the introduction of deer or horse, whereas carnivores gathered ibexs and chamois (Blasco 1995, 1997; Altuna et al. 2002). In other sites with the same chronology, such as Cova Beneito (Valencia), a similar situation is mentioned (Villaverde and Martínez Valle 1995). Furthermore, at sites such as Boquete de Zafaraya (Gerads 1997) or Grotte de Buraca Escura (Aubry et al. 2001), in which ibex are common, the presence of this species may be associated with carnivore consumption (lynx in the latter case).

Nevertheless, the aforementioned examples do not imply that carnivore agency for small-sized prey and human introduction of medium-sized prey is a universal feature. Counterexamples include sites such as Esquilleu (Yravedra 2005) or Valdegoba (Díez 2006), where the available taphonomic evidence indicates an anthropogenic role in the accumulation of small ungulates such as ibex or chamois.

Discussion

Summing up, the data presented in this paper show a persistence of hunting activities in both the Middle and Upper Paleolithic (Figs. 6.3, 6.4, 6.6). This is evidenced in prey selection, particularly in the Mousterian and the Aurignacian-Gravettian, where medium- and large-sized ungulates such as horse, deer and aurochs dominate the record. This situation tends to change in the Solutrean, when the human use of small ungulates becomes frequent, as seen in the cut and percussion mark data from Amalda, Castillo, Ruso, Cobalejos, Hornos de la Peña and Morín (Dari 2003; Landry and Burke 2006; Yravedra 2006a, 2007, 2009, 2010a, b; Yravedra et al. 2010).

Although the possibility that human were utilizing small ungulates cannot be dismissed for the earlier periods under consideration, the available evidence does not support human interaction with small ungulates before the end of the Upper Paleolithic. Taphonomic analyses indicated that carnivores were the main agent responsible for the accumulation of small ungulates. These predators would probably have used these caves as shelter when humans were absent and introduced ibex and chamois carcasses at these times. It further suggests that human occupations in the area during the Middle and Upper Paleolithic were rather short, facilitating an alternating occupation of the sites by human and carnivores. Other sites in the Iberian Peninsula with Middle and Upper Paleolithic levels show a similar pattern. At Amalda, human agency was focused on large-sized ungulates whereas carnivores consumed ibex and chamois (Yravedra 2007, 2009, 2010a, b). In Moros de Gabasa (Blasco 1995, 1997) and at Cova Beneito (Villaverde and Martínez Valle 1995), human activity was identified on deer and horse bones, with hyena and wolf tooth marks on ibex. Similarly, ibex in Grotte de Buraca Escura were introduced by lynx, while medium-sized ungulates were introduced by humans (Aubry et al. 2001).

Nevertheless, this change of focus from large-sized prey to smaller prey demands further explanation. We cannot enlarge on this topic here due to its complexity, but a number of hypotheses may be postulated: (1) Neanderthals may have lacked the technology and knowledge needed to hunt small-size ungulates; (2) human populations may have preferred different resources at different times; (3) territorial strategies may have changed from a selection of valleys and plains to the incorporation of the rocky and steep environments where chamois and ibex lived.

Briefly, we may easily discard the first hypothesis due to the large number of caves with evidence of Neanderthals hunting ibex and chamois. For instance, at Esquilleu (Yravedra 2005, 2006b), Valdegoba (Díez 2006), Covalejos, and Morín (Yravedra 2005, 2006b, 2011; Yravedra and Gómez-Castanedo 2010a) small ungulates were introduced by humans during the Middle Paleolithic.

Unfortunately, there is not yet enough data to evaluate the second and third hypotheses. The patterning could be a result of a number of variables, such as the reduced meat content of smaller-sized ungulates. Another factor may be the difficulties posed by the environments in which these animals lived. Future taphonomic studies in other contexts, alongside detailed topographic analyses of the site region—including sites located in different ecological zones—would probably help to address these issues.

Final Remarks

An accurate analysis of subsistence strategies requires the incorporation of data from a number of disciplines. In this paper, zooarchaeological and taphonomic analyses of faunal remains from a series of sites were discussed in order to explore some aspects of human subsistence across the Middle-Upper Paleolithic transition in northern Spain. From the new and revised data presented here, we can postulate the persistence of subsistence strategies across the transition. In all of the sites studied, the same prey types were hunted in both periods. Moreover, the sites were always inhabited for short time periods, being used by carnivores during periods of abandonment. The rise of the Solutrean brought a change in this behavior, with the incorporation of smaller-sized ungulates and the use of sites for longer periods of time.

Acknowledgments I thank Jamie Clark and John D. Speth for the invitation to participate in Session S5-4. Variability in human hunting behavior during Oxygen Isotope Stages (OIS) 4/3: implications for understanding modern human origins in ICAZ 2010. I thank Raúl Gutierrez, Pedro Fernández Vega and MUPAC (Museo Provincial de Arqueología de Cantabria), Javier Baena (Director of the Esquilleu project) and Jesus Altuna, Koro Mariexcurrena and the Sociedad de Ciencias de Aranzadi for assistance during the completion of our study. I appreciate the comments made by Manuel Domínguez-Rodrigo on the drawing of Fig. 6.2, and the translation made by Aixa Vidal. I finally want to acknowledge the suggestions made by three anonymous reviewers.

References

Altuna, J. (1971). Fauna de Mamiferos de los Yacimiento Prehistórico de Morín, Santander. In J. González Echegaray & L. Freeman (Eds.), *La Cueva Morin excavaciones 1966–68* (pp. 367–398). Santander (España): Publicaciones Patronato Cuevas Prehistóricos de Santander VI.

Altuna, J. (1972). *Fauna de Mamíferos de los Yacimiento Prehistórico de Guipuzcua* (pp. 1–404). XXIV: Munibe.

Altuna, J. (1973). Fauna de Mamiferos de la Cueva Morín, Santander. In J. González Echegaray & L. Freeman (Eds.), *La Cueva Morin excavaciones 1969* (pp. 281–292). Santander: Publicaciones Patronato Cuevas Prehistóricos de Santander X.

Altuna, J. (1989). Subsistencia d'origene animale pendant le Mousterien dans la Región Cantabrique (Spagne). In M. Otte. (Ed.), *L'Homme de Néandertal vol. 6, Le Subsistence* (pp. 31–44). Liege.

Altuna, J. (1990). Caza y alimentación procedente de macromamíferos durante el Paleolítico de Amalda. In J. Altuna, A. Baleron, & R. Mariezkurrena (Eds.), *La cueva de Amalda (P. Vasco) ocupaciones paleolíticas y postpaleolíticas* (pp. 149–192). Sociedad de Estudios Vascos serie B4. Bilbao: España.

Altuna, J. (1992). Le Paleolithique Moyen de la Region Cantábrique. *L'Antropologie, 96*(1), 87–102.

Altuna, J., & Mariezkurrena, K. (2010). Tafocenosis en yacimientos del País vasco con predominio de grandes carnívoros. *Consideraciones sobre el yacimiento de Amalda. Zona Arqueológica, 13,* 214–228.

Altuna, J., Mariezkurrena, K., & Elorza, M. (2002). Arqueología de los animales paleolíticos de la cueva de Abauntz (Arraiz Navarra). *Saldvie, II,* 1–26.

Arrizabalaga, A., Altuna, J., Areso P., Falgueres C., Iriarte, M. J., Mariezkurrena, K., et al. (2005). Retorno a Lezetxiki (Arrasate, País Vasco): Nuevas perspectivas de la investigación. In M. Santonja, A. Pérez González, & M. J. Machado (Eds.), *Geoarqueología y patrimonio en la Península Ibérica y el entorno mediterráneo* (pp. 81–98). Almazan, Soria: ADEMA.

Aubry, T., Brugal, J. P., Chauviere, P. X., Figueiral, I., Moura, M. H., & Plisson, H. (2001). Modalités d'occupations au Paleolithique Supérieur dans la Grotte de Buraca Escura (Rendinha, Pombal, Portugal). *Revista Portuguesa de Arqueologia, 4*(2), 19–47.

Baena, J., Carrión, E., Ruiz, B., Sesé C., Yravedra, J., & Jordá, J. (2005). Paleoecología y comportamiento humano durante el Pleistoceno Superior en la comarca de Liébana: La secuencia de la Cueva del Esquilleu (Occidente de Cantabria, España). In R. Montes & M. Laceras (Eds.), *Neandertales cantábricos estado de la cuestión. El Paleolítico Medio cantábrico* (pp. 461–487). Santillana del Mar: Monografías del Museo Nacional y Centro de Investigación de Altamira n° 20.

Barba, R., & Domínguez-Rodrigo, M. (2005). The taphonomic relevance of the analysis of bovid long limb bone shaft features and their application to element identification: Study of bone thickness and morphology of the medullary cavity. *Journal of Taphonomy, 3*(1), 29–42.

Bernaldo de Quirós, F. (1980). *Notas sobre la economía del Paleolítico Superior Cantábrico. Centro de Investigaciones y Museo de Altamira.* Santander: Memorias.

Bernaldo de Quirós, F. (1982). The early upper Palaeolithic in Cantabria Spain. *Eraul 13*(2), 65–78 (Nitra, Liege).

Binford, L. R. (1981). *Bones: Ancient men, modern myths*. New York: Academic Press.

Blasco, M. F. (1995). *Hombres y Fieras. Estudio Zooarqueológico y Tafonómico del Yacimiento del Paleolítico Medio de la Cueva de Gabasa 1 (Huesca).* Zaragoza: Monografías de la Universidad de Zaragoza.

Blasco, M. F. (1997). Cave site of Gabasa in the Spenish Pyrinees. *Journal of Anthropological Research, 53*, 177–218.

Blumenschine, R. (1986). *Early hominid scavenging opportunities. Implications of carcass availability in the Serengeti and Ngrongoro ecosystems.* B.A.R. International Series 283. Oxford: Archaeopress.

Blumenschine, R. (1988). An experimental model of the timing of hominid and carnivore influence on archaeological bone assemblages. *Journal of Archaeological Science, 15*, 483–502.

Blumenschine, R. J., & Selvaggio, M. M. (1988). Percussion marks on bone surfaces as a new diagnostic of hominid behaviour. *Nature, 333*, 763–765.

Blumenschine, R. (1995). Percussion marks, tooth marks and the experimental determinations of the timing of hominid and carnivore access to long bones at FLK Zinjanthropus, Olduvai Gorge, Tanzania. *Journal of Human Evolution, 29*, 21–51.

Brain, C. K. (1969). The contribution of Namib desert Hottentot to understanding of Australopithecus bone accumulations. *Scientific Papers in Namibian Desert Research Station, 32*, 1–11.

Breuil, H., & Obermaier, H. (1912). Les premiers travaux de l'Institut de Paléontologie Humaine. *L'Anthropologie,* XXIII, 1–27.

Bunn, H. (1982). *Meat-eating and human evolution: Studies on the diet and subsistence patterns of Plio-Pleistocene hominids in East Africa.* Ph.D. dissertation, University of California, Berkeley.

Castaños, P. (2003). Estudio arqueozoológico de la fauna de Cofresnedo. In J. Ruiz Cobo & P. Smith (Eds.), *La Cueva de Cofresnedo en el Valle de Matienzo. Actuaciones Arquológicas 1996–2001* (pp. 118–121). Consejería de Cultura, Turismo y Deporte, Santander: Cantabria.

Castaños, P. (2005). Revisión actualizada de las faunas de macromamíferos del Würn antiguo en la Región cantábrica. In R. Montes & J. A. Laceras (Eds.), *Neandertales cantábricos estado de la cuestión. El Paleolítico Medio cantábrico* (pp. 201–207). Santillana del Mar: Monografías del Museo Nacional y Centro de Investigación de Altamira n° 20.

Castaños, P., & Castaños, J. (2007). Estudio de la fauna de la Cueva del Cuco. In E. Muñoz Fernández & R. Montes (Eds.), *Intervenciones Arqueológicas en Castro Urdiales. Arqueología y Arte rupestre Paleolítico en las cavidades de el Cuco o Sobera y la Lastrillla* (pp. 161–171). Excelentísimo Ayuntamiento de Castro Urdiales. Cantabria: España.

Coutourier, M. A. (1962). Le bouquetis des Alpes. Grenoble: chez l'auteur.

Dari, A. (2003). *Comportement de subsistence pendant le transition paaleolithique moyen- Paleolithique superieur en Cantabria à partir de l'etude archaeozooologique des restes osseaux des grands mamiferes de la Grotte d'el Castillo (Espagne).* Ph.D. dissertation, Museun Nationale D'histoire Naturalle, Paris.

Díez, J. C. (2006). Huellas de descarnado en el Paleolítico Medio: La cueva de Valdegoba (Burgos). *Zona Arqueológica, 7*(1), 304–317.

Domínguez-Rodrigo, M. (1997). Meat eating by early hominids at FLK Zinj 22 Site, Olduvai Gorge Tanzania: An experimental approach using cut-mark data. *Journal of Human Evolution, 33*, 669–690.

Domínguez-Rodrigo, M., & Barba, R. (2005). A study of cut marks on small-sized carcasses and its application to the study of cut marked bones from small mammals at the FLK Zinj site. *Journal of Taphonomy, 3*(3), 121–134.

Fernandez, H. (2001). *Osteologie comparee des petties ruminants eurasiatiques sauvages et domestiques (genres Rupicapra, Ovis, Capra et Capreolus): diagnose differentialle du squelette apendiculaire.* Universite de Geneva, Facultat de Ciencies.

Freeman, L. G. (1973). The significance of mammalians faunas from Palaeolithic occupations in Cantabrian Spain. *American Antiquity, 38*, 3–44.

Fuentes, C. (1980). Estudio de la Fauna del Pendo en González Echegaray, El Yacimiento de la Cueva del Pendo Exc. 1953.57. *Biblioteca Prehistórica Hispana, 17*, 215–238.

Gerads, D. (1997). La grande faune associee aux desniers neandertaliens de Zafarraya (Andalusie Espagne). Systematique et essai de interpretation. *C. R. Acad. Sci de Paris Sciencies de la Terre et des Planetes/Earth y Plantary Sciencies, 325*, 725–731.

González Echegaray, J., & Freeman, L. G. (1978). *Vida y Muerte en Cueva Morin.* Santander (España): Instituto de Cultura Cantábrica.

González Echegaray, J., García Guinea, M. A., & Begines Ramirez, M.A. (1966). *Cueva del Otero Servicio Nacional de Excavaciones Arqueológicas,* Madrid: España.

Klein, R., & Cruz Uribe, K. (1994). The Palaeolithic mammalian fauna from the 1910–14 excavations at El Castillo cave (Cantabria). *Museo y Centro de Investigaciones de Altamira, 17*, 141–158.

Landry, G., & Burke, A. (2006). El Castillo: The Obermaier faunal collection. *Zona Arqueológica, 7*, 104–113.

Lavocat, R. (1966). *Faunes et Flores prehistoriques de L'Europe Occidentale*. Paris: Collection L'homme et ses Origenes.

Lupo, K. D., & O'Connell, J. F. (2002). Cut and tooth mark distributions on large animal bones: Ethnoarchaeological data from the Hadza and their implications for current ideas about early human carnivory. *Journal of Archaeological Science, 29*, 85–109.

Lyman, R. L. (1994). *Vertebrate taphonomy*. Cambridge: Cambridge University Press.

Maillo, J. M., Valladas, H., Cabrera, V., & Bernardo de Quirós, F. (2001). Nuevas dataciones para el Paleolítico superior de Cueva Morín (Villanueva, Villaescusa, Cantabria). *Espacio, Tiempo y Forma,* 14, 145–150.

Mallye, J. B., Boudai, M., Costamagno, S., Mourre, V., Prucca, A., Thiebaut, C., et al. (2010). La faune de la Grotte du Noissier (Fréchet-Aure, Hautes Pyrénées): Une histoire taphonomique complexe. In J. D. Vigne, M. E. Patou-Mathis, & C. Lefèvre (Eds.), *Session 7-1. New Perspectives in Taphonomy. ICAZ 2010.* 11 International Conference of Archaeozoology.

Marín-Arroyo, A. B., Fosse, P., & Vigne, J. D. (2009). Probable evidences of bone accumulation by Pleistocene bearded vulture at the archaeological site of El Mirón Cave (Spain). *Journal of Archaeological Science, 36*, 284–296.

Martínez, J. (1998). *El modo de vida Neandertal: Una reflexión en torno a la ambigüedad en la interpretación de la subsistencia durante el Paleolítico Medio Cantábrico.* Ph.D. dissertation, Universidad Autónoma de Barcelona, España.

Montes R., Sanguino J., Martín P., Gómez A. J., & Morcillo C. (2005). La secuencia estratigráfica de la Cueva del Pendo (Escobedo de Camargo, Cantabria): Problemas geoarqueológicos de un referente cronocultural. In M. Santonja, A. Pérez González, & M. J. Machado (Eds.), *Geoarqueología y patrimonio en la Península Ibérica y el entorno mediterráneo* (pp. 139–159). Almazán Soria, España: ADEMA.

Muñoz Fernández, E. (1991). Excavaciones arqueológicas en la Cueva del Ruso I. Avance preliminar. *Arquenas, 1*, 1–157.

Muñoz Fernández, E., & Montes, R. (2007). *Intervenciones Arqueológicas en Castro Urdiales. Arqueología y Arte rupestre*

Paleolítico en las cavidades de el Cuco o Sobera y la Lastrillla. Cantabria: Excelentísimo Ayuntamiento de Castro Urdiales.

Muñoz Fernández, E., & Serna, A. (1994). *Los niveles solutrenses de la Cueva del Ruso I* (pp. 161–176). VI: Sautuola.

Pales, L., & Lambert, C. (1971). *Atlas ostéologique pour servir à la identification des mammifères du quaternaire*. Paris: Institut de Paleontologie Humaine.

Pérez Ripoll, M. (1988). Estudio de la secuencia del desgaste de los molares de Capra pyrenaica de los yacimientos prehistóricos. *Archivo de Prehistoria Levantina, 18*, 83–128.

Pike Tay, A., Cabrera, V., & Bernaldo de Quirós, F. (1999). Seasonal variation of the Middle- Upper Palaeolithic transition at El Castillo, Cueva Morin and El Pendo (Cantabria Spain). *Journal of Human Evolution*, 36, 283–317.

Potts, R., & Shipman, P. (1981). Cutmarks made by stone tools from Olduvai Gorge, Tanzania. *Nature, 291*, 577–580.

Ruiz, J., & Smith, P. (2003). *La Cueva de Cofresnedo en el Valle de Matienzo. Actuaciones Arquológicas 1996–2001*. Santander: Gobierno de Cantabria.

Sanguino, J., & Montes, R. (2005). Nuevos datos para el conocimiento del Paleolítico Medio en el centro de la región cantábrica: La cueva de Covalejos. In R. Montes & J. M. Laceras (Eds.), *Neandertales cantábricos estado de la cuestión. El Paleolítico Medio cantábrico* (pp. 443–460). Santillana del Mar: Monografías del Museo Nacional y Centro de Investigación de Altamira n° 20.

Soto Barreiro, M. J. (2003). *Cronología, radiometría, ecología y clima del Paleolítico cantábrico*. Santillana del Mar, Santander: Museo Nacional y Centro de Investigación de Altamira.

Steele, T. E. (2002). *Red deer: Their ecology and how they were hunted by late Pleistocene Hominids in Western Europe*. Ph.D. dissertation, Stanford University, Stanford.

Straus, L. G. (1976). Análisis de la fauna arqueológico del norte de la Península Ibérica. *Munibe, 28*, 277–285.

Straus, L. G. (1977). Of deer slayers and mountain men: Palaeolithic faunal exploitation in Cantabrian Spain. In L. Binford (Ed.), *For theory building in archaeology* (pp. 41–78). New York: Academic Press.

Straus, L. G. (1982). Carnivores and cave sites in Cantabrian Spain. *Journal of Anthropological Research, 1*, 75–96.

Straus, L. G. (1992). *Iberia before the Iberians. The Stone Age Prehistory of Cantabrian Spain*. Albuquerque: University of New México Press.

Villaverde, V., & Martínez Valle, R. (1995). Características culturales y económicas del final del Paleolítico superior en el Mediterráneo español. In V. Villaverde (Ed.), *Los últimos cazadores: transformaciones culturales y económicas durante el Tardiglaciar y el inicio del Holoceno en el ámbito mediterráneo* (pp. 79–118). Instituto Alicantino Juan Gil-Albert. España.

Yravedra, J. (2002a). Implicaciones Tafonómicas en el debate de la caza y el carroñeo en el Paleolítico de la Península Ibérica. *Nivel Cero, 10*, 7–22.

Yravedra, J. (2002b). Subsistencia en la transición del Paleolítico Medio al Paleolítico Superior de la Península Ibérica. *Trabajos de Prehistoria, 59*(1), 9–29.

Yravedra, J. (2005). *Patrones de aprovechamiento de recursos animales en el Pleistoceno Superior de la Península Ibérica: estudio tafonómico y zooarqueológico de los yacimientos del Esquilleu, Amalda, cueva Ambrosio y la Peña de Estebanvela.* Ph.D. dissertation, Universidad Nacional de Educación a Distancia, Madrid.

Yravedra, J. (2006a). Acumulaciones biológicas en yacimientos arqueológicos: Amalda VII y Esquilleu III-IV. *Trabajos de Prehistoria, 62*(2), 55–78.

Yravedra, J. (2006b). Aportes biológicos y antrópicos en la Cueva del Esquilleu (Cantabria, España). *Zona Arqueológica,* 7(1), 280–289.

Yravedra, J. (2007). Nuevas contribuciones en el comportamiento cinegético de Amalda. *Munibe, 58*, 43–88.

Yravedra, J. (2009). Tafonomía en la Cueva de Amalda. La intervención de carnívoros. *Zona Arqueológica, 13*, 144–154.

Yravedra, J. (2010a). A taphonomic perspective on the origins of the faunal remains from Amalda Cave (Spain). *Journal of Taphonomy, 8*(4), 301–334.

Yravedra, J. (2010b). Zooarqueología y tafonomía del yacimiento de Hornos de la Peña (San Felices de Buelna, Cantabria). *Complutum, 21*, 69–86.

Yravedra, J. (2011). *Análisis tafonómico de los macromamíferos del yacimiento paleolítico de Covalejos (Velo de los Piélagos, Cantabria, España).*

Yravedra, J., & Domínguez-Rodrigo, M. (2009). The shaft-based methodological approach to the quantification of long limb bones and its relevance to understanding hominid subsistence in the Pleistocene: Application to four Palaeolithic sites. *Journal of Quaternary Science, 24*(1), 85–96.

Yravedra, J., & Gómez-Castanedo, A. (2010a). Tafonomía en Cueva Morín. Resultados preliminares de un estudio necesario. *Zephirus, LXVII*, 69–90.

Yravedra, J., & Gómez-Castanedo, A. (2010b). Zooarqueología y Tafonomía de un viejo conocido la Cueva del Otero. *Espacio, Tiempo y Forma, 10*(3), 21–38.

Yravedra, J., Baena, J., Arrizabalaga, A., & Iriarte M. J. (2005). El empleo de material óseo como combustible durante el Paleolítico Medio y Superior en el Cantábrico. Observaciones experimentales. In R. Montes & J. M. Laceras (Eds.), *Neandertales cantábricos estado de la cuestión. El Paleolítico Medio cantábrico* (pp. 369–383). Santillana del Mar: Monografías del Museo Nacional y Centro de Investigación de Altamira n° 20.

Yravedra, J., Muñoz, E., & Gómez-Castanedo, A. (2010). Estrategias de Subsistencia en el yacimiento del Ruso (Igollo, Camargo, Cantabria España). *Espacio, Tiempo y Forma, 10*(3), 39–58.

Chapter 7
Iberian Archaeofaunas and Hominin Subsistence during Marine Isotope Stages 4 and 3

Lawrence Guy Straus

Preamble

What follows is the result of a fairly exhaustive synthesis of the available archaeofaunal record for Iberia done by a prehistoric archaeologist who is a "consumer" of such information. The author, while he produces faunal data by excavating and collaborates with archaeozoologists, is not a faunal specialist himself. This review seeks to gather into one place the data as they stand published in 2010–2011 in a variety of forms (at best NISP and/or MNI), in diverse publication outlets (some very hard to access outside Spain and Portugal), and representing stratigraphic units ("levels") of very widely divergent nature, albeit mainly from more-or-less modern excavations. It is extremely difficult to statistically compare faunal "assemblages" from different sites in whose excavations different criteria for defining "levels" may have been used and different methods (wet vs. dry) and mesh grades of screening, recovery and curation, undoubtedly were applied, and for the study of whose faunas different standards may have been used for identification (e.g., what was defined as being "unidentifiable"?) and quantification, as well as for "assemblage" creation (i.e., the lumping of finds from stratigraphic entities that may have represented palimpsests of greater or lesser temporal formation magnitude). Thus, these data sets are presented (hopefully for further—albeit cautious—manipulation by archaeozoological specialists) in an effort to expose the known facts to a wider audience and to suggest some broad, apparent trends. Such tentative conclusions are based on global—and only semi-quantitative—comparisons at the level of major blocks of cultural time (i.e., early and late Middle Paleolithic, Early Upper Paleolithic [Châtelperronian and Aurignacian] and Gravettian), covering the period before, during and immediately after the so-called Middle-Upper Paleolithic transition. The author has resisted reviewer suggestions to do further quantitative analysis and the interpretations have been kept modest. Thus the intention of this paper is to contribute facts (admittedly of unequal value) to the ongoing debate on the nature of this supposed cultural-adaptive revolution, with hints of both continuity and change in subsistence that must be interpreted in light of the Iberian environments and demography, as well as the possible dietary needs and capacities of the different hominin populations that likely were involved.

Introduction

The Iberian Peninsula was one of the last places where Neanderthals managed to survive (until sometime around 30–27 ka) and it was also a subcontinental region where early "Aurignacian" artifact assemblages seem to have been coeval with late "Mousterian" ones after ca. 40 ka. Current debates revolve around such issues as the exact timing of the extinction of the last Neanderthals, especially in Gibraltar, Andalucía and southern Portugal, the possible existence of late Mousterian "enclaves" in northeastern and north-central Spain, the hypothesis of an in situ development of an initial Aurignacian from the terminal Mousterian (notably at El Castillo Cave in Cantabria) (Cabrera et al. 2006, with references), and the controversial idea that certain anatomical traits of the Lagar Velho (Portugal) juvenile skeleton in a burial of Gravettian style with a radiocarbon age of 25 ^{14}C kBP may be suggestive of the presence of Neanderthal genes among anatomically modern humans in Iberia (despite the notable lack of *any* hominin finds associated with the Iberian Aurignacian) (Zilhão and Trinkaus 2002). This possibility has recently been strengthened by the finding of small percentages of Neanderthal genes among modern Eurasians (Green et al. 2010). In this context, interesting arguments are being made about, on the one

L. G. Straus (✉)
Department of Anthropology, University of New Mexico, Albuquerque, NM 87131, USA
e-mail: lstraus@unm.edu

Jamie L. Clark and John D. Speth (eds.), *Zooarchaeology and Modern Human Origins: Human Hunting Behavior during the Later Pleistocene*, Vertebrate Paleobiology and Paleoanthropology, DOI: 10.1007/978-94-007-6766-9_7,

hand, whether or not there were special ecological conditions in southern Iberia that favored the survival of the Neanderthals (e.g., Finlayson et al. 2006; d'Errico and Sanchez-Goñi 2003) until either environments or subsistence strategies may have caused faunal changes; and, on the other hand, whether or not there was some critical niche separation between the Iberian Neanderthals and the (putative) modern humans involving a nutritionally mediated reproductive advantage for the latter that ultimately allowed them to out-compete and replace the Neanderthals (Hockett and Haws 2005).

Much hinges on whether one can find evidence for significant differences in subsistence between the two populations that would directly or indirectly lead to the final success of "moderns" at least by Gravettian times and the demise of the "archaics". Put simply, the question here is whether or not there was a sharp break in human subsistence between the Mousterian cultures of Neanderthals in Marine Isotope Stage (MIS) 4 and early MIS 3 and those of the Early Upper Paleolithic (various facies of Aurignacian–hominin makers unknown in Iberia—plus Gravettian) in late MIS 3. The null hypothesis is that there was substantial continuity, as has been convincingly demonstrated for southwestern France by Grayson and Delpech (2002, 2003, 2006, 2008).

In examining this question, it is important to filter out aspects of change in the faunal record that might be due to evolutionary trends for which any hominin involvement is likely to have had no more than incidental impact. It is also important to keep in mind the role of raptors, bears and other carnivores in the creation and modification of ungulate and lagomorph bone assemblages in a site sample that is entirely composed of caves, especially in the earlier time periods under consideration, when the human presence and role was arguably relatively minor in many cases. A survey of the faunal evidence across the so-called Middle-Upper Paleolithic transition in Iberia is seriously hampered by the paucity of analyzed assemblages demonstrably dating to MIS 4 and early MIS 3, as well as by the geographically uneven distribution of assemblages even from later MIS 3. In many cases, there are so few sites with analyzed faunal assemblages that, if one or a small number thereof happens to be heavily dominated by a single species, the pooled faunal spectrum for the entire region and time period may be totally skewed, falsely giving the impression of regional subsistence specialization. (A case in point is the large quantity of chamois in Amalda Cave, the agency of whose deposition—hominin or felid—is current under debate.)

Although a credible argument can be mounted for human settlement of the high, relatively hostile environments of the Iberian interior during much of the Upper Paleolithic, this was clearly not always the case during the Middle Paleolithic. But studied faunal assemblages from those regions are simply absent at least at present, so the record for all periods under consideration here is essentially a peripheral one, based on sites that are coastal or peri-coastal, generally no more than a few tens of kilometers from the present (interglacial) shore. It is also a record that is very uneven. Although most of the faunal collections included here are from relatively recent excavations, surface areas or sediment volumes dug and methodologies of recovery and analysis all vary widely (some excavations were small pits, others large blocks). The degrees of expertise and effort of the many archaeozoologists/paleontologists who analyzed the collections reported here clearly varied, as did their methodologies. These facts lead to problems of inter-site comparability when comparing faunal assemblages.

I have made a modest attempt to standardize taxonomic names where possible (e.g., among the Rhinocerotidae and Equidae), opting for the nomenclature most generally used at present in the Iberian Peninsula. At some sites, many remains were only identified to family level, although one can generally assume that "Cervidae" mainly means red deer and "Capridae", mainly ibex. The basic faunal data (derived from the primary references for each site, as listed below and in the References), on which the tables and discussion are based, are given in Appendices 7.1, 7.2, 7.3 and 7.4.

MIS 3 and 4 Paleoenvironments

The Neanderthals evolved in Europe from *Homo heidelbergensis* in the period between ca. 300 and 200 ka and they survived the extreme environmental vicissitudes of the late Middle and early Late Pleistocene, including major, extreme glacials in MIS 8 and 6. MIS 5 was a generally warm, but highly variable interglacial that, after a long, saw-tooth cooling trend, terminated in a short and moderately cold glacial, MIS 4, between ca. 71 and 57 ka. MIS 3 was neither a classic glacial nor a full-fledged interglacial, but rather a highly variable interstadial, the last millennia of which were characterized by the onset at ca. 25 ka of a sharp climatic downturn that culminated in the Last Glacial Maximum early in MIS 2. The Neanderthals disappeared at the beginning of that downturn. A recent synthesis of paleobotanical evidence from Spain and Portugal by González-Sampériz et al. (2010, with extensive references; but see d'Errico and Sanchez-Goñi 2003 for alternative views on vegetation reconstruction) serves as the basis for the following characterization of MIS 4 and 3 vegetation in the peripheral regions of the Peninsula.

The narrow northern Atlantic strip of northwestern and north-central Spain (together with a small northwestern corner of Portugal and a thin band south of the Pyrenean crestline including extreme northeastern Catalonia) is the only region of the Iberian Peninsula that is part of the Eurosiberian biogeographic or ecological zone. From west to east, it consists of the modern cultural/administrative regions of Galicia (where there are no Middle or Early Upper Paleolithic sites with faunal remains), Asturias, Cantabria and Euskadi (Basque Country). The rest of Spain and (despite its bordering the Atlantic Ocean) most of Portugal (the southern and central regions being the only ones with sites of relevance to this discussion) fall within the Mediterranean biogeographic zone. This division also had a significant degree of relevance during Pleistocene times, both glacial and interglacial/interstadial.

Limited pollen evidence from Galicia shows the presence of heath and grasslands, but with both some deciduous trees (including several rather temperate taxa) and conifers during MIS 4, while charcoal from El Castillo Cave (Cantabria) revealed the presence of both Scots pine and birch and the more thermophile beech in the high-relief, north-coastal Atlantic region. MIS 3 witnessed pulses of woodland expansion in this region, punctuated by episodes of arboreal contraction into regional micro-refugia. The more wooded phases included the spread of deciduous forms of *Quercus* or birch, presumably depending on temperatures. Increases in deciduous tree and shrub taxa came at the expense of Scots pine. The downturn that led to MIS 2 saw increases in juniper and pine, as well as birch, but willow continued to grow along water courses. The vegetation was always a mosaic of open heath and grasslands and tree stands or woods of varying importance. This Atlantic oceanic region was always relatively humid even during colder phases, in striking contrast to much of the rest of the Peninsula.

For Portuguese Estremadura (south-central Portugal) we have very limited data only from MIS 3 (from wood charcoal at archaeological sites). They suggest the presence of open steppes and heaths, but with stands of deciduous oaks, as well as maritime or stone pine.

Turning to the Mediterranean regions of Spain, there is pollen evidence for MIS 4 and early MIS 3 for the northeast from the Abric Romaní (Barcelona, Catalonia). Pine dominates throughout, but always accompanied by juniper, birch and a variety of rather temperate deciduous trees. Arboreal pollen fluctuates between 40 and 60 %. MIS 3 evidence from L'Arbreda (Gerona) and Lake Banyoles confirms the substantial presence of trees, as well as a fluctuating extent of open vegetation in Catalonia. The evidence from several sites in Aragón (interior northeastern Spain's Ebro River Basin)—including Gabasa Cave (Huesca) indicates MIS 3 arboreal vegetation dominated by Scots pine and juniper, but with both evergreen and deciduous oaks, as well as a wide variety of other deciduous trees (including some quite temperate Mediterranean taxa such as olive). The vegetational mosaics of this period included varying amounts of steppe-like grasses, weeds and shrubs, including *Artemisia,* which often characterizes the cold, dry glacial phases in Mediterranean Spain. For the eastern and southern sectors (the regions of Valencia and Andalucía), the most substantial records are palynological and come from the Padul bog core and the cave of Carihuela, both in Granada, complemented by a few other natural and archaeological loci. MIS 4 shows the co-presence of both steppe plants and trees that are dominated by pines and junipers, but also include numerous more temperate taxa (various deciduous trees plus scrub taxa). The Mediterranean taxa clearly survived in refugia in these meridional, high-relief regions, despite the presence of cold, dry-loving plants. Such mosaics continued throughout MIS 3 (including in the record for the Cova Beneito archaeological site in Valencia, which also has wood charcoal assemblages, dominated by Scots pine and traces of juniper). In the far south, there is an overall trend for alternation between more steppe (with pockets of trees) and episodes of greater woodland cover, as attested in both pollen (mainly Padul and Carihuela) and charcoal records (including such key archaeological sites as Zafarraya, Bajondillo and Nerja in Málaga and Gorham's Cave in Gibraltar. "Disharmonious" floras—including both "cold" and "warm", non-Mediterranean and Mediterranean taxa—characterize the southern Spanish refugia during MIS 3 and 4. This is corroborated by pollen records from marine cores in both the Alborán Sea (extreme western Mediterranean) and Atlantic margin of Portugal. Apparently there were no dramatic shifts between MIS 3 and 4 in Mediterranean Spain or Portugal, and even in North Atlantic (Cantabrian) Spain the differences in vegetation were only matters of degree in terms of open versus wooded environments rather than a sharp reversal. However d'Errico and Sanchez-Goñi (2003) have argued that there was a significant increase in desert-steppe vegetation in southern Iberia during the Heinrich 4 event, possibly making this region unattractive to herd ungulate grazers and hence to modern humans, leaving Neanderthals in place at least temporarily.

The Early Mousterian Faunal Record

There are very few archaeofaunal assemblages that can be credibly argued to date to MIS 4. These may include Level III in Teixoneres Cave (see Fig. 7.1 for locations of main sites with faunal assemblages), 40 km north of the city of Barcelona (Rosell et al. 2010). The base of this stratum is dated by uranium-series to 94.6 ± 3.2 ka, but the top is unconstrained (capping flowstone Level I is 14–16 ka). The others are Levels VI and V in Cova Negra (Valencia),

Fig. 7.1 Map of the Iberian Peninsula showing the Eurosiberian and Mediterranean ecological zones, the principal sites mentioned in the text and relevant regions of Spain and Portugal. *1* Amalda, *2* Labeko, *3* Lezetxiki, *4* Axlor, *5* Arrillor, *6* Morín, *7* Venta Laperra, *8* El Pendo & Covalejos, *9* El Castillo, *11* Gabasa, *12* Teixoneres, *14* L'Arbreda, *16* Els Ermitons, *17* Romaní, *19* Cova Negra, *20* Les Mallaetes, *21* Beneito, *22* Zafarraya, *25* Gorham's & Vanguard, *26* Columbeira, *28* Figueira Brava, *29* Val Boi, *30* Picareiro, *31* Anecrial, *33* Caldeirão. All are caves except *17* (*rockshelter*) and *29* (*collapsed rockshelter & open-air talus slope*)

which are argued on stratigraphic and geological grounds to have been formed under cold conditions (with gelifraction) corresponding to traditional Würm II (Villaverde et al. 1996), and El Castillo Levels 22 and 21 with ESR dates of ca. 70 and 69 ka respectively (Dari 1999). The rather vague term "Early Mousterian" is used here as a rough proxy for levels probably formed during MIS 4 (Table 7.1).

The faunally richer Early Mousterian stratum at El Castillo (Level 22)—from Obermaier's pre-World War I excavation, with all the caveats entailed by the inclusion of such an old collection along with more modern ones from other sites—is overwhelmingly dominated by horse, followed by red deer, with a scattering of other taxa [including a single element of Merck's rhinoceros, as identified by R. Vaufrey and cited by Cabrera (1984) [see also Landry and Burke (2007) and Klein and Cruz-Uribe (1994) for analyses of Obermaier's collections], but probably misidentified as an unlikely hippopotamus by Dari (1999) and—*if* it is the same bone in the IPH collection (a radio-ulna)—re-identified as bovine by Stuart and Lister (2007: 289)]. The sole carnivores are traces of cave lion and wolf (Table 7.1). The Teixoneres ungulate assemblage is nearly evenly divided, in order of importance, between red deer and horse, followed by ass, bovines/aurochs. Cave bear is well represented, as is lynx, and there are traces of hyena, beaver and porcupine. The two assemblages from Cova Negra both have very small numbers of remains from as many as 10 ungulate taxa. Wolf and lynx are present (one element each) in Level VI and both assemblages have relatively large numbers of rabbit remains, whose human agency would have to be demonstrated. There is really nothing in these collections suggestive of particularly cold conditions or the quantitatively significant persistence of "archaic" faunas. Undated Middle Paleolithic levels (said only to be of Late Pleistocene age and which overlie late Acheulean levels) in Cueva Hora (Granada) are dominated by horse (albeit in small

Table 7.1 Relative frequencies of ungulates and carnivores and ratio of ungulates to carnivores, based on NISP for combined Early Middle Paleolithic assemblages in Vasco-Cantabria and Mediterranean Spain

	Mediterranean Spain		Vasco-Cantabria	
	NISP	%	NISP	%
Ungulates				
Sus	11	1.4	0	0.0
Cervus	55	6.9	238	15.2
Capreolus	7	0.9	3	0.2
Bovini	29	3.6	84	5.4
Rupicapra	46	5.8	45	2.9
Capra	265	33.3	3	0.2
Equus	269	33.8	1194	76.1
Rhinocerotidae	1	0.1	1	0.1
Other	112	14.1	0	0.0
Total ungulates	**795**	**100.0**	**1568**	**100.0**
Carnivores				
Canis	15	21.1	2	20.0
Ursus	26	36.6	1	10.0
Hyaenidae	15	21.1	0	0.0
Vulpes	0	0.0	0	0.0
Felis	1	1.4	0	0.0
Lynx spp.	7	9.9	0	0.0
Panthera spp.	1	1.4	7	70.0
Mustelidae	0	0.0	0	0.0
Other	6	8.5	0	0.0
Total carnivores	**71**	**100.0**	**10**	**100.0**
Total ungulates/total carnivores	**11.20**		**156.80**	

Table 7.2 Early Middle Paleolithic: lagomorph NISP counts from Mediterranean Spain sites of Teixoneres, Gabasa and Cova Negra; ratio of ungulates to lagomorphs based on NISP for Mediterranean Spain

	Mediterranean Spain			
	III Teixoneres	18a + b Gabasa	VI Cova Negra	V Cova Negra
Leporidae	18	0	0	0
Oryctolagus	0	647	75	41
Lepus	0	0	0	0
Total	**18**	**647**	**75**	**41**
Total ungulates/total lagomorphs	**1.017925736**			

numbers), followed by ass. There are even smaller numbers of red deer and ibex remains, also distributed widely among the levels. Finally there are a few, scattered remains of bovines, rhinoceros, wolf and lynx (Martín Penela 1986). Calculated globally, the Mediterranean Early Mousterian non-carnivore fauna spectrum is divided evenly (50/50 %) between ungulates and lagomorphs in NISP (Table 7.2).

The Late Mousterian Faunal Record

Late Mousterian levels (as defined here) pertain to the first 30 millennia of MIS 3. There are 20 assemblages from Vasco-Cantabria. Most are dominated by red deer remains, often followed by horse (Table 7.3; Altuna 1978, 1989; Altuna and Mariezkurrena 1988). One site, Amalda

Table 7.3 Relative frequencies of ungulates and carnivores and ratio of ungulates to carnivores, based on NISP for combined Late Middle Paleolithic assemblages in Vasco-Cantabria, Mediterranean Spain and Portugal

	Mediterranean Spain		Vasco-Cantabria		Portugal	
	NISP	%	NISP	%	NISP	%
Ungulates						
Sus	46	0.4	11	0.1	10	1.6
Cervus	3654	29.0	4847	53.0	317	51.4
Capreolus	58	0.5	133	1.5	2	0.3
Bovini	398	3.2	1763	19.3	41	6.6
Rupicapra	325	2.6	720	7.9	5	0.8
Capra	4899	38.9	751	8.2	135	21.9
Equus	2697	21.4	822	9.0	68	11.0
Rhinocerotidae	78	0.6	89	1.0	29	4.7
Other	425	3.4	5	0.1	10	1.6
Total ungulates	**12580**	**100.0**	**9141**	**100.0**	**617**	**100.0**
Carnivores						
Canis	281	18.5	31	13.1	6	3.7
Ursus	601	39.6	95	40.1	60	37.0
Hyaenidae	114	7.5	8	3.4	55	34.0
Vulpes	77	5.1	47	19.8	9	5.6
Felis	136	9.0	3	1.3	3	1.9
Lynx spp.	117	7.7	0	0.0	18	11.1
Panthera spp.	27	1.8	8	3.4	9	5.6
Mustelidae	9	0.6	0	0.0	0	0.0
Other	157	10.3	45	19.0	2	1.2
Total carnivores	**1519**	**100.0**	**237**	**100.0**	**162**	**100.0**
Total ungulates/total carnivores	**8.28**		**38.57**		**3.81**	

(in Guipúzcoa), has large numbers of chamois remains (Altuna 1990), but it has been argued that these may have been killed by carnivores (Yravedra 2007; *pace* Altuna and Mariezkurrena 2010), which are indeed very diverse and abundant in the level in question (VII). Other sites, Lezexiki (Level VI) and Morín (Level 17), are dominated by bovines, which are also very numerous in El Castillo Level 20 and Arrillor (Smk-I) (Castaños 2005; Martínez-Moreno 2005). Ibex is relatively common at Esquilleu (Cantabria), though the numbers are very small (Yravedra 2006). It is also dominant (and absolutely somewhat more abundant) in Venta Laperra, Axlor, Arrillor and Amalda. There are traces of rhinoceroses in several of the Late Mousterian assemblages (Covalejos, Morín, Arrillor, Axlor, Lezetxiki). Cave bear [possibly a facultative omnivore, although some (contested) stable isotope studies (e.g., Bocherens et al. 1994 vs. Hilderbrand et al. 1996; see Pacher and Stuart 2008 with references, for discussion) suggest it was mainly herbivorous] and numerous carnivores—notably wolf, fox and occasionally hyena and leopard—are also present in the Late Mousterian assemblages. There are very small numbers of hare remains in a few of the assemblages (Table 7.4); other rare small mammals include marmot.

The Late Mousterian of Eastern Spain (i.e., Aragon, Catalonia and Valencia) includes some assemblages that *might* pertain to late MIS 4 (i.e., Gabasa Levels g and h), but this is unproven so they are included here with MIS 3. All the Gabasa (Huesca, in pre-Pyrenean Aragón) levels have relatively large numbers of horse and ibex remains, and most (except the uppermost ones) have similar amounts of red deer (Blasco 1995). Chamois is constantly represented, but by only relatively small quantities. Very small amounts of boar, roe deer and aurochs are found in all levels, and traces of ass and rhinoceroses in most. Carnivores are fairly numerous and very diverse; they notably include wolf, hyena and lynx. The latter is interesting given the surprisingly high (for the Mousterian) representation of rabbit in all the levels. Based on taphonomic studies, including cut and gnaw mark analyses, it is likely that both hominins and carnivores (principally wolves and hyenas) were agents of accumulation and that carnivores also scavenged from

Table 7.4 Late Middle Paleolithic: lagomorph NISP counts from Cantabrian site of Esquilleu, Mediterranean Spain/Gibraltar sites of Zafarraya, Vanguard, Gorham's, Romaní, L'Arbreda, Ermitons, Beneito and Cova Negra, and Portuguese sites of Caldeirão and Figueira Brava; Ratio of ungulates to lagomorphs based on NISP for Vasco-Cantabria, Mediterranean Spain and Portugal

	Mediterranean Spain													
	Gabasa						L'Abreda		Beneito		Cova Negra			
	a + b	d	e	f	g	h	I	H	D2	D1	IV	IIIB	IIIA	II
Leporidae	0	0	0	0	0	0	0	0	0	0	0	0	0	0
Oryctolagus	647	277	211	489	453	81	0	0	169	113	368	337	94	151
Lepus	0	0	0	0	0	0	39	8	0	0	0	0	0	0
Total	**647**	**277**	**211**	**489**	**453**	**81**	**39**	**8**	**169**	**113**	**368**	**337**	**94**	**151**
Total Ungulates/total Lagomorphs	**3.65910413**													

	Vasco-Cantabria						Portugal	
	Esquielleu		Covalejos	Axlor	Arrillor		Caldeirão	F. Brava
	IV	III	J	B-C	Amk	Lmc	N-K	2
Leporidae	2	8	0	0	0	0	0	0
Oryctolagus	0	0	0	0	0	0	806	100
Lepus	0	0	5	1	1	1	0	0
Total	**2**	**8**	**5**	**1**	**1**	**1**	**806**	**100**
Total ungulates/total lagomorphs	**507.8333333**						**0.681015453**	

human-hunted carcasses (Blasco 1995). The seven early-mid MIS 3 Mousterian levels from Abric Romaní (Barcelona) whose ungulate faunal assemblages have been studied are all dominated by either red deer or horse (Rosell et al. 2012; Fernández-Laso et al. 2010; Cáceres et al. 1998). Aurochs is generally present, but usually in only moderate amounts. Small numbers of rhinoceros remains are found in three levels and traces of chamois in only two. The mammalian faunas from late Late Mousterian Levels I and H in L'Arbreda Cave (Gerona) seem to have been more thoroughly studied and thus include carnivores, mustelids and lagomorphs (Estevez 1987). Red deer, with only moderate numbers, dominate both levels. Horse and ass also are substantially present and bovines (probably aurochs) are found in one. There are a few proboscidean remains. Cave bear is very abundant in the lower level (I) and there are especially many other carnivores in this level (wolf, hyena, fox, wild cat, lion). There are small numbers of hare remains, but no rabbit, despite the Mediterranean environment. The Mousterian strata (VI and IV) of montane Els Ermitons Cave (Gerona) are lacking in red deer and mainly (and logically) contain ibex, though the quantities are small (Maroto et al. 1996). Cave bear is very abundant and there are several remains of hyena, wolf, fox, leopard and lynx, again suggesting an alternation in use of this cave by Neanderthals, bears and carnivores.

Further south in Mediterranean Spain, the Late Mousterian assemblages of Cova Beneito are dominated by ibex, followed by red deer and horse, but all the counts are rather small (Iturbe et al. 1993). Rabbit bones are relatively abundant, but carnivores are virtually absent. In the MIS 3 levels of Cova Negra (most of whose ungulate remains were only identified to genus or family level), thar (*Hemitragus*) and other caprines (probably mainly ibex), horse and cervids (probably mainly red deer) are dominant in that approximate descending order (Villaverde et al. 1996). There are also some bovines (probably aurochs). Remains of rabbits are very abundant, but carnivores again are scarce. There are traces of rhinoceros.

The Late Mousterian data from Andalucía are dominated by Zafarraya (Málaga), which has some complex problems of stratigraphic mixing at least in its upper layers (Barroso et al. 2006a, b). The site is located high on a very steep cliff-side, so not surprisingly its assemblages are dominated by ibex remains, with only small numbers of chamois and red deer, and occasionally some aurochs, plus traces of ass and horse in one level (UD). On the other hand, carnivores are numerous and diverse [abundant leopard (also represented by coprolites) and dhole, plus hyena, wildcat, lynx, mustelids and some fox, plus brown bear in most levels]. Both carnivores and Neanderthals may have been the agents of ungulate accumulation and they alternated their occupation of the cave in each stratigraphic layer (Barroso et al. 2006a). Rabbit remains are abundant (NISP = 7,309; MNI = 118), but they are reported globally, not by level, and they are said to have been accumulated mostly by small-medium carnivores and owls, not by humans, at least in the Mousterian strata (Barroso et al. 2006b). The small

Late Mousterian assemblage from Vanguard Cave (Gibraltar—a quintessentially steep, rocky habitat) is also dominated by ibex, followed by red deer from the then-dry coastal plain directly in front of the cave (Finlayson et al. 2006). There are also several carnivores and bear. Level IV in adjacent Gorham's Cave (ca. 28 ^{14}C kBP) also has abundant ibex remains followed by red deer (NISP = 205 and 89; MNI = 13 and 7 respectively), huge numbers (NISP = 1,620; MNI = 97) of rabbit bones (not mentioned for Vanguard, perhaps for lack at present of a published study). There are small numbers of other ungulates and a wide variety of carnivores (notably hyena and lynx) and bear (Rodríguez et al. 2010). Obviously, the relative roles of Neanderthals and carnivores in the accumulation of rabbits in each site will have to be determined by careful taphonomic studies. Indeed, much work remains to be done to sort out the relative roles of hominins, carnivores and raptors in the accumulation of the remains of other mammals, birds and fish in caves such as those of Gibraltar. For example, Ibex Cave, high on the east-facing cliff of "the Rock" has Mousterian stone tools associated with uncalibrated radiocarbon dates of 35–40 ^{14}C kBP and a mammalian fauna dominated by ibex remains, along with rabbits, red deer, wolf, birds, voles, etc., but, unlike the cases of Gorham's and Vanguard, there is "(n)o evidence of human activity on any of the… large, medium or small mammal remains" according to taphonomists Fernández-Jalvo and Andrews (2000: 174). Wolves, along with rock-falls, seem to have been the ibex assassins of Ibex Cave!

Presence/absence data from ≥54 ka (TL) Mousterian levels VII, VI a, VI (and possibly culturally poor level V) in Higueral de Valleja Cave (in the interior of Cadiz Province, north of Gibraltar) show a continuous presence of rabbit (also in the Gravettian and Solutrean levels). Red deer and horse are also ubiquitous and one Mousterian level each has traces of ibex, hare, wolf and wild cat (Jennings et al. 2009). Unfortunately there is no discussion of taphonomy in the publication, so one cannot judge whether the rabbits in any of the levels of this site were caught, processed or consumed by humans. Overall, the summed Mediterranean Late Mousterian non-carnivore faunas contain 79 % ungulate remains versus 21 % lagomorph ones (Table 7.4).

The Final Mousterian in Portugal has only three published faunal assemblages. Figueira Brava Cave is dominated by cervids (probably red deer), followed closely by ibex, with smaller equal numbers of aurochs and horse (Antunes 2000a). There are traces of boar, rhino and mammoth. Once again, carnivore remains are fairly numerous and diverse, including especially hyena. Rabbit remains are abundant and seem to have been consumed by Neanderthals in at least some cases. There are only four remains of land tortoise and two of pond tortoise in Figueira Brava (Antunes 2000b). Tortoises are not at all common in Iberian archaeofaunas—Middle or Upper Paleolithic—in contrast to some assemblages from the central and eastern Mediterranean basin studied by Stiner (Stiner 2001; Stiner et al. 1999).

Another Portuguese Late MP site is Caldeirão (Levels N-K), which is dominated by red deer (though the absolute numbers are fairly small), followed by horse, with even smaller numbers ibex and traces of several other ungulates (Davis 2002). There are small numbers of a wide variety of carnivores and a very large number of rabbit remains. Gruta Nova da Columbeira, Level 8, is also dominated by red deer, with small numbers of ibex, horse, aurochs and rhinoceros (Hockett and Haws 2009). Hyena is relatively abundant and here are traces of lynx, wild cat and wolf, and brown bear. The presence of rabbit in the Columbeira Mousterian is undocumented in the sources available to me at least. Overall, the Portuguese Late Mousterian non-carnivore spectrum is heavily dominated in terms of NISP by lagomorphs (92 %) versus ungulates (8 %). Rabbits as supplementary hominin food clearly preceded the UP.

Marine Resource Exploitation in the Late Mousterian

While it is true that many MIS 4 coastal plain sites are now drowned as a consequence of interglacial sea level transgression, there is no meaningful evidence of Mousterian exploitation of marine resources at coastal sites during MIS 3 in Vasco-Cantabria. [Such exploitation actually seems to have begun on a significant basis ironically during the Last Glacial Maximum (Solutrean period), when a key site for such evidence, La Riera Cave (eastern Asturias), would have been at least a couple of hours' walk from the shore as opposed to the present-day half-hour (Straus and Clark 1986). This is paralleled by the sequence at Nerja Cave in Málaga (Aura et al. 2001).] A case has been made by Finlayson et al. (2006) and Stringer et al. (2008; see also Carrión et al. 2008; Fernández-Jalvo and Andrews 2000) that Neanderthals in Gorham's and Vanguard Caves in Gibraltar exploited marine resources, namely mollusks, fish, seals and dolphins. This sort of argument had been made long ago at nearby Devil's Tower by its excavator, Garrod et al. (1928), although doubt has been cast by Freeman (1981) on the anthropogenic origin of at least most of the marine mollusks found in (Last Interglacial) beach sands at that site, while noting that some are burned. Likewise, Klein and Steele (2008) contest the significance of the total of 149 shells, five seal bones, three dolphin bones and three fish bones from Gorham's and Vanguard, arguing that even if collected (and in some cases of shells, burned) by Neanderthals, the remains are so few as to be nutritionally meaningless. There is no evidence that the seal and dolphin remains were not simply picked up as curiosities, even if they may have had some attached meat at the time. At any rate, these caves, currently directly on the shore, were never very far from the littoral [about 3–4 km during the Late Mousterian (Barton 2000) even if some of the mollusks came from an estuary in the present Bay of Algeciras], so the presence of at least many of these remains (and those of birds, which are diverse and abundant in all the Gibraltar caves) could be "natural". The Humo Caves, directly on the northern shore of the Bay of Málaga and very near present sea level, have also yielded relatively abundant marine mollusks from undated Mousterian levels (Cortés 2007a: 48; Cortés et al. 2008: 2183). Their human agency remains to be demonstrated. Similarly, there are some marine molluscs in Last Interglacial deposits of Bajondillo Cave, near sea level on the western shore of the Bay of Málaga (Cortés et al. 2008: 2183)—agency unknown.

The Late Mousterian horizon (Level 2, dated to 30 ^{14}C kBP) of Figueira Brava Cave, right on the present Atlantic shore at the mouth of the Sado River estuary, yielded numerous mussel and limpet shells and crabs, as well as smaller numbers of a variety of other mollusks. Evidence of breakage is interpreted by the analyst (Callapez 2000, see also Antunes 1990–91) as indicative of human exploitation. A few marine mammal remains (one ringed seal ulna and 6 vertebrae from a common dolphin, both of which of course could have been beached animals) have gnaw- or cut-marks (Antunes 2000c). Given the site's location, the molluscan collection is deserving of quantification and taphonomic re-analysis. A few marine mollusks have also been found with poorly known Mousterian materials in the Ibn Ammar caves on the Portimão estuary of the Algarve (southernmost Portugal) and (also from old excavations) in Furninha Cave on the Peniche Peninsula of western Portugal (Bicho and Haws 2008). Small animal foods were clearly of some significance (albeit limited in absolute terms) in Neanderthal diet in Portugal probably in the form of (seasonal?) pulses. This pattern seems to have been widespread throughout the eastern Mediterranean basin and in advance of the Middle-Upper Paleolithic transition, although all r-selected animal foods (shellfish, lagomorphs) obviously increased in relative importance with higher Upper Paleolithic human populations (see Stiner 1994, 2001).

The Aurignacian and Other Early Upper Paleolithic

Clearly, the general interest of this survey is to see if there are any notable differences in archaeofaunas between those (presumably at least in part) obtained by Neanderthals and those obtained by the earliest Upper Paleolithic people, who are generally assumed to have been *Homo sapiens sapiens* (though this has not been demonstrated on the Iberian Peninsula, and not really well elsewhere in Europe either, for that matter).

The Early Aurignacian levels of El Castillo (18c and 18b of the new excavations) have very large numbers of red deer remains, swamping the nonetheless substantial amounts of chamois, and aurochs (Dari and Renault-Miskovsky 2001). There are smaller numbers of roe deer, ibex, horse, rhinoceros, mammoth and a trace of boar in the latter level. Carnivores are not common—just a few wolf, hyena and leopard remains in one or both levels—plus small amounts of brown bear. The dramatic quantity of and apparent specialization in red deer is obviously a novelty compared with the Mousterian faunal assemblages from the Cantabrian region (Table 7.5). Whether this is the result of a decline in the relative abundance of horse and an increase in red deer in the region due to climatic and vegetational changes in late MIS 3 and/or changes in human hunting strategies and practices is unknown. This is a key question to be resolved. Red deer is also relatively well-represented in all the Châtelperronian and Aurignacian levels at nearby Cueva Morin, but in nothing like the numbers found in El Castillo (Altuna and Mariezkurrena 1988). Small numbers of roe deer, bovine and horse remains round out the Morín EUP assemblages (Altuna 1972; Quesada 2006). There are virtually no carnivores (one item each of fox and wild cat in only Level 6, the uppermost Aurignacian). The Basque Country site of Labeko Koba, with a modern excavation and full publication, yielded a massively red deer-dominated Châtelperronian level (IX inf.), which also had large numbers of horse and bovine remains (Altuna and Mariezkurrena 2000). There are a few reindeer remains. Hyena is relatively abundant and there are traces of fox and cave bear. The same red deer-dominant pattern holds in the culturally indeterminate level (IX sup.), but there are small numbers of other ungulates [*Megaloceros* (extinct giant elk), roe deer, boar], a substantial number of rhinoceros remains, a few of mammoth and the same carnivores (hyena being even more numerous). The Proto-Aurignacian and Early Aurignacian levels at Labeko have far fewer red deer and horse, but many bovines and consistent presence of chamois. Some of the levels have traces of wooly rhinoceros and others of mammoth. The oldest Proto-Aurignacian level (VII) has very large quantities of cave bear and hyena remains. Fox is well-represented throughout, and some levels have traces of wild cat. The cave clearly saw alternating use among humans, bears and hyenas and the latter must have been carcass (i.e., bone) accumulation agents. Humans were not yet permanently in control of at least this cave.

In Catalonia, at L'Arbreda Cave, red deer bones (albeit all in rather small quantities) alternate with bovines, horse and ass for the "number one" position in terms of NISP in

Table 7.5 Relative frequencies of ungulates and carnivores and ratio of ungulates to carnivores, based on NISP for combined Early Upper Paleolithic assemblages in Vasco-Cantabria, Mediterranean Spain and Portugal

	Mediterranean Spain		Vasco-Cantabria		Portugal	
	NISP	%	NISP	%	NISP	%
Ungulates						
Sus	3	0.4	8	0.0	17	19.5
Cervus	170	20.2	12434	77.6	41	47.1
Capreolus	1	0.1	225	1.4	1	1.1
Bovini	29	3.4	1515	9.5	0	0.0
Rupicapra	2	0.2	663	4.1	4	4.6
Capra	524	62.2	16	0.1	24	27.6
Equus	105	12.5	954	6.0	0	0.0
Rhinocerotidae	1	0.1	137	0.9	0	0.0
Other	7	0.8	74	0.5	0	0.0
Total ungulates	**842**	**100.0**	**16026**	**100.0**	**87**	**100.0**
Carnivores						
Canis	0	0.0	10	1.1	0	0.0
Ursus	98	45.6	397	44.7	9	45.0
Hyaenidae	14	6.5	357	40.2	4	20.0
Vulpes	3	1.4	119	13.4	1	5.0
Felis	50	23.3	3	0.3	0	0.0
Lynx spp.	23	10.7	0	0.0	4	20.0
Panthera spp.	0	0.0	3	0.3	2	10.0
Mustelidae	6	2.8	0	0.0	0	0.0
Other	21	9.8	0	0.0	0	0.0
Total carnivores	**215**	**100.0**	**889**	**100.0**	**20**	**100.0**
Total ungulates/total carnivores	**3.92**		**18.03**		**4.35**	

the two Late Aurignacian levels and there are traces of rhinoceros and mammoth in one level each (Maroto et al. 1996). The older of the two levels has a very large quantity of rabbit remains (with far fewer in the upper Aurignacian level). There are small numbers or at least traces of fox, lynx and hyena. The upper level yielded a large number of cave bear remains. Again, this cave may have been occupied by humans and cave bears on a "time-sharing" basis. The data are not all systematically presented for the three Aurignacian levels of Cova Beneito (Iturbe et al. 1993; Pérez and Martínez 2001). Globally, the dominant species is ibex, followed by red deer and then horse. There are traces of roe deer and boar. The number of rabbit remains rises steadily from the oldest to the youngest of these levels, attaining an impressive number (NISP = 1,534) in the Late Aurignacian one (B) (only to more than double again in the Gravettian level (B7) (Table 7.6). Together with the rabbits are small numbers of lynx remains in the upper two

Table 7.6 Early Upper Paleolithic: ratio of ungulates to lagomorphs based on NISP for Vasco-Cantabria, Mediterranean Spain and Portugal; lagomorph NISP counts from Cantabrian site of Morín (n = 1), Mediterranean Spain sites of L'Arbreda, Beneito and Mallaetes, and Portuguese site of Caldeirão

	Mediterranean Spain						Vasco-Cantabria	Portugal
	L'Abreda		Beneito			Mallaetes	Morin	Caldeirão
	Gbc	Ga	C	B9	B8	11	5c	
Leporidae	0	0	0	0	0	0	0	0
Oryctolagus	766	39	68	385	1534	55	0	1551
Lepus	0	0	0	0	0	0	1	2
Total	**766**	**39**	**68**	**385**	**1534**	**55**	**1**	**1553**
Total ungulates/total lagomorphs	**0.295542296**						**16026**	**0.056020605**

Aurignacian (and Gravettian) levels. This might suggest that humans were not the only rabbit-killers at the site in the EUP. Rabbit is also numerically dominant in the Aurignacian level (11) of Les Mallaetes, while ungulates are very scarce (small numbers of red deer, ibex and horse) (Davidson 1989). Carnivores are absent from the list.

The ambiguous (possibly mixed and contradictorily dated) EUP (?) levels in Zafarraya are heavily dominated by ibex (not surprising given the site's cliff-side location), with only traces of red deer, chamois and horse (Barroso et al. 2006a, b). (The large numbers of ibex remains from Zafarraya swamp and thus distort the combined ungulate assemblages for both the Late MP and EUP of Mediterranean Spain, which is unfortunate given the chronological ambiguities of some of its levels.) Once again, carnivores (leopard, wild cat, dhole, hyena, fox) are relatively common, though bears are now absent. The cave continued to serve as a carnivore lair when not being used by humans. There are no other data from Andalucia—and none at all that are clearly Aurignacian, with the exception of Bajondillo Cave in Torremolinos (Málaga). This site has an apparent late Aurignacian component, radiocarbon dated to ~34 ^{14}C kBP, with small numbers of marine mollusks (as in the underlying Mousterian and overlying Gravettian levels) (Cortés 2007a, b). No other faunal information has yet been published for this important site—possibly the southernmost Aurignacian locality in Western Europe.

The same is true for Portugal. Caldeirao Cave (Estremadura) has an indeterminate EUP level (J), whose small assemblage is dominated by red deer, followed by ibex and boar, plus traces of boar, roe deer and chamois (Davis 2002). There is a very large number of rabbit remains (1,551), plus a large variety of carnivores and brown bear, all represented by very few remains. Human agency for the rabbit bones is possible, *if* one extrapolates back from what is known from the LUP assemblages (Hockett and Haws 2002). A possible EUP level in Picareiro Cave, also in south-central Portugal, is said by Hockett and Haws (2009) to have red deer, rabbit and hedgehog, but no quantities are yet published for this important, carefully excavated site.

Observations on Rabbits as Human Food

Globally, the EUP ungulate/lagomorph ratios for Mediterranean Spain and Portugal are 28/77 and 8/92 respectively. Portugal continued to be "the land of the rabbit", as it had been in the late Mousterian. Can one presume the existence of nets and rabbit drives? Even so, obviously it took many rabbits (especially with their lean meat) to equal a single red deer in terms of nutritional value to the hunter-gatherers (Speth and Spielmann 1983: 3, 4; but see Hockett and Haws 2002; Hockett and Bicho 2000; see also Broughton et al. 2011 for a theoretical discussion of the relative importance of large body size prey relative to small ones like lagomorphs, with examples from the American Great Basin). A major practical problem, especially among Iberian sites that are almost all caves (not kill-sites), is the likely undercounting of large mammals by NISP since these game were field-butchered before only certain selected parts were brought back to residential sites (whether long- or short-term). This contrasts with the probable complete transport of rabbit carcasses back to such sites for processing and consumption, thus "inflating" the rabbit counts relative to the ungulate ones based on NISP. This is a case where comparison between animals of such widely divergent body sizes could be done more accurately by using MNI, which is unfortunately not often given for rabbits. Naturally, a secondary use for rabbits would have been their pelts. The point here is that in those Iberian regions where they were abundant (and perhaps red deer less abundant than in humid, plant food-rich Vasco-Cantabria), rabbits were a secondary food resource for hominins from at least Late MP times onward, though increasing in the Middle and Late Upper Paleolithic. Whether this increase reflected increased human subsistence stress and/or the development of more efficient methods and technologies for mass rabbit slaughter remains to be shown.

The Gravettian

Late MIS 3 is represented by a number of Gravettian levels in the various Iberian regions, beginning ca. 28 ka. Some of these assemblages (the most recent ones) can have been formed near the onset of the Last Glacial Maximum, some 12,000–19,000 years after the so-called Middle-Upper Paleolithic transition. There are modern-quality, published Gravettian faunal assemblages from only two sites in Vasco-Cantabria (Table 7.7). At Cueva Morín the three Gravettian assemblages are heavily dominated by red deer, with moderate numbers of roe deer, bovines, and horse, plus traces of boar, chamois and mammoth (1 item) (Altuna 1972, 1978; Quesada 2006). There are traces of wolf, fox, hyena, wild cat, leopard, but no bear. There are also a few hare remains. The Amalda Late Gravettian levels are very heavily dominated by chamois (argued in a reanalysis by Yravedra [2002; but see Altuna and Mariezkurrena 2010] to have been accumulated by carnivores, but nonetheless very strongly swamping and distorting the published Gravettian ungulate record from Vasco-Cantabria), with large numbers of ibex and red deer, plus traces of boar, roe deer and reindeer. The older of the two levels (VI) has a large number of horse remains, while the younger one (IV) has

Table 7.7 Relative frequencies of ungulates and carnivores and ratio of ungulates to carnivores, based on NISP for combined Gravettian assemblages in Vasco-Cantabria, Mediterranean Spain and Portugal

	Mediterranean Spain		Vasco-Cantabria		Portugal	
	NISP	%	NISP	%	NISP	%
Ungulates						
Sus	4	0.5	1	0.0	18	2.3
Cervus	322	37.4	1170	19.6	513	65.8
Capreolus	0	0.0	193	3.2	1	0.1
Bovini	24	2.8	173	2.9	20	2.6
Rupicapra	15	1.7	3795	63.4	4	0.5
Capra	196	22.8	437	7.3	94	12.1
Equus	295	34.3	190	3.2	130	16.7
Rhinocerotidae	0	0.0	0	0.0	0	0.0
Other	5	0.6	24	0.4	0	0.0
Total ungulates	**861**	**100.0**	**5983**	**100.0**	**780**	**100.0**
Carnivores						
Canis	1	0.5	50	13.7	2	4.4
Ursus	169	83.7	128	35.2	9	20.0
Hyaenidae	6	3.0	13	3.6	4	8.9
Vulpes	6	3.0	157	43.1	10	22.2
Felis	4	2.0	1	0.3	0	0.0
Lynx spp.	8	4.0	2	0.5	15	33.3
Panthera spp.	1	0.5	6	1.6	5	11.1
Mustelidae	0	0.0	3	0.8	0	0.0
Other	7	3.5	4	1.1	0	0.0
Total carnivores	**202**	**100.0**	**100.0**	**100.0**	**45**	**100.0**
Total ungulates/total carnivores	**4.26**		**16.44**		**17.33**	

only a few. Carnivores are diverse and sometimes (wolf and especially fox) abundant. The impressive roster of carnivores plus bears strongly does suggest a major role for these agents of accumulation, with humans as only part-time residents of the cave and hunters of some of the ungulates. There are only traces of hare.

In Catalonia, horse dominates one of the three Gravettian levels (E) in L'Arbreda, followed by red deer (Maroto et al. 1996). All the Gravettian levels have those species plus small numbers or traces of ass, boar, chamois and aurochs. There is one proboscidean remain. One of the levels (F) has a large number of cave bear remains and all of them have traces of a variety of carnivores. Rabbits are very abundant and increase through time, certainly due to human predation. South of Catalonia, in Valencia, Cova Beneito Level B7—Gravettian—is roughly equally dominated by ibex and red deer, followed by horse (Iturbe et al. 1993; Pérez and Martínez 2001). There are only traces of a few carnivore species, but rabbits are very well represented (NISP = 3,625), no doubt killed by humans. The very small Mallaetes (Valencia) Gravettian (Level 10) assemblage is dominated by red deer and rabbit (Davidson 1989). The Gravettian levels in Les Cendres Cave (Alicante) have abundant rabbit bones with considerable evidence of butchery by humans (Table 7.8; Pérez Ripoll 2006).

Nerja Cave, on the shore of Málaga, saw its first marine and terrestrial mollusks (mostly *Iberus*) appear during the pre-Magdalenian (ca. 24–17.5 ^{14}C kBP) levels of its Vestibule area. There are small numbers of limpet and mussel shells in the Gravettian levels dated between 25 and 21 ^{14}C kBP (Cortés et al. 2005). Rabbit remains are also present in the late EUP and Solutrean levels. Human agency for these is claimed on the basis of some taphonomic analyses (Aura et al. 2002). Nerja was never more than 5–6 km from the glacial shore. But it was in the Magdalenian that full-scale, ocean fishing began (Aura et al. 2001). The pre-Magdalenian and Magdalenian ungulate faunas of Nerja are overwhelmingly dominated by ibex—not surprising given its location at the foot of a 1,500 m-high

Table 7.8 Gravettian: ratio of ungulates to lagomorphs based on NISP for Vasco-Cantabria, Mediterranean Spain and Portugal; lagomorph NISP counts from Cantabrian sites of Morín and Amalda (n = 1–4), Mediterranean Spain sites of L'Arbreda, Beneito and Mallaetes, and Portuguese sites of Caldeirão, Val Boi and Anecrial

	Mediterranean Spain					Vasco-Cantabria				Portugal			
	L'Abreda			Beneito	Mallaetes	Morin	Amalda			Caldeirão	Val Boi	Anecrial	
	F	Eb	E	B7	10	5a	IV	VI	IV	J	pooled	2inf	2sup
Leporidae	0	0	0	0	0	0	0	0	0	0	0	0	0
Oryctolagus	142	155	571	3625	28	0	0	0	0	1551	0	743	744
Lepus	0	0	4	2	0	1	2	3	4	2	2802	0	0
Total	**142**	**155**	**575**	**3627**	**28**	**1**	**2**	**3**	**4**	**1553**	**2802**	**743**	**744**
Total ungulates/total lagomorphs	**0.19019218**					**598.3**				**0.133515919**			

mountain chain that plunges straight down to the Mediterranean shore. There are also some marine molluscs in ~24 ^{14}C kBP, cold-climate Gravettian Level 10 of Bajondillo Cave in Torremolinos (Málaga) (Cortés et al. 2008).

In extreme southwestern Portugal at the rather unique open-air, coastal plain site of Val Boi, the combined Gravettian assemblage (22–27 ^{14}C kBP) is dominated by red deer, followed by horse, then aurochs and ass, plus traces of boar, ibex, and a few carnivores (notably lynx)—possibly trapped for fur by humans (Manne and Bicho 2009; Stiner 2003). There are a very large number of rabbit remains (NISP = 2,802) which are anthropogenic in terms of their accumulative agency and intensive breakage. The Val Boi Gravettian component is distinguished by the presence of a very large number of marine mollusk remains (NISP = 8,286, with an MNI of 1,054), overwhelmingly dominated by limpets (*Patella* sp.), at a time when sea level was falling but had not reached its LGM low when the shore would be 15–20 km from the site (Manne and Bicho 2009, 2011). There is also a vertebra fragment from a small cetacean (Manne and Bicho 2009)—probably scavenged or collected as a curiosity on the shore.

Picareiro Cave and Anecrial Cave in central Portugal show evidence of human exploitation of rabbits with marrow extraction. The combined Gravettian of Picareiro has produced >3,000 leporid bones (and 220 bird bones) (Hockett and Haws 2009). In Anecrial Level I the rabbit NISP is 1,601 and in Gravettian Level J there is a hearth full of burned leporid bones (Hockett and Haws 2002). The Gravettian of Lagar Velho is also rich in leporid remains (NISP = 1,336; MNI = 76) (Hockett and Haws 2009). Rabbit drives were obviously growing in importance.

Overall, for Mediterranean Spain and Portugal respectively the ungulate/lagomorph ratios are 16/84 and 12/88 in terms of NISP. Throughout all time, there was a dramatic difference between the Eurosiberian (i.e., Vasco-Cantabrian) and Mediterranean eco-zones in terms of the abundance of rabbits and thus their exploitation by humans—whether Neanderthal or Cro-Magnon.

Discussion and Conclusions

The watchword for Iberian archaeofaunas throughout MIS 3 and 4 in Iberia is "continuity". There are no major breaks either between the Middle and Early Upper Paleolithic or between each of the two phases of either major phase (i.e., Early and Late Mousterian or Aurignacian and Gravettian). The Iberian Peninsula, south of the Pyrenees and south of the Cantabrian Sea/Gulf of Gascony was and is significantly different from France, never having been a land inhabited by many "arctic" taxa during glacials. Even the narrow, oceanic Vasco-Cantabrian strip, the only region of Spain that belongs to the Eurosiberian ecological zone (the rest of the Peninsula being in the Mediterranean zone, as noted earlier), never saw large numbers of reindeer, woolly mammoth or rhino, arctic fox, etc. Muskoxen, saiga antelope [one bone in the early Magdalenian of Abauntz in Navarra was probably carried there across the Pyrenees as a curiosity from southwestern France (Altuna and Mariezkurrena 1996)] do not seem to have lived here ever or hardly ever (Altuna 1996; Alvarez-Lao and García 2010). Ironically, MIS 4 archaeological deposits have virtually no cold-climate ungulate fauna. For late MIS 3, ca. 35–25 ka—also ironically—there are mammoth remains in both archaeological and non-archaeological contexts in the South (respectively in Figueira Brava Cave and the Padul bog in Granada), as well as in a handful of sites in Vasco-Cantabria and Catalonia. Naturally, there is a caveat in that we have very few purely "paleontological" sites; almost all the large mammal faunal evidence comes from archaeological sites (in caves), where human selection was operative, although almost certainly other carnivores (hyenids, canids, felids) were also involved to varying extent as agents of carcass accumulation (Straus 1982; Lindly 1988; Blasco 1995; Yravedra 2002). Besides Padul, for example, there have been a few purely paleontological finds of isolated mammoths from other loci in Spain, but in LGM contexts. Mammoth could have been present in the open-vegetation environments of Last Glacial central Iberia, but the paucity of sites makes it impossible to judge its relative abundance. The big steppe-tundra beasts of Ice Age France "visited" Spain and Portugal only rarely in both Middle and Early Upper Paleolithic times—reindeer a bit more frequently, but only in the far north of the Peninsula—but all were rare sights for the hominin residents whoever they were. And it is likely that only the reindeer were their prey in any meaningful way, however slight and sporadic. On the other hand, as with plants, Iberia was clearly a refugium or reservoir for species such as boar, red and roe deer that recolonized France during temperate times.

Although there is often more variability between sites of the same period than between different periods, it is generally the case that Mousterian sites have relatively more large and very large ungulates (bovines, horses and rarely rhinos) than do the EUP sites. There is some question, however, as to whether hominins were actually butchering whole rhinos, since the numbers of their remains are always extremely few, leaving open the possibility that rhino bones and teeth were collected (already "bare" of any meat) as oddities. The tendency toward specialization in red deer and/or ibex hunting [which became overwhelming in the LUP—Solutrean and Magdalenian (Straus 1977, 1992; Freeman 1973, 1981; Marín-Arroyo 2009a, 2010)] is already beginning to be manifested in a few MP and EUP assemblages (e.g., El Castillo). That specialization often becomes quite clear by Gravettian times, though there may still have been confounding factors (i.e., non-hominin carnivore agency) in final MIS 3 times.

There are a couple of clear biogeographic aspects to the record that separate Atlantic Vasco-Cantabria from the Mediterranean remainder of the Peninsula. One is the presence of ass in many Mediterranean sites of various periods, though it is never very abundant as human prey. The other is vastly more important: rabbits—never present in the Eurosiberian zone, but omnipresent (when included in the published faunal reports) in Mediterranean Spain and southern and central Portugal. *Oryctolagus cuniculus* is present as early as the Early Mousterian, although hominin agency needs to be rigorously demonstrated in each case [see, for example, the virtual exclusion of human agency in Zafarraya after taphonomic analysis (Barroso et al. 2006a, b)]. Late Mousterian and EUP levels in Eastern Spain often have very large numbers of rabbit remains and this species becomes even more important quantitatively in Gravettian levels, no doubt (despite their small mass and the leanness of their meat) contributing a critical part of hominin diet, perhaps during seasons of scarcity of red deer and ibex. Growing numbers of studies in Mediterranean Spain and Portugal demonstrate that rabbits were butchered (and presumably hunted in drives or other types of mass kills, using nets, rabbits sticks, etc.) and consumed by people, with ample evidence of cut marks and burning (e.g., Pérez Ripoll 2001; Hockett and Haws 2002). Such a supplementary specialization in rabbit slaughter may have been motivated by regional human population pressure and/or over-exploitation of larger game, despite the relatively low nutritional return from these lagomorphs vis à vis large-medium ungulates (mainly red deer, ibex). The environmental conditions of the Mediterranean eco-zone may have been less favorable to high red deer densities than those of the Eurosiberian zone, while favoring prolific rabbit populations, with their high rate of reproduction. Given the high potential returns (in terms of meat, fat, marrow, organs, hides and—from red deer stags—antler) from the hunting of *Cervus* and *Capra*, it is hard to understand why humans would invest a lot of time and effort in killing many *Oryctolagus* if the ungulate populations were large and accessible enough to fully satisfy human food (and other) needs year-round. All these animals are, after all, *fast* and require considerable planning, skills and specialized technologies for killing *en masse*. The growing focus on rabbits in Mediterranean Spain and Portugal throughout the late Middle and Upper Paleolithic suggests that the reverse may have been the case, as well as the obvious, namely that fast-breeding rabbits were very abundant in these environments.

There is no clear-cut evidence of a break in hominin subsistence patterns between the Middle and Upper Paleolithic or indeed between MIS 4 and 3 in the Iberian Peninsula. The archaeofaunas of the Early and Late Mousterian, Aurignacian and Gravettian show considerable continuity within each ecological region of the Peninsula. It was biogeography that was mostly driving the observed patterns of human hunting. As I have argued several times before (e.g., Straus 1977, 1992; Straus and Clark 1986), the really significant changes in subsistence seem to have taken place with the Last Glacial Maximum and Tardiglacial in the Late Upper Paleolithic (Solutrean and Magdalenian periods) and may have been responses to increased regional population densities more than to climatic/environmental changes (see Marín-Arroyo 2009b; Stiner 2001). The interest of this overview is the clear evidence that hominins [both Neanderthals, followed (presumably) by anatomically modern humans] were already beginning to exploit rabbits (and perhaps shellfish) in small quantities in Mediterranean Spain and Portugal. This was probably not demographically driven, although the vast increase in small-animal exploitation (terrestrial and marine, as well as birds) in the Solutrean and Magdalenian periods probably was. The parallelism in the significant development of marine resource exploitation during the Solutrean in both north Atlantic Spain and Mediterranean Spain and in Portugal is also clear. The significant differences between Iberia as a whole and France in terms of both Mousterian and Early Upper Paleolithic subsistence are also clear: Iberia—the southwestern refugium of Europe—was not a land of reindeer, woolly mammoths or rhinos. It was always (and still would be, without the vast ecological changes wrought by the spread of agricultural lifeways) the land of red deer and ibex, accompanied by roe deer, boar, chamois, aurochs, horse, ass, and rabbit, depending on the region. Though Iberia witnessed occasional incursions of "glacial" fauna from the North, it was from the Iberian refugium that France etc. were repeatedly recolonized by "temperate" faunas—and hominins.

There is clear evidence neither of "superior" EUP subsistence practices nor of a more nutritious EUP diet relative to the subsistence of Mousterian Neanderthals in Iberia, at least as one can perceive from the (admittedly low-resolution) faunal evidence. The "wild card" could conceivably be the use of plant foods, but there is certainly no EUP lithic technology suggestive of an increase in that aspect of diet. Intensification surely began in Gravettian times in some regions (notably southern Portugal), as indicated by bone grease rendering and shellfish collection at Val Boi, for example (Bicho et al. 2010a, b; Manne and Bicho 2009). By then [and based on the Lagar Velho child burial (Zilhão and Trinkaus 2002)] the human inhabitants of Iberia definitely were anatomically modern (though possibly Neanderthal-"tainted") humans, and in some regions, such as Portugal and Mediterranean Spain—as in Italy or the Czech Republic—their numbers were growing. This is the crux of the story that would lead to the major changes that marked the second half of the Upper Paleolithic in Iberia (e.g., Straus 1993), with a ratcheting-up of regional subsistence intensification that included situational specialization and overall diversification of mammalian, molluscan, piscine and avian species exploited and sometimes over-exploited, heavy carcass and bone processing, as humans scrambled to feed more hungry stomachs.

Acknowledgments Jamie Clark and John Speth kindly invited this dirt archaeologist to attempt the impossible, namely to synthesize the faunal evidence on hominin subsistence in Iberia during MIS 4-3. Thanks to Jesús Altuna, Nuno Bicho, Paco Giles, Jonathan Haws, Jordi Rosell and Isabel Cáceres for sending me data, manuscripts, and articles. *Muchas gracia tambien* to Joshua Vallejos for database construction, taxonomic research and graph production. All errors of interpretation are (as a "consumer" of faunal data—not a specialized archaeozoologist) solely mine. This work is offered as another first approximation of a synthesis, just like my 1977 chapter, "Of deerslayers and mountain men", in the late Lew Binford's book, *For Theory Building in Archaeology.* Les Freeman and Altuna set me on a course of trying to bring faunal evidence into equal consideration along with the artifacts found at European Paleolithic sites. For the two decades we were "next-door office neighbors" in the Anthro Annex at UNM, Lew was always provoking me to think about how to explain the differences (and similarities) between Middle and Upper Paleolithic behavior. I and the profession will miss his genius, stimulus and impetus to understand the archaeological record. In various ways, Jane Wheeler, Richard Klein, Achilles Gautier, Don Grayson, Mary Stiner and Ana Belen Marín-Arroyo (the latter two of whom I count as partly "my" students) have also influenced my imperfect appreciation of archaeozoology. Three reviewers and Jamie Clark provided numerous helpful corrections and suggestions, as well as criticisms, only some of which I have been able to act on in the time allocated for revision. The map was professionally drafted by Ron Stauber. After this was written, I learned about the discovery of humanly modified (cut, broken, burnt) tortoise carapaces and bones from Level IV (dating between ca. 121 kya and ca. 225 kya) in Bolomor Cave (Valencia) (Blasco 2008). This would be evidence of at least local exploitation of small, slow prey by Neanderthals in the early Middle Paleolithic. Published after this was written is an article by E. L. Jones (2012) that discusses rabbit exploitation in the Upper Paleolithic of Mediterranean Spain.

Appendix

Appendix 7.1 Faunal data from MP and EUP sites in Vasco-Cantabria

	Esquilleu	Esquilleu	Covalejos	Covalejos	Covalejos	El Pendo	El Castillo	El Castillo	El Castillo	El Castillo	El Castillo	Morin
Level	IV	III	J	I	C	XVI	22*	21*	20	18c	18b	22
14C age (ka)	<34	<34	38 (TL)	–	33		70 (ESR)	69 (ESR)	43–45	40	38	
Culture	Mou	Mou	Mou	Mou	Mou	Mou	Mou	Mou	Mou	Aur	Aur	Mou
Sus scrofa				3\							5\3	
Cervidae												
Cervus elaphus	6\	5\	801\	321\	480\	164\6	228\	10\	1572\34	4981\86	5882\89	
Rangifer tarandus			1\									
Megaloceros												
Capreolus			21\	16\	11\		3\		34\4	39\7	95\10	
Dama dama												
Bovini						95\6	84\		544\21	190\6	152\5	\2
Bos primigenius												
Caprini												
Hemitragus												
Rupicapra	14\	27\		2\		13\2	45\		28\7	155\13	458\29	
Capra pyrenaica	34\	115\				8\2	3\		2\2	6\2	6\2	
Equus sp.												
E. hydruntinus												
E. caballus			120\	23\	12\	26\3	1152\	42\	347\16	19\3	22\3	\1
Rhinocerotidae									54\12			1\
Coelodonta												
S. kirchbergensis							1(+)\			11\3	13\3	
Stephanorhinus sp.												
S hemitoechus			2\	2\	1\							1\1
Mammuthus										11\	1\	
Elephas sp.												
Canis lupus			4\				2\			4\1	3\1	
Vulpes			4\									
Alopex												
Cuon												

(continued)

Appendix 7.1 (continued)

	Esquilleu	Esquilleu	Covalejos	Covalejos	Covalejos	El Pendo	El Castillo	El Castillo	El Castillo	El Castillo	El Castillo	Morin
Ursus sp.												
U. arctos									1\1	1\1	17\3	
U. spelaeus			1\					1\	4\1			
Crocuta		3\		1\							3\1	
Hyaenidae												
Felis sylvestris		1\	1\									
Panthera spelaea							6\	1\				
P. leo												
P. pardus										3\1		
Lynx spelaea												
L. pardina												
Erinaceus												
Leporidae	2\	8\										
Oryctolagus												
Lepus			5\									
Marmota			8\									
Castor												
Hystrix												
Meles												
Mustela putorius												
M. nivalis												
Martes												

	Morin	Morin	Morin	Morin	Morin	Morin	Morin	Morin	Morin	Morin	Venta Laperra	Axlor	Axlor
Level	17	15	10	8	7	6	5c	5b	5a	4		D	B-C
14C age (ka)			36	36.5	28	31.5			20			42	
Culture	Mou	Mou	Ch.	Aur	Aur	Aur	Aur	Gr	Gr	Gr	Mou	Mou	Mou
Sus scrofa	2\1			1\1					1\1		3\	1\	
Cervidae													
Cervus elaphus	142\8	6\1	3\2	14\3	95\7	108\7	121\7	143\8	388\12	271\11	28\	279\	
Rangifer tarandus													
Megaloceros													
Capreolus	15\3	2\2		11\2	26\2	26\6	21\5	31\5	95\6	49\4	7\		

(continued)

Appendix 7.1 (continued)

	Morin	Morin	Morin	Morin	Morin	Morin	Morin	Morin	Morin	Morin	Venta Laperra	Axlor	Axlor
Dama dama													
Bovini	215\10	2\1	23\2	6\2	26\2	23\5	9\1	17\3	29\3	19\2	8\	258\	41\
Bos primigenius													
Caprini													
Hemitragus													
Rupicapra							1\1		3\1	1\1	1\	7\	1\
Capra pyrenaica	1\1					1\1	3\1	8\1	8\3	13\3	136\	204\	41\
Equus sp.													
E. hydruntinus													
E. caballus	124\11	1\1	1\1	13\1	8\2	37\4	6\1	12\2	30\4	39\4	1\	42\	31\
Rhinocerotidae	4\2												
Coelodonta													
S. kirchbergensis													
Stephanorhinus sp.													
S hemitoechus												5\	
Mammuthus										1\1			
Elephas sp.													
Canis lupus	\1						3\1			1\1	1\		
Vulpes						1\1	2\1		1\1	\1	1\		2\
Alopex													
Cuon													
Ursus sp.													
U. arctos											1\		
U. spelaeus													
Crocuta	1\1							1\1	1\1				
Hyaenidae													
Felis sylvestris						1\1				1\1			
Panthera spelaea													
P. leo													
P. pardus								1\1				1\	
Lynx spelaea													
L. pardina										1\1			
Erinaceus													

(continued)

Appendix 7.1 (continued)

	Morin	Morin	Morin	Morin	Morin	Morin	Morin	Morin	Morin	Morin	Venta Laperra	Axlor	Axlor
Leporidae													
Oryctolagus													
Lepus							1\1		1\1	2\1			1\
Marmota													
Castor													
Hystrix													
Meles													
Mustela putorius													
M. nivalis													
Martes											1\		
	Arrillor	Arrillor	Arrillor	Amalda	Amalda	Amalda	Lezetxiki	Labeko K	Labeko K	Labeko K	Labeko K	Labeko K	Labeko K
Level	Amk	Smk-l	Lmc	VII	VI	IV	VI	IX inf.	IX sup.	VII	VI	V	IV
14C age (ka)	45.5	43	37		19	19		34.2	29.8	31.5, 26.9	30.6		21.7
Culture	Mou	Mou	Mou	Mou	late Gr	late Gr	Mou	Ch	(?)	Proto-Aur	Proto-Aur	early Aur	early Aur
Sus scrofa							2\1		1\	1\			
Cervidae					3\2	18\1							
Cervus elaphus	164\	103\	600\	150\5	274\6	94\5	26\2	792\16	367\8	79\7	59\	29\5	13\
Rangifer tarandus					2\1			14\	13\		1\		1\
Megaloceros							4\1		5\4	2\			1\
Capreolus	11\		5\	3\3	17\2	1\1	8\2		4\	2\			1\
Dama dama													
Bovini	146\	125\	45\	58\3	99\4	9\3	226\1	143\5	206\5	111\7	175\6	290\7	161\5
Bos primigenius													
Caprini													
Hemitragus													
Rupicapra			91\	536\16	2769\59	1022\35	13\2		7\1	23\2	4\2	7\2	8\2
Capra pyrenaica	19\	128\		61\5	236\7	172\8	2\1						
Equus sp.								210\6	305\8	183\9	36\4	61\6	53\4
E. hydruntinus													
E. caballus	16\	6\	5\	48\4	101\6	8\3	20\2						
Rhinocerotidae													
Coelodonta									92\7	21\4			
S. kirchbergensis													

(continued)

Appendix 7.1 (continued)

	Arrillor	Arrillor	Arrillor	Amalda	Amalda	Amalda	Lezetxiki	Labeko K	Labeko K	Labeko K	Labeko K	Labeko K	Labeko K
Stephanorhinus sp.							3\1						
S hemitoechus	3\		13\										
Mammuthus									8\1	6\		6\	5\
Elephas sp.													
Canis lupus	3\		6\	17\3	37\3	12\2							
Vulpes	2\		9\	29\2	133\3	22\2		2\	3\	22\	67\	6\	16\
Alopex					1\1								
Cuon				1\1	1\1	2\1							
Ursus sp.													
U. arctos													
U. spelaeus	17\	6\	7\	58\5	103\7	25\4		3\2	14\	338\	20\		4\
Crocuta				3\2	11\1			55\	117\	128\	43\	2\	9\
Hyaenidae													
Felis sylvestris			1\							1\		1\	
Panthera spelaea													
P. leo						1\1							
P. pardus			4\	3\1	3\1	1\1							
Lynx spelaea						1\1							
L. pardina													
Erinaceus													
Leporidae													
Oryctolagus													
Lepus	1\		1\		3\1	4\1							
Marmota			11\		1\1								
Castor													
Hystrix													
Meles			1\										
Mustela putorius						2\1							
M. nivalis						1\1							
Martes					1\1								

Ages based on uncalibrated radiocarbon dates unless otherwise indicated (*TL* Thermoluminescence, *ESR* Electron Spin Resonance). *Mou* Mousterian, *Ch* Châtelperronian, *Aur* Aurignacian, *Gr* Gravettian. The numbers in the cells (n\n) refer to NISP and MNI counts, respectively. References are found in the chapter text

Appendix 7.2 Faunal data from MP and EUP sites in Aragon, Catalunya, and Greater Valencia

	Gabasa	Gabasa	Gabasa	Gabasa	Gabasa	Gabasa	Romani	Romani	Romani	Romani	Romani	Romani
Level	a + b	d	e	f	g	h	M	L	K	Jb	Ja	I
14C age (ka)	>46		>52, 46.5		>51		54	52	51	49	48.5	48
Culture	Mou	Mou	Mou	Mou	Mou	Mou	Mou	Mou	Mou	Mou	Mou	Mou
Sus scrofa	9\2	1\1	1\1	3\1	7\1	5\2						
Cervidae												
Cervus elaphus	66\9	75\6	348\25	359\25	511\22	159\9	479\9	96\5	335\8	96\4	497\12	117\6
Rangifer tarandus												
Megaloceros												
Capreolus	5\3	2\1	7\3	15\3	22\4							
Dama dama												
Bovini	9\3	17\3	26\4	15\4	13\2	17\2						
Bos primigenius							15\3	6\1	75\1	9\2	88\5	5\1
Caprini												
Hemitragus												
Rupicapra	46\7	44\7	83\10	48\6	40\5	4\2				2\1	6\5	
Capra pyrenaica	265\27	330\24	434\29	399\21	654\46	272\14						
Equus sp.												
E. hydruntinus			11\1	1\1	5\2	2\1						
E. caballus	206\18	162\12	374\23	249\17	380\23	263\14	58\6	34\4	56\4	139\4	351\15	86\9
Rhinocerotidae	1\1		1\1	5\1	6\1	1\1						
Coelodonta												
S. kirchbergensis												
Stephanorhinus sp.												
S. hemitoechus										27\2	33\3	
Mammuthus												
Elephas sp.												
Canis lupus	14\2	32\2	62\6	14\3	84\6	39\2						
Vulpes	15\2	13\2	6\2	2\1	13\1	8\1						
Alopex												
Cuon		1\1	1\1		2\1							
Ursus sp.												
U. arctos												
U. spelaeus	1\1	3\2	2\1	2\2	20\1	1\1						

(continued)

Appendix 7.2 (continued)

	Gabasa	Gabasa	Gabasa	Gabasa	Gabasa	Gabasa	Romani	Romani	Romani	Romani	Romani	Romani
Crocuta	11\2	18\2	24\2	7\3	26\3	7\1						
Hyaenidae												
Felis sylvestris	3\2	2\1	1\1	1\1	3\1	1\1						
Panthera spelaea					1\1							
P. leo			6\1									
P. pardus			2\1	1\1	4\1							
Lynx spelaea	7\1	3\1	16\3	2\1	23\4	26\2						
L. pardina												
Erinaceus												
Leporidae												
Oryctolagus	647\30	277\14	211\43	489\28	453\28	81\7						
Lepus			1\1									
Marmota												
Castor												
Hystrix												
Meles				1\1		1\1						
Mustela putorius					1\1							
M. nivalis												
Martes												

	Romani	Teixoneres	L'Arbreda	L'Arbreda	L'Arbreda	L'Arbreda	L'Arbreda	L'Arbreda	L'Arbreda	Ermitons	Ermitons	Beneito
Level	H	III	I	H	Gbc	Ga	F	Eb	E	VI	IV	D2
14C age (ka)	46.5	<94 > 36(U)	40	38	22.1	22.1			20.1		34	
Culture	Mou	Mou	Mou	Mou	late Aur	late Aur	Gr	?	Gr	Mou	final Mou	Mou
Sus scrofa		1\1							2\			2\
Cervidae			4\1			5\						
Cervus elaphus	141\2	47\24	127\5	60\4	42\	14\	7\	4\	114\			34\
Rangifer tarandus												
Megaloceros												
Capreolus		2\2	4\1	1\1								
Dama dama												
Bovini			68\3		14\	14\	4\	4\	15\	1\1		
Bos primigenius		10\7										5\
Caprini		3\3										

(continued)

Appendix 7.2 (continued)

	Romani	Teixoneres	L'Arbreda	L'Arbreda	L'Arbreda	L'Arbreda	L'Arbreda	L'Arbreda	L'Arbreda	Ermitons	Ermitons	Beneito
Hemitragus												
Rupicapra				18\4			4\		11\			1\
Capra pyrenaica					1\			1\	5\	c.110\6	c. 110\6	44\
Equus sp.			57\4	34\4	6\	14\	9\	3\	220\	1\1		
E. hydruntinus		15\9	10\	3\1	16\	6\	4\	1\	6\			
E. caballus	56\2	36\4	10\		19\	3\	3\					8\
Rhinocerotidae												
Coelodonta												
S. kirchbergensis												
Stephanorhinus sp.					1\							
S. hemitoechus	1\1											
Mammuthus												
Elephas sp.				7\1		2\		1\				
Canis lupus			6\1	2\1						4\1	2\1	
Vulpes			6\1	7\2	1\		1\		4\		1\1	
Alopex												
Cuon												
Ursus sp.					3\							
U. arctos												
U. spelaeus		25\11	377\21	5\1		95\	160\	9\		c. 11\2	c. 158\7	
Crocuta		4\2	1\1	1\1	14\		5\				c. 8\3	
Hyaenidae												
Felis sylvestris			5\2									
Panthera spelaea			2\1									
P. leo												
P. pardus									1\	2\1	1\1	
Lynx spelaea			7\1	7\1	1\	1\	1\					
L. pardina										1\1	2\1	1\
Erinaceus		1\1	8\1	12\2								1\
Leporidae		18\13										
Oryctolagus					766\	39\	142\	155\	571\			169\
Lepus			39\7	8\1					4\			
Marmota												
Castor												

(continued)

Appendix 7.2 (continued)

	Romani	Teixoneres	L'Arbreda	L'Arbreda	L'Arbreda	L'Arbreda	L'Arbreda	L'Arbreda	L'Arbreda	Ermitons	Ermitons	Beneito
Hystrix		3\1										
Meles		2\1										
Mustela putorius												
M. nivalis					6\							
Martes												

	Beneito	Beneito	Beneito	Beneito	Beneito	Cova Negra	Cova Negra	Cova Negra	Cova Negra	Cova Negra	Cova Negra	Mallaetes	Mallaetes
Level	D1	C	B9	B8	B7	VI	V	IV	IIIB	IIIA	II	11	10
14C age (ka)	30 or 39	26 or 34				MIS 4	MIS 4	MIS 3	MIS 3	MIS 3	MIS 3	30.5	
Culture	Mou	Aur	late Aur	late Aur	Gr	Mou	Mou	Mou	Mou	Mou	final Mou	Aur	Gr
Sus scrofa			1\	2\			1\	1\	2\				2\1
Cervidae						5\	12\	36\	12\	8\	10\		
Cervus elaphus	19\	46\	45\	46\	159\	1\	7\	13\	4\	2\	2\	6\1	38\2
Rangifer tarandus													
Megaloceros													
Capreolus				1\					2\				
Dama dama						6\	8\	11\	2\	1\	2\		
Bovini		1\				5\	4\	8\	6\	1\	1\		
Bos primigenius	1\						1\						1\1
Caprini					4\	2\	6\	28\	74\	8\	57\		
Hemitragus						2\	2\	24\	58\	13\	70\		
Rupicapra													
Capra pyrenaica	46\	90\	90\	91\	177\				1\	1\	2\	4\1	13\1
Equus sp.						9\	2\	37\	38\	12\	23\		
E. hydruntinus													
E. caballus	20\	13\	12\	13\	48\		1\	2\					1\1
Rhinocerotidae													
Coelodonta													
S. kirchbergensis													
Stephanorhinus sp.									1\	2\			
S. hemitoechus													
Mammuthus													
Elephas sp.													
Canis lupus	1\				1\	1\		8\	5\		6\		

(continued)

Appendix 7.2 (continued)

	Beneito	Beneito	Beneito	Beneito	Beneito	Cova Negra	Cova Negra	Cova Negra	Cova Negra	Cova Negra	Cova Negra	Mallaetes	Mallaetes
Vulpes					1\				3\	1\	1\		
Alopex													
Cuon											4\		
Ursus sp.											1\		
U. arctos									1\		6\		
U. spelaeus													
Crocuta													
Hyaenidae	4\							2\					
Felis sylvestris		4\			1\						3\		3\1
Panthera spelaea													
P. leo													
P. pardus						1\				1\			
Lynx spelaea													
L. pardina			6\	11\	7\					2\	1\		
Erinaceus					2\	1\							
Leporidae													
Oryctolagus	113\	68\	385\	1534\	3625\	75\	41\	368\	337\	94\	151\	55\7	28\6
Lepus				2\	2\								
Marmota													
Castor										2\	2\		
Hystrix													
Meles													
Mustela putorius													
M. nivalis													
Martes													

Ages based on uncalibrated radiocarbon dates unless otherwise indicated (*U* Uranium Series). *Mou* Mousterian, *Aur* Aurignacian, *Gr* Gravettian. The numbers in the cells (n\n) refer to NISP and MNI counts, respectively. References are found in the chapter text

Appendix 7.3 Faunal data from MP and EUP sites in Andalucia (including Gibraltar)

	Zafarraya	Zafarraya	Zafarraya	Zafarraya	Zafarraya	Zafarraya	Zafarraya	Vanguard	Gorham's
Level	UG	UF	UE	UD	UC	UB	UA	C + D	IV
^{14}C age (ka)	46.5		48–46	47–41	19 or 53		19	>43	~28
Culture	Mou	Mou	Mou	Mou	Mou	UP	UP	Mou	Late Mou
Sus scrofa					4\2			11\	3\2
Cervidae									
Cervus elaphus	7\1	5\1	22\4	3\1	37\10	6\1	3\1	40\	89\7
Rangifer tarandus									
Megaloceros									
Capreolus									
Dama dama									
Bovini									
Bos primigenius		1\1	6\2		5\4				3\2
Caprini									
Hemitragus									
Rupicapra	7\2	2\2	14\6	1\1	9\2	2\2	?\5		
Capra pyrenaica	437\21	99\3	592\3	80\3	902\63	213\12	35\5	121\	205\13
Equus sp.									
E. hydruntinus				4\2					
E. caballus				5\3		3\2			4\3
Rhinocerotidae									
Coelodonta									
S. kirchbergensis									
Stephanorhinus sp									
S. hemitoechus									
Mammuthus									
Elephas sp.									
Canis lupus								2\	2\1
Vulpes			1\1			1\1	1\1		1\1
Alopex									
Cuon	9\2	2\1	21\1	2\1	61\11	13\2	8\1		
Ursus sp.									
U. arctos	5\2		2\2	1\1	3\1			2\	4\1
U. spelaeus									

(continued)

Appendix 7.3 (continued)

	Zafarraya	Zafarraya	Zafarraya	Zafarraya	Zafarraya	Zafarraya	Zafarraya	Vanguard	Gorham's
Crocuta								5\	19\3
Hyaenidae									
Felis sylvestris	17\1	14\2	31\4	6\1	45\5	39\4	7\1	4\	2\1
Panthera spelaea									
P. leo									
P. pardus								7\	4\1
Lynx spelaea									
L. pardina	4\1		1\1	1\1	13\5	4\2			12\2
Erinaceus									
Leporidae									
Oryctolagus									1620\97
Lepus									
Marmota									
Castor									
Hystrix									
Meles									
Mustela putorius									
M. erminea					1\1				
M. nivalis	3\3	2\1	2\2						
Martes									

Ages based on uncalibrated radiocarbon dates. *Mou* Mousterian, *UP* Upper Paleolithic. The numbers in the cells (n\n) refer to NISP and MNI counts, respectively. References are found in the chapter text

Appendix 7.4 Faunal data from MP and EUP sites in Portugal

	Val Boi	Picareiro	Picareiro	Picareiro	Picareiro	Picareiro	Picareiro	Picareiro	Picareiro	Caldeirao	Caldeirao	Anecrial	Anecrial	F. Brava	Columbeira
Level	Pooled Gr	HH	GG	FF	EE	BB	Z	U	T	N–K	J	2 inf.	2 sup.	2	8
^{14}C age (ka)	27–22			28.6				22.5	20.5	>27	27–22	23.5	21.5	30.5	29
Culture	Gr	Mou	Mou	Late Mou or EUP	Gr	Gr	Gr	Late Gr	Late Gr	Mou	EUP	Late Gr	Late Gr	Mou	Mou
Sus scrofa	1\				present					3\	17\			7\	
Cervidae															
Cervus elaphus	472\	Present	Present	Present			Present	Present	Present	50\	41\			116\	151\
Rangifer tarandus															
Megaloceros															
Capreolus								Present		1\	1\				1\
Dama dama															
Bovini															
Bos primigenius	22\									1\				25\	15\
Caprini															
Hemitragus															
Rupicapra								Present		5\	4\				
Capra pyrenaica	4\		Present			Present		Present	Present	12\	24\	75\1	15\1?	101\	22\
Equus sp.										24\					
E. hydruntinus	15\														
E. caballus	115\								Present					28\	16\
Rhinocerotidae															
Coelodonta															
S. kirchbergensis															
Stephanorhinus sp.															
S. hemitoechus														14\	15\
Mammuthus														10\	
Elephas sp.															
Canis lupus	2\									3\				1\	2\
Vulpes	9\									1\	1\			8\	
Alopex															
Cuon															
Ursus sp.															
U. arctos										1\	9\			3\	1\
U. spelaeus															
Crocuta										6\	4\			25\	24\
Hyaenidae															
Felis sylvestris														2\	1\
Panthera spelaea															
P. leo	3\										2\			1\	

(continued)

Appendix 7.4 (continued)

	Val Boi	Picareiro	Picareiro	Picareiro	Picareiro	Picareiro	Picareiro	Picareiro	Picareiro	Caldeirao	Caldeirao	Anecrial	Anecrial	F. Brava	Columbeira
P. pardus										1\				7\	
Lynx spelaea	11\														
L. pardina										11\	4\				7\
Erinaceus				Present				Present	Present					2\	
Leporidae															
Oryctolagus	2802\	Present	Present	Present	Present	Present	Present	Present	Present	806\	1551\	743\7	744\8	>100\	
Lepus											2\				
Marmota															
Castor															
Hystrix															
Meles															
Mustela putorius															
M. nivalis															
Martes															

Ages based on uncalibrated radiocarbon dates. *Mou* Mousterian, *EUP* Early Upper Paleolithic, *Gr* Gravettian. The numbers in the cells (n\n) refer to NISP and MNI counts, respectively. References are found in the chapter text

References

Altuna, J. (1972). Fauna de mamíferos de los yacimientos prehistóricos de Guipúzcoa. *Munibe, 24*, 1–464.

Altuna, J. (1978). Los mamíferos de Cueva Morin. In J. González Echegaray & L. G. Freeman (Eds.), *Vida* y *Muerte en Cueva Morín* (pp. 201–209). Santander: Institución Cultural de Cantabria.

Altuna, J. (1989). Subsistance d'origine animale pendant le Moustérien dans la région cantabrique. In M. Otte (Ed.), *L'Homme de Néandertal*, Vol. 6, *La Subsistance* (pp. 31–43). Liège: ERAUL 33.

Altuna, J. (1990). Caza y alimentación procedente de macromamíferos durante el Paleolítico de Amalda. In J. Altuna, A. Baldeón, & K. Mariezkurrena (Eds.), *La Cueva de Amalda* (pp. 149–192). San Sebastián: Sociedad de Estudios Vascos.

Altuna, J. (1996). Faunas de clima frío en la Península Ibérica durante el Pleistoceno superior. In C. Rodríguez, P. Ramil-Rego, & M. Rodríguez (Eds.), *Biogeografia Pleistocena-Holocena de la Península Ibérica* (pp. 13–39). Santiago de Compostela: Xunta de Galicia.

Altuna, J., & Mariezkurrena, K. (1988). Les macromammifères du Paléolithique moyen et supérieur ancien dans la région cantabrique. *Archaeozoologia, I*(2), 179–196.

Altuna, J., & Mariezkurrena, K. (1996). Primer hallazgo de restos óseos de antilope Saiga (*Saiga tatarica L.*) En la Península Ibérica. *Munibe, 48*, 3–6.

Altuna, J., & Mariezkurrena, K. (2000). Macromamíferos del yacimiento de Labeko Koba. In A. Arrizabalaga & J. Altuna (Eds.), *Labeko Koba* (*Munibe* 52), (pp. 107–181). San Sebastian: Sociedad de Ciencias Aranzadi.

Altuna, J., & Mariezkurrena, K. (2010). Tafocenosis en yacimientos del País Vasco con predominio de grandes carnivoros. Consideraciones sobre el yacimiento de Amalda. *Zona Arqueológica, 13*, 213–228.

Alvarez-Lao, D., & García, N. (2010). Chronological distribution of Pleistocene cold-adapted large mammal faunas in the Iberian Peninsula. *Quaternary International, 212*, 120–128.

Antunes, M. T. (2000a). *Ultimos Neandertais em Portugal: Evidencia, Odontologica e Outra*. Lisbon: Academia das Ciências de Lisboa.

Antunes, M. T. (2000b). The Pleistocene fauna from Gruta da Figueira Brava: A synthesis. In M. T. Antunes (Ed.), *Ultimos Neandertais em Portugal* (pp. 259–282). Lisbon: Academia das Ciências de Lisboa.

Antunes, M. T. (2000c). Pleistocene marine animals. In M. T. Antunes (Ed.), *Ultimos Neandertais em Portugal* (pp. 246–257). Lisbon: Academia das Ciências de Lisboa.

Aura, J. E., Jordá, J., Pérez Ripoll, M., & Rodrigo, M. J. (2001). Sobre dunas, playas y calas. Los pescadores prehistóricos de la Cueva de Nerja (Málaga) y su expresión arqueológica en el tránsito Pleistoceno-Holoceno. *Archivo de Prehistoria Levantina* 24, 9–39.

Aura, J. E., Villaverde, V., Pérez Ripoll, M., Martínez Valle, R., & Guillem, P. (2002). Big game and small prey: Paleolithic and Epipaleolithic economy form Valencia. *Journal of Archaeological Method and Theory* 9, 215–268.

Barroso, C., Bailon, S., El Guennouni, K., & Desclaux, E. (2006a). Les lagomorphes (Mammalia, Lagomorpha) du Pléistocène supérieur de la Grotte du Boquete de Zafarraya. In C. Barroso & H. de Lumley (Eds.), *La Grotte du Boquete de Zafarraya* (Vol. 2, pp. 893–926). Sevilla: Junta de Andalucía.

Barroso, C., Riquelme, J. A., Moigne, A.-M., & Banes, L. (2006b). Les faunes de grands mammifères du Pléistocène supérieur de la Grotte du Boquete de Zafaraya. Etude paléontologique, paléoécologique et archéologique. In C. Barroso & H. de Lumley (Eds.), *La Grotte du Boquete* de *Zafarraya* (Vol. 2, pp. 675–891). Seville: Junta de Andalucía.

Barton, R. N. E. (2000). Mousterian hearths and shellfish: Late Neanderthal activities on Gibraltar. In C. Stringer, R. N. E. Barton, & J. C. Finlayson (Eds.), *Neanderthals on the Edge* (pp. 211–220). Oxford: Oxbow Books.

Bicho, N., & Haws, J. (2008). At the land's end: Marine resources and the importance of fluctuations in the coastline in the prehistoric hunter-gatherer economy of Portugal. *Quaternary Science Reviews, 27*, 2166–2175.

Bicho, N., Manne, T., Cascalheira, J., Mendoça, C., Evora, M., Gibaja, J. & Pereira, T. (2010a). O Paleolitico superior do sudoeste da Península Ibérica: o caso do Algarve. In *El Paleolítico Superior Peninsular*: Novedades del Siglo XXI, pp. 215–234. Barcelona.

Bicho, N., Gibaja, J. F., Stiner, M., & Manne, T. (2010b). Le Paléolithique supérieur au sud du Portugal: le site de Val Boi. *L'Anthropologie, 114*, 48–67.

Blasco, M. F. (1995). *Hombres*, *Fieras* y *Presas*: *Estudio Arqueozoológico* y *Tafonómico del Yacimiento del Paleolítico Medio de la Cueva de Gabasa* I (Monografías Arqueológicas 38). Zaragoza: Universidad de Zaragoza.

Blasco, R. (2008). Human consumption of tortoises at Level IV of Bolomor Cave (Valencia, Spain). *Journal of Archaeological Science, 35*, 2839–2848.

Bocherens, H., Fizet, M., & Mariotti, A. (1994). Diet, physiology and ecology of fossil mammals as inferred from stable carbon and nitrogen isotope biogeochemistry: Implications for Pleistocene bears. *Palaeogeography, Palaeoclimatology, Palaeoecology, 107*, 213–225.

Broughton, J., Cannon, M., Bayham, F., & Byers, D. (2011). Prey body size and ranking in zooarchaeology: Theory, empirical evidence and applications from the northern Great Basin. *American Antiquity, 76*, 403–428.

Cabrera, V. (1984). *El Yacimiento de la Cueva de "El Castillo"* (Bibliotheca Praehistoria Hispana 22). Madrid: CSIC.

Cabrera, V., Bernaldo de Quirós, F., & Maillo, J. M. (2006). La Cueva de El Castillo: las nuevas excavaciones. In V. Cabera, F. Bernaldo de Quirós, & J. M. Maillo (Eds.), *En el Centenario de la Cueva del Castillo* (pp. 349–365). Santander: Universidad Nacional de Educación a Distancia.

Cáceres, I., Rosell, J., & Huguet, R. (1998). Séquence d'utilisation de la biomasse animale dans le gisement de l'Abric Romani. *Quaternaire, 9*, 379–383.

Callapez, P. (2000). Upper Pleistocene marine invertebrates from Gruta da Figueira Brava. In M. Telles Antunes (Ed.), *O Ultimos Neandertais em Portugal* (pp. 84–101). Lisbon: Academia das Ciências de Lisboa.

Carrión, J. S., Finlayson, C., Fernández, S., Finlayson, C., Allué, E., López-Sáez, J., et al. (2008). A coastal reservoir of biodiversity for Upper Pleistocene human populations: Palaeoecological investigations in Gorham's Cave (Gibraltar) in the context of the Iberian Peninsula. *Quaternary Science Reviews, 27*, 2118–2135.

Castaños, P. (2005). Revisión actualizada de las faunas de macromamíferos del Würm antiguo en la región cantábrica. In R. Montes & J. A. Lasheras (Eds.), *Neanderthales Cantábricos*, *Estado de la Cuestión* (pp. 201–207). Santander: Museo Nacional y Centro de Investigación de Altamira, Monografías 20.

Cortés, M. (2007a). *El Paleolítico Medio y Superior en el Sector Central de Andalucía*. Madrid: Monografias del Museo de Altamira 22.

Cortés, M. (2007b). *Cueva Bajondillo (Torremolinos)*. Málaga: Diputación de Málaga.

Cortés, M, Simón, M., Fernández, E., Gutierrez, C., Lozano-Francisco, M., Morales, A., et al. (2005). Algunos datos sobre el Paleolítico Superior de la Cueva de Nerja. In J. L. Sanchidrián, A. M. Márquez, & J. M. Fullola (Eds.), *La Cuenca Mediterránea durante el*

Paleolítico Superior (pp. 298–315). Málaga: Fundación Cueva de Nerja.

Cortés, M., Morales, A., Simón, M., Bergadà, M., Delgado, A., López, P., et al. (2008). Palaeoenvironmental and cultural dynamics of the coast of Málaga (Andalusia, Spain) during the Upper Pleistocene and early Holocene. *Quaternary Science Reviews, 27*, 2176–2193.

Dari, A. (1999). Les grands mammifères du site Pleistocène supérieur de la grotte du Castillo. *Espacio, Tiempo y Forma. Serie I, Prehistoria y Arqueología, 12*, 103–107.

Dari, A., & Renault-Miskovsky, J. (2001). Etudes paléoenvironnementales dans la grotte "El Castillo". *Espacio, Tiempo y Forma. Serie I, Prehistoria y Arqueología, 14*, 121–144.

Davidson, I. (1989). *La Economía del Final del Paleolítico en la España Oriental*. Valencia: Servicio de Investigación Prehistórica, Trabajos Varios 85.

Davis, S. (2002). The mammals and birds from the Gruta do Caldeirão, Portugal. *Revista Portuguesa de Arqueologia, 5*, 29–98.

d'Errico, F., & Sanchez-Goñi, M. F. (2003). Neanderthal extinction and the millennial scale climatic variability of OIS 3. *Quaternary Science Reviews, 22*, 769–788.

Estevez, J. (1987). La fauna de l'Arbreda (sector Alfa) en el conjunt de faunes del Plistocè Superior. *Cypsela, 6*, 73–87.

Fernández-Jalvo, Y., & Andrews, P. (2000). The taphonomy of Pleistocene caves, with particular reference to Gibraltar. In C. Stringer, R. N. E. Barton, & J. C. Finlayson (Eds.), *Neanderthals on the edge* (pp. 171–182). Oxford: Oxbow Books.

Fernández-Laso, M. C., Rivales, F., & Rosell, J. (2010). Intra-site changes in seasonality and their consequences on the faunal assemblages from Abric Romani. *Quaternaire, 21*, 255–263.

Finlayson, C., Giles, F., Rodríguez-Vidal, J., Fa, D., Gutierrez, J., Santiago, A., et al. (2006). Late survival of Neanderthals at the southernmost extreme of Europe. *Nature* 443, 850–853 (with supplementary on-line information).

Freeman, L. G. (1973). The significance of mammalian faunas from Paleolithic occupations in Cantabrian Spain. *American Antiquity, 38*, 3–44.

Freeman, L. G. (1981). The fat of the land: Notes on Paleolithic diet in Iberia. In R. S. O. Harding & G. Teleki (Eds.), *Omnivorous primates* (pp. 104–165). New York: Columbia University Press.

Garrod, D., Buxton, L., Smith, G. E., & Bate, D. (1928). Excavation of a Mousterian rock-shelter at Devil's Tower, Gibraltar. *Journal of the Royal Anthropological Institute, 58*, 33–113.

González-Sampériz, P., Leroy, S., Carrión, J., Fernández, S., García-Antón, M., Uzquiano, P., et al. (2010). Steppes, savannahs, forests and phytodiversity reservoirs during the Pleistocene in the Iberian Peninsula. *Journal of Palaeobotany & Palynology*. doi: 10.1016/j.revpalbo.2010.03.009.

Grayson, D., & Delpech, F. (2002). Specialized Early Upper Palaeolithic hunters in Southwestern France? *Journal of Archaeological Science, 29*, 1439–1449.

Grayson, D., & Delpech, F. (2003). Ungulates and the Middle-Upper Paleolithic transition at Grotte XVI (Dordogne, France). *Journal of Archaeological Science, 30*, 1633–1648.

Grayson, D., & Delpech, F. (2006). Was there increasing dietary specialization across the Middle-to-Upper Paleolithic transition in France? In N. Conard (Ed.), *When Neanderthals and Modern Humans Met* (pp. 377–417). Tübingen: Kerns Verlag.

Grayson, D., & Delpech, F. (2008). The large mammals of Roc de Combe (Lot, France): The Châtelperronian and Aurignacian assemblages. *Journal of Anthropological Archaeology, 27*, 338–362.

Green, R. et al. (2010). A draft sequence of the Neandertal genome. Science 328: 710–725.

Hilderbrand, G., Farley, S., Robbins, C., Hanley, T., Titus, K., & Servheen, C. (1996). Use of stable isotopes to determine diets of living and extinct bears. *Canadian Journal of Zoology, 74*, 2080–2088.

Hockett, B., & Bicho, N. (2000). The rabbits of Picareiro Cave: Small mammal hunting during the late Upper Palaeolithic in the Portuguese Estremadura. *Journal of Archaeological Science, 27*, 715–723.

Hockett, B., & Haws, J. (2002). Taphonomic and methodological perspectives of leporid hunting during the Upper Paleolithic of the Western Mediterranean Basin. *Journal of Archaeological Method and Theory, 9*, 269–302.

Hockett, B., & Haws, J. (2005). Nutritional ecology and the human demography of Neanderthal extinction. *Quaternary International, 137*, 21–34.

Hockett, B., & Haws, J. (2009). Continuity in animal resource diversity in Late Pleistocene human diet of Central Portugal. *Before Farming 2009/2.2*, 1–13.

Iturbe, G., Fumanal, M. P., Carrión, J. S., Cortell, E., Martínez, R., Guillem, P., et al. (1993). Cova Beneito: una perspectiva interdisciplinar. *Recerques del Museu d'Alcoi, 2*, 23–88.

Jennings, R., Giles, F., Barton, R., Collcutt, S., Gale, R., Gleed-Owen, C., et al. (2009). *Quaternary Science Reviews*. doi: 10.1016/j.quascirev.2008.11.014.

Jones, E.L. (2012). Upper Paleolithic rabbit exploitation and landscape patchiness: the Dordogne vs. Mediterranean Spain. *Quaternary International 265*, 52–60.

Klein, R., & Cruz-Uribe, K. (1994). The Paleolithic mammalian fauna from the 1910-14 excavations at El Castillo Cave. In J. A. Lasheras (Ed.), *Homenaje al Dr. Joaquín González Echegaray* (pp. 141–158). Madrid: Museo y Centro de Investigación de Altamira, Monografías 17.

Klein, R., & Steele, T. (2008). Gibraltar data are too sparse to inform on Neanderthal exploitation of coastal resources. *Proceedings of the National Academy of Sciences of the United States of America, 105*, E115.

Landry, G., & Burke, A. (2007). El Castillo: The Obermaier faunal collection. In J. M. Maillo & E. Baquedano (Eds.) *Miscelánea en Homenaje a Victoria Cabrera* (Vol. I, pp. 104–112). Alcalá: Zona Arqueológica 7.

Lindly, J. (1988). Hominid carnivore activity at Middle and Upper Paleolithic cave sites in eastern Spain. *Munibe, 40*, 45–70.

Manne, T., & Bicho, F. (2009). Val Boi: Rendering new understandings of resource intensification & diversification in southwestern Iberia. *Before Farming 2009/2.1*, 1–21.

Manne, T., & Bicho, N. (2011). Prying new meaning from limpet harvesting at Vale Boi during the Upper Paleolithic. In N. Bicho, J. Haws, & L. Davis (Eds.), *Trekking the Shore* (pp. 273–289). New York: Springer.

Marín-Arroyo A. B. (2009a). The human use of the montane zone of Cantabrian Spain during the Lalte Glacial: Faunal evidence from El Mirón Cave. *Journal of Anthropological Research, 65*, 69–102.

Marín-Arroyo A. B. (2009b). Economic adaptations during the Late Glacial in northern Spain. A simulation approach. *Before Farming 2009/2.3*, 1–18.

Marín-Arroyo A. B. (2010). *La fauna de Mamíferos en el Cantábrico Oriental durante el Magdaleniense y Aziliense*. Santander: Ediciones de la Universidad de Cantabria.

Maroto, J., Soler, N., & Fullola, J. M. (1996). Cultural change between Middle and Upper Palaeolithic in Catalonia. In E. Carbonell & M. Vaquero (Eds.), *The last Neanderthals, The first antatomically modern humans* (pp. 219–250). Tarragona: Universitat Rovira i Virgili.

Martín Penela, A. (1986). Los grandes mamíferos del yacimiento pleistoceno superiorde Cueva Hora (Darro, Granada, España). *Antropología y Paleoecología Humana, 4*, 107–128.

Martínez-Moreno, J. (2005). Una aproximación zooarqueológica al estudio de los patrones de subsistencia del Paleolítico Medio Cantábrico. In R. Montes & J. A. Lasheras (Eds.), *Neanderthales Cantábricos, Estado de la Cuestión* (pp. 209–230). Santander: Museo Nacional y Centro de Investigación de Altamira, Monografías 20.

Pacher, M., & Stuart, A. J. (2008). Extinction chronology and palaeobiology of the cave bear (*Ursus spelaeus*). *Boreas*. doi: 10.1111/j.1502-3885.2008.00071.x.

Pérez Ripoll, M. (2001). Marcas antrópicas en los huesos de conejo. In V. Villaverde (Ed.), *De Neanderthales a Cromañones: El Inicio del Poblamiento Humano en las Tierras Valencianas* (pp. 119–124). Valencia: Universitat de Valencia.

Pérez Ripoll, M. (2006). Caracterización de las fracturas antrópicas y sus tipologías en huesos de conejo procedentes de los niveles gravetienses de la Cova de les Cendres. *Munibe, 57*(1), 239–254.

Pérez Ripoll, M., & Martínez R. (2001). La caza, el aprovechamiento de las presas y el comportamiento de las comunidades cazadores prehistóricas. In V. Villaverde (Ed.), *De Neanderthales a Cromañones: El Inicio del Poblamiento Humano en las Tierras Valencianas* (pp. 73–98). Valencia: Universitat de Valencia.

Quesada, J. M. (2006). Faunas del Auriñaciense y Gravetiense cantábrico. In J. Maillo & E. Baquedano (Eds.), *Miscelánea en Homenaje* a *Victoria Cabrera* (Vol. 1, pp. 406–421). Alcalá: Zona Arqueológica 7.

Rodríguez, J., Giles, F., & Riquelme, J. (2010). El registro fósil de hienas en las cuevas de Gorham's y Vanguard (Gibraltar): contexto paleogeográfico. *Zona Arqueológica, 13*, 354–364.

Rosell, J., et al. (2010). A stop along the way: The role of Neanderthal groups at Level III of Teixoneres Cave (Moià, Barcelona, Spain). *Quaternaire, 21*, 239–253.

Rosell, J., Blasco, R., Huguet, R., Caceres, I., Saladie, P., Rivals, F., et al. (2012). Occupational patterns and subsistence strategies in Level J of Abric Romani. In E. Carbonell (Ed.), *High resolution archaeology and Neanderthal behavior: Time and space in Level J of Abric Romaní (Capellades, Spain)* (pp. 313–372). New York: Springer.

Speth, J., & Spielmann, K. (1983). Energy source, protein metabolism, and hunter-gatherer subsistence strategies. *Journal of Anthropological Archaeology, 2*, 1–31.

Stiner, M. (1994). *Honor among thieves: A zooarchaeological study of Neanderthal ecology*. Princeton: Princeton University Press.

Stiner, M. (2001). Thirty years on the "Broad Spectrum Revolution" and Paleolithic demography. *Proceedings of the National Academy of Sciences of the United States of America, 98*, 6993–6996.

Stiner, M. (2003). Zooarchaeological evidence for resource intensification in Algarve, Southern Portugal. *Promontoria, 1*, 27–61.

Stiner, M., Munro, N., Surovell, T., Tchernov, E., & Bar-Yosef, O. (1999). Paleolithic population growth pulses evidenced by small animal exploitation. *Science, 283*, 190–194.

Straus, L. G. (1977). Of deerslayers and mountain men: Paleolithic faunal exploitation in Cantabrian Spain. In L. R. Binford (Ed.), *For theory building in archaeology* (pp. 41–76). New York: Academic Press.

Straus, L. G. (1982). Carnivores and cave sites in Cantabrian Spain. *Journal of Anthropological Research, 38*, 75–96.

Straus, L. G., & Clark, G. A. (1986). La Riera Cave. *Tempe: Anthropological Research Papers 36.*

Straus, L. G. (1992). *Iberia before the Iberians*. Albuquerque: University of New Mexico Press.

Straus, L. G. (1993). Upper Paleolithic hunting tactics and weapons in Western Europe. In. G. L. Peterkin, H. Bricker, & P. Mellars (Eds.), *Hunting and animal exploitation in the Later Palaeolithic and Mesolithic of Eurasia* (pp. 83–93). Washington: Archaeological Papers of the American Anthropological Association 4.

Stringer, C., Finlayson, J. C., et al. (2008). Neanderthal exploitation of marine mammals in Gibraltar. *Proceedings of the National Academy of Sciences of the United States of America, 105*, 14319–14324.

Stuart, A. J., & Lister, A. (2007). Patterns of Late Quaternary megafaunal extinctions in Europe and northern Asia. *Courier Forschungsinstitut Senckenberg, 259*, 287–297.

Villaverde, V., Martínez-Valle, R., Guillem, P., & Fumanal, M. P. (1996). Mobility and the role of small game in the Middle Paleolithic of the central region of the Spanish Mediterranean: A comparison of Cova Negra with other Paleolithic deposits. In E. Carbonell & M. Vaquero (Eds.), *The last Neanderthals the first anatomically modern humans* (pp. 267–288). Tarragona: Universitat Rovira i Virgili.

Yravedra, J. (2002). Subsistencia en la transición del Paleolítico Medio al Paleolítico Superior de la Peninsula Ibérica. *Trabajos de Prehistoria, 59*, 9–28.

Yravedra, J. (2006). Acumulaciones biológicas en yacimientos arqueológicos: Amalda VII y Esquilleu III-IV. *Trabajos de Prehistoria, 63*, 55–78.

Yravedra, J. (2007). Nuevas contribuciones en el comportamiento cinegético de la Cueva de Amalda. *Munibe, 58*, 43–88.

Zilhão, J., & Trinkaus, E. (2002). *Portrait of the Artist as a Child: The Gravettian Human Skeleton from the Abrigo do Lagar Velho and its Archaeological Context*. Lisbon: Trabalhos de Arqueologia 22.

Chapter 8
Animal Exploitation Strategies during the Uluzzian at Grotta di Fumane (Verona, Italy)

Antonio Tagliacozzo, Matteo Romandini, Ivana Fiore, Monica Gala, and Marco Peresani

Introduction

The intense debate regarding the Middle Paleolithic to Early Upper Paleolithic transition in Europe includes efforts to reconstruct subsistence strategies, particularly relating to the differences and similarities between the food supplies of Neanderthals and anatomically modern humans, and the alleged broadening of the diet by the first *sapiens* (Chase 1989; Clark 1997; Stiner 1999; Marean 2005; Bocherens 2009; Gaudzinski and Niven 2009). Of crucial relevance to resolving these questions is evidence from the major transitional cultural complexes specific to certain regions of the continent. These include the Châtelperronian in the west, the Neronian in southeast France, the Lincombian in the far north, the Szeletian in Central Europe and the Uluzzian in the south (Conard 2006; Slimak 2008). Considered, sometimes with considerable uncertainty, as a creation of the last Neanderthals (Bar-Yosef and Bordes 2010), these complexes have provided important information showing that subsistence was focused on meat from various species of ungulates.

The picture is still suffering from limited and discontinuous data, especially concerning the Uluzzian, a cultural complex widespread on the Italian peninsula and in Greece (Ronchitelli et al. 2009) and recently attributed to anatomically modern humans on the basis of a thorough re-examination of some teeth (Benazzi et al. 2011). In southern Italy, Uluzzian hunting strategies and food sourcing are relatively well known thanks to data from Cavallo Cave layer EIII5 that attest to the use of aurochs and secondly, deer and horse (Boscato and Crezzini 2012). However, similar information is still scanty in central and northern Italy, where the Uluzzian technocomplex has been identified at the foot of the Alps (Peresani 2008). This paper contributes to the dataset regarding Uluzzian hunting and food sourcing patterns in northern Italy by presenting results of the archaeozoological and taphonomic study of a sample of an estimated 37,000 bone remains recovered in the Uluzzian occupation levels (Layer A3-A4) at Fumane Cave (Grotta di Fumane). The aim here is to reconstruct subsistence practices and to provide a basis for further study of the relationship between economic changes and the new human population through comparison with evidence for the latest Mousterian occupations at this important site.

A. Tagliacozzo (✉) · I. Fiore · M. Gala
Soprintendenza al Museo Nazionale Preistorico Etnografico "L. Pigorini", Sezione di Paleontologia del Quaternario e Archeozoologia, P. le Marconi 14, 00144 Rome, Italy
e-mail: antonio.tagliacozzo@beniculturali.it

I. Fiore
e-mail: iva_fiore@yahoo.it

M. Gala
e-mail: monarix@yahoo.it

M. Romandini · M. Peresani
Università degli Studi di Ferrara, Dipartimento di Studi Umanistici, Sezione di Quaternario, Preistoria e Antropologia, Corso Ercole I d'Este, 32, 44100 Ferrara, Italy
e-mail: matteo.romandini@unife.it

M. Peresani
e-mail: marco.peresani@unife.it

Site Description

Grotta di Fumane lies 350 m above sea level at the southern fringe of the Venetian Pre-Alps in a landscape characterized by long, narrow valleys and peaks with elevations above 1,500 m. The cave is part of a complex karst system composed of several cavities that allowed for the formation of a sedimentary sequence over 12 m thick (Fig. 8.1).

The site has produced a dated sequence from the Middle to Early Upper Palaeolithic spanning the period from MIS 5 to MIS 2 (Martini et al. 2001; Peresani et al. 2008; Higham et al. 2009). Since 1988, excavations have been carried out yearly, for periods of variable duration, beyond the

Jamie L. Clark and John D. Speth (eds.), *Zooarchaeology and Modern Human Origins: Human Hunting Behavior during the Later Pleistocene*, Vertebrate Paleobiology and Paleoanthropology, DOI: 10.1007/978-94-007-6766-9_8,

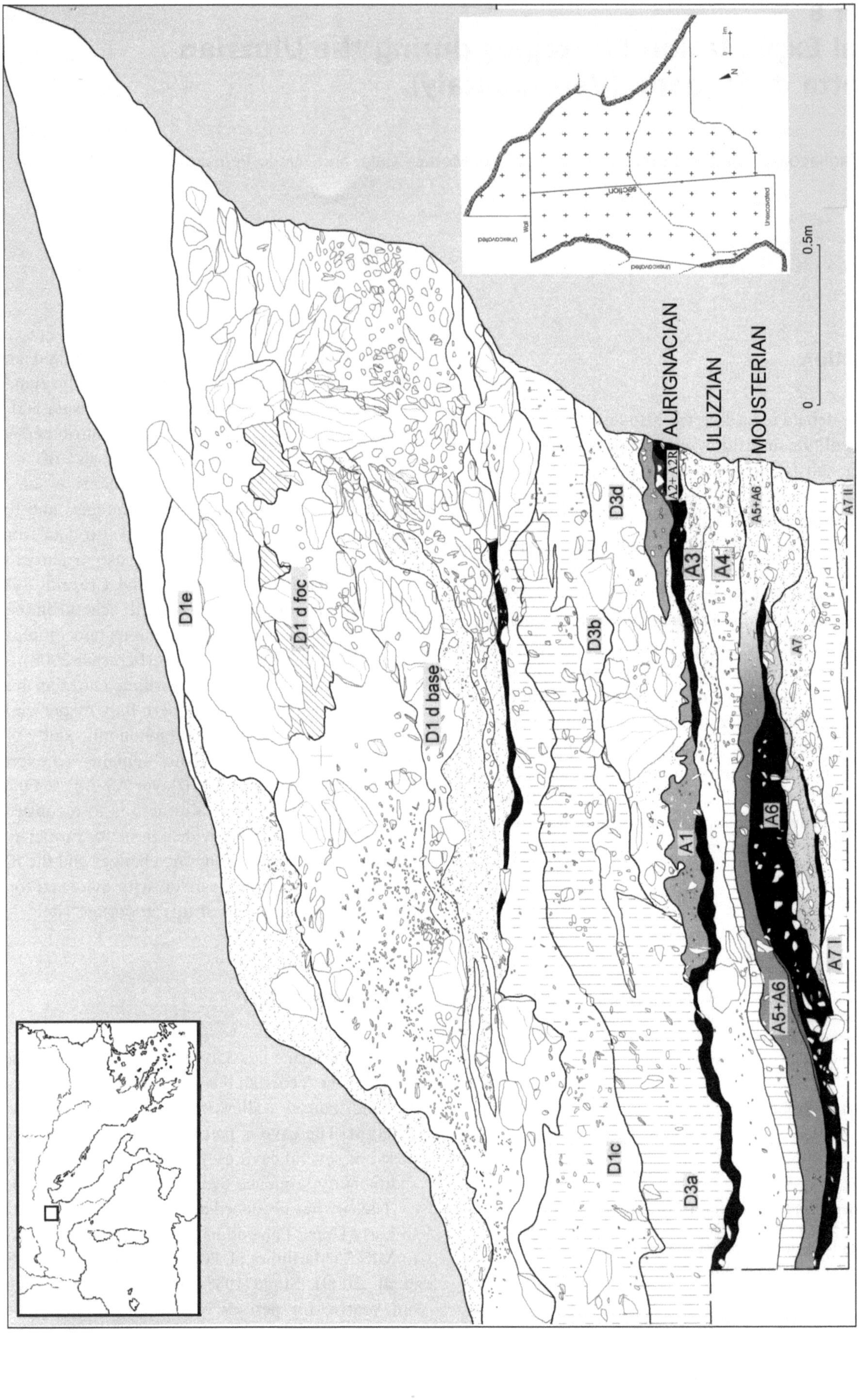

Fig. 8.1 Stratigraphy of Grotta di Fumane indicating the late Mousterian (A6-A5), Uluzzian (A4-A3), and Aurignacian (A2-A2R) layers

present-day drip-line and at the cave entrance. In this area, sediments from the final Mousterian to the Aurignacian derive primarily from frost-shattered slabs in which sand and aeolian dust are present in varying amounts, the former being prevalent in the western zone and the latter increasing from the entrance to the exterior. Lithics, faunal remains, hearths and other structures are densely scattered on the ground, particularly in layers A11, A10, A9, A6, A5 (Mousterian) and A2, A1 (Aurignacian), whereas lighter densities have been noted in A4, A3 (Uluzzian), D3, and D1 (Aurignacian) (Broglio et al. 2006; Peresani et al. 2011a).

Some osteological analyses were carried out on remains from the Mousterian-Aurignacian levels from the 1988–1991 and 1992–1996 excavations (Cassoli and Tagliacozzo 1994; Cremaschi et al. 2005; Gala and Tagliacozzo 2006). Results show relevant ecological and environmental changes from the Mousterian (cool-temperate, with prevailing red deer and roe deer) to the Aurignacian (cool, with prevailing ibex and chamois; evidence for cool climatic conditions also provided by carnivores and birds).

Uluzzian layers have been excavated at various intervals since 1989, with more extensive explorations undertaken during the last 2 years in a 20 m^2 area at the left side of the cave's mouth. In these most recent excavations, the layers were excavated in 33 $\times$ 33 cm squares and bones and lithic material were either directly positioned using the total station or recovered through flotation and wet sieving. Dwelling structures with hearths and a toss-zone—namely a concentration of discarded knapped by-products, charcoal and few bones—have been brought to light; flakes and bones have also been found scattered on the soil and in proximity to the combustion structures (Peresani 2008). The lithic technological and typological features of layers A4 and A3 are clearly distinct from those of the late Mousterian (A6, A5) and the directly overlying Aurignacian levels A2 and A1 (Peresani 2012; Broglio et al. 2005, 2009). The Uluzzian, in fact, coincides with a group of innovations in technological procedures like blade- and bladelet-production and in the use of new inventories of tools. The Levallois is still the most commonly used method in the initial phase, but it then disappears. Levallois cores for the extraction of polygonal or fan-shaped flakes were made using the centripetal rather than the unidirectional recurrent modality. Modelled ages of the Uluzzian based on radiocarbon measurements lie between 43.6–43.0 and 41.2–40.4 cal kBP (Higham et al. 2009).

Materials and Methods

Taxonomic and skeletal element identifications made in this study are based on the reference collection of the Laboratorio di Paleontologia del Quaternario e Archeozoologia of the Museo Nazionale Preistorico Etnografico "L. Pigorini" (Rome, Italy). Microscopic analyses of the bone surfaces were carried out using Nikon 1000 (Rome) and Leica S6D Green Ough (Ferrara) stereomicroscopes with a 20–220 magnification range. In order to identify the nature of the surface alterations on bones, and to distinguish human traces from animal traces, trampling abrasion, and modern mechanical modifications produced by excavation tools, reference was made to the well-established taphonomic literature (Binford 1981; Brain 1981; Potts and Shipman 1981; Shipman 1981; Shipman and Rose 1984; Blumenschine and Selvaggio 1988; Capaldo and Blumenschine 1994; Lyman 1994; Blumenschine 1995; Fisher 1995). The degree of combustion was evaluated employing the methodology developed by Stiner et al. (1995).

Sex and age at death were determined in order to reconstruct strategies of exploitation of the different species (Aitken 1974; Mariezkurrena 1983; Vigal and Machordom 1985; d'Errico and Vanhaeren 2002; Fiore and Tagliacozzo 2006). Measurements were taken following von den Driesch (1976). In order to evaluate species abundance, the following methods were used: number of identified specimens (NISP) (Grayson 1984), minimum number of elements (MNE) (Binford 1981; Klein and Cruz-Uribe 1984; Stiner 1994), and the estimate of the minimum number of individuals (MNI) (Bökönyi 1970). Estimation of deer MNI is based on the eruption and dental wear. Fragmentation indices (Binford 1981; Brain 1981; Lyman 1994) were calculated to evaluate the skeletal representation of the different animals and the skeletal survival rate.

Results

Bone Assemblage

Only 1,188 out of 36,944 bone remains (3 % of the faunal assemblage) have been identified to the taxonomic level (Table 8.1). Seventeen genera and nineteen species belonging to five orders of mammals (lagomorphs, rodents, carnivores, perissodactyls and artiodactyls) have been recognized. Some of these mammals are represented by only very few specimens (*Ursus spelaeus*, Mustelids, *Crocuta crocuta*, *Lynx lynx*, *Panthera pardus*, *Coelodonta antiquitatis*), while others are represented by a large quantity of bone elements (*Cervus elaphus*, *Capra ibex*, *Capreolus capreolus*, foxes). Among the layers considered, differences in faunal composition are minimal, with red deer always prevalent over ibex and roe deer. However, in layer A3 red deer decreases compared to A4, while conversely ibex, chamois and bison increase. Also in A3, there are woolly rhino, glutton, ermine and lynx, which suggest colder

Table 8.1 Mammals NISP (number identified specimens), NISP %, MNI (minimum number individuals) and MNI % for the Uluzzian layers A4-A3

Fumane Cave—Taxa	A4		A3		Total (layers A4-A3)			
	NISP	%	NISP	%	NISP	%	MNI	%
Lepus sp.			3	*0.5*	3	*0.3*	1	*1.4*
Marmota marmota			2	*0.3*	2	*0.2*	1	*1.4*
Canis lupus	11	*1.8*	21	*3.7*	32	*2.7*	4	*5.5*
Vulpes vulpes	61	*9.9*	36	*6.3*	97	*8.2*	4	*5.5*
Vulpes/Alopex	5	*0.8*	6	*1.0*	11	*0.9*	2	*2.7*
Ursus arctos	11	*1.8*	7	*1.2*	18	*1.5*	4	*5.5*
Ursus spelaeus	2				2	*0.2*	2	*2.7*
Ursus sp.	2		2	*0.3*	4	*0.3*		
Mustela erminea			1	*0.2*	1	*0.1*	1	*1.4*
Mustela nivalis	1		2	*0.3*	3	*0.3*	1	*1.4*
Gulo gulo	1		3	*0.5*	4	*0.3*	1	*1.4*
Crocuta crocuta			3	*0.5*	3	*0.3*	1	*1.4*
Lynx lynx			1	*0.2*	1	*0.1*	1	*1.4*
Panthera pardus	2				2	*0.2*	1	*1.4*
Carnivora indet.	5	*0.8*	7	*1.2*	12	*1.0*		
Coelodonta antiquitatis			1	*0.2*	1	*0.1*	1	*1.4*
Megaloceros giganteus	12	*1.9*	8	*1.4*	20	*1.7*	2	*2.7*
Cervus elaphus	242	*39.3*	169	*29.5*	411	*34.6*	13	*17.8*
Capreolus capreolus	54	*8.8*	50	*8.7*	104	*8.8*	6	*8.2*
Cervidae indet.	29	*4.7*	33	*5.8*	62	*5.2*		
Bison priscus	5	*0.8*	5	*0.9*	10	*0.8*	5	*6.8*
Bos/Bison	16	*2.6*	24	*4.2*	40	*3.4*		
Capra ibex	82	*13.3*	116	*20.3*	198	*16.7*	14	*19.2*
Rupicapra rupicapra	31	*5.0*	34	*5.9*	65	*5.5*	8	*11.0*
Caprinae indet.	13	*2.1*	12	*2.1*	25	*2.1*		
Ungulata indet.	31	*5.0*	26	*4.5*	57	*4.8*		
Total NISP	**616**	***100***	**572**	***100***	**1188**	***100***	**73**	***100***
Indeterminate								
Mammals small size	11	*0.1*	7	*0.0*	18	*0.1*		
Mammals medium-small size	4	*0.0*	3	*0.0*	7	*0.0*		
Mammals medium size	375	*1.9*	352	*2.1*	727	*2.0*		
Mammals large-medium size	63	*0.3*	91	*0.6*	154	*0.4*		
Mammals large size	171	*0.9*	165	*1.0*	336	*0.9*		
Fragm. indet.	18715	*96.8*	15799	*96.2*	34514	*96.5*		
Total indet.	**19339**	***96.9***	**16417**	***96.6***	**35756**	***96.8***		
Total NISP	**616**	***3.1***	**572**	***3.4***	**1188**	***3.2***		
Total NR	**19955**	***100***	**16989**	***100***	**36944**	***100***		
Fragm. burned	5187	*26.0*	2840	*16.7*	8027	*21.7*		
Fragm. calcined	2134	*10.7*	1883	*11.1*	4017	*10.9*		
Total burned	**7321**	***36.7***	**4723**	***27.8***	**12044**	***32.6***		

environmental conditions (Table 8.1). Of the 575 bird remains, 47.1 % have been identified at genera or species levels, 6.4 % are undeterminable and 46.4 % are in the process of being identified (Table 8.2).

Bones are mostly fragmented and over 95 % of the specimens are smaller than 3 cm. Among the analyzed remains, 32 % are burned: over 67 % of these show a brown or black color (burning temperature c. 200–400 °C),

Table 8.2 Birds NISP (number identified specimens), NISP %, MNI (minimum number individuals) and MNI % for the Uluzzian layers A4-A3

Fumane Cave—Taxa	A4				A3				Total (layers A4-A3)			
	NISP	%	MNI	%	NISP	%	MNI	%	NISP	%	MNI	%
Anas platyrhynchos	1	*0.7*	1	*3.3*					1	*0.4*	1	*1.8*
Aquila chrysaetos	1	*0.7*	1	*3.3*					1	*0.4*	1	*1.8*
Falco tinnunculus					4	*3.1*	1	*3.8*	4	*1.5*	1	*1.8*
Falco subbuteo	1	*0.7*	1	*3.3*	1	*0.8*	1	*3.8*	2	*0.7*	2	*3.6*
Falco sp.	1	*0.7*	1	*3.3*					1	*0.4*	1	*1.8*
Lagopus mutus					1	*0.8*	1	*3.8*	1	*0.4*	1	*1.8*
Tetrao tetrix	28	*19.4*	3	*10.0*	24	*18.9*	4	*15.4*	52	*19.2*	7	*12.5*
cfr. *Tetrao tetrix*	2	*1.4*			2	*1.6*			4	*1.5*		
Alectoris graeca	1	*0.7*	1	*3.3*					1	*0.4*	1	*1.8*
Perdix perdix	3	*2.1*	1	*3.3*	1	*0.8*	1	*3.8*	4	*1.5*	2	*3.6*
Coturnix coturnix					2	*1.6*	1	*3.8*	2	*0.7*	1	*1.8*
Rallus aquaticus	1	*0.7*	1	*3.3*					1	*0.4*	1	*1.8*
Crex crex	24	*16.7*	4	*13.3*	25	*19.7*	4	*15.4*	49	*18.1*	8	*14.3*
Scolopax rusticola	1	*0.7*	1	*3.3*					1	*0.4*	1	*1.8*
cfr. *Strix aluco*					1	*0.8*	1	*3.8*	1	*0.4*	1	*1.8*
Asio otus	6	*4.2*	3	*10.0*	1	*0.8*	1	*3.8*	7	*2.6*	4	*7.1*
Asio flammeus	2	*1.4*	2	*6.7*	2	*1.6*	2	*7.7*	4	*1.5*	4	*7.1*
Asio sp.	2	*1.4*			2	*1.6*			4	*1.5*		
Ptyonoprogne rupestris	1	*0.7*	1	*3.3*					1	*0.4*	1	*1.8*
Pica pica	2	*1.4*	2	*6.7*	1	*0.8*	1	*3.8*	3	*1.1*	3	*5.4*
Nucifraga caryocatactes					1	*0.8*	1	*3.8*	1	*0.4*	1	*1.8*
Pyrrhocorax graculus	44	*30.6*	3	*10.0*	48	*37.8*	5	*19.2*	92	*33.9*	8	*14.3*
cfr. *Pyrrhocorax graculus*	5	*3.5*			5	*3.9*			10	*3.7*		
Pyrrhocorax pyrrhocorax	1	*0.7*	1	*3.3*					1	*0.4*	1	*1.8*
Corvus monedula	1	*0.7*	1						1	*0.4*	1	*1.8*
Corvus corone	1	*0.7*	1	*3.3*					1	*0.4*	1	*1.8*
Corvidae indet.	3	*2.1*			1	*0.8*			4	*1.5*		
Montifringilla nivalis					1	*0.8*	1	*3.8*	1	*0.4*	1	*1.8*
Carduelis cannabina	1	*0.7*	1	*3.3*					1	*0.4*	1	*1.8*
Pyrrhula pyrrhula					1	*0.8*	1	*3.8*	1	*0.4*	1	*1.8*
Passeriformes indet.	11	*7.6*			3	*2.4*			14	*5.2*		
Total identified birds	**144**	***100***	**30**	***97***	**127**	***100***	**26**	***100***	**271**	***100***	**56**	***100***
Identified birds	144	40.0			127	*59.1*			271	*47.1*		
Unidentified birds	19	5.3			18	*8.4*			37	*6.4*		
Aves being identified	197	54.7			70	*32.6*			267	*46.4*		
Total bird remains	**360**	***100***			**215**	***100***			**575**	***100***		

33 % are calcined to a grey-whitish color (T > 700 °C). Calcined bones mostly include small diaphysis fragments and approximately 18 % are fragments of spongy bone. These data may suggest that bones were used as fuel (Costamagno and Thèry-Parisot 2005).

Bone surfaces are well preserved: only a few remains have root marks while micro-fissures on other fragments resulted from weathering phenomena and/or weathering cracks. Manganese staining, concretions, run-off surfaces and exfoliations are also present. Rodent gnaw marks are almost absent (Number of Remains, or NR 15), while carnivore marks, in the form of scoring, pits, or gastric juice erosion, were detected only on 149 specimens. Hence, carnivores did not play a significant role in accumulating these bone remains inside the cave, even if some large digested bones may suggest occasional use of the cave by hyenas.

Human modifications have been observed both on identified specimens and unidentifiable fragments. As far as the identified specimens are concerned, human traces include striae and impacts. Many of the butchering traces on unidentifiable remains are percussion cones and prove that bone fragmentation was carried out for the purpose of marrow extraction. In sum, then, the high fragmentation observed in the studied bone assemblage is attributable to butchering activities, burning and weathering.

Birds

The avifaunal assemblage is dominated by the Alpine chough, *Pyrrhocorax graculus* (Linnaeus, 1766), black grouse, *Tetrao tetrix* (Linnaeus, 1758) and corncrake, *Crex crex* (Linnaeus, 1758) (Table 8.2). Many specimens belong to the Passeriformes, mainly Corvids. Bones attributed to diurnal and nocturnal raptors are also frequent; these primarily include the common kestrel, *Falco tinnunculus* (Linnaeus, 1758), and the long-eared owl, *Asio otus* (Linnaeus, 1758). Galliformes are represented by rock partridge, *Alectoris graeca* (Meisner, 1804), gray partridge, *Perdix perdix* (Linnaeus, 1758) and quail, *Coturnix coturnix* (Linnaeus, 1758). Important ecologic indicators, such as the mallard, *Anas platyrhynchos* (Linnaeus, 1758) and the water rail, *Rallus aquaticus* (Linnaeus, 1758), suggest that humid zones existed in the cave surroundings while the Alpine chough proves the presence of rocky environments, black grouse alpine forest habitat, and corncrake grasslands indicate hill and mountain zones (Spagnesi and Serra 2003). Choughs and black grouse currently live in the Alps, while corncrake is a regular migrant, breeding but largely absent in the winter (Spina and Volponi 2008). Bones of young corncrake may suggest that capture occurred during the summer season.

Of the whole bird assemblage made up of 25 species, birds of eastern-Palearctic chorology (mostly Mediterranean-middle Eurasian but also Euro-Siberian, boreal-Alpine and Eurasian) prevail. Some of these nest in the alpine zone (besides choughs and black grouse): rock ptarmigan, *Lagopus mutus* (Montin, 1781), spotted nutcracker, *Nucifraga caryocatactes* (Linnaeus, 1758), white-winged snow finch, *Montifringilla nivalis* (Linnaeus, 1766) and Eurasian bullfinch, *Pyrrhula pyrrhula* (Linnaeus, 1758).

With regard to the most abundant species, the black grouse is mainly represented by claw bones (tarsometatarsus and tibiotarsus), the Alpine chough by wing bones (carpometacarpus and ulna) and the corncrake by humeri and coracoids. This bias in the skeletal composition may suggest that various predators contributed to the accumulation of these remains.

Anthropogenic modifications have been observed on a handful of specimens: four bones are cut-marked, *arrachement* is present on another five bones, impact points on two, and peeling on 11 identified specimens. Cut-marks have been observed on one golden eagle proximal humerus, one left proximal humerus of Alpine chough, on the distal articulation of one black grouse tibiotarsus and on one unidentified long bone diaphysis (Alpine chough size). Cut-marks suggest that birds were disarticulated and defleshed. Several fresh bone fractures have been found (NR 35), often associated with other modification, especially peeling (eight) and sometimes small impacts (two).

A small sample of bones is affected by scorings and pits (15 bones) and by digestive juice erosion (six). Several pieces show grooves typical of roots and other thin isolated striae caused by sedimentary particles. A complete taphonomic analysis of the Uluzzian bird bone assemblage will be carried out as part of future work.

Mammals

Lagomorpha
Leporidae G. Fischer, 1817
Lepus timidus Linnaeus, 1758 or *Lepus europaeus* Pallas, 1778

The only three hare teeth (P_4, P^x/M^x, I^x) belong to a single individual. The scarcity of remains and the absence of diagnostic elements do not favor a distinction between Alpine (*L. timidus*) or common (*L. europaeus*) hare.

Rodentia
Sciuridae (Hemprich, 1820)
Marmota marmota (Linnaeus, 1758)

Marmot bones include one proximal femur and one distal metacarpal. Tunnels that also run through the final Mousterian levels prove that the cave was used by this species.

Carnivora
Canidae G. Fischer, 1817
Canis lupus Linnaeus, 1758

Wolf bones (NISP = 32) can be ascribed to four individuals of different age (young, young-adult, adult, adult-old). Most are teeth (13), one hyoid, four vertebrae, one shaft of radius and two distal epiphyses of tibia and fibula. The distal limb extremities are phalanxes (six), sesamoids (two) and one metacarpal, carpal and astragalus.

Six bones (18.7 % of the determined specimens) are cut-marked. One cut-mark on the hyoid, disposed transversally to the *epihyoideum*, could be proof of slaughter. Skinning has been inferred from striae on the ventral face of three first phalanges and on a sesamoid, and on the articular proximal face of the V/VI caudal vertebra when the tail was removed. The lack of third phalanges and the scarcity of second phalanges may suggest that the limb extremities and the tail were left attached to the skin (Fig. 8.2, 1A–B).

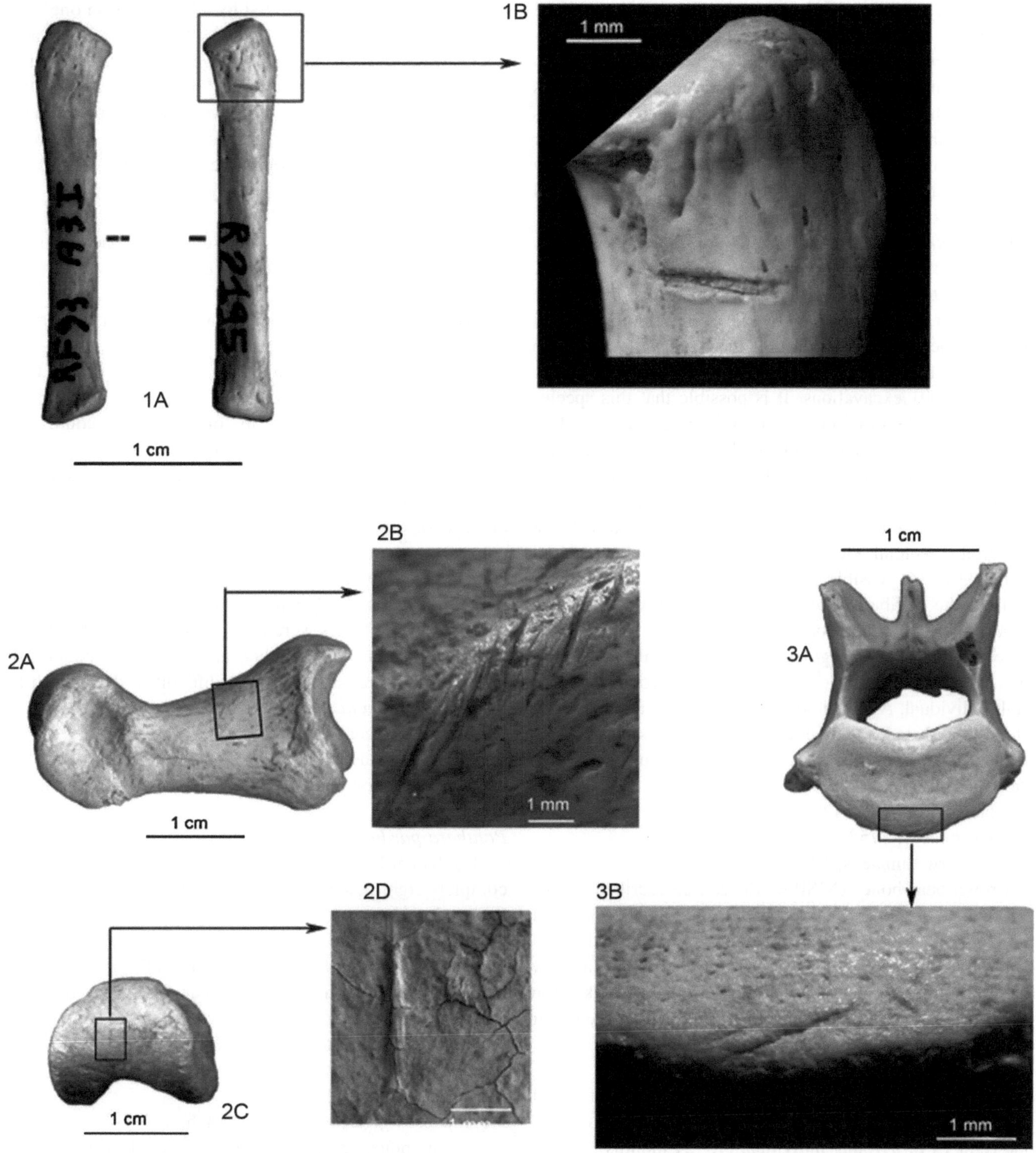

Fig. 8.2 Cut-mark on a *Canis lupus* hyoid of the (*1A*), detail of the stria (*1B*). Cut-marks on an *Ursus arctos* phalanx II. Striae on the dorso-lateral face for skinning (*2A*) and one stria on the distal articular face for disarticulation (*2C*). Detail of the striae (*2B*, *2D*). Cut-marks on a *Vulpes vulpes* lumbar vertebra made during disarticulation (*3A*), detail of the striae (*3B*)

Vulpes vulpes (Linnaeus, 1758)

The fox bone assemblage (NISP = 97) includes at least four individuals of different age (one young-adult, two adults and one senile). There are 30 cranial remains, of which most are isolated teeth (22), then vertebrae (22) and phalanges (22), while ribs are scarce (four) as are the long bones (four proximal fragmented ulnae and one distal fragment of a fibula). Carpals, tarsals, sesamoids, metacarpals and metatarsals are also poorly represented.

Butchering traces have been observed on four bones (4.1 % of the determined specimens). Skinning cut-marks are on a mandible, a third metatarsal and a rib. One lumbar vertebra shows striae on the ventral face, both on the body and the distal articular surface, made during de-fleshing and disarticulation (see Fig. 8.2, 3A, 3B).

Vulpes vulpes (Linnaeus, 1758) or *Alopex lagopus* (Linnaeus, 1758)

The existence of the polar fox (*Alopex lagopus*) in the cave was already documented by one of the authors (Cassoli and Tagliacozzo 1994) during the study of the bone assemblage found in the lowermost Aurignacian layers from the 1988–1991 excavations. It is possible that this species may have reached the north of Italy in a period of cooling during which wolverine, snowy owl, and other species also spread to the foothills of the Alps. Osteological distinction between the polar and the common fox can only be made on the basis of the features of specific cranial bones or the size of certain skeletal elements (Poplin 1976; Compagnoni et al. 1997). A possible coexistence of these two species restricts the attribution of some fox remains (11) to the genus level: 5 cranial bones (mandibles and teeth), three vertebrae and three fragments of metacarpal, fibula and tibia. These remains belong at least to one young and one adult individual. A mandible fragment with P_2 and M_1 teeth, which is still under taxonomic study, matches the features of the polar fox and bears several grouped cut-marks produced during skinning.

Ursidae G. Fischer, 1817

Ursus Linnaeus, 1758

Ursus arctos Linnaeus, 1778

Brown bear bones (NISP = 18) can be ascribed to two young (13–24 months) and two adult individuals. Bones of the hand and foot (phalanxes, sesamoids, tarsal and metacarpal) prevail in number over other types of bones, namely the cranium (teeth and zygomatic bone), the axial skeleton (two ribs, one of which is complete), one fragmentary femur of an adult male and one penis bone.

Ursus spelaeus Rosenmüller, 1794

Cave bear bones are constituted by the root of a left upper large canine, probably of a male adult individual, and one right P_4 of a young individual (13–24 months).

Ursus sp.

Another four remains of uncertain taxonomical attribution must be added to the bones described above: two milk canines, one fragmentary M^1 and a third phalanx of two young and very young individuals.

Only brown bear bones show traces of human exploitation (5 specimens out of 18). Three cut-marked second phalanges suggest possible skinning (Fig. 8.2, 2A, 2B) and disarticulation, as corroborated by a few striae on one distal articular face (Fig. 8.2, 2C, 2D). The oblique and axial traces observed on the dorso-lateral face at the distal extremity of an almost complete rib should not exclude possible recovery of meat. Moreover, the use as a retoucher of a fragmentary femur marks a unique occurrence throughout the whole stratigraphic succession.

Mustelidae G. Fischer, 1817

Mustela nivalis Linnaeus, 1766

A tiny adult weasel is represented by one canine, a proximal portion of humerus and a distal tibia.

Mustela erminea Linnaeus, 1758

Ermine has been identified by a single right humerus.

Gulo gulo (Linnaeus, 1758)

Only four bones coming from the A3 layer are taxonomically attributed to wolverine: a P_3, tarsal cuboid, calcaneum and possibly a rib of a single adult. The morphometric features of the two tarsals and the premolar (low, massive, rough and grooved enamel surfaces) strengthen the attribution to wolverine instead to badger or other carnivores of similar size. Other wolverine bones come from the uppermost Aurignacian layers.

Hyaenidae Gray, 1821

Crocuta crocuta (Erxleben, 1777)

Three hyena bones were found in layer A3: a left P_3 and a right P_4 of an old individual and one right phalanx; the lack of cranial remains makes it difficult to distinguish the massive *C. crocuta spelaea* from *C. crocuta*.

Felidae G. Fischer, 1817

Lynx lynx (Linnaeus, 1758)

One single first phalanx from A3 testifies the presence of an adult lynx.

Panthera pardus (Linnaeus, 1758)

The leopard is exclusive of A4 and is represented by one complete right patella and one left coxal of a single adult.

Perissodactyla

Rhinocerotidae Gray, 1821

Coelodonta antiquitatis Blumenbach, 1807

One fragmented upper milk molar (D^2) is the only evidence of the presence of this pachyderm in the stratigraphic series of Fumane (Fiore et al. 2004)

Artiodactyla

Cervidae Goldfuss, 1820

Megaloceros giganteus (Blumenbach, 1799)

Twenty bones are attributed to the giant deer. In size they exceed the range of the red deer, a species whose large males can reach considerable size. From the teeth and mandibles (five specimens) we estimated the presence of at least two individuals, an adult and a juvenile under the age of three (hemi-mandible with D_4 still inserted). The post-cranial skeleton is mostly represented by appendicular parts of the limbs (six phalanges, one malleolar bone and one sesamoid). Most of the diaphyses come from metapodials (five). Long bones include one proximal extremity of a radius and one mesial diaphysis of a tibia.

Cut-marks are preserved on five bones (25 %). Skinning striae on metapodials (two metacarpal diaphyses and one metatarsal diaphysis) are obliquely arranged, in mesial and lateral position. One of these also bears a percussion mark for marrow extraction. One fragmentary hemi-mandible bears traces in the proximity of the coronoid process, suggesting defleshing or disarticulation from the cranium. The lateral tuberosity of the proximal part of the radius is marked by two short, deep and transverse striae probably produced during disarticulation (Fig. 8.3, 2A, 2B).

Cervus elaphus Linnaeus, 1758

Red deer is the most represented ungulate in terms of number of remains (NISP = 411); the MNI, instead, is represented only by 13 individuals of different ages: three young (8, 10–16 and 17–22 months); one young-adult of 2–3 years; six adults (3–4 years, probable female, 3–5, 4–5, 5–7, 6–9 and 7–9 years); three senile, one of 9–12 years and two represented by two atrophic canines of different-sized females, over 11–12 years old. Considering that hinds give birth at the end of spring, it is possible to estimate the capture period for the three young red deer: one winter, one autumn-winter and one spring-summer.

The deer remains are mostly teeth and fragmented diaphyses of long bones; except for rare cases, epiphyses are underrepresented. Phalanges are relatively numerous compared to the carpal and tarsal bones, which are almost absent. The NISP/MNI ratio indicates each deer is represented by about 32 bone fragments. As a whole, the head (25.5 %) and the distal limb bones (27.5 %) have comparable frequencies, compared to the hind limb and the frontal/fore limb bones which are more (33.8 %) and less frequent (15 %), respectively. The rarity of antler and the presence of two fragmentary hyoids are also notable, as is the scarcity of trunk elements. The absence of the atlas and the axis is of relative importance, because these bones may remain with the cranium or still connected with the other vertebrae when the head is disarticulated from the body. A single rib has been attributed to this species, but among the indeterminate ribs we counted 36 fragments whose size associates them with the deer. Also significant is the rarity of specimens attributed to coxae and scapulae. The most represented element of the long bones is the metatarsal, followed by the metacarpal, tibia and femur, while the other elements are lower in number. The patella is completely absent, and the carpal and tarsal bones are very underrepresented. The bones of the distal limbs are represented by fragmented phalanges, all three rudimental phalanxes, and sesamoids.

In light of this anatomical composition, it seems plausible that the deer trunk and girdle were left at the kill site and that only selected portions with meat were transported into the cave. Considering the high number of metapodials, it remains difficult to interpret the scarcity of carpal and tarsal bones, which are usually highly resistant to post-depositional phenomena, while it is plausible that phalanges were often discarded or left attached to the skin.

Considering the MNE, our assessments of the deer anatomical composition remain substantially unchanged (Table 8.3). The most represented anatomical elements belong to the hind limbs: tibiae, followed by femurs and metatarsals. The survival rate reveals that parts of the skeleton are highly underrepresented. As a whole, the cranium represents all 13 individuals recognized, which correspond to a survival rate of 100 %, but each single head component, such as the hemi-mandible (34.6 %), the rest of the cranium (46.2 %) and the hyoid (15.4 %), is less completely represented. The parts of the forelimb record low survival rates and only the metacarpal and humerus reach values similar to the cranium. The tibia has the highest value, followed by the femur and metatarsal. The bones of the trunk and the extremities of the limbs have very low survival rates.

The fragmentation index (MNE/NISP) reveals a high incidence of fractured elements except for short and compact bones of the extremities, which have values that are close to 1. Other elements may have overestimated values due to the small number of remains (such as the scapula or coxae). The long bones, which are particularly subject to fracture, have values for the forelimb ranging from 0.35 (metacarpal, intensively fragmented), to 0.53 (radius), 0.63 (humerus) and 0.88 (ulna, less fractured). The hind leg appears to have been subject to increased fracturing as indicated by the value of the metatarsal (0.2), tibia (0.47) and femur (0.58).

Numerous traces of anthropic modifications were identified on over 42 % of the red deer bones, while the number of bones that show evidence of gnawing by carnivores or digestion traces is low. Butchering traces like cut-marks and percussion marks have been detected on over 167 fragments, while a few flakes were used as retouchers (Table 8.4; Fig. 8.4). All the long bones of the limbs show cut-marks: radius, ulna, metacarpal, femur and metatarsal show very high percentages, above 60 %, and over 40 % of the remaining elements also bear cut-marks. Also, cut-marks have been observed on all the short end-bones of the limbs with the exception of third phalanges, with a high incidence over the first rudimental (75 %) and the second phalanges (with one case of an impact point). Bones may bear only striae or percussion marks or both, but percussion cones are also sometimes associated with striae. In some cases, fragments of diaphyses of long bones show multiple percussion marks which, in rare cases, have not resulted in the fracturing of the bones due to their strength and resistance to fracture. In a few instances detachments were also produced by the counterblows.

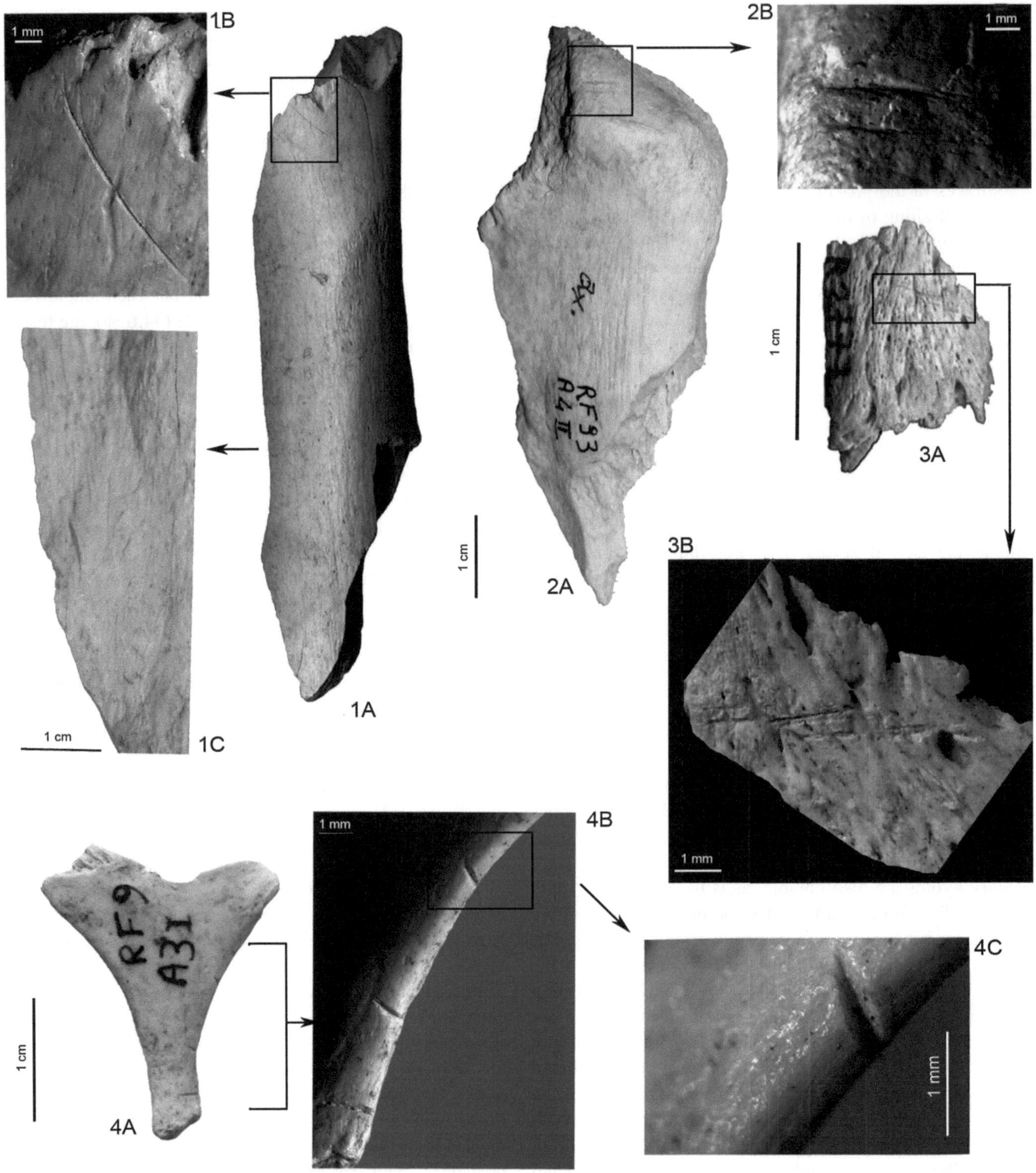

Fig. 8.3 Butchering marks on the diaphysis of the radius of *Bison priscus* (*1A*). Detail of the stria from de-fleshing (*1B*), detail of the impact (*1C*). Cut-marks on the proximal part of the radius of the *Megaloceros giganteus* with striae for disarticulation (*2A*), detail of the striae (*2B*). Cut-marks on the fragment of horn core of the *Rupicapra rupicapra* for the recovery of the horny sheath or skinning (*3A*), detail of the striae (*3B*). Cut-marks on the fragment of hyoid of the *Capreolus capreolus* from the cutting of the throat or recovery of the tongue (*4A*), detail of the striae (*4B*, *4C*)

Table 8.3 Anatomical elements of *Cervus elaphus* and *Capra ibex*

Anatomical element	*Cervus elaphus* (Layers A4-A3)					*Capra ibex* (Layers A4-A3)				
	NISP	MNE	NEE	Skeletal surv. rate (%)	Fragm. index MNE/NISP	NISP	MNE	NEE	Skeletal surv. rate (%)	Fragm. index MNE/NISP
Cranium	19	6	13	*46.2*	*0.32*	22	6	14	*42.9*	*0.27*
Hemimandible	27	9	26	*34.6*	*0.33*	41	13	28	*46.4*	*0.32*
Tooth indet.	38					10				
Hyoid	2	2	13	*15.4*	*1*			14		
Total cranium	**86**	**17**	**52**	***32.7***	***0.20***	**73**	**19**	**42**	***45.2***	***0.26***
Atlas-axis			26					28		
Vertebra	4	4	351	*1.1*	*1*	1	1	378	*0.3*	*1*
Rib	1	1	338	*0.3*	*1*			364		
Total trunk	**5**	**5**	**715**	***0.7***	***1***	**1**	**1**	**770**	***0.1***	***1***
Scapula	3	2	26	*7.7*	*0.67*	1	1	28	*3.6*	*1*
Humerus	16	10	26	*38.5*	*0.63*	8	4	28	*14.3*	*0.50*
Radius/Ulna						8	5	28	*17.9*	*0.6*
Radius	15	8	26	*30.8*	*0.53*					
Ulna	8	7	26	*26.9*	*0.88*					
Carpals	2	2	156	*1.3*	*1*	13	13	168	*7.7*	*1*
Metacarpal	34	12	26	46.2	*0.35*	7	4	28	*14.3*	*0.57*
Metacarpal rudim.	1	1	26	3.8	*1*	1	1	28	*3.6*	*1*
Total frontal limb	**79**	**42**	**312**	***13.5***	***0.53***	**38**	**28**	**308**	***9.1***	***0.74***
Coxal	3	2	26	7.7	*0.67*	4	2	14	*14.3*	*0.50*
Femur	26	15	26	57.7	*0.58*	12	6	28	*21.4*	*0.50*
Patella			26					28		
Tibia	34	16	26	61.5	*0.47*	10	6	28	*21.4*	*0.60*
Malleolar Bone	2	2	26	7.7	*1*					
Calcaneum	1	1	26	3.8	*1*					
Astragalus	1	1	26	3.8	*1*					
Tarsals	2	2	78	2.6	*1*	2	2	168	*1.2*	*1*
Metatarsal	70	14	26	53.8	*0.20*	6	4	28	*14.3*	*0.67*
Total hind limb	**139**	**53**	**286**	***18.5***	**0.38**	**34**	**20**	**294**	**6.8**	**0.59**
Metapodial	19	21			*1.11*	6	3			*0.50*
First phalanx	11	7	104	6.7	*0.64*	12	9	112	*8.0*	*0.75*
Second phalanx	25	21	104	20.2	*0.84*	7	5	112	*4.5*	*0.71*
Third phalanx	9	9	104	8.7	*1*	3	3	112	2.7	*1*
First phalanx rudim.	4	4	104	3.8	*1*					
Second phalanx rudim.	12	12	104	11.5	*1*					
Third phalanx rudim.	5	5	104	4.8	*1*					
Sesamoid	17	17	208	8.2	*1*	24	24	336	*7.1*	*1*
Total indet. limb	**102**	**96**	**832**	***11.5***	*0.94*	**52**	**44**	**672**	***6.5***	***0.85***
Total	**411**	**213**	**2210**	***9.6***	***0.52***	**198**	**112**	**2058**	***5.4***	***0.57***

NISP number identified specimens, *MNE* Minimum number elements, *NEE* number expected elements, % skeletal survival rate, and breakage index

Table 8.4 Number of anatomical elements of *Cervus elaphus* and *Capra ibex* with anthropogenic modifications and carnivore marks

Cervus elaphus - Layers A4-A3												*Capra ibex* - Layers A4-A3							
Anatomical element	NISP	CM	PM	PC	C + PM	C + PC	BM	% BN	R	G	B	NISP	CM	PM	C + PM	BM	% BM	G	B
Cranium	19	1					**1**	***5.3***		1	2	22							
Hemimandible	27	5					**5**	***18.5***			1	41	2			**2**	***4.9***		1
Tooth indet.	38											10							
Hyoid	2	1					**1**	***50.0***											
Total cranium	**86**	**7**					**7**	***8.1***		**1**	**3**	**73**	**2**			**2**	***2.7***		1
Atlas-axis																			
Vertebra	4	1					**1**	***25.0***				1							
Rib	1																		
Total trunk	**5**	**1**					**1**	***20.0***				**1**							
Scapula	3										1	1							
Humerus	16	6			1		**7**	***43.8***	1		1	8	3	1		**4**	***50.0***	1	
Radius/Ulna												8	1			**1**	***12.5***	2	
Radius	15	3		1	5		**9**	***60.0***											
Ulna	8	5			1		**6**	***75.0***											
Carpals	2											13						3	1
Metacarpal	34	11	5	1	5	1	**23**	***67.6***	2	1		7	4		1	**5**	***71.4***		
Metacarpal rudim.	1											1							
Total frontal limb	**79**	**25**	**5**	**2**	**12**	**1**	**45**	***57.0***	**3**	**1**	**2**	**38**	**8**	**1**	**1**	**10**	***26.3***	**6**	1
Coxal	3											4							
Femur	26	7	3		7		**17**	***65.4***	2	1		12	4	1	1	**6**	***50.0***	1	1
Patella																			
Tibia	34	10	1	1	4		**16**	***47.1***	3		1	10	1	1	1	**3**	***30.0***	1	1
Malleolar Bone	2	1					**1**	***50.0***											
Calcaneum	1		1				**1**	***100.0***											
Astragalus	1																		
Tarsals	2											2							1
Metatarsal	70	12	12	4	14	3	**45**	***64.3***			6	6	3	1		**4**	***66.7***	1	
Total hind limb	**139**	**30**	**17**	**5**	**25**	**3**	**80**	***57.6***	**5**	**1**	**7**	**34**	**8**	**3**	**2**	**13**	***38.2***	**3**	3
Metapodial	19	6	2				**8**	***42.1***			2	6	1			**1**	***16.7***	2	
First phal.	11	2					**2**	***18.2***				12	3			**3**	***25.0***	1	
Second phal.	25	12	1				**13**	***52.0***		1	3	7	1			**1**	***14.3***		1

(continued)

Table 8.4 (continued)

Cervus elaphus - Layers A4-A3												*Capra ibex* - Layers A4-A3							
Anatomical element	NISP	CM	PM	PC	C + PM	C + PC	BM	% BN	R	G	B	NISP	CM	PM	C + PM	BM	% BM	G	B
Third phal.	9										1	3							
First phal. rudim.	4	3					3	*75.0*											
Second phal. rudim.	12	5					5	*41.7*			2								
Third phal. rudim.	5																		
Sesamoid	17	3					3	*17.6*				24	5			5	*20.8*	3	
Total indet. limb	**102**	**31**	**3**				**34**	**33.3**		**1**	**8**	**52**	**10**			**10**	**19.2**	**6**	1
Total	**411**	**94**	**25**	**7**	**37**	**4**	**167**	***40.6***	**8**	**4**	**20**	**198**	**28**	**4**	**3**	**35**	***17.7***	**15**	**6**

CM Cut-marks, *PM* Percussion marks, *PC* Percussion cones, *C* + *PC* Cut-marks + Percussion cones, *BM* Total butchery marks, *% BM* % NISP with butchery marks, *R* Retoucher, *G* Carnivore Gnaw marks, *B* burned

Cut-marks are primarily related to stripping (long, sparse, straight and/or sinuous, oblique and longitudinal striae, scrapings) and the cutting of tendon and muscle insertions (short, repeated, numerous, transverse and/or slightly oblique striae) and are located on the diaphyses of long bones and on five fragmentary hemi-mandibles (Fig. 8.4, 2). Skinning (longitudinal oblique, isolated and/or densely spaced striae) is equally well documented by cuts on the phalanges (first, second and rudimentals), metapodials, sesamoids, one incisor and one caudal vertebra (Fig. 8.4, 3). Traces of disarticulation are rare, due to the scarcity of articular portions: they are more frequent over some proximal portions of metapodials, over both the articulations of the second phalanges, as well as on one ulna proximal extremity, one malleolus, and one calcaneus.

Capreolus capreolus (Linnaeus, 1758)

Roe deer is the third most represented (NISP = 104) species among the ungulates. The estimated minimum number of individuals based on the teeth (24, including mandibles and jaws) reveals the presence of at least six individuals: two young (4–6 and 8–10 months); three adults, two of 5–6 years and one of 6–7 years, one senile (older than 7–8 years). The age of the two young ones indicates that they were hunted at the end of summer and during winter. The fragmentary antlers cannot be ascribed to this species. We report the presence of two hyoid bones.

Metacarpals (10), metatarsals (17) along with indeterminate metapodial fragments (seven) are the most represented skeletal remains. This observation, however, is certainly influenced by the ease of recognition of these bones even from small fragments. The limb extremities are mostly represented by phalanges (23) and by a few carpal, tarsal and sesamoid bones. Among the long bones the humerus and femur are absent, while radius (four), tibia (three) and ulna (one) are scarce. The trunk is completely missing while two coxae fragments are the only bones belonging to the girdle.

Butchering traces have been observed on 15 out of 104 identified bones (14.4 %) and concentrate mainly on the extremities of the limbs: five metatarsals, thee metacarpals, two metapodials, one calcaneus, one second phalanx and one second rudimentary phalanx. Striae on the metapodials and phalanges relate primarily to skinning and recovery of tendons. Moreover, percussion marks are visible only on these specimens. Transverse cut-marks on the upper face of a calcaneus may suggest the removal of muscles or disarticulation. One stylohyoid (*stylohyoideum*) is grooved by deep, short and transverse striae which may have been produced during the cutting of the throat or recovery of the tongue (Fig. 8.3, 4A–C). Stripping of flesh is suggested by the presence of short and oblique cut-marks on the diaphysis of a radius.

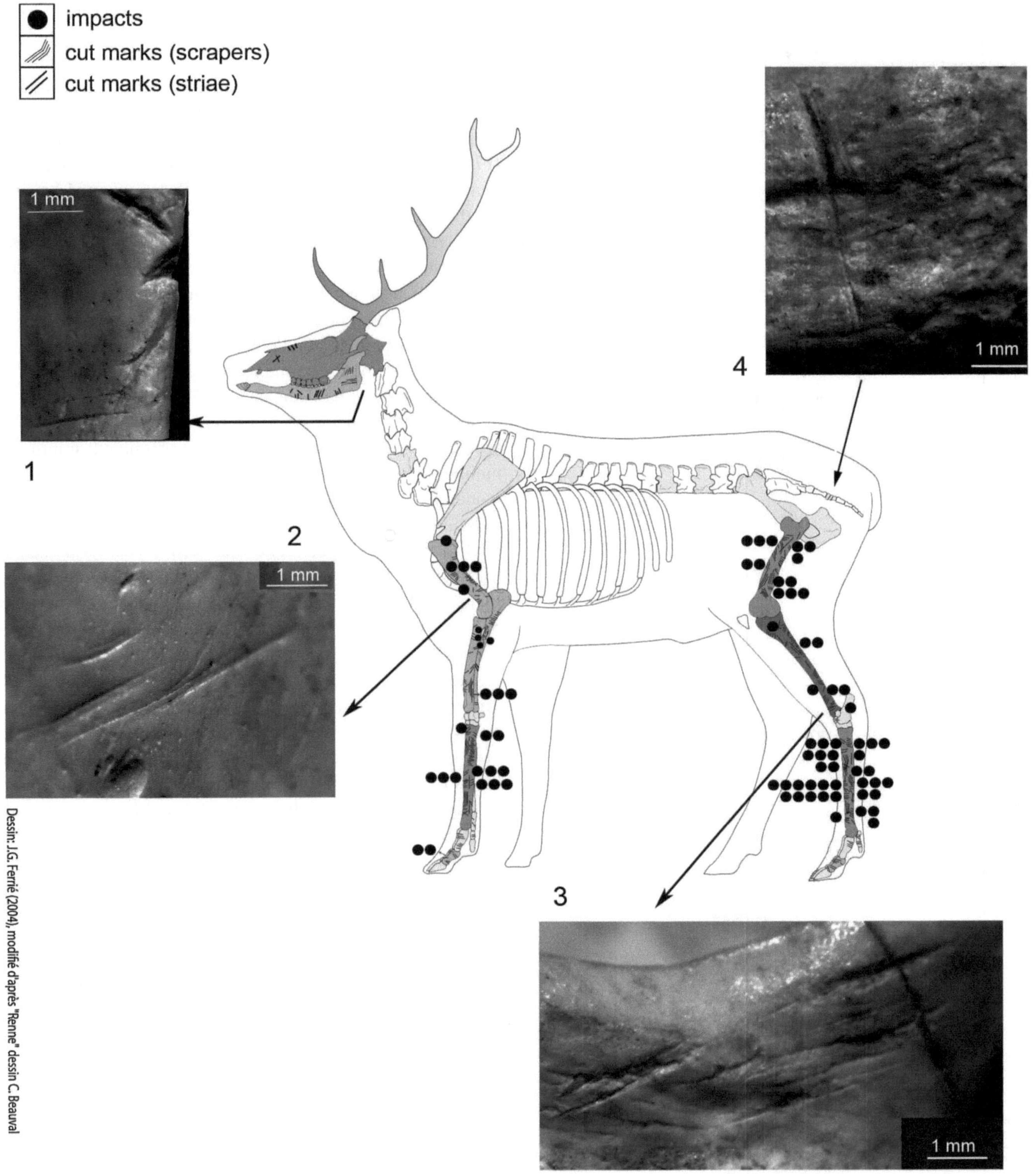

Fig. 8.4 Distribution of the butchering marks on the skeletal elements of the *Cervus elaphus* and coefficient of survival (in *gray-scale*). Details of striae on the hyoid (*1*), on the distal humerus (*2*), on the caudal vertebra (*3*), on the distal tibia (*4*)

Cervids
There are some remains (62) that could be only attributed to the general "cervid" level, although it is very likely that the major part belongs to red deer (Table 8.1). Many of these are small fragments of tooth crown, antler fragments or tip of antler tines. Furthermore, some skeletal fragments of large-sized deer (six) may belong to either red or giant deer.
Bovidae Gray, 1821
Bovinae Gray, 1821
Bison priscus (Bojanus, 1827) and *Bos primigenius* Bojanus, 1827

Apart from the handful of ten bones taxonomically attributed to bison, 40 additional remains have been labelled at the general level as *Bos* or *Bison* as a consequence of the absence of diagnostic anatomical portions. Moreover, it must be noted that across the Fumane stratigraphy, the presence of *Bos primigenius* is scanty and has been recorded only in the Aurignacian layers. Thus, there were at least five bison, if we take into account the teeth (three isolated teeth and two mandibles with the teeth still inserted). There is at least one individual younger than 16 months, one young-adult and three adults, two of which are represented by mandibles with P2-M1 and M1-M3.

Among the whole sample of large bovids, the cranial remains are eight teeth, five mandibles and six fragments of horn core. Long bones (13) and metapodials (six) are mostly represented by diaphyses. The distal limb elements are scarce (four phalanges, one sesamoid, one pisiform) and there are five rib fragments, while vertebrae, including atlas and axis, coxae, carpal and tarsal elements, are completely lacking.

Out of 50 elements, 17 (34 %) of the bones belonging to large bovids are cut-marked and eight bones show evidence of percussion marks. Cut-marked long bones are mostly fragments of diaphyses (Fig. 8.3, 1A–B). The transverse or twisted arrangement of the traces, which are sometimes also ordered in series, suggests an intense exploitation of meat. Twisted-transverse striae on the dorsal and ventral surfaces of four ribs likely suggest the recovery of meat also from the trunk, the former being attributed to flesh removal, the latter to evisceration. Ribs may also have been disarticulated from the backbone, as suggested by two striae found in proximity of the articular head. Several percussion marks (Fig. 8.3, 1A, 1C) are evidence of marrow removal. Skinning is indicated by several traces, grouped or isolated, short and transverse, on a fragmentary metatarsal. Skinning and disarticulation is also suggested by transverse and oblique cut-marks on the ventral, lingual and buccal surfaces of two mandibles.
Caprinae Gray 1821
Capra ibex Linnaeus, 1758

Ibex is the second most represented ungulate of this assemblage (NISP = 198) and at least 14 individuals are present (Table 8.1): one young less of 1 year, one young adult (20–30 months), ten adults (three over 3–4 years, one of 4–6 years, one of 5–7 years, two of 8–9 years, one of 8–10 years, and two of 9–10 years) and two senile of 12–14 years old.

The NISP/MNI ratio shows that each ibex is represented by 14 bones, a value lower than deer (32 bones). In fact, while the number of remains representing the deer is more than double that of the ibex, there are fewer individual deer (MNI 13). Of the overall skeleton, the bones from the head (36.8 %) are numerous, either as a consequence of the fragmentation of jaws and mandibles or the number of isolated teeth. Horn fragments are very scarce (Table 8.3). Limb extremities (26.2 %) include numerous sesamoids (24), phalanges (22) and carpals (13). Forelimbs and hindlimbs are equally represented. Among the long bones, the most common are femur (12), then tibia, humerus and radio/ulna, metacarpal and metatarsal. Except for a few rare cases, epiphyses are underrepresented. Bones from the trunk are almost completely missing, except for one thoracic vertebra. The absence of the atlas and the axis is notable. This evidence is further confirmed by the scarce and fragmentary vertebrae, the thickness and size of which is comparable to those of the ibex. Ribs of medium-sized ungulates include 87 fragments, many of them ascribable to *Capra ibex*. Given these data, we infer that ibex carcasses were dismembered at the location of kill and the fleshier parts were carried to the cave. Similarly to deer, trunk and girdle were thrown away, but portions of the rib cage were kept. This is also confirmed by the MNE (Table 8.3). Excluding the small bones of the limb extremities, tibiae, femur and radio-ulna are more represented than the humerus, metacarpals and metatarsals.

The survival pattern described here reveals the ways in which certain skeletal elements are clearly and selectively underrepresented (Table 8.3). The highest values are in cranial remains, as shown by the hemi-mandibles (13 MNE/28 NEE, 46.4 %) and the other cranial bones (6 MNE/14 NEE, 42.9 %). Forelimb bones have lower rates, between 14 and 17 %. The tibia and femur have the highest values (21 %). Lastly, the lowermost values are those of the limb extremities and the trunk bones.

The fragmentation index (MNE/NISP) has an average value, if we exclude the compactness of the bones at the limb extremities and other elements, the values of which are overestimated as a consequence of the low number of remains (scapulae and vertebrae). Long bones range from 0.5 (humerus and femur), to 0.57 (metacarpal) and 0.6 (radio/ulna, tibia and metatarsals).

Cut-marks and percussion marks have been observed, sometimes together (Table 8.4; Fig. 8.5), on 17 % of the

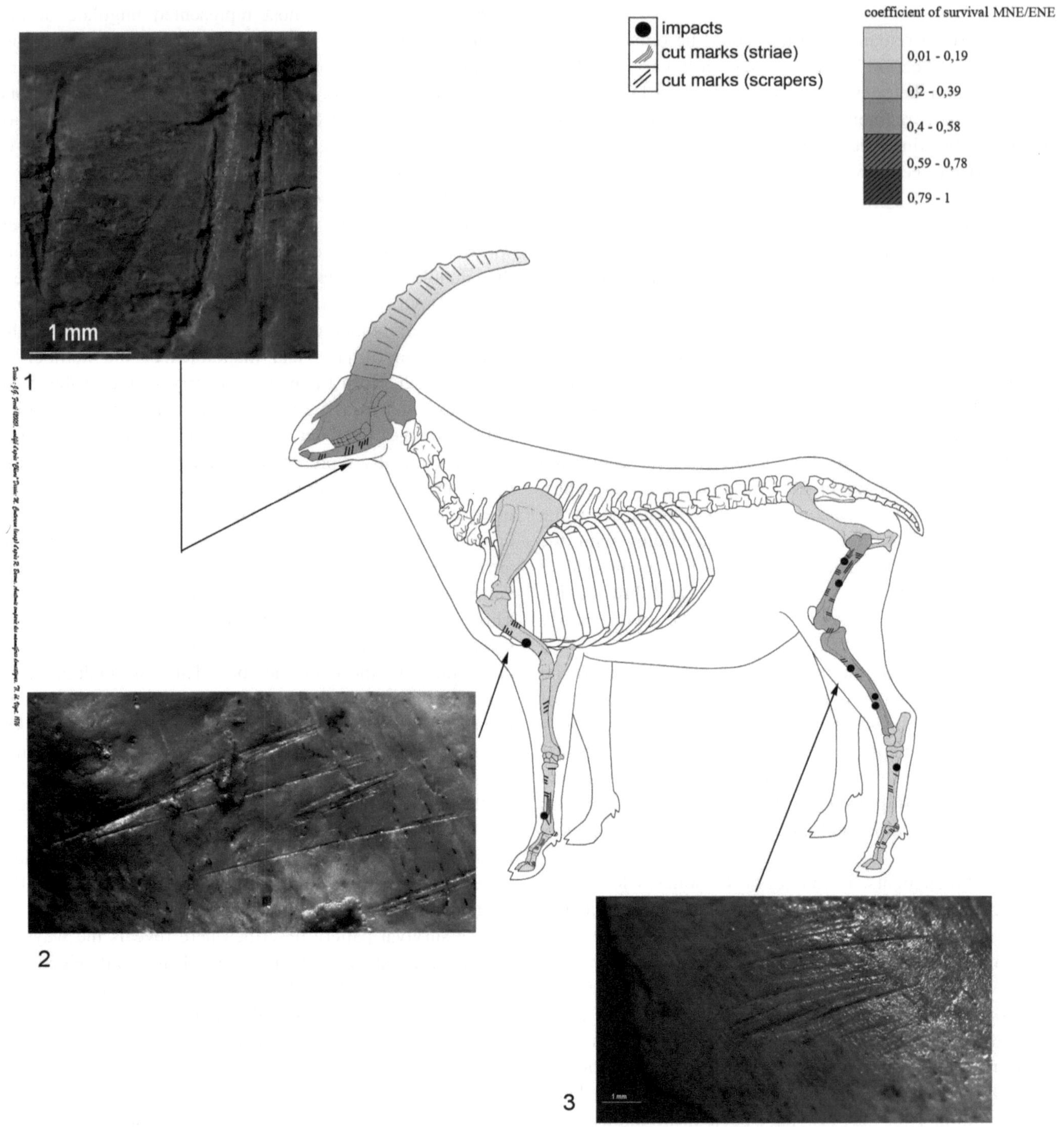

Fig. 8.5 Distribution of the butchering marks on the skeletal elements of the *Capra ibex* and coefficient of survival (in *gray-scale*). Details of striae on the mandible (*1*), on the humerus shaft (*2*), on the tibia shaft (*3*)

overall assemblage, while carnivore gnawing and digestion traces are few. Certain long bones like metacarpals, metatarsals, humerus and femur were more intensively subject to these human modifications; the extremities of the limbs bear traces on the first and second phalanges and the sesamoids. Impact scars have been identified only on the long limb bones. Striae are preserved over several fragments. Cutmarks suggest defleshing and the cutting off of insertions of muscles and tendons, and mostly affect the diaphyses of the long bones (Fig. 8.5, 2–3). Skinning is suggested by cutmarks on two hemi-mandibles (Fig. 8.5, 1) and on phalanges, metapodials and sesamoids. Traces of disarticulation are almost completely lacking as a consequence of the near absence of articular portions of bones, with the exception of a lateral condyle of femur and a proximal portion of a second phalanx.

Rupicapra rupicapra (Linnaeus, 1758)

The chamois is represented by 65 remains and the MNI (8) is based primarily on the teeth, which are the most abundant remains: one very young (first phalanx not completely ossified), two young-adult (17–30 months), four adults (one of 3–4 years, two of 4–5 years and one of 7–10 years) and one senile individual. The newborn suggests that hunting took place during the summer.

Cranial bones are the most numerous (21), followed by the limb extremities (12 phalanges, five tarsals, four carpals, three sesamoids, one calcaneus). The axial skeleton is represented by two vertebrae (axis and thoracic) and the girdle from one coxa. With regard to the limbs, both diaphyses and epiphyses are proportionally represented. Metapodials are more frequent (six) than the bones from the hind limb (four femura, one tibia, one patella) and the forelimb (two radius, one humerus, one scapula).

There are 11 bones with butchering marks (16.9 %). A basal fragment of horn shows three oblique striae proving either the recovery of the horny sheath or skinning (Fig. 8.3, 3A–B). Traces of defleshing have been observed on the spine of a scapula, on one diaphysis of a radius, the shaft of the ilium of a coxa and two femur diaphyses. Skinning is indicated by oblique and longitudinal striae observed on two metatarsal diaphyses and by short and transverse striae on the proximal part of a first phalanx. Cut-marks on a single capitate trapezoid confirm the disarticulation of the distal extremity of the limb. Deliberate fragmentation of bone has been inferred from two percussion marks and notches on a single radius diaphysis.

Discussion

The Uluzzian faunal assemblage of the Grotta di Fumane includes many ungulates, carnivores and birds from various habitats and climatic contexts (Tables 8.1, 8.2). Among the ungulates there are deer, ibex and roe deer and the less numerous chamois, giant deer and bison. The assemblage also includes one tooth of a woolly rhino. Most of the carnivore remains belong to the fox, but wolf and brown bear are also present. Cave bear, mustelids (ermine, weasel and wolverine), feline (lynx and leopard) and hyena bones have also been identified. Hare and marmot are very scarce. Among the birds, the most abundant species are Alpine chough, black grouse and corncrake, and various corvids, galliforms, diurnal and nocturnal raptors, mallard and water rail. The birds are associated with various habitats: rocky, Alpine forest and Alpine grasslands, but also wet zones close to the cave.

Broadly, the faunal spectrum indicates close woody environments, with conifers and broad-leave trees in a context of temperate to cool climatic conditions. These ecological conditions alternated with colder temperatures and a drier climate that caused the lowering of the local timberline (mainly in A3). However, there are also indications of temperate forest fauna (red deer, roe deer, brown bear, black grouse) living in the area surrounding the cave with species typical of cold (wolverine) and open habitats (ibex, chamois, ermine, marmot, Alpine chough) above the tree line. Given the landscape around the cave, we can infer that all these different habitats existed in proximity to the shelter.

Some differences in the faunal composition between the Late Mousterian, Uluzzian and Aurignacian layers (Table 8.5) are worth noting. Bones from the Mousterian A5-A6 layers show a reduced number in the identified species of mammals, especially among the carnivores, where felines, wolverine and other mustelids are lacking. The clearly dominant incidence of deer over the other ungulates (over 70 %) and the co-presence of elk suggest the existence of wet zones in the land close to the cave. The most prevalent birds in the Late Mousterian are comparable to those of the Uluzzian levels, even if the corncrake is more prevalent than the black grouse. Among the species typical of wet zones in the Uluzzian the mallard replaces the common moorhen *Gallinula chloropus* (Linnaeus, 1758).

Hence, the Uluzzian records the onset of a series of ecological and economical modifications that lead to a colder climate and steppe-like conditions in the Aurginacian occupation (A2-A2R) and thus to the distinct prevalence of ibex over deer, chamois, roe deer and giant deer. Carnivores provide a more detailed picture of the available Uluzzian biome. Wolf and fox are the most represented, and some fox remains present qualities characteristic of the polar species. Hyenas, mustelids and feline also increase. Again, the avifauna changes both in quantity (of bones and species) and in the composition of species, with the appearance of birds typical of cold climatic and Nordic environments like the rough-legged buzzard, *Buteo lagopus* (Pontoppidam, 1763) and the white-backed woodpecker, *Picoides leucotos* (Bechstein, 1803).

During the Uluzzian occupation, red deer and ibex were the most hunted species, but the giant deer, roe deer, bison and chamois were also pursued. Carnivores like wolf, fox and brown bear were hunted as well, for the recovery of fur as suggested by the traces of skinning, although it is possible that their meat was also consumed. The range of the exploited carnivores is wider in these layers and in the Aurignacian layers (fox, wolf and lynx) than in the Mousterian layers (only brown bear). Marmot and leporid bones

Table 8.5 Mammals composition in the Late Mousterian, Uluzzian and Aurignacian layers at Grotta di Fumane

Fumane Cave—Taxa	Mousterian		Uluzzian		Aurignacian	
	A6-A5		A4-A3		A2-A2R	
	NISP	%	NISP	%	NISP	%
Lepus sp.	1	*0.1*	3	*0.3*	1	*0.1*
Lepus cfr. timidus					4	*0.4*
Marmota marmota	1	*0.1*	2	*0.2*	2	*0.2*
Castor fiber					1	*0.1*
Total rod-lagom	**2**	***0.1***	**5**	***0.5***	**8**	***0.9***
Canis lupus	11	*0.6*	32	*3.1*	38	*4.2*
Alopex lagopus					2	*0.2*
Vulpes vulpes	45	*2.3*	97	*9.4*	28	*3.1*
Alopex/Vulpes	3	*0.1*	11		7	*0.8*
Ursus arctos	9	*0.5*	18	*1.7*	1	*0.1*
Ursus spelaeus	4	*0.2*	2	*0.2*	1	*0.1*
Ursus sp.	21	*1.1*	4	*0.4*	1	*0.1*
Mustela erminea			1	*0.1*	2	*0.2*
Mustela nivalis	1	*0.1*	3	*0.3*	1	*0.1*
Mustela putorius					1	*0.1*
Gulo gulo			4	*0.4*	1	*0.1*
Crocuta crocuta			3	*0.3*	15	*1.7*
Lynx lynx			1	*0.1*	3	*0.3*
Panthera leo					1	*0.1*
Panthera pardus			2	*0.2*		
Total carnivora	**94**	***4.7***	**178**	***17.2***	**102**	***11.3***
Coelodonta antiquitatis			1	*0.1*		
Sus scrofa	1	*0.1*				
Megaloceros giganteus	38	*1.9*	20	*1.9*	34	*3.8*
Cervus elaphus	1403	*70.3*	411	*39.9*	170	*18.8*
Capreolus capreolus	235	*11.7*	104	*10.1*	37	*4.1*
Alces alces	5	*0.2*				
cfr. *Bos primigenius*					2	*0.2*
Bison priscus	2	*0.1*	10	*1*	6	*0.7*
Bos/Bison	24	*1.2*	40	*4*	28	*3.1*
Capra ibex	100	*4.9*	198	*19.2*	447	*49.5*
Rupicapra rupicapra	91	*4.5*	65	*6.3*	71	*7.8*
Total ungulata	**1899**	***95.2***	**849**	***82.3***	**795**	***87.8***
Total	**1995**	***100***	**1032**	***100***	**905**	***100***

NISP number identified specimens, and NISP %

do not bear any clear evidence of butchery and therefore it is not certain that these species were introduced by humans into the cave. Some birds (golden eagle, Alpine chough and black grouse) were also disarticulated and defleshed for food purposes, as in the case of the black grouse, or for the recovery of bones, perhaps to be used as ornaments, as suggested in the Mousterian levels (Fiore et al. 2004; Peresani et al. 2011b). It also seems that birds were more subject to carnivore exploitation than mammals.

The bone assemblage studied here indicates that red deer and ibex were first dismembered at the kill site, and that the trunk and girdle were left behind when the most nutritionally valuable portions were selected to be consumed inside the cave. The very low number of vertebrae and ribs suggest that carcasses were occasionally introduced into the cave to be processed (from skinning to marrow exploitation). Percussion marks on the long bones are more frequent on deer than ibex, certainly due to differences in tissue strength. It is also possible that the absence or low number of certain bones in addition to the high degree of fragmentation could result from the selective cleaning of the shelter and the removal of bulky, fragmented bones. The identification of percussion flakes inside the cave may indirectly support this possibility. The high incidence of burned remains suggests that bone was used as fuel for the hearths. This might also explain the absence or scarcity of the carpal and tarsal small bones, which surely entered the cave still joined to ungulate articulations. The nature of the traces of burning does not accord with the results of cooking. The scarcity of other artiodactyls does not allow us to outline the methods of carcass treatment more broadly.

The scanty data regarding the age of death of juveniles (three deer, two roe deer, one ibex and one chamois) are not specific or abundant enough to demonstrate a pattern of a seasonal human presence in the Uluzzian. Rather, as such, they seem to show that the cave was used year round. From the thin sections of the two teeth found in A4, we inferred that deer were killed during the summer (Facciolo and Tagliacozzo 2006). In the Aurignacian layers, deer and ibex teeth indicate a wider interval in which death could have occurred, spanning late spring to the onset of winter. Unlike in the Aurignacian, deer in the Uluzzian were killed at any age while ibexes were subject to more selective hunting of adults.

Conclusion

The faunal data from the Uluzzian period at Grotta di Fumane take on considerable importance for the information they provide regarding the human exploitation of animals and butchering practices in a critical transitional phase from the Middle to the Early Upper Paleolithic. In northern Italy, the Grotta di Fumane is the only site from which faunal data are presently available from a Mousterian—Uluzzian—Aurignacian sequence. Other sites with the same stratigraphic sequence and with faunal data are found in central and southern Italy: Grotta la Fabbrica (Toscana), Grotta di Castelcivita and Grotta La Cala (Campania), and Grotta del Cavallo (Puglia). These sites are widely distributed across the Italian peninsula and are located in diverse environments. As such, their assemblages include a variety of different taxa. Implications of these differences are detectable in evidence for both the quantity and the exploitation of faunal resources.

From Uluzzian layer 2 at Grotta la Fabbrica (Pitti et al. 1976), only 68 remains of four taxa have been identified: two ungulates—horse is predominant and deer is scarce—and scanty remains of two carnivores (hyena and fox). The Uluzzian occupation at Grotta di Castelcivita (Masini and Abbazzi 1997) is represented by multiple layers (rsi-tg18, pie, rpi) in which 286 remains of 18 taxa have been identified. Of these, in addition to hare, nine are ungulates (horse, boar, bison, aurochs, ibex, chamois, deer, fallow deer and roe deer), and eight are carnivores (bear, wolf, fox, leopard, lynx, wildcat, hyena and least weasel). Between one layer and the other, the Uluzzian occupation is characterized by variations in species present and in the relative quantities of species. The remains of carnivores in the lowest layer (rsi-tg18) make up only 8 % of the total; this value increases in the middle and highest layers to 15–18 %. In the lowest layer, there is a predominance of cervids, but in the middle and upper layers, cervids decline while horse increases. Bovids are present but scarce throughout the sequence and caprids decline at the end of the sequence. Remains of ichthyofauna are frequent (Cassoli and Tagliacozzo 1997). For Uluzzian levels, this seems to document fishing activity that took place year round, with eel and chub (*Leuciscus cephalus*) harvested in winter months and trout in summer and autumn months. Compared to Mousterian layers, the abundant avifaunal remains indicate climatic cooling for which there is even stronger evidence in the Aurignacian.

At Grotta La Cala (Benini et al. 1997), the Uluzzian fauna is represented by 364 identified remains divided among nine species of ungulate (rhinoceros, horse, boar, aurochs, ibex, chamois, deer, fallow deer and roe deer), and six *taxa* of carnivore (fox, bear, marten, wildcat, lynx and leopard), whose 46 remains make up 12.6 % of the assemblage. Hare is also present. Of the ungulate remains, fallow deer constitutes 43 %, deer 16.7 %, boar 13.2 %, and roe deer 9.8 %. The other ungulate species are minimally represented. Among the carnivore remains, leopard and marten are predominant.

At Grotta del Cavallo (Boscato and Crezzini 2012), in the Uluzzian level EIII5, 204 remains belonging to eight taxa were identified. There are five ungulate taxa (NISP 196): horse, *Equus hydruntinus*, boar, deer and aurochs. Of these, aurochs remains make up 43.9 % of the total, indicating that it was the most hunted ungulate. Horse and deer follow, with values of around 27 %, while boar and *Equus hydruntinus*, are rare. Carnivores are represented by seven remains (3.4 %) identified as wolf, fox and hyena. One bone of a lagomorph was also identified.

In general, at all of these sites a distinct predominance of ungulate bones relative to carnivore bones is notable. The faunal assemblages indicate the exploitation of different environments from site to site, the most common being: coastal plain and prairies with herbaceous and sparse vegetation (horse and aurochs); arboreal landscapes with brush (fallow deer, deer, boar) or more dense woods (roe deer, bear, wildcat and lynx); and environments without arboreal vegetation preferred by ibex and chamois.

It is difficult to compare faunal data from these central and southern Italian sites to those from Fumane, both due to different methods employed in collection and study and, even more so, to the environmental diversity among them. The quantity of remains identified as those of large mammals (NISP 1,032) from the Uluzzian levels of Grotta di Fumane is greater than that at the other deposits. However, the ungulate/carnivore relationship, both in terms of number of species and quantity of remains, seems similar to those of Grotta della Cala and Grotta di Castelcivita. Meanwhile, the numbers of species and remains of carnivores at Grotta del Cavallo are significantly lower, not to mention the small sample from Grotta La Fabbrica. As at Fumane, the abundance of cervids and caprids records the exploitation of both forested and alpine prairie environments at all the other sites.

The comparison of taphonomic data with the other Uluzzian sites further demonstrates the importance of the faunal remains from Fumane for the reconstruction of the exploitation of ungulates and carnivores. Impacts and cut-marks made by lithic tools have otherwise only been noted at Grotta del Cavallo on aurochs, deer and horse bones. At Grotta di Fumane, all ungulates present clear traces of human exploitation. Moreover, at present Fumane is the only Italian site where traces of butchering have also been identified on remains of carnivores (wolf, fox, bear) and birds. At Grotta La Cala, the recovery of carnivore furs has been hypothesized on the basis of the presence of bones from the extremities of leopard limbs.

Comparison between the composition of skeletal elements from Fumane and those from Grotta La Cala and Grotta del Cavallo is especially productive for information regarding the survival of elements from the extremities of ungulates' limbs. In their analysis of bones from Grotta del Cavallo, Boscato and Crezzini (2012) noted in particular the differences in the composition of skeletal elements from the extremities of ungulate limbs that could reflect different strategies for ungulate exploitation between Mousterian and Uluzzian occupations. The percentages of these bones rise from Mousterian layers (carpals and tarsals, between 0.5 and 1.6 %; phalanges and sesamoids, 2.7–7.9 %) and the Uluzzian (carpals and tarsals, 15.3 %; phalanges and sesamoids, 30.1 %). The highest quantity of these elements in Uluzzian levels could indicate the recovery and a more systematic exploitation of tendons and bone marrow. This development seems to be confirmed by data sets from other deposits in southern Italy between the Middle and Upper Paleolithic (Boscato and Crezzini, 2012: Table 2). At Grotta La Cala this increase is observable from the Uluzzian layers (carpals and tarsals, 9.4 %; phalanges and sesamoids, 12.3 %) and the Aurignacian layers (carpals and tarsals, 22.3 %; phalanges and sesamoids, 13.6 %). At Grotta di Fumane, between the Mousterian, Uluzzian and Aurignacian layers, there is a notable increase in sesamoids and phalanges together; the percentages are 12.9, 19.4 and 25.6 %, respectively. Meanwhile, carpal and tarsal bones show constant values between 22 and 26 % throughout this same period. Human modifications, both points of impact and cut-marks, recur on the bones studied from all three cultural horizons.

The changes in percentages between these elements are dramatic in deposits from southern Italy. However, they are more subtle at Grotta di Fumane, where already by the Mousterian their levels reach nearly those of the Uluzzian from comparable sites discussed here. Moreover, at Fumane, it is possible to define differences between the two predominant species, the deer and the ibex. This is probably the result of other factors (among which different skeletal compositions, recognizability of certain elements, size of the animal, distance from kill site) and was not only the result of differential treatment of the carcass.

In addition to providing useful paleoenvironmental information, the faunal data from the Uluzzian occupation at Fumane represents what is at present the most important indication of subsistence strategies during the critical transition from the Middle to the Upper Paleolithic in the North of Italy.

Acknowledgments Research at Fumane is supported by the Superintendent for Archaeological Heritage of Veneto, the Venetian Region, the Cariverona Foundation, the Comunità Montana della Lessinia, the Fumane Municipality and several private companies (Roberto Gardina, Albino Armani winegrovers, Banca di Credito Cooperativo della Valpolicella). The authors are grateful to E. M., M. V. and A. J. S. for their help with the English translation of the manuscript. Most of the data achieved in this research are taken from the Masters thesis of M. Romandini (2007–2008).

References

Aitken, R. J. (1974). Delayed implantation in roe deer (*Capreolus capreolus*). *Journal of Reproduction and Fertility, 39*, 225–233.

Bar-Yosef, O., & Bordes, J.-G. (2010). Who were the makers of the Châtelperronian culture? *Journal of Human Evolution, 59*(5), 586–593.

Benazzi, S., Douka, K., Fornai, C., Bauer, C. C., Kullmer, O., Svoboda, J., et al. (2011). Early dispersal of modern humans in Europe and implications for Neanderthal behaviour. *Nature, 479*, 525–528.

Benini, A., Boscato, P., & Gambassini, P. (1997). Grotta della Cala (Salerno): industrie litiche e faune uluzziane e aurignaziane. *Rivista di Scienze Preistoriche, 48*, 37–95.

Binford, L. R. (1981). *Bones, ancient men and modern myths*. New York: Academic Press.

Blumenschine, R. J. (1995). Percussion marks, tooth marks, and experimental determinations of the timing of hominid and carnivore ace to long bones at FLK Zinjanthropus, Olduvai Gorge, Tanzania. *Journal of Human Evolution, 27*, 197–213.

Blumenschine, R. J., & Selvaggio, M. M. (1988). Percussion marks on bone surfaces as a new diagnostic of hominid behavior. *Nature, 333*, 763–765.

Bocherens, H. (2009). Neanderthal dietary habits: Review of the isotopic evidence. In J. J. Hublin & M. P. Richards (Eds.), *The evolution of Hominin diets* (pp. 241–250). Germany: Springer.

Bökönyi, S. (1970). A new method for determination of the number of individuals in animal bone material. *American Journal of Archaeology, 74*, 291–292.

Boscato, P., & Crezzini, J. (2012). Middle-Upper Palaeolithic transition in Southern Italy: Uluzzian macromammals from Grotta del Cavallo (Apulia). *Quaternary International, 252*, 90–98.

Brain, C. K. (1981). *The hunters or the hunted? An introduction to African Cave Taphonomy*. Chicago: University of Chicago Press.

Broglio, A., Bertola, S., De Stefani, M., Marini, D., Lemorini, C., & Rossetti, P. (2005). La production lamellaire et les armatures lamellaires de l'Aurignacien ancien dans la Grotte de Fumane (Monts Lessini, Vénétie). In F. Le Brun-Ricalens (Ed.), *Productions lamellaires attribuées à l'Aurignacien.* Actes du XIVe congrès de l'UISPP (Liege, 2001). Archéologiques 1. (pp. 415–436). Luxemburg: MNHA.

Broglio, A., De Stefani, M., Gurioli, F., Pallecchi, P., Giachi, G., Higham, T., et al. (2009). L'art aurignacien dans la décoration de la Grotte de Fumane. *L'Anthropologie, 113*, 753–761.

Broglio, A., Tagliacozzo, A., De Stefani, M., Gurioli, F., & Facciolo A. (2006). Aurignacian dwelling structures, hunting strategies and seasonality in the Fumane Cave (Lessini Mountains). In *The Early Upper Paleolithic of Eurasia: General trends, local developments (the Materials of International Conference devoted to the 125-th Anniversary of the Paleolithic Investigations in Kostenki, 2004, August, 23–26)*. Russian Academy of Sciences. Proceedings of Kostenki—Borschevo archaeological expedition (Vol. 4, pp. 263–278). Saint-Petersburg: Nestor-Historia Publishers.

Capaldo, S. D., & Blumenschine, R. J. (1994). A quantitative diagnosis of notches made by hammerstone percussion and carnivore gnawing on bovid long bones. *American Antiquity, 59*, 724–748.

Cassoli, P. F., & Tagliacozzo, A. (1994). Considerazioni paleontologiche, paleoecologiche e archeozoologiche sui macromammiferi e gli uccelli dei livelli del Pleistocene superiore del Riparo di Fumane (VR) scavi 1988–91. *Bollettino del Museo Civico di Storia Naturale di Verona, 23*, 85–117.

Cassoli, P. F., & Tagliacozzo, A. (1997). Avifauna e Ittiofauna di Grotta di Castelcivita: considerazioni ecologiche ed inquadramento crono-stratigrafico. In P. Gambassini (Ed.), *Il Paleolitico di Castelcivita. Culture e ambiente* (pp. 60–74). Napoli: Electa.

Chase, P.-G. (1989). How different was Middle Paleolithic subsistence? A zooarchaeological perspective on the Middle to Upper Paleolithic transition. In P. Mellars & C. Stringer (Eds.), *The human revolution: Behavioral and biological perspectives on the origins of modern humans* (pp. 321–337). Princeton: Princeton University Press.

Clark, G. A. (1997). The Middle-Upper Paleolithic transition in Europe: An American perspective. *Norwegian Archaeological Review, 30*, 25–53.

Compagnoni, B., Curci, A., & Tagliacozzo, A. (1997). Exploitation of fox in the Epigravettian levels of Grotta Romanelli (Apulia, Italy). In M. Kokabi & J. Wahl (Eds.), *Proceedings of the 7th ICAZ Conference (Konstanz, 26/09/1994-01/10/1994). Anthropozoologica, 25-26*, 319-328.

Conard, N. (2006). *When Neanderthals and modern humans met. International workshop*. Tübingen: KernsVerlag.

Costamagno, S., & Thèry-Parisot, I. (2005). Propriétés combustibles des ossements. *Gallia Préhistoire, 47*, 235–254.

Cremaschi, M., Ferraro, F., Peresani, M., & Tagliacozzo, A. (2005). La Grotta di Fumane: nuovi contributi su stratigrafia, cronologia, fauna a macromammiferi e industrie musteriane. In A. Broglio & G. Dalmeri (Eds.), *Pitture paleolitiche nelle Prealpi Venete. Grotta di Fumane e Riparo Dalmeri*, Atti del Simposio (Verona, 19-20 giugno 2003), Memorie del Museo Civico di Storia Naturale di Verona, 2. serie, Sezione Scienze dell'Uomo, 9. Preistoria Alpina, NR. Speciale, (pp. 12–22) Verona: Comune di Verona.

D'Errico, F., & Vanhaeren, M. (2002). Criteria for identifying red deer (*Cervus elaphus*) age and sex from their canines: Application to the study of Upper Palaeolithic and Mesolithic ornaments. *Journal of Archaeological Science, 29*, 211–232.

Facciolo, A., & Tagliacozzo, A. (2006). L'occupazione stagionale di Grotta di Fumane(VR) durante l'Aurignaziano attraverso l'analisi delle sezioni sottili dei denti di cervo e stambecco. In G. Malerba & P. Visentini (Eds.), *Atti del 4° Convegno Nazionale di Archeozoologia, Pordenone, 13-15 novembre 2003.* Quaderni del Museo Archeologico del Friuli Occidentale, 6 (2005) (pp. 43–52). Pordenone: Comune di Pordenone Editore.

Fiore, I., & Tagliacozzo, A. (2006). Lo sfruttamento dello stambecco nel Tardiglaciale di Riparo Dalmeri (TN): il livello 26c. In U. Tecchiati & B. Sala (Eds.), *Archaeozoological studies. In honour of Alfredo Riedel* (pp. 59–76). Bolzano: Ripartizione Beni Culturali.

Fiore, I., Gala, M., & Tagliacozzo, A. (2004). Ecology and subsistence strategies in the Eastern Italian Alps during the Middle Palaeolithic. *International Journal of Osteoarchaeology, 14*, 273–286.

Fisher, W. J. (1995). Bone surface modifications in zooarchaeology. *Journal of Archaeological Method and Theory, 2*, 7–68.

Gala, M., & Tagliacozzo, A. (2006). L'avifauna dei livelli aurignaziani di Grotta di Fumane (VR). Risultati preliminari dello studio tafonomico. In G. Malerba & P. Visentini (Eds.), *Atti del 4° Convegno Nazionale di Archeozoologia, Pordenone, 13-15 novembre 2003.* Quaderni del Museo Archeologico del Friuli Occidentale, 6 (2005) (pp. 53–57). Pordenone: Comune di Pordenone Editore.

Gaudzinski, S., & Niven, L. (2009). Hominin subsistence patterns during the Middle and Late Paleolithic in Northwestern Europe. In J. J. Hublin & M. P. Richards (Eds.), *The evolution of Hominin diets* (pp. 99–112). Dordrecht: Springer.

Grayson, D. K. (1984). *Quantitative zooarchaeology: Topics in the analysis of archaeological Faunas.* Orlando: Academic Press.

Higham, T., Brock, F., Peresani, M., Broglio, A., Wood, R., & Douka, K. (2009). Problems with radiocarbon dating the Middle and Upper Palaeolithic transition in Italy. *Quaternary Science Reviews, 28*(13–14), 1257–1267.

Klein, R. G., & Cruz-Uribe, K. (1984). *The analysis of animal bones from archaeological sites*. Chicago: University of Chicago Press.

Lyman, L. D. (1994). *Vertebrate taphonomy.* Cambridge: Cambridge University Press.

Marean, C. W. (2005). From the tropics to the colder climates: Contrasting faunal exploitation adaptation of modern humans and Neanderthals. In F. d'Errico & L. Backwell (Eds.), *From tools to symbols. From early hominids to modern humans* (pp. 333–371). Johannesburg: Wits University Press.

Mariezkurrena, K. (1983). Contribution al conocimiento del desarrollo de la dencicion y el esqueleto postcraneal de *Cervus elaphus*. *Munibe, 35*, 149–202.

Martini, M., Sibilia, E., Croci, S., & Cremaschi, M. (2001). Thermoluminescence (TL) dating of burnt flints: Problems, perspectives and some example of application. *Journal of Cultural Heritage, 2*, 179–190.

Masini, F., Abbazzi, L., (1997). L'associazione di mammiferi della Grotta di. Castelcivita. In P. Gambassini (Ed.), *Il Paleolitico di Castelcivita. Culture e ambiente* (pp. 3–59). Napoli: Electa.

Peresani, M. (2008). A new cultural frontier for the last Neanderthals: The Uluzzian in Northern Italy. *Current Anthropology, 49*(4), 725–731.

Peresani, M. (2012). Variability in lithic technology from the recent Middle Palaeolithic to the Uluzzian across the 50ky record of Fumane Cave. *Quaternary International, 247*, 125–150.

Peresani, M., Cremaschi, M., Ferraro, F., Falanguères, C., Bahain, J. J., Grupponi, G., et al. (2008). Age of the final Middle Palaeolithic and Uluzzian levels at Fumane Cave, Northern Italy, using ^{14}C, ESR, ^{234}U/^{230Th} and thermoluminescence methods. *Journal Archaeological Science, 35*, 2986–2996.

Peresani, M., Chravzez, J., Danti, A., De March, M., Duches, R., Gurioli, F., et al. (2011a). Fire-places, frequentations and the environmental setting of the final Mousterian at Grotta di Fumane: A report from the 2006–2008 research. *Quartär, 58*, 131–151.

Peresani, M., Fiore, I., Gala, M., Romandini, M., & Tagliacozzo, A. (2011b). Late Neandertals and the intentional removal of feathers as evidenced from bird bone taphonomy at Fumane cave 44ky BP, Italy. *Proceedings of the National Academy of Sciences of the United States of America, 108*(10), 3888–3893.

Pitti, C., Sorrentino, C., & Tozzi, C. (1976). L'industria di Tipo Paleolitico Superiore Arcaico della Grotta La Fabbrica (Grosseto). *Nota preliminare. Atti della Società Toscana di Scienze Naturali, A, 83*, 173–206.

Poplin, F. (1976). *Les grands vertébrés de Gönnersdorf:fouilles 1968. Der Magdalenien Fundplatz Gönnersdorf* (Vol. 2, pp. 1–212). Wiesbaden: Steiner.

Potts, R., & Shipman, P. (1981). Cutmarks made by stone tools on bones from Olduvai Gorge, Tanzania. *Nature, 291*, 577–580.

Romandini, M., (2007–2008). Sfruttamento del cervo e della volpe nei livelli uluzziani della Grotta di Fumane (Monti Lessini, VR). Facoltà di Scienze MM.FF.NN., Corso di Laurea Specialistica in Scienze Preistoriche.

Ronchitelli, A., Boscato, P., & Gambassini, P. (2009). Gli ultimi neandertaliani in Italia: aspetti culturali. In F. Facchini & M. G. Belcastro (Eds.), *La lunga storia di Neandertal. Biologia e comportamento* (pp. 257–287). Milano: Jaca Book.

Shipman, P. (1981). *Life history a fossil. An introduction to taphonomy and paleoecology*. Harvard: Harvard University Press.

Shipman, P., & Rose, J. (1984). Cutmark mimics on modern fossil bovid bones. *Current Anthropology, 25*, 116–177.

Slimak, L. (2008). The Neronian and the historical structure of cultural shifts from Middle to Upper Palaeolithic in Mediterranean Europe. *Journal of Archaeological Science, 35*, 2204–2214.

Spagnesi, M., & Serra, L. (2003). *Uccelli d'Italia.* Quaderni di Conservazione della Natura, 16. Roma—Ozzano dell'Emilia: Ministero dell'Ambiente—Istituto Nazionale Fauna Selvatica.

Spina, F., & Volponi, S. (2008). *Atlante della Migrazione degli Uccelli in Italia. 1. Non-Passeriformi*. Roma: Ministero dell'Ambiente e della Tutela del Territorio e del Mare, Istituto Superiore per la Protezione e la Ricerca Ambientale (ISPRA).

Stiner, M. C. (1994). *Honor among thieves. A zooarchaeology study of Neandertal ecology*. Princeton, NJ: Princeton University Press.

Stiner, M. C. (1999). Paleolithic population growth pulses evidenced by small animal exploitation. *Science, 283*, 190–194.

Stiner, M. C., Kuhn, S. L., Weiner, S., & Bar-Yosef, O. (1995). Differential burning, recrystallization and fragmentation of archaeological bone. *Journal of Archaeological Science, 22*, 223–237.

Vigal, C. R., & Machordom, A. (1985). Tooth eruption and replacement in the Spanish wild goat. *Acta Theriologica, 30*, 305–320.

von den Driesch, A. (1976). *A guide to the measurement of animal bones from archaeological sites*. Peabody Museum Bulletin, 1. Cambridge Mass: Peabody Museum of Archaeology and Ethnology, Harvard University.

Chapter 9
A Diachronic Evaluation of Neanderthal Cervid Exploitation and Site Use at Pech de l'Azé IV, France

Laura Niven

Introduction

Several decades of zooarchaeological research focusing on Neanderthal hunting and exploitation of large ungulates have provided a great deal of insight on the dynamic interplay among hominins, their prey, and their natural and social landscapes. Such knowledge helps us better understand the subsistence strategies of Neanderthals and their implications for such factors as site use, mobility, and life history. Sites containing multi-layer stratigraphic sequences of Neanderthal deposits with associated Mousterian artifacts and archaeofaunas that span long ranges of time are particularly good sources for evaluating change and stability in subsistence across this hominin's long existence in Eurasia (e.g., Chase 1986, 1999; Stiner 1994, 2005; Gaudzinski 1995; Blasco 1997; Pike-Tay et al. 1999; Burke 2000; Conard and Prindiville 2000; Beauval 2004; Morin 2004; Rabinovich and Hovers 2004; Speth and Clark 2006; Speth and Tchernov 2007; Steele et al. 2009; Rivals et al. 2009).

The site of Pech de l'Azé IV, France is another such example of an extensive Mousterian sequence that yielded an exceptionally rich archaeological record of Neanderthal occupations (e.g., Bordes 1954; Turq et al. 2011). With three meters of deposits spanning much of the Mousterian, we have roughly 60 kyr of Neanderthal life to evaluate. The study presented here is based on the zooarchaeological remains recovered from the recent excavation campaign and focuses on the Neanderthals' treatment of large cervid prey in three large archaeological horizons that cover most of the stratigraphic sequence. The results show that environmental conditions influenced the abundance of the two dominant prey species across the sequence—red deer and reindeer—but factors relating to nutritional yield likely influenced how Neanderthals transported and processed these prey. The zooarchaeological results are then evaluated in conjunction with data from the lithics and site features in order to understand the use of this site over time and through changing climate regimes.

The Site and Research Background

The site of Pech de l'Azé IV (hereafter Pech IV) is a collapsed cave located in the Perigord region of southwest France. Located at the base of a cliff, Pech IV is one of four caves containing Lower and Middle Palaeolithic deposits that have been excavated over the past century and a half (e.g., McPherron et al. 2001; Soressi et al. 2007). Pech IV was discovered and tested by François Bordes in 1952 (Bordes 1954) and excavated by him between 1970 and 1977. A second campaign of excavations at the site was undertaken by Dibble and McPherron (Turq et al. 2011) between 2000 and 2003 with the aim of clarifying the stratigraphy, collecting samples for dating the sequence, and better understanding site formation processes.

The archaeological sequence at Pech IV is 3 m deep at its thickest point. At the base is Layer 8, which is characterized by Typical Mousterian lithic assemblages and evidence for the extensive use of fire by Neanderthals during these occupations (Dibble et al. 2009). The faunal spectrum is dominated by red deer (*Cervus elaphus*) and roe deer (*Capreolus capreolus*), indicating temperate, wooded environs. Thermoluminescence (TL) dates from burned flints place the occupations in Layer 8 in MIS 5c (Richter et al. 2010). This is followed by Layer 6 (Levels 6A–6B) containing a lithic industry that is infrequently seen, named the Asinipodian, marked by the production of small flakes (Dibble and McPherron 2006, 2007). Environmental conditions similar to Layer 8 persist through Layer 6 as evidenced by the fauna. The upper level 6A was dated by TL to

L. Niven (✉)
Department of Human Evolution, Max Planck Institute for Evolutionary Anthropology, Deutscher Platz 6, 04105 Leipzig, Germany
e-mail: laura@lauraniven.com

Jamie L. Clark and John D. Speth (eds.), *Zooarchaeology and Modern Human Origins: Human Hunting Behavior during the Later Pleistocene*, Vertebrate Paleobiology and Paleoanthropology, DOI: 10.1007/978-94-007-6766-9_9,

the end of MIS 5 (Richter et al. 2010). Above this are several layers of scraper-rich assemblages (5-4), including rich Quina Mousterian occupations in Level 4A. Scraper production was also markedly high in Layer 4. Reindeer (*Rangifer tarandus*) replaces red deer as the dominant taxon in Layers 5-4, although roe deer still persists in small amounts in Layer 5. Finally, the upper sequence comprises Mousterian of Acheulian Tradition (MTA) inventories lacking scrapers (Layer 3) with small faunal assemblages containing primarily reindeer and *Bos/Bison*.

A study of the fauna from Bordes' first campaign at Pech IV was undertaken by Laquay (1981). His analysis covered most of the same sequence as ours, except for the uppermost MTA horizons. Laquay's results on species representation across the sequence provided a valuable framework for understanding the climatic conditions under which the Pech de l'Azé caves and other regional sites such as Combe Grenal were occupied by Neanderthals (e.g., Mellars 1996). The recent zooarchaeological study of the Pech IV fauna shows the same general pattern of species representation found by Laquay.

Materials and Methods

Approximately 23,000 piece-plotted (>2 cm) faunal remains were recovered from the 2000–2003 excavations at Pech IV. A portion of this material forms the basis of the archaeozoological study presented here, with coarse waterscreen fauna excluded from the analysis except for a sample from Layer 8. Due to extensive fragmentation of the faunal remains overall, the level of identifiability in many layers was less than ideal and as a result many specimens could be placed only in general categories by family or body size.

Bone surface preservation is generally good for the faunal remains across the sequence. Cancellous bone portions and axial skeletal parts are strongly underrepresented in every layer, with the denser elements including teeth tending to preserve well. Statistically significant relationships between bone density and skeletal abundance (element and portion of element) are evident for the red deer and reindeer assemblages from Levels 4A, 4C and 6A. Long bones are overwhelmingly represented by mid-shaft portions and unidentifiable shaft splinters. These specimens consistently show a high degree of fresh breakage, presumably the result of marrow processing, and hammerstone impact marks are common. Stone tool cutmarks are present and in some layers exceptionally high. Every bone specimen was thoroughly examined for the presence of cutmarks using a strong primary light source and 10× magnification. Only clear and unambiguous evidence of bone modifications was recorded.

Evidence for carnivore gnawing is minimal (0.002 %) across the entire sequence. In addition, only 10 carnivore bone or tooth specimens were identified amongst all of the assemblages, indicating that bone collecting carnivores were simply not a factor in the faunal accumulations at Pech IV. Therefore, the assemblages at this site are assumed to be anthropogenic accumulations. The high frequencies of bone modifications such as cutmarks, hammerstone impacts and fresh breakage support this scenario.

The abundance of fauna varied significantly across the sequence of Pech IV, as reflected in the number of identified specimens (NISP) and minimum number of individuals (MNI) per layer (Table 9.1). Since only some of the layers

Table 9.1 Summary of faunal remains from Layers 4, 6 and 8 at Pech de l'Azé IV, expressed as NISP and MNI [in brackets]

	4A	4B	4C	6A	6B	8
Lepus sp. (hare)						1 [1]
Castor fiber (beaver)				3 [1]	1 [1]	3 [1]
Canis lupus (wolf)			2 [1]			2 [1]
Vulpes/Alopex (fox)				1 [1]		
Ursus sp. (indeterminate bear)				1 [1]	4 [1]	
Equus ferus (horse)	6 [1]	2 [1]	5 [1]	79 [3]	104 [3]	8 [1]
Cervus elaphus (red deer)	5 [1]	4 [1]	7 [1]	870 [13]	876 [19]	269 [10]
Rangifer tarandus (reindeer)	207 [4]	95 [2]	828 [19]	22 [1]	3 [1]	3 [1]
Capreolus capreolus (roe deer)	9 [1]	13 [1]	35 [2]	428 [7]	425 [7]	41 [2]
Bos/Bison (aurochs/bison)	1 [1]		1 [1]	26 [1]	17 [1]	
Capra ibex (ibex)				1 [1]		1 [1]
Sus scrofa (boar)				15 [3]	24 [2]	3 [3]
Aves						1 [1]
	228	114	878	1,446	1,454	331

Specimens placed in body size or family categories excluded from this list contained adequate faunal samples for evaluating Neanderthal subsistence, the study presented here will be limited to the assemblages of Layers 8, 6 and 4. Moreover, these layers provide contrasting pictures of environmental conditions and Neanderthal site use. In terms of subsistence, the bulk of this paper focuses on the data for reindeer and red deer only, since these samples were the largest, but evidence for site seasonality also includes data from boar.

Lying on bedrock at the bottom of the sequence is Layer 8, representing occupations that look different from subsequent ones in a number of important ways (Dibble et al. 2009). What sets apart Layer 8 most significantly is the evidence for the controlled use of fire, as indicated by numerous ash lenses and hearth features along with high frequencies of burned lithics and bone. We see indications that shallow hearths were used for fires, cleaned out of their content, reused, cleaned out again, and so on, with the burned refuse subsequently trampled by human occupants. These indications are visible on a macro- and microscale, though the trampling is best discerned in geomorphological thin sections where in situ bone breakage is clearly shown (Dibble et al. 2009: 188–189 and Fig. 7c). In addition to wood and plant remains, animal bones were burned for fuel. As a result of burning and trampling, much of the bone was highly fragmented and thus the majority could only be placed in body size categories. Interestingly, the presence of fire is virtually absent from the entire rest of the Pech IV sequence (Dibble et al. 2009). The fauna indicates temperate conditions with mixed deciduous and conifer forests and is dominated by red deer and roe deer, with small amounts of boar and beaver present as well.

Layer 6 (Levels 6A and 6B) in the lower part of the Pech IV sequence is rich in stone and bone assemblages. The faunal spectrum here is virtually the same as in Layer 8; horse and *Bos/Bison* are more numerous in Layer 6 than anywhere else in the sequence, but are still sparsely represented compared to the cervids. Environmental conditions are presumably similar to those reflected in Layer 8.

Reindeer replaces red deer and roe deer as the dominant taxon higher up in the sequence in Layer 4 (Levels 4A, 4B, 4C). The faunal assemblages from each of the levels are smaller than in Layer 6 and the richness of animal taxa represented is also much lower. Cold and dry environments are indicated by the reindeer, while the presence of horse and a hint of *Bos/Bison* shows that these taxa are present in a variety of climatic conditions.

In this paper, the discussion focuses only on Neanderthal utilization of the predominant species, red deer and reindeer. Although small assemblages of roe deer and sparse remains of large bovids and horse are also represented in these layers, they are not included here due to small sample sizes.

Zooarchaeological Results

In order to understand how Neanderthals exploited large cervids across the Pech IV sequence, and in turn, to begin addressing broader questions about subsistence change and site use, several zooarchaeological datasets derived from the assemblages in Layers 4, 6 and 8 are evaluated. Information on sex and age ratios, site seasonality, skeletal element abundances, and butchery modifications on bones provide insight on the ecological influence on Neanderthal prey choice as well as situational factors that may have played a role in carcass processing decisions.

Sex Ratios

Limited information on the sex ratios of red deer were gleaned from small diagnostic assemblages in Layers 6 and 8, despite the general lack of diagnostic skeletal parts or articular ends for osteometrics across the sequence overall. For example, one neonate and one fetal red deer indicate the presence of at least two females among the 10 adult individuals in Level 6A, while a few fragments of antler point to at least one male. Level 6B yielded a small number of antler fragments but no fetal remains; skeletal elements in this level were generally more robust than in the surrounding levels, probably reflecting at least some males, but there are no quantifiable data to determine more precise sex ratios. The best data come from Layer 8, where red deer canine teeth indicate six females and one male (following methods in d'Errico and Vanhaeren 2002). At least three fetal individuals also support the scenario of a female-dominated assemblage.

Seasonality and Age

Information on season of occupation at Pech IV comes from ageable fetal bone and teeth. An assemblage (NISP = 21) of fetal remains belonging to red deer and boar from Layer 8 and a smaller number of fetal red deer from Layer 6 (NISP = 6) were preserved. These specimens (mostly long bones) were aged as precisely as possible using two sets of comparative fetal red deer skeletons, one of which was approximately half-term (4.5–5.0 months) and the other nearly full-term (7–8 months). Based on the gestation schedule of ~250 days for *Cervus elaphus* with the mating season taking place in September–October and birth in May–June (Habermehl 1985), the comparative skeletons allowed age estimations of the Pech IV specimens within a seasonal timeframe. A single fetal boar radius from Layer 8 was aged using osteometric data from domestic pig long bones at various stages of gestation (Habermehl 1975: Table 14). Both wild boar and domestic pig have gestation periods of 110–120 days and give birth to 5–6 young in spring (Habermehl 1975, 1985). Metrics on the Pech IV fetal specimen indicate a nearly full-term individual, i.e., pointing to death in early spring.

Season of death for reindeer from Layer 4 was determined by means of dental eruption and wear stages of partial toothrows. Deciduous and permanent teeth from the maxilla and mandible were evaluated. The Pech IV specimens were compared to the extensive sample of precisely aged dentition from modern caribou found in Miller (1974). Because *Rangifer* have seasonally restricted mating (autumn) and birthing (late May–early June) periods, their season of death can be accurately determined using comparative specimens of known age. Individuals ranging from 2 to 3 months old (MNI = 1), 12 months old (MNI = 1) and 20–24 months old (MNI = 6) all indicate death in spring and perhaps early summer (the newborn and yearling). The presence of two or more premolars and molars from the same toothrow of these individuals allowed for well-defined season of death data in this layer.

Evaluated together, the season of death data from several prey taxa show that Neanderthal occupation of Pech IV, as well as their ungulate hunting strategies, varied throughout the sequence (Table 9.2). Data from Layer 8 show the site being visited during all seasons of the year, with locally available red deer being hunted during most of these occupations. Both levels in Layer 6 indicate that Neanderthals were elsewhere during the fall but occupying the cave in winter and spring. Similar to Layer 8, it looks as if the prey taxa were available year-round and hunted in most seasons. The cold period Layer 4 shows a much different picture, with occupations and reindeer hunting limited to a seasonally restricted window. These data likely show that reindeer were available locally only during this time period in spring and perhaps early summer, i.e., during their migration. Pleistocene *Rangifer* did indeed migrate long distances like their modern counterparts (Britton et al. 2009) and the fact that other Quina Mousterian sites (e.g., Costamagno et al. 2006; Britton et al. 2011) have yielded reindeer assemblages showing hunting episodes restricted to reindeer migration times in spring or fall suggests that Neanderthals took advantage of the local abundance of this taxon during restricted seasonal periods. This in turn implies that Pech IV was occupied for shorter durations in cold periods, including the Quina Mousterian Level 4A. This scenario is supported by the sparser lithic assemblages in Layer 4, particularly Levels 4A and 4B (Turq et al. 2011).

Table 9.2 Seasonality evidence across the sequence at Pech de l'Azé IV, based on eruption and wear of teeth and fetal bone

	Spring	Summer	Fall	Winter
4A	*Rangifer*	*Rangifer*		
4B	*Rangifer*			
4C	*Rangifer*	*Rangifer*		
6A	*Sus/Cervus*	*Sus/Cervus*		*Cervus*
6B	*Sus/Cervus*	*Sus scrofa*		*Cervus*
8	*Sus scrofa*	*Cervus*	*Cervus*	*Sus/Cervus*

Ageable skeletal specimens are relatively sparse amongst the Pech IV assemblages, both in terms of teeth as well as unfused skeletal bone. Layer 6 is an exception with a substantial sample of cervid teeth (NISP = 91), in contrast to Layer 8, which contains just four. Layer 4 yielded smaller tooth assemblages (NISP = 65) but they were nonetheless valuable for age and season information.

Mandibular and maxillary teeth of red deer and reindeer were aged based on eruption state and/or wear stages (Miller 1974; Carter 2006) and placed into three age groups: juvenile (from time of birth to prior to loss of deciduous teeth), prime adult (reproductive years of life), and old adult (tooth crown is >50 % worn) (following Stiner 1990). Grouping the data in this way allows us to construct a mortality profile for evaluating whether Neanderthals targeted certain age groups of prey and in turn, if their hunting strategies remained consistent or varied across the sequence at Pech IV.

Both red deer assemblages in Layer 6 show a predominance of prime adults with one or two juveniles and just one old adult in Level 6A (Table 9.3). Based on the evidence we have on multiple seasons of red deer hunting in these levels, we can presume that this taxon was locally available most or all of the year. Pleistocene red deer were not a highly migratory species like reindeer, but instead made limited seasonal movements in the regional landscape (Pellegrini et al. 2008). Perhaps the reliability of this prey animal in the vicinity of Pech IV influenced the more frequent targeting of prime adult individuals by Neanderthals, some of whom had calves younger than a month old.

The data from Layer 4 are quite varied, with more juveniles in Level 4A, all prime adults in 4B, and no clear pattern is visible in 4C. Because the samples are small (3–4 individuals in each level), we should interpret a "focus" on any age group with caution. That said, the varying patterns

Table 9.3 Summary of age groups for reindeer (Layer 4) and red deer (Layers 6) based on teeth (Layer 8 contained no ageable teeth)

MNI by level	4A	4B	4C	6A	6B	8
Juvenile	3	0	2	2	1	Nd
Prime adult	1	0	1	5	5	Nd
Old adult	0	3	1	1	0	Nd

in each level might be specific to hunting this highly migratory taxon. If Pleistocene *Rangifer* migrations were of similar character to those of today, then they would have proceeded in several stages lasting one to two months each, incorporating a calving period, pulses of movement, pauses, the rut, and occasional splits in herd structure (males separating from females with calves) (Burch 1972; Enloe and David 1997). Using these modern data as a framework, the assemblages from Level 4A and 4C, which show a spring through summer seasonal signal, might represent hunting of females and calves (4A) and mixed herds (4C). The prime adults hunted in summer from Level 4B could reflect a migratory phase when the females and calves were elsewhere and Neanderthals encountered barren females and males in the vicinity of Pech IV. Overall, the results from Layer 4 show much more variability than we see in Layer 6, which seems to be the persistent trend in these cold, reindeer-dominated levels.

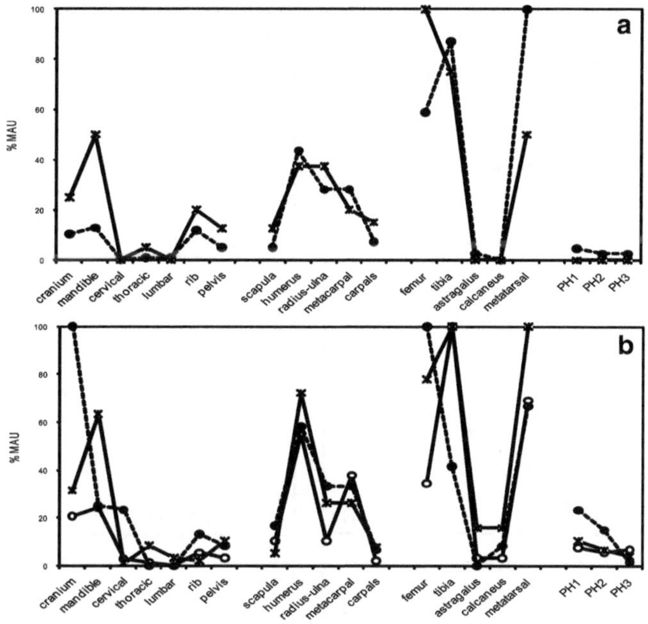

Fig. 9.1 Skeletal element abundance expressed as ratio MAU (minimum animal units). Reindeer from Levels 4A and 4C are plotted in (**a**) and red deer from Layers 6 and 8 are plotted in (**b**)

Skeletal Element Frequencies

The cave of Pech IV was presumably the place to which Neanderthals brought prey animals for processing and consumption, as opposed to the actual kill site. Being the end-point of prey transport (Stiner 1994), caves such as Pech IV therefore provide an opportunity for evaluating transport decisions by hominins. Although the ultimate goal is assumed to have been the transport of as much of the prey animal as possible, we know from ethnographic studies of modern hunter-gatherers that people must sometimes selectively transport certain carcass portions or elements based on nutritional yield (e.g., Binford 1978; Bunn et al. 1988; O'Connell et al. 1988; Bartram 1993). Such decisions are influenced by various factors including the body size of prey, distance to the site, weather or topography and the number of hunters involved (Bunn et al. 1988). In order to evaluate prey transport strategies of Neanderthals at Pech IV, the relative abundances of skeletal elements from red deer and reindeer are considered.

Skeletal frequencies tabulated as MNE (minimum number of elements) are expressed as ratio MAU (minimum number of animal units; Binford 1978) (Fig. 9.1). Because Level 4B is a small assemblage, it has been excluded from the graph. Overall, the posterior axial skeleton (vertebrae, ribs, pelvis) is strongly underrepresented in all levels. Considering that the assemblages from Levels 4A, 4C and 6A were influenced by bone density-mediated attrition, as seen in significant and positive correlations between bone density values and abundance, the lack of the vertebrae and ribs could be purely taphonomic. However, this pattern might reflect the decision to leave the bulky rib and spinal segments behind after having removed the meat. The abundance of heads varies but mandibles are more frequent than crania except in Layer 8.

Whether the discrepancies in skeletal element abundances are the result of transport decisions based on nutritional value can be explored through utility indices. Ratio MAU data from Pech IV are plotted against values for standardized food utility ([S]FUI; Metcalfe and Jones 1988) and standardized marrow cavity volume (Binford 1978). The food utility index averages values for meat, marrow and overall nutritional gain of skeletal elements for this taxon while the marrow index is based on the actual volume of each marrow-yielding long bone shaft; both indices are based on data averaged from modern *Rangifer*. Because the vertebrae and ribs are so poorly represented in the Pech IV assemblages, we limit our analysis to the "high survival set" of skeletal elements (Marean and Cleghorn 2003). These include elements containing thick cortical bone but lacking fragile cancellous bone and include all of the long bone shafts, mandibles, and cranium (Marean and Cleghorn 2003: 34). Because of their better tendency to preserve in the fossil record, these skeletal parts represent the best opportunity for evaluating human subsistence strategies.

A significant and positive correlation exists between overall food utility and skeletal abundance of reindeer in Levels 4A and 4C and red deer in Level 6A (Fig. 9.2). No statistical relationship is seen between marrow cavity volume and ratio MAUs for the long bones and mandible in any level. This is somewhat surprising since the overall frequency of these elements is high, particularly the tibia

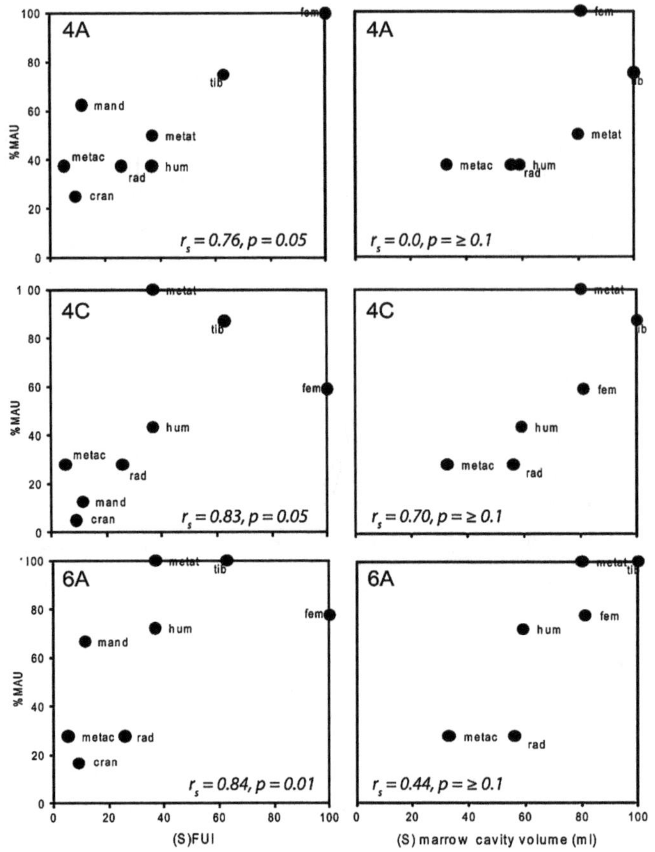

Fig. 9.2 Relation between ratio MAU and the standardized food utility index ([S]FUI) and standardized (S) marrow cavity volume for reindeer in Levels 4A and 4C; and red deer in Level 6A

and the non-meat bearing metatarsal (see Fig. 9.1). This might be in part due to weaknesses in the return rate data on which these utility indices are based, which do not account for every variable involved in the transport of prey (Egeland and Byerly 2005). Therefore, in the case of Pech IV, it is useful to explore more closely the nutritional qualities of marrow specific to skeletal element and time of year (Binford 1978; Morin 2007).

In a reevaluation of Binford's marrow index, Morin (2007) confirmed that the overall quantity of unsaturated fatty acids was the most valuable factor in whether an element was chosen for marrow processing by the Nunamiut, the Alaskan native group that Binford studied (e.g., Binford 1978). With the relevant data recalculated, Morin (2007: Table 4) established the Unsaturated Marrow Index (UMI) to be applied in place of marrow cavity volume. When plotted against the Pech IV ratio MAUs, the UMI shows significant correlations for the assemblages in Level 4C ($r_s = 0.84$, $p = 0.05$) and 6A ($r_s = 0.88$, $p = 0.05$), which likely relate to the higher rank of the femur in the UMI. However, neither index helps us understand the pattern in Level 4C, where the frequencies of tibiae (MNE = 34) and metatarsals (MNE = 39) are extraordinarily high—by a factor of two times or more—in comparison to equally robust humeri (MNE = 17) and radii (MNE = 11). These numbers suggest that the richest marrow bones were selectively transported to the cave in greater quantity than the upper limb elements for further processing. This might relate to the poor physical condition of some of the reindeer, in light of their having been hunted in spring when they tend to be undernourished (Burch 1972; Spiess 1979). Fat stores such as marrow are depleted in such situations, beginning with the upper limb bones (humerus and femur) (Speth 1983). The tibia and metatarsal retain fat longer (Binford 1978; Morin 2007), which may explain the much higher numbers of these elements among the Level 4C reindeer assemblage. Similar fat depletion processes affect other ungulates (e.g., Speth 1983) at least in terms of the proximal-to-distal sequence, so we assume that similar economic-based decisions influenced the higher abundances of tibiae (MNE = 29) and metatarsals (MNE = 20) from red deer in Level 6B versus the humerus (MNE = 16).

Finally, the abundance of heads is important to consider, since they are high-survival body parts but low in nutritional utility and bulky to transport. The reindeer assemblages in Layer 4 show low frequencies of heads; mandibles are the reason for the 50.0 ratio MAU value in Level 4A, although only four individuals are represented here, so this abundance might not be as significant as the value in Level 4C where we have 19 reindeer. Mandibles are frequent in Level 6A, suggesting that the small amount of marrow available in this element was exploited along with the other marrow-rich elements in this level. Moreover, this fits the general pattern in 6A where overall food utility was a factor in subsistence decisions. Crania and mandibles are about equally (and poorly) represented in Level 6B, but interestingly, Level 8 shows both crania and tibiae as most abundant. Perhaps the active use of fire in this level allowed for more efficient processing of the crania for their small amounts of nutrition.

In summary, there appears to be consistency in the relative abundances of the high-survival set of skeletal elements (i.e., the long bones and heads) across the Pech IV assemblages. Even the somewhat robust (but still considered low-survival; Marean and Cleghorn 2003) scapula and pelvis are poorly represented, suggesting that these elements were rarely transported. Carpals, tarsals and patellae are not abundant. It is unlikely that their underrepresentation is taphonomic or a result of collection bias during excavation, since they are generally robust (exception being the patella), diagnostic, and large enough to have been piece-plotted and collected separately during excavation. Therefore, their abundance probably relates to anthropogenic factors, namely that they rarely "rode" along articulated to their respective long bones because the limbs were previously disarticulated and defleshed before being transported to the cave. This corroborates the evidence for the levels in which we see the selective transport of the best marrow bones in much higher frequencies. Level 4C shows the most distinctive discrepancy between relative abundance of certain skeletal elements, pointing to different strategies of carcass transport during these cold period occupations.

Carcass Butchery and Utilization

Neanderthals' exploitation of ungulate prey across the Pech IV sequence is evaluated by means of frequencies of stone tool cutmarks, percussion marks and fresh (i.e., green or spiral) breakage. In order to have comparable datasets, the results are presented for the long bones of red deer and reindeer only (Fig. 9.3; Tables 9.4, 9.5). The actual number of cutmark striations or percussion marks was not counted but instead, the data presented here involve the NISP bearing one or more modifications.

Cutmarks are more than twice as frequent on the reindeer assemblages of Layer 4 in comparison to the lower layers. Since articular ends of long bones are scarce in the Pech IV assemblages overall, the majority of cutmarks are found on the shaft portions, indicating the removal of meat and tissue

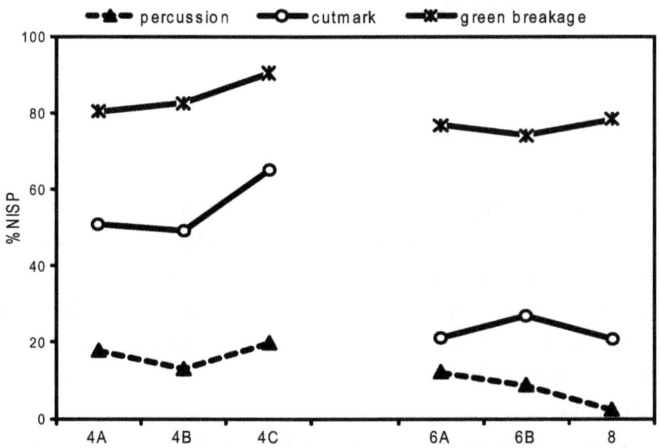

Fig. 9.3 Summary of bone modifications, expressed as %NISP, for reindeer (Layer 4) and red deer (Layers 6 and 8) long bones

(e.g., Binford 1981; Nilssen 2000). The reasons behind the abundant cutmarked bone in Layer 4 is difficult to explain, particularly because cutmark production and abundance are not fully understood (e.g., Egeland 2003). If we compare the Pech IV Quina Mousterian (Level 4A) cutmark frequencies to two other known Quina Mousterian reindeer assemblages, the overall pattern is that the numbers at Pech IV are generally high; for example, Jonzac exhibits 20–25 % cutmarks on the long bones (Steele et al. 2009). However, the cutmark frequencies in Level 9 of Les Pradelles range between 30 and 60 % (Costamagno et al. 2006) and are comparable on some long bones to those seen in Level 4A at Pech IV.

A further evaluation of cutmark frequencies focuses on discrepancies between specific skeletal elements or between the meatier upper limbs and meat-poor lower limbs. Similar, overall abundances of cutmarks characterizes the red deer assemblages in Layers 6 and 8 (Fig. 9.4c). This pattern also applies to the reindeer in Level 4C but not in the other two levels (Fig. 9.4a). Levels 4A and 4B reflect differences in utilization of the meatier upper elements (humerus, femur) versus the lower limb elements that hold little muscle mass; perhaps these variations reflect carcass condition (e.g., stiff versus supple) or alternatively, the differences in morphology among each element that lend them to more or fewer cutmarks regardless of butchery effort (Egeland 2003).

Percussion impact marks are similarly abundant across the cervid assemblages (Figs. 9.3, 9.4b–d). These distinctive modifications are produced when a butcher uses a hard hammer such as a cobblestone to break open the long bone for accessing the marrow, although some of the percussion marks are likely lost or obscured by fracture surfaces in the bone. This may be a factor in the low frequency of marks in the Layer 8 assemblage, which was heavily fractured by burning and trampling. The lower frequencies in Layer 6 could be from the loss of preserved diagnostic marks from subsequent breakage but they might also reflect different strategies of processing red deer long bones for marrow. Excluding Layer 8 with its burning and trampling influences, the assemblages in Layer 6 show the smallest mean fragment size (50–52 mm) of green-broken long bones across the Pech IV sequence (for comparison, Level 4A

Table 9.4 Comprehensive summary of bone modifications across the six long bones, expressed as NISP and %(modification)

Level	NISP	Cut	%cut	Percussion	%perc	Green brk	%green
4A	118	60	50.8	21	17.8	95	80.5
4B	69	34	49.3	9	13.0	57	82.6
4C	581	378	65.1	114	19.6	522	90.5
6A	510	108	21.2	62	12.2	392	76.9
6B	508	137	26.9	46	9.1	377	74.2
8	293	61	20.8	7	2.4	230	78.5

Table 9.5 Summary of cutmark and percussion frequencies by skeletal element (long bones only) expressed as NISP and %(modification)

Level	4A			4B			4C			6A			6B			8		
	NISP	%c	%p	NISP	%c	%p	NISP	%c	%p	NISP	%c	%p	NISP	%c	%p	NISP	%c	%p
Humerus	11	27.0	55.0	9	67.0	22.0	50	60.0	26.0	64	28.0	20.0	81	27.0	15.0	51	22.0	0.0
Femur	20	60.0	30.0	11	91.0	27.0	59	61.0	27.0	66	26.0	12.0	65	32.0	9.2	52	23.0	3.8
Tibia	20	40.0	15.0	10	70.0	10.0	108	65.0	32.0	78	22.0	15.0	104	38.0	6.7	60	25.0	5.0
Radius-ulna	28	54.0	3.6	14	36.0	7.1	100	62.0	20.0	66	20.0	4.5	68	28.0	5.9	52	35.0	0.0
Metacarpal	7	71.0	57.0	3	67.0	33.0	42	67.0	31.0	39	13.0	28.0	85	19.0	7.1	27	7.4	0.0
Metatarsal	32	53.0	3.1	22	18.0	4.5	222	69.0	7.7	197	19.0	7.6	105	19.0	11.0	51	5.9	3.9

mean = 62 mm; Level 4C mean = 58 mm). Because ethnoarchaeological data on ungulate carcass processing from the Kua in Botswana revealed a distinct pattern where long bone fragmentation increased with time spent at the processing site (Bartram 1993: 86), we should reconsider the importance of fragmentation in conjunction with bone surface modification data to better understand issues such as carcass utilization and site use (Egeland and Byerly 2005).

Fresh, spiral breakage of long bones is consistent across the Pech IV cervid assemblages but again, Layer 4 shows a slightly different pattern. The frequency of green breakage is higher and long bone fragment length is slightly larger (see above). Spiral breakage is generally attributed to marrow processing activities but it might also result from disarticulation in smaller ungulates (e.g., Stiner 2005). It is interesting to note that the frequencies of spiral breaks, cutmarks and impact marks rise and fall more or less together in the Layer 4 assemblages (Fig. 9.4). What these potential correlations mean is less clear, but the bone modification data together show that butchering and processing of reindeer in Layer 4 differed from these activities on red deer in Layers 6 and 8. The significantly higher frequencies of cutmarks in Levels 4A and particularly 4C may reflect more intensive processing of elements for their nutritional yield, but the time spent on these activities was brief in comparison to the lower layers if we consider that larger fragment size implies less time investment for processing (Bartram 1993).

Conclusion

The long stratigraphic sequence at Pech IV provides a unique opportunity for evaluating Neanderthal subsistence behavior and site use over a long period of time and within variable climate regimes. The faunal spectrum at the base of the sequence (Layers 8 through 6) indicates site occupations during generally temperate conditions with forested landscapes.

Zooarchaeological results from these layers show that the cave was occupied repeatedly during most or all of the year; several temperate ungulate taxa were hunted, although red deer was the predominant prey in these layers. Based on the seasonality data, we presume that this taxon was locally available during much of the year. Marrow-rich long bones are the most abundant skeletal parts in all of these assemblages and were cracked and processed thoroughly for their within-bone nutrients. The lithic inventories in Layer 8 indicate the production of tools on local raw materials. In conjunction with evidence for the controlled use of fire in this layer as well as the faunal data, it looks as if occupations were of longer duration. Minus the evidence of fire,

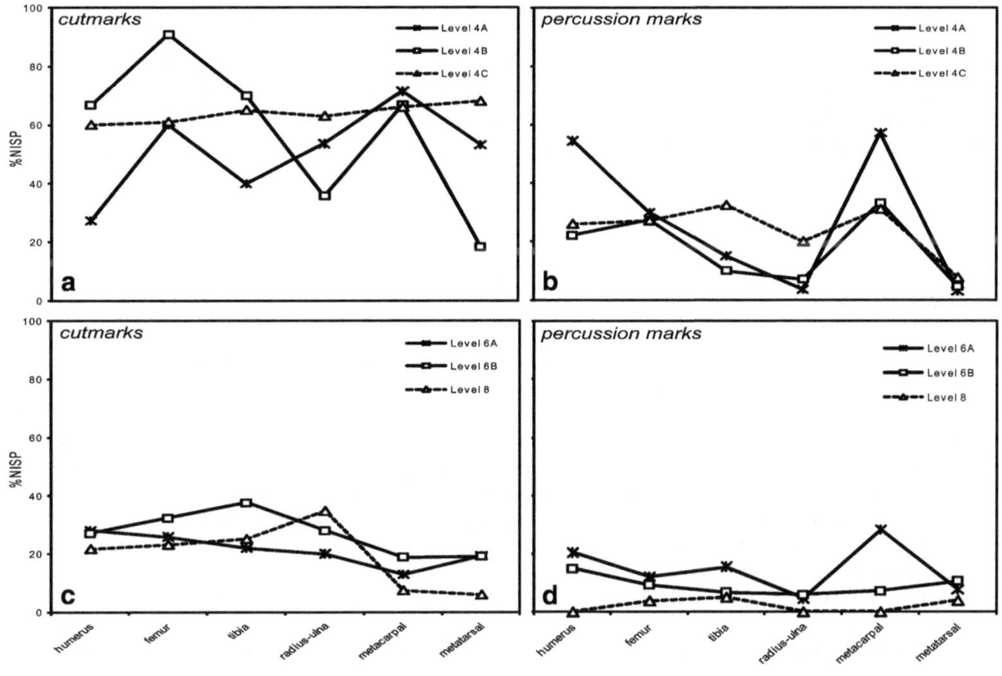

Fig. 9.4 Cutmark and percussion mark frequencies, expressed as %NISP, for each long bone of reindeer in Layer 4 (**a–b**) and red deer in Layers 6 and 8 (**c–d**)

Layer 6 shows a similar pattern of occupation, with on-site production of small flakes and extensive exploitation of cervid prey. Overall, the temperate layers of Pech IV show remarkable stability in the way the cave was used, suggesting that this locality played an important role in the local settlement system.

The cold period Layer 4 at the end of MIS 4 and beginning of MIS 3 reflects a different character of occupations by Neanderthals. The faunal spectrum is quite limited, with reindeer overwhelmingly predominant, most likely due to climatic factors as opposed to human choice (Grayson and Delpech 2005). Occupations and hunting of reindeer took place in seasonally-restricted windows, likely when this prey species was available locally as it moved through the area on its spring migration. The evidence for focused predation on this seasonal ungulate during its migrations from Layer 4 at Pech IV adds to the number of cold period Mousterian or Quina Mousterian sites in southwestern France showing this subsistence strategy (e.g., Les Pradelles (Costamagno et al. 2006); Jonzac (Steele et al. 2009)). This does not suggest that Neanderthals followed the herds during the year, but instead were knowledgable about the seasonal abundance of a preferred prey animal and took advantage of these opportunities. At other times of the year, Neanderthals presumably hunted the less migratory and more predictable animals such as horse and bison. It remains unclear where hominin groups based themselves during the seasons they were not occupying Pech IV, but all of the evidence from Layer 4 indicates higher mobility. A higher percentage of exotic raw materials is seen in the Quina Mousterian Level 4A, and the overall trend seen in the zooarchaeological data is that reindeer were processed intensively in this level as well as the Typical Mousterian Levels 4B and 4C, with selective transport of the best marrow bones (distal limb). In summary, the picture gleaned from the datasets in Layer 4 shows much less stability in site use in comparison to the earlier occupations, which reflects adaptations by Neanderthals to such factors as climate, ecological conditions, and prey availability.

The record from Pech de l'Azé IV provides valuable new insight on Neanderthal occupations of this site. The extent of its stratigraphic sequence makes Pech IV particularly interesting and further results from the faunal, lithic and geological analyses are forthcoming. In terms of subsistence behavior, the zooarchaeological results presented here support those from numerous other Neanderthal sites (e.g., Boyle 2000; Burke 2004; Speth and Tchernov 2007; Rendu 2010), showing the flexibility and adaptability of this hominin's hunting strategies and utilization of its landscape.

Acknowledgments Thank you to Jamie Clark and John Speth for the invitation to participate in their stimulating session of the 11th ICAZ Congress in Paris. Thanks to the Pech IV team, particularly Harold Dibble and Shannon McPherron, for giving me the opportunity to work on the fauna and for many great discussions about the stones and bones. Thank you also to other team members Paul Goldberg, Dennis Sandgathe, and Alain Turq for their help and discussions. Thanks to Charles Egeland, Sabine Gaudzinski-Windheuser, Jamie Hodgkins, Olaf Jöris, Lutz Kindler, Claire Letourneux, Natalie Munro, Isabelle Thèry-Parisot, Will Rendu, Daniel Richter, Utsav Schurmans, Uta Schwarz, Marie Soressi, Teresa Steele, Martin Street and Elaine Turner for their contributions to this research. I am also grateful to Mark Diab for his help with this paper; and to Donald Grayson and two anonymous reviewers for their valuable comments and suggestions for improving the manuscript. Support for the excavation and analysis of Pech IV came from the National Science Foundation (0073856; 0551927), the Leakey Foundation, the University of Pennsylvania Museum of Archaeology and Anthropology, the Department of Human Evolution at the Max Planck Institute for Evolutionary Anthropology, and the Forschungsbereich Altsteinzeit—RGZM, Neuwied/Mainz.

References

Bartram, L. E. (1993). *An ethnoarchaeological analysis of Kua San (Botswana) bone food refuse*. Ph.D. Dissertation, University of Wisconsin.

Beauval, C. (2004). La faune des niveaux Mousteriens de "Chez-Pinaud" (Jonzac, Charente-Maritime, France). Première analyse. In J. Airvaux (Ed.), *Le site paléolithique de Chez-Pinaud à Jonzac, Charente-Maritime* (pp. 125–156). Préhistoire du Sud Ouest.

Binford, L. R. (1978). *Nunamiut ethnoarchaeology*. New York: Academic Press.

Binford, L. R. (1981). *Bones: Ancient men and modern myths*. New York: Academic Press.

Blasco, M. F. (1997). The pursuit of game: The Mousterian cave site of Gabasa 1 in the Spanish Pyrenees. *Journal of Anthropological Research, 53*, 177–217.

Bordes, F. (1954). Le gisement de Pech de l'Azé IV: note préliminaire. *Bulletin de la Société Préhistorique Francaise, 72*, 293–308.

Boyle, K. V. (2000). Reconstructing Middle Palaeolithic subsistence strategies in the South of France. *International Journal of Osteoarchaeology, 10*, 336–356.

Britton, K., Grimes, V., Dau, J., & Richards, M. P. (2009). Reconstructing faunal migrations using intra-tooth sampling and strontium and oxygen isotope analyses: A case study of modern caribou (*Rangifer tarandus granti*). *Journal of Archaeological Science, 36*, 1163–1172.

Britton, K., Grimes, V., Niven, L., Steele, T., McPherron, S., Soressi, M., et al. (2011). Strontium isotope evidence for migration in late Pleistocene *Rangifer*: Implications for Neanderthal hunting strategies at the Middle Palaeolithic site of Jonzac, France. *Journal of Human Evolution, 61*, 176–185.

Bunn, H. T., Bartram, L. E., & Kroll, E. M. (1988). Variability in bone assemblage formation from Hadza hunting, scavenging, and carcass processing. *Journal of Anthropological Archaeology, 7*, 412–457.

Burch, E. S. (1972). The caribou/wild reindeer as a human resource. *American Antiquity, 37*, 339–368.

Burke, A. (2000). The view from Starosele: Faunal exploitation at a Middle Palaeolithic site in Western Crimea. *International Journal of Osteoarchaeology, 10*, 325–335.

Burke, A. (2004). The ecology of Neanderthals: A preface. *International Journal of Osteoarchaeology, 14*, 155-161.

Carter, R. J. (2006). A method to estimate the ages at death of red deer (*Cervus elaphus*) and roe deer (*Capreolus capreolus*) from developing mandibular dentition and its application to Mesolithic NW Europe. In D. Ruscillo (Ed.), *Recent advances in ageing and sexing animal bones* (pp. 40–61). Oxford: Oxbow Books.

Chase, P. G. (1986). The *hunters of Combe Grenal: Approaches to Middle Palaeolithic subsistence in Europe*. Oxford: British Archaeological Reports 286.

Chase, P. G. (1999). Bison in the context of complex utilization of faunal resources: a preliminary report on the Mousterian zooarchaeology of La Quina (Charente, France). In J.-P. Brugal, F. David, J. Enloe, & J. Jaubert (Eds.), *Le Bison: Gibier et Moyen de Subsistence des Hommes du Paléolithique aux Paléoindiens des Grandes Plains* (pp. 159–184). Antibes: Éditions APDCA.

Conard, N. J., & Prindiville, T. J. (2000). Middle Palaeolithic hunting economies in the Rhineland. *International Journal of Osteoarchaeology, 10*, 286–309.

Costamagno, S., Meignen, L., Beauval, C., Vandermeersch, B., & Maureille, B. (2006). Les Pradelles (Marillac-le-Franc, France): A Mousterian reindeer hunting camp? *Journal of Anthropological Archaeology, 25*, 466–484.

D'Errico, F., & Vanhaeren, M. (2002). Criteria for identifying red deer (*Cervus elaphus*) age and sex from their canines. Application to the study of Upper Palaeolithic and Mesolithic ornaments. *Journal of Archaeological Science, 29*, 211–232.

Dibble, H. L., & McPherron, S. P. (2006). The missing Mousterian. *Current Anthropology, 47*, 777–803.

Dibble, H. L., & McPherron, S. P. (2007). Truncated-faceted pieces: Hafting modification, retouch or cores? In S. P. McPherron (Ed.), *Tools versus cores: Alternative approaches to stone tool analysis* (pp. 75–90). Newcastle: Cambridge Scholars Publishing Newcastle.

Dibble, H. L., Berna, F., Goldberg, P., McPherron, S. P., Mentzer, S., Niven, L., et al. (2009). A preliminary report on Pech de l'Azé IV, Layer 8 (Middle Paleolithic, France). *PaleoAnthropology*, 182–219.

Egeland, C. P. (2003). Carcass processing intensity and cutmark creation: An experimental approach. *Plains Anthropologist, 48*, 39–51.

Egeland, C. P., & Byerly, R. M. (2005). Application of return rates to large mammal butchery and transport among hunter-gatherers and its implications for Plio-Pleistocene hominid carcass foraging and site use. *Journal of Taphonomy, 3*, 135–157.

Enloe, J. G. & David, F. (1997). *Rangifer* herd behavior: Seasonality of hunting in the Magdalenian of the Paris Basin. In L. J. Jackson & P. T. Thacker (Eds.), *Caribou and reindeer hunters of the Northern Hemisphere* (pp. 52–68). Hampshire: Avebury.

Gaudzinski, S. (1995). Wallertheim revisited: A re-analysis of the fauna from the Middle Palaeolithic site of Wallertheim (Rheinessen/Germany). *Journal of Archaeological Science, 22*, 51–66.

Grayson, D. K., & Delpech, F. (2005). Pleistocene reindeer and global warming. *Conservation Biology, 19*, 557–562.

Habermehl, K.-H. (1975). *Die Altersbestimmung bei Haus- und Labortieren*. Berlin: Verlag Paul Parey.

Habermehl, K.-H. (1985). *Altersbestimmung bei Wild- und Pelztieren*. Hamburg: Verlag Paul Parey.

Laquay, G. (1981). *Recherches sur les faunes du Würm I en Périgord*. Ph.D. Dissertation, Université de Bordeaux I.

Marean, C. W., & Cleghorn, N. (2003). Large mammal skeletal element transport: Applying foraging theory in a complex taphonomic system. *Journal of Taphonomy, 1*, 15–42.

McPherron, S. P., Soressi, M., & Dibble, H. L. (2001). Deux nouveaux programmes de recherche au Pech-de-l'Aze I et IV (Dordogne, France). *Bulletin Préhistoire du Sud-Ouest, 8*, 11–30.

Mellars, P. (1996). *The Neanderthal legacy: An archaeological perspective from Western Europe*. Princeton: Princeton University Press.

Metcalfe, D., & Jones, K. T. (1988). A reconsideration of animal body-part utility indices. *American Antiquity, 53*, 486–504.

Miller, F. L. (1974). *Biology of the Kaminuriak population of barren-ground caribou Part 2*. Ottawa: Canadian Wildlife Service.

Morin, E. (2004). *Late Pleistocene population interaction in western Europe and modern origins: New insights based on the faunal remains from Saint-Césaire, Southwestern France*. Ph.D. Dissertation, University of Michigan.

Morin, E. (2007). Fat composition and Nunamiut decision-making: A new look at the marrow and bone grease indices. *Journal of Archaeological Science, 34*, 69–82.

Nilssen, P. J. (2000). *An actualistic butchery study in South Africa and its implications for reconstructing hominid strategies of carcass acquisition and butchery in the Upper Pleistocene and Plio-Pleistocene*. Ph.D. Dissertation, University of Cape Town.

O'Connell, J. F., Hawkes, K., & Blurton Jones, N. (1988). Hadza hunting, butchering, and bone transport and their archaeological implications. *Journal of Anthropological Research, 44*, 113–161.

Pellegrini, M., Donahue, R. E., Chenery, C., Evans, J., Lee-Thorp, J. A., Montgomery, J., et al. (2008). Faunal migration in Late-Glacial central Italy: Implications for human resource exploitation. *Rapid Communications in Mass Spectometry, 22*, 1714–1726.

Pike-Tay, A., Valdés, V. C., & de Quirós, F. B. (1999). Seasonal variations of the Middle-Upper Paleolithic transition at El Castillo, Cueva Morín and El Pendo (Cantabria, Spain). *Journal of Human Evolution, 36*, 283–317.

Rabinovich, R., & Hovers, E. (2004). Faunal analysis from Amud cave: Preliminary results and interpretations. *International Journal of Osteoarchaeology, 14*, 287–306.

Rendu, W. (2010). Hunting behavior and Neanderthal adaptability in the Late Pleistocene site of Pech-de-l'Aze′ I. *Journal of Archaeological Science, 37*, 1798–1810.

Richter, D., Dibble, H., Goldberg, P., McPherron, S., Niven, L., Sandgathe, D., et al. (2010). New chronometric data for Pech de l'Azé IV and its place in the Middle Palaeolithic of southwestern France. *PaleoAnthropology*, A27.

Rivals, F., Moncel, M.-H., & Patou-Mathis, M. (2009). Seasonality and intra-site variation of Neanderthal occupations in the Middle Palaeolithic locality of Payre (Ardèche, France) using dental wear analyses. *Journal of Archaeological Science, 36*, 1070–1078.

Soressi, M., Jones, H. L., Rink, W. J., Maureille, B., & Tillier, A. M. (2007). The Pech-de-l'Azé I Neandertal child: ESR, uranium-series, and AMS 14C dating of its MTA type B context. *Journal of Human Evolution, 52*, 455–466.

Speth, J. D. (1983). *Bison kills and bone counts: Decision making by ancient hunters*. Chicago: University of Chicago Press.

Speth, J. D., & Clark, J. L. (2006). Hunting and overhunting in the Levantine Late Middle Palaeolithic. *Before Farming, 3*, 1–42.

Speth, J. D., & Tchernov, E. (2007). The Middle Paleolithic occupations at Kebara Cave: A faunal perspective. In O. Bar-Yosef & L. Meignen (Eds.), *Kebara Cave Mt Carmel, Israel: The Middle and Upper Paleolithic archaeology Part I* (pp. 165–260). Cambridge: Peabody Museum of Archaeology and Ethnology.

Spiess, A. E. (1979). *Reindeer and caribou hunters: An archaeological study*. San Francisco: Academic Press.

Steele, T. E., Niven, L., Britton, K., Grimes, V., Mallye, J.-B., Rendu, W., et al. (2009). Neandertal exploitation of reindeer in the Quina Mousterian at Chez Pinaud, Jonzac (France): Zooarchaeological and isotopic evidence. *PaleoAnthropology*, A34.

Stiner, M. C. (1990). The use of mortality patterns in archaeological studies of hominid predatory adaptations. *Journal of Anthropological Archaeology, 9*, 305–351.

Stiner, M. (1994). *Honor among thieves: A zooarchaeological study of Neandertal ecology*. Princeton: Princeton University Press.

Stiner, M. C. (2005). *The faunas of Hayonim Cave, Israel: A 200,000-year record of Paleolithic diet, demography and society*. American School of Prehistoric Research Bulletin 48, Peabody Museum, Harvard University.

Turq, A., Dibble, H. L., Goldberg, P., McPherron, S. P., Sandgathe, D., Jones, H., et al. (2011). Les Fouilles Recentes du Pech de L'Azé IV (Dordogne). *Gallia Préhistoire, 53*, 1–58.

Chapter 10
Neanderthal to Neanderthal Evolution: Preliminary Observations on Faunal Exploitation from Mousterian to Châtelperronian at Arcy-sur-Cure

James G. Enloe

Introduction

The karst system at Arcy-sur-Cure provides substantial information pertinent to the transition from the Middle to Upper Paleolithic (Leroi-Gourhan 1988; Farizy 1990; Baffier and Girard 1997). One of the most important stratigraphic sequences is found in the Grotte du Renne (Fig. 10.1), with a sequence from Mousterian through Châtelperronian to Aurignacian. The Châtelperronian includes a fairly large sample of Neanderthal fossil material, including teeth (Bailey and Hublin 2006) and a distinctive temporal fragment (Hublin et al. 1996). The neighboring Grotte du Bison includes a stratigraphic sequence spanning from the Mousterian through the Châtelperronian (Leroi-Gourhan and Leroi-Gourhan 1964) and provides the data discussed in this paper.

While much attention has been paid to the transition from Neanderthals to anatomically modern humans (e.g., Discamp et al. 2011), relatively little has been paid to the ecological or economic contexts of the transition from flake to blade industries *within* the context of a pre-modern hominid species, the Neanderthals. This study attempts a preliminary exploration of indications of continuity or change in environment or subsistence practices *prior to* the transition from the Châtelperronian to the Aurignacian, focusing on potential differences or similarities between the Mousterian and the Châtelperronian at the Grotte du Bison at Arcy-sur-Cure. There are clearly important technological innovations or changes in the transitional industries such as the Châtelperronian and the Uluzzian (Neruda and Nerudová 2011; Perisani 2012). Some have argued that because the species lists of prey are very similar for both the Mousterian and early Upper Paleolithic, the ecological niches occupied by their producers were identical; this does not address whether changes in technology might have had a more profound effect not only on the weaponry that was used to acquire prey but also on the organization of other aspects of the exploitation of prey species (Enloe 2001, 2003a, b). While environment and subsistence are clearly linked in terms of the availability of prey and the techniques necessary to procure and process game, a discussion of the interactions between technology and prey choice, acquisition and processing is beyond the scope of this paper.

Grotte du Bison

At Arcy-sur-Cure, the Cure River cuts through a coral massif amidst softer limestone sedimentary marine deposits. This harder reef formed a barrier to the flow of the river, resulting in large east-west bends in the northward flowing Cure (Fig. 10.1). This resulted in rather spectacular south facing cliffs perforated by karstic galleries leading through the massif north to the next bend in the river. These galleries offer openings at several different levels above the river which were sequentially exposed and dry as the river entrenched itself into the valley according to variations in the climatic regime. Among the upper caves are the Grotte du Renne and the Grotte du Bison. Today, a mere 5 m separates these two collapsed cave mouths, and they were certainly interconnected at the time of their occupation. The Grotte du Bison includes a stratigraphic sequence from the Mousterian through the Châtelperronian. Two levels in the lengthy geological sequence were considered for this analysis (Fig. 10.2). Of primary interest is Level D, which corresponds to Level VIII in the adjacent Grotte du Renne. It includes material from the Châtelperronian and is dated to 34,050 ± 750 ^{14}C BP and 33,670 ± 450 ^{14}C BP (David et al. 2006: 11). In recent excavation of the earlier Mousterian Level I, the discovery of new Neanderthal fossil

J. G. Enloe (✉)
Department of Anthropology, University of Iowa, Iowa City, IA 52242, USA
e-mail: james-enloe@iowa.edu

Jamie L. Clark and John D. Speth (eds.), *Zooarchaeology and Modern Human Origins: Human Hunting Behavior during the Later Pleistocene*, Vertebrate Paleobiology and Paleoanthropology, DOI: 10.1007/978-94-007-6766-9_10,

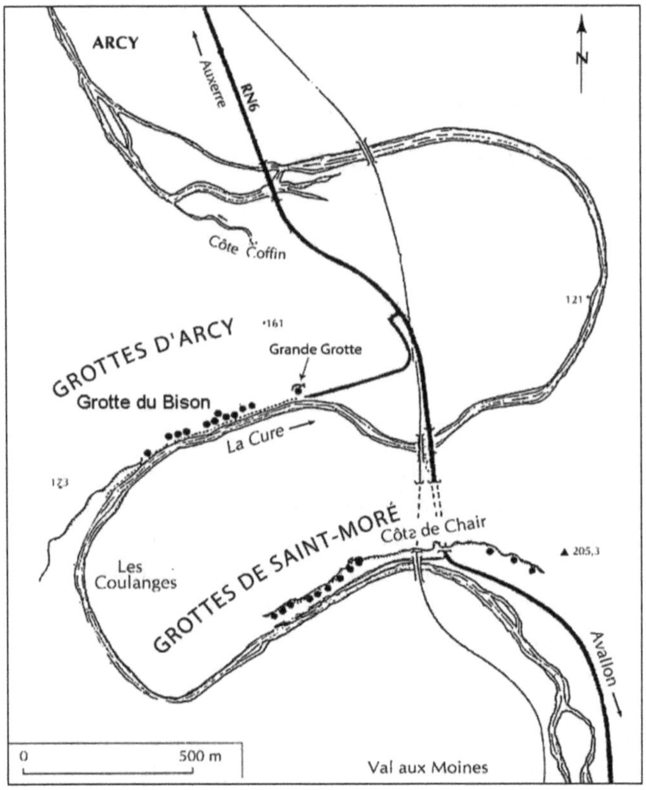

Fig. 10.1 Location of Arcy-sur-Cure caves

specimens adds to knowledge about the authorship of final Middle Paleolithic industries (David et al. 2009a, b).

Faunal materials from the Grotte du Bison have been recovered and recorded using a variety of methods, reflecting the long period of time over which excavations were performed. The site was discovered in 1958 as a lateral gallery during the excavation of the Grotte du Renne by the team lead by Andre Leroi-Gourhan. From 1961 through 1963 excavation of test pits and trenches established a stratigraphic sequence of seven geological levels (D through J) which contained archaeological evidence of human occupation (Leroi-Gourhan 1961; Leroi-Gourhan and Leroi-Gourhan 1964). In 1995, a team led by Francine David began a series of more comprehensive excavations. Primary goals included: (1) correlation of the stratigraphy of the Grotte du Bison with that of the Grotte du Renne and to acquire modern chronometric dates for the levels, (2) investigation of the evolution of the cave and of the climatic conditions, and (3) expansion of the excavated surface to allow observation of spatial patterning. This last goal dictated the excavation method of décapage procedures pioneered by Leroi-Gourhan, privileging the examination of horizontal surfaces rather than vertical stratigraphic cuts. It included vertical photo cartography of each square meter for recording provenience, and retrieval of each archaeological specimen and fine-mesh water screening of all sediment. Beginning in 2006, a total station theodolite was used to record three-dimensional data on potentially identifiable material. The following year, that three-dimensional recording was expanded to include the small, unidentifiable splinters, primarily bone fragments (David et al. 2010). Results of those measurements are indicated in Table 10.1, but are not divided by occupation levels. Thus, the material was recorded and analyzed at differing degrees of intensity, precluding across-the-board comparisons for such things as anatomical part representation and taphonomic observations.

Faunal Assemblages, Levels I and D

Data from Level I (Table 10.2) were chosen for examination of the earlier Neanderthal occupation of the Grotte du Bison. This is the most recently excavated part of the cave, which has been made even more interesting by the discovery of new Neanderthal fossil specimens (David et al. 2009a, b). The lithic assemblage is a denticulate Mousterian (Lhomme et al. 2004). We are still awaiting results from several dating techniques, but overlying levels E and F have been dated to 38 and 40 ka (David et al. 2006: 12). Level I probably dates to around 50 ka or more. This level has had the most comprehensive collection and recording over the history of investigations in the Grotte du Bison. The faunal assemblage included 2,616 identified faunal specimens (NISP) reported from the 1961–1963 excavations. In addition to these, a total of 9,949 faunal remains (NR) were plotted and recovered from Level I during the 2006–2009 field seasons. Of these, 3,225 were numbered; the remainder (6,724) was largely unidentified diaphysis fragments smaller than 2.5 cm in length.

Microfauna were recorded separately in 2007 through 2009, totaling 1,234 specimens that were not included in the macrofaunal count. Due to the frequently porous nature of the *éboulis* fill, it was deemed impossible to distinguish between more recent intrusive burrowers and those specimens that might have been deposited during the Pleistocene. Microfaunal remains were dominated by incisors and molars of rodents, primarily recovered in 2 mm water screening, but also included cranial and postcranial elements from shrews and bats. A substantial portion probably derive from ancient or recent rejection pellets from hawks and owls. None of these remains were taken into account for this study.

The faunal assemblage of Level I (Table 10.2; Fig. 10.3) is dominated by horse and reindeer, together representing over 65 % of the NISP, with MNI counts of twelve each. Reports of the earlier faunal analysis from the 1961–1963 excavations preclude comparisons of skeletal element representation, as those data were not published in the

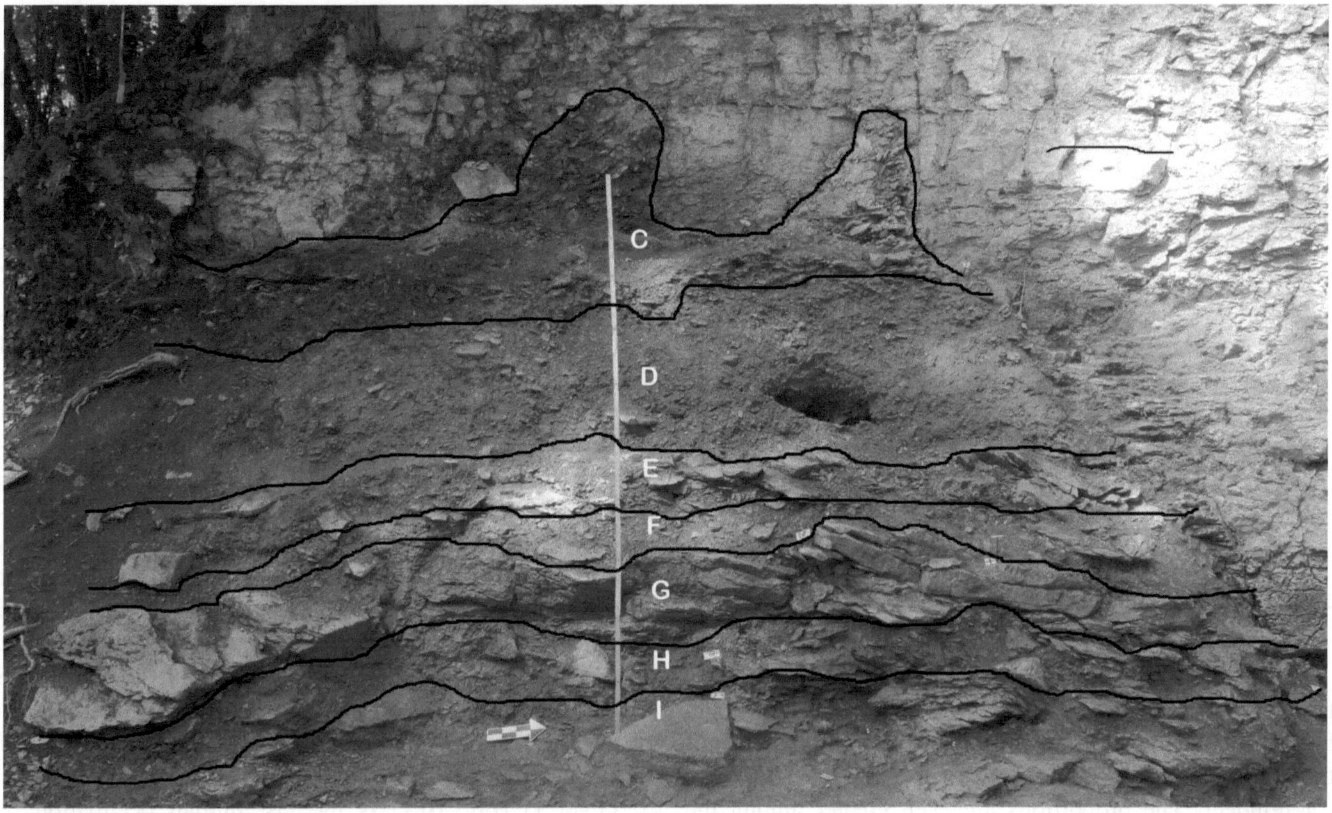

Fig. 10.2 Stratigraphic sequence, as preserved in the witness block against the west wall of the cave prior to excavation of those units (photo J. G. Enloe)

Table 10.1 Three-dimensionally recorded artifacts, Level I

Remains/year	2006	2007	2008	2009	2010	Total
Total	1,045	4,032	4,964	2,866	3,608	16,515
Bone	1,000	3,336	4,064	2,006	3,120	13,526
Numbered bone	1,000	458	1,147	777	1,583	4,920
Bone splinters	0	2,878	2,917	1,229	1,538	8,562
Burned bone	0	194	35	29	128	386
Microfauna	0	417	488	345	156	1406
Coprolite	0	223	190	392	117	1,922
Lithic industry	24	27	30	42	87	210
Lithic splinters	0	12	26	10	29	77
Cobbles	6	7	103	49	73	838
Carbon	0	7	16	4	6	33
Ochre	0	1	10	12	14	37
Miscellaneous	15	9	2	6	1	33

excavation report (Leroi-Gourhan 1961) and the assemblages were not available for examination. Notable among the remains in the recent excavations are reindeer antler and largely intact horse metapodials, neither of which offers significant nutritional utility. Other species, in decreasing importance by respective NISP counts are hyena, bear, fox, aurochs or bison, and wolf, with MNI counts ranging from 5 to 10. Mammoth, red deer, chamois, marmot, hare, rhinoceros and a large felid are represented by small NISP counts and minimum numbers of individuals of one.

Table 10.2 Taxonomic diversity, Level I

Level I	NISP	NISP (%)
Horse	453	31.37
Reindeer	494	34.21
Hyena	123	8.52
Fox	80	5.54
Bear	125	8.66
Bovid	67	4.64
Wolf	47	3.25
Red deer	8	0.55
Mammoth	21	1.45
Chamois	8	0.55
Marmot	5	0.35
Hare	10	0.69
Large feline	1	0.07
Rhinoceros	2	0.14
NISP	1,444	99.99

In Level I, a great proportion of all specimens identified to species were teeth. This was true for both carnivores and herbivores. Among the herbivores, teeth represent 50 % of the identified specimens for horse and red deer, and 25 % for reindeer. The proportions are even greater among the carnivores, with 68 and 67 % for wolf and hyena, 50 % for fox, and 25 % for bear (David et al. 2006: 39–40). These figures suggest significant taphonomic effects on the preservation and identifiability of the faunal assemblage. Of 13,526 bone specimens, only 1,444 (10.7 %) were identifiable, with small (<2.5 cm) diaphysis fragments representing the overwhelming majority of the assemblage. One factor that may account for this pattern is mechanical crushing by the extremely large roof block falls in Level I, but carnivore attrition is likely an even larger factor.

Carnivore to herbivore ratios may suggest who had predominance or exclusivity of occupation of the cave, hominins or other carnivores, and who might be responsible for the accumulation of the faunal assemblage. Of the specimens identified to species (NISP) for level I, the ratio is 1:2.56, suggesting a considerable proportion of carnivores. This is emphasized by the large number of coprolites—505—largely constituted of very fine bone fragments and attributed to hyenas. In summary, a varied faunal spectrum is probably the product of numerous agents of accumulation and modification. The relative paucity of human artifactual remains suggests that other large carnivores were largely responsible for the faunal assemblage. Even the human fossils show evidence of carnivore gnawing (David et al. 2009a: 808). And yet, there is arguably a significant portion of the faunal assemblage in Level I that was contributed by Mousterian Neanderthals, as suggested in a spatial study by Enloe (2011), in which no single species focus could be discerned. These mixed results make it very difficult to assign authorship for the faunal assemblage and render it ambiguous for discovering patterns of human hunting.

Level D (Fig. 10.2) is the last human occupation of the Grotte du Bison, which corresponds to Level VIII in the adjacent Grotte du Renne. It includes material from the Châtelperronian and is dated to 34,050 ± 750 BP and 33,670 ± 450 BP (David et al. 2006: 12). Most of this level was excavated during 1997 and 1998 (David et al. 2006). An additional 7 m^2 had been left along the west wall of the cave as a witness block for stratigraphic control and long-term dosimeter placement for dating; level D of this block

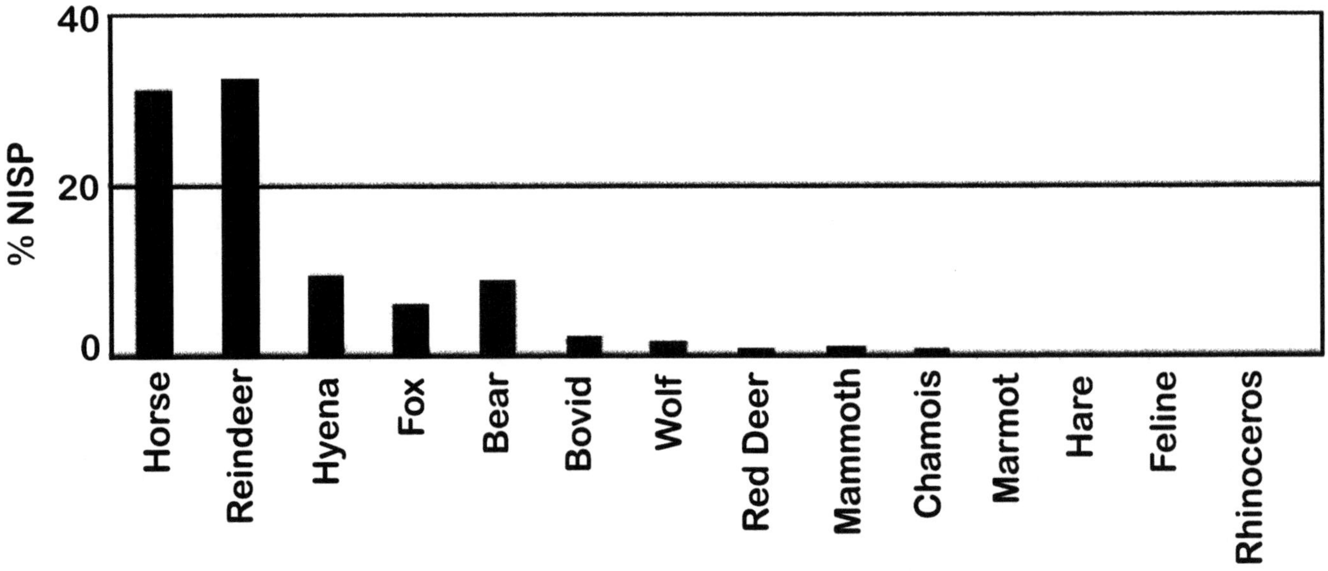

Fig. 10.3 Proportional representation of taxonomic diversity of Level I

Table 10.3 Taxonomic diversity, Level D, with and without bear counts

Level D	NISP	No bear	NISP (%)
Bear	1,169	0	0
Horse	126	126	30.66
Reindeer	199	199	48.42
Hyena	18	18	4.38
Fox	12	12	2.92
Bovid	32	32	7.79
Wolf	4	4	0.97
Red deer	10	10	2.43
Mammoth	0	0	0
Chamois	7	7	1.7
Marmot	2	2	0.49
Hare	1	1	0.24
Feline	0	0	0
Rhinoceros	0	0	0
NISP	1,580	411	100

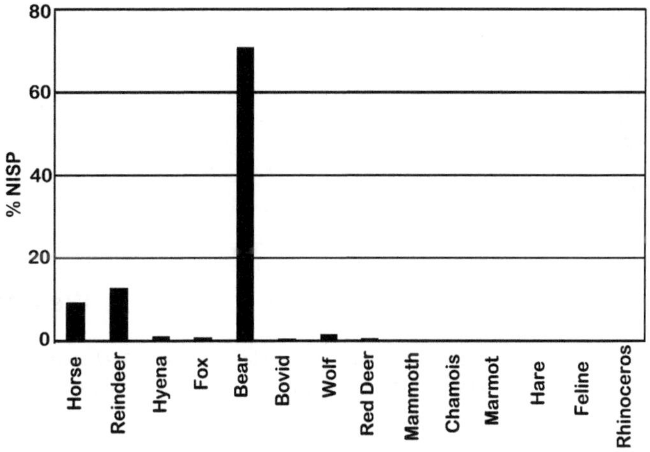

Fig. 10.4 Proportional representation of taxonomic diversity of Level D

was excavated in 2009 by the author (Enloe and David 2010). The faunal spectrum (Table 10.3; Fig. 10.4) is almost as varied as that of Level I, lacking only rhinoceros, mammoth and the large felid. Again, reindeer and horse are the most frequent herbivores, followed by bovid, red deer and chamois. Carnivores include bear, hyena, fox and wolf.

Level D had a carnivore to herbivore ration of 1:0.31, but this may be misleading, as the entire faunal assemblage is dominated by bear remains, particularly fetal bones and deciduous teeth, which are very characteristic of an occupation exclusively for winter hibernation (Baryshnikov and David 2000; David 2002). Level D is the thickest geological stratum in the Grotte du Bison, generally over 50 cm thick, as can be seen in Fig. 10.2. David et al. (2006) anecdotally reported that most of the bear remains from the 1997 and 1998 excavations were found in the upper portion of the level, while reindeer and horse were found at the bottom, suggesting potential differences in the agents of accumulation according to depth in Level D. Figure 10.5 portrays a back plots of the elevations of artifactual material recovered during the 2009 excavation of the western witness block in square N and O, 6 through 10, for which detailed three-dimensional coordinates were recorded with the total station. Three-dimensional data were not systematically recorded for level D during the earlier Leroi-Gourhan or David excavations. A total of 195 object locations were measured, including 2 ochre fragments, 11 cobble fragments, 24 lithic artifacts, 11 burned bones and 148 other faunal remains. The NISP of this faunal sample was 47, of which 35 were identified as bear, including 33 teeth.

Figure 10.5 presents the west-facing coordinates of 6 through 10. A small hearth was discovered in square O8, consisting of a roughly 30 cm diameter circular concentration of ash, burned bone and specks of charcoal, at the base of Level D, on the edge of a projecting limestone bench along the west wall. In Fig. 10.5, non-bear and unidentified faunal specimens are represented by small points; bear remains are represented by "b" symbols; cultural remains, including flint, chert, fire-cracked cobbles, burned bone and charcoal, are represented by "c" symbols. As is evident in Fig. 10.5, almost all of the bear remains came from the upper portions of that stratigraphic unit, the vast majority (33 of 35) above +0.05 m elevation. About 45 % of the unidentified faunal specimens, principally small splinter of diaphyses, were dispersed through the upper 0.55 m of level D above +0.05 m elevation; the remaining 55 % were concentrated in the lower portion, between +0.05 m and −0.10 m elevation. Most of the other identified species' remains and the Châtelperronian cultural materials came from that same thin elevation range at the very bottom of Level D, at the same elevation as the O8 hearth. This suggests that it was after the last Châtelperronian occupation at the base of Level D that the use of the cave passed from intermittent human occupation in the lower Mousterian levels and more exclusive human occupation in Level D, to bear hibernation for the duration of the accumulation of the upper portion of that thick geological layer, until the final roof collapse occurred in level C (David et al. 2006).

When the bear remains are excluded from the Châtelperronian occupation (Table 10.3; Fig. 10.6), a radical change in proportional representation by NISP is produced. Reindeer make up almost half of the remains, horse a third, and all of the other species are insignificant. The carnivore to herbivore ratio drops to 1:11.09, which is substantially different from that of Level I. I have argued elsewhere (Enloe and David 2010) that spatial analyses suggest greater

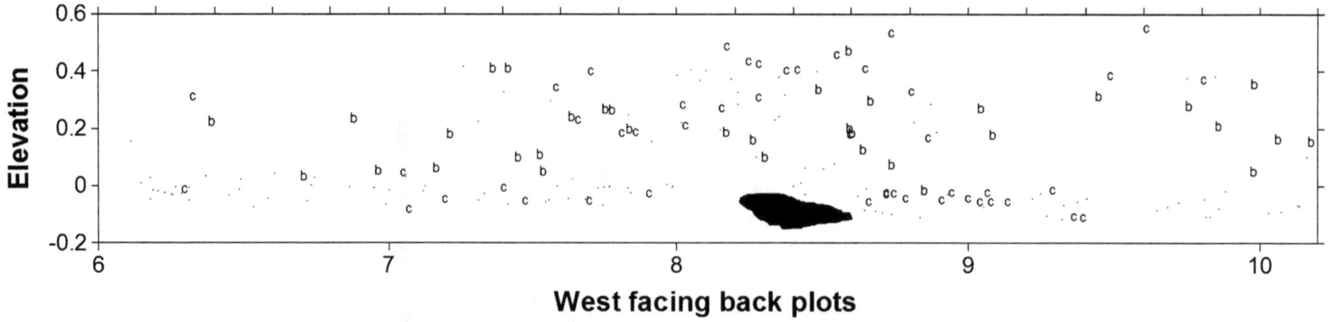

Fig. 10.5 West facing back plots of the elevation of bear (*b*), cultural (*c*) and other fauna (.) from the total station controlled provenience data of level D from 2009 excavations. Hearth O8 indicated by dark mass in lower part of the stratum

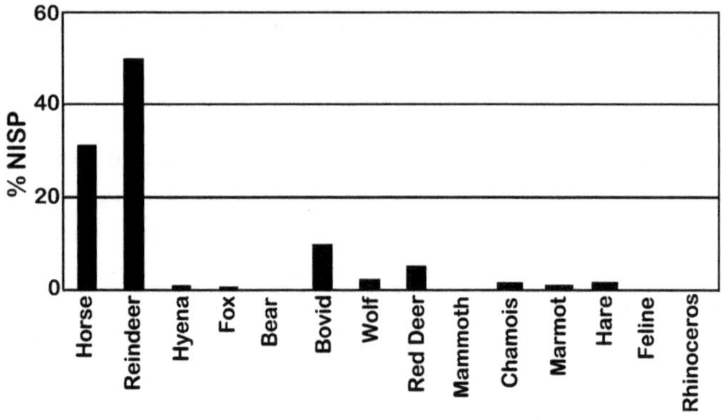

Fig. 10.6 Proportional representation of taxonomic diversity of Level D without bear

exclusivity of occupation by the Châtelperronian inhabitants of the Grotte du Bison, with a more restricted prey spectrum, almost exclusively reindeer and horse. This may well have been the case for the earlier Mousterian occupation of Level I, but the palimpsest deposition obscures clear patterning in taxonomic representation.

Taphonomy

Any consideration of taxonomic representation must commence with a taphonomic assessment. The relatively low 10.7 % of identifiable remains already mentioned for level I is similarly reflected in a 10.3 % NISP for Level D. Clearly, large and robust species are generally better preserved and more identifiable in the paleontological and archaeological record (Grayson 1984; Lyman 1994), but a more precise assessment might be found in the element representation of two of the dominant species for which sample sizes allow the strongest statistical evaluation, reindeer and horse, in Level D (Tables 10.4, 10.5). Anatomical landmarks and portion and segment overlaps were used to derive minimum numbers of element (MNE) frequencies, which were converted into % survivorship proportions, based on minimum numbers of individuals (MNI) for those two well-represented species. For reindeer, the MNI = 3 was derived from MNE counts of metacarpi and tibiae. For horse, the MNI = 2 was derived from counts of crania, mandibles and radii-ulnae. Those proportions were compared with mineral density derived from Lam et al. (1999) calculated for the respective elements of the two species. While % survivorship and mineral density may both qualify as interval-scale variables, we cannot assume a bivariate normal distribution. Therefore, a nonparametric statistic would be most appropriate to evaluate that relationship, so in this case Spearman's Rank Order Correlation Coefficient was employed to assess the significance of the relationship between mineral density and element survival in these samples. For horse ($n = 26$), Spearman's rho = 0.420*, $p = 0.033$. For reindeer, Spearman's rho = 0.546**, $p = 0.004$. In both cases, skeletal element representation is significantly correlated with bone mineral density. Density mediated attrition may be a result of numerous agencies; it is very difficult to distinguish among multiple causes of equifinality such as carnivore ravaging, differential transport or human processing (Enloe et al. 2000; Enloe 2004).

These are small sample sizes and there are a considerable number of zero cells for each data set, which could potentially affect the correlation. As an exploratory data analysis approach, the correlations were run without the zero cells for those elements entirely missing in the assemblage, in an attempt to discern patterning that may be indicative of what might be affecting differential element representation. When the zero cells for horse are removed from the calculation, Spearman's rho = 0.629**, $p = 0.009$, an even stronger correlation. Clearly, with or without zero cells, skeletal element representation for horse indicates density mediated attrition.

Mid-density elements are well represented in reindeer. The zero cells are primarily the low-density elements, although the highest density of all non-tooth elements, the cranium, is also missing. When the zero cells are removed

Table 10.4 Density mediated attrition of reindeer in Level D (MNI = 3)

Element	MNE	Mineral density	Survivorship (%)
Cranium	0	1.29	0
Mandible	2	1.07	33.33
Atlas	0	0.49	0
Axis	0	0.62	0
Cervical	0	0.45	0
Thoracic	0	0.38	0
Lumbar	2	0.45	4.76
Sacrum	0	0.37	0
Caudal	0	0.43	0
Rib	0	0.47	0
Scapula	1	1.01	16.67
Humerus	3	0.62	50
Radius-ulna	4	1.08	66.66
Carpals	3	0.69	10
Metacarpal	5	0.68	83.33
Coxal	1	0.64	16.67
Femur	4	0.74	66.6
Patella	0	0.57	0
Tibia	5	0.73	83.33
Astragalus	0	0.7	0
Calcaneum	2	0.94	33.33
Navicular	0	0.62	0
Cuneiform	0	0.71	0
Metatarsal	3	0.71	50
Phalanx 1	3	0.61	12.5
Phalanx 2	3	0.48	12.5

Table 10.5 Density mediated attrition of horse in Level D (MNI = 2)

Element	MNE	Mineral density	Survivorship (%)
Cranium	3	1.25	75
Mandible	3	0.98	75
Atlas	0	0.64	0
Axis	1	0.53	50
Cervical	0	0.5	0
Thoracic	2	0.32	7.14
Lumbar	0	0.42	0
Sacrum	0	0.36	0
Rib	2	0.36	3.57
Scapula	1	1.03	25
Humerus	2	1.05	50
Radius-ulna	3	1.04	75
Metacarpal	1	0.6	25
Coxal	1	0.65	25
Femur	2	0.99	50
Patella	0	0.4	0
Tibia	2	0.45	50
Astragalus	0	0.64	0
Calcaneum	0	0.69	0
Navicular	0	0.71	0
Cuneiform	0	0.6	0
Metatarsal	2	0.6	50
Phalanx 1	0	0.67	0
Phalanx 2	0	0.62	0
Phalanx 3	1	0.57	25
Styloid	1	0.69	12.5

from the calculation, the picture is not quite the same for reindeer as it was for horse. Spearman's rho = 0.499, p = 0.069, which is not significant, in strong contrast to the very high significance of the same statistic for horse. Without the zeros, the low and high density elements are represented by low to moderate proportions. Axial elements are poorly represented irrespective of their density. What appears to render the relationship insignificant is the comparatively high representation of mid-density elements. These were plotted on a log/log basis which would emphasize the linearity of the relationship and more clearly show what fell outside of that forced linearity. Figure 10.7 plots the 95 % confidence intervals for reindeer skeletal elements. The only elements over-represented are appendicular elements, both forelimbs and hindlimbs. While these may represent substantial meat packages, such as carcass quarters that were differentially transported to the cave, it is important to note that not only the major meat-bearing elements of the humerus, femur and tibia (Binford 1978) are highly represented, but also that the low meat utility elements of the lower limbs, the metacarpals and metatarsals, are highly represented, while carpals, tarsals and phalanges are poorly represented or absent. This suggests that the metapodials are not merely carried along as riders to the major limb bones (since the carpals, tarsals and phalanges were not attached), but were actively selected and transported. These metapodials have very low meat utility; their primary value was as a marrow source (Morin 2007). In fact, most of the major meat-bearing elements in question have higher marrow utility values than meat utility values (Binford 1978: 21, 27). The conclusion appears to be that carcass portions which are present beyond the proportions expected based on to bone mineral density were selected for transport to the Grotte du Bison primarily for the caloric value of their marrow (Speth and Spielmann 1983). This is quite consistent with the discovery of six "pockets" of bones on the main occupation surface at the base of level D reported by David et al. (2006: 42). These were exclusively reindeer diaphysis fragments on which all of the cut marks from the entire level were concentrated.

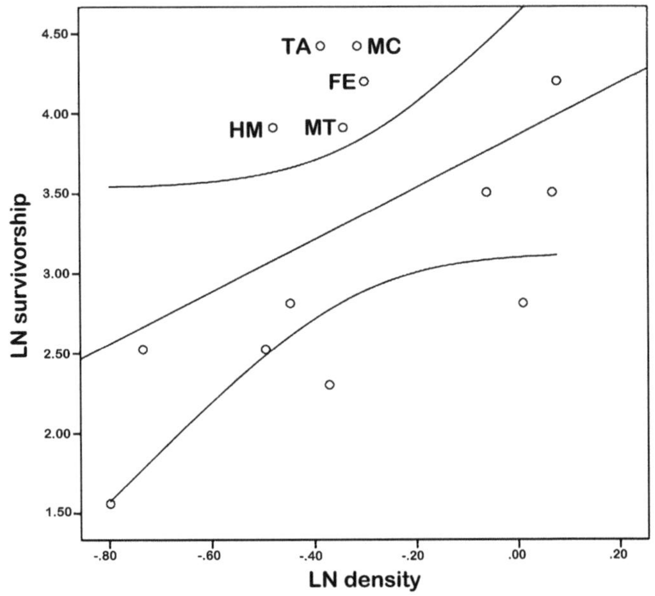

Fig. 10.7 95 % confidence interval of regression of log bone mineral density with log survivorship of reindeer in level D, Grotte du Bison, Arcy-sur-Cure. Over-represented outlier elements include tibia, metacarpal, femur, metatarsal and humerus

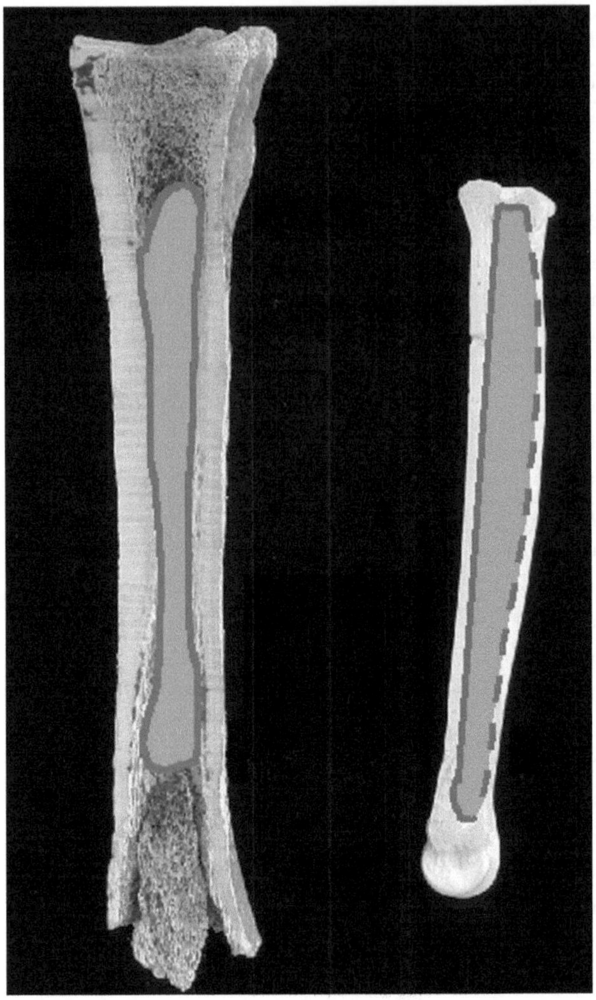

Fig. 10.8 Relative marrow cavity volume of horse (*left*) and reindeer (*right*) metapodials (photo J. G. Enloe)

They correspond to putative marrow processing areas on that level identified by Enloe and David (2010). Blumenschine and Madrigal (1993) have demonstrated that, despite substantial differences in body size, cervid metapodials can yield more than ten times the volume of marrow than can equid metapodials, due to differences in wall thickness affecting medullar cavity volume and proportion of cancellous trabecular tissue (see Fig. 10.8). It is substantially easier to extract the marrow mechanically from reindeer (see Binford 1978: 152–157), in contrast to horse, which requires boiling to extract the maximum of fatty tissues from the metapodial (Enloe 2007). A greater and more systematic focus on nutritionally rich marrow, particularly in the increased exploitation of reindeer, may have been particularly significant for Châtelperronian adaptation compared to that of earlier Mousterian Neanderthals at Arcy-sur-Cure.

Conclusions: Ecology and Taphonomy

Many attempts have been made to compare or contrast the Middle and Upper Paleolithic, whether to demonstrate continuity in patterns of faunal exploitation, and thus basic continuity in ecological niche of Neanderthals and anatomically modern humans, or to demonstrate radical differences to explain differential evolutionary success. Zooarchaeologists typically report species representation, numbers of individuals, element representation, and utility indices to try to enlighten ourselves about the nature of food acquisition, preparation and consumption. While these may be appropriate ways of examining subsistence among modern populations, I am not sure if our traditional ways of looking at faunal assemblages are the best for exploring the potential evolutionary significance of the Middle to Upper Paleolithic transition.

The complicating issue is always taphonomic. Most Middle Paleolithic cave occupations contain a considerable contribution by other carnivores (e.g., Stiner 1994; Stiner et al. 1996). This makes it difficult to assign agency for the accumulation and modification of any faunal materials recovered from such deposits. It is exceedingly problematic to recognize signature patterning of overall assemblages attributed to the various agents. There is a substantial literature debating the morphology, frequency, location and superposition of cut marks and tool marks, but such evidence is infrequently sufficiently represented in the totality of an assemblage to resolve the recognition problem.

We need additional information about pure Neanderthal subsistence ecology because at least some portion of our database consists of assemblages that may be substantially mixed with the residues of other carnivore inhabitants of the sites. While the goal of this research was to compare the Mousterian and Châtelperronian at Arcy-sur-Cure, the results are not entirely clear. The Level I faunal assemblage is not entirely anthropogenic, but at least a portion of it can be attributed to Mousterian Neanderthals (Enloe 2011), although it is not internally spatially structured as argued for the Châtelperronian Level D at the Grotte du Bison (Enloe and David 2010).

While Neanderthal sites are generally lacking in structures or much else in the way of spatial organization, some earlier Middle Paleolithic sites appear to have some elements of differential space use and organization of space, such as in the Mousterian levels at the Abric Romaní (Vaquero and Pastó 2001; Vaquero et al. 2001) or at the Grotta di Fumane (Perisani et al. 2011). The picture for the Châtelperronian seems perhaps a bit clearer. Spatially structured site organization is more evident through differential space use and maintenance of cleaned space at Arcy-sur-Cure, in both the Grotte du Renne's "huts" (Leroi-Gourhan 1961) and in the processing and discard areas in the Grotte du Bison (Enloe and David 2010). These appear to occur in more of an exclusive occupation of the site, where evidence of other carnivores is absent or minimal, or where they become part of the species exploited by humans rather than independent contributors to the faunal assemblage, as demonstrated by David (2004) for the Grotte du Renne. In this case, there also appears to be a substantial focus on the acquisition and processing of marrow-rich skeletal elements from reindeer, seen in taxonomic and element representation, as well as in cut mark and spatial distribution. With those late Neanderthal cases, we can perhaps begin to build a clearer picture of human hunting behavior, and to identify subsistence practices that had greater significance for modern human origins and evolutionary success. These practices may not be evident in the proportions of different species as much as in how we can see differences in the ways species were exploited. They might be functions of seasonal or nutritional characteristics of the prey species that formed the bases for new behavioral patterns among the human exploiters of those species, visible in carcass processing or social mechanisms for cooperative acquisition and consumption, as have been demonstrated for carcass processing and food sharing in the Late Upper Paleolithic (e.g. Enloe 1999, 2003a, b, 2007).

Acknowledgments The author would like to thank Francine David for her endless generosity and support and for her sharing of data. I would like to express my appreciation also to all of the excavation crews over the years at the Grotte du Bison, who made this project possible, and to François de la Varende for his gracious permission and assistance for excavation at the Grotte du Bison. I would like to thank Matthew Hill of the Department of Anthropology at the University of Iowa for his advice and assistance in the statistical analysis of skeletal element representation, as well as Sara Lane and Katie Hove at the University of Iowa for their assistance in preparing data tables and figures. I acknowledge the Service Régional de l'Archéologie de Bourgogne, the LSB Leakey Foundation for Anthropological Research and the University of Iowa's Arts and Humanities Initiative for their contributions to the support for the excavation and analysis. I would like to express my appreciation to the anonymous reviewers, whose suggestions helped make this a stronger paper.

References

Baffier, D., & Girard, M. (1997). Le karst d'Arcy-sur-Cure (Yonne) et ses occupations humaines paléolithiques. *Quaternaire, 8*, 245–255.

Bailey, S. E., & Hublin, J. J. (2006). Dental remains from the Grotte du Renne at Arcy-sur-Cure (Yonne). *Journal of Human Evolution, 50*, 485–508.

Baryshnikov, G., & David, F. (2000). Les ours des cavernes à Arcy-sur-Cure (Yonne, France)—*Ursus* (*Spelearctos*) *spelaeus* Rosenmüller et Heinroth, 1794. *Quaternaire, 11*, 65–79.

Binford, L. R. (1978). *Nunamiut ethnoarchaeology*. New York: Academic Press.

Blumenschine, R. J., & Madrigal, T. C. (1993). Variability in long bone marrow yields of East African ungulates and its zooarchaeological implications. *Journal of Archaeological Science, 20*, 555–587.

David, F. (2002). Les ours du Châtelperronien de la Grotte du Renne à Arcy-sur-Cure (Yonne). In T. Tillet & L. R. Binford (Eds.), *L'Ours et l'Homme* (pp. 185–192). Etudes et Recherches Archéologiques de l'Université de Liège (ERAUL) 100. Liège: Université de Liège.

David, F. (2004). Note sur l'hyène des cavernes (*Crocuta crocuta spelaela* G.) dans trois secteurs de la grotte du Renne à Arcy-sur-Cure (Yonne): Moustérien de la Galerie Schoepflin et de la Rotonde, Châtelperronien du Xc. *Revue de Paléobiologie, 23*(2), 761–769.

David, F., Connet, N., Girard, M., Miscovsky, J.-C., Mourer-Chauviré, C., & Roblin-Jouve, A. (2006). Les niveaux du Paléolithique supérieur à la Grotte du Bison (Arcy-sur-Cure, Yonne): Couches A à D. *Revue Archéologique de l'Est, 54*, 5–50.

David, F., D'Iatchenko, V., Enloe, J. G., Girard, M., Hardy, M., Lhomme, V., et al. (2009a). New Neandertal remains from the Grotte du Bison at Arcy-sur-Cure, France. *Journal of Human Evolution, 57*, 805–809.

David, F., D'Iatchenko, V., Enloe, J. G., Girard, M., Hardy, M., Roblin-Jouve, A., et al. (2009b). Nouvelles découverts de dents néandertaliennes dans la grotte du Bison à Arcy-sur-Cure (Yonne). *Bulletin de la Société Préhistorique Française, 106*, 375–390.

David, F., Connet, N., D'Iatchenko, V., Enloe, J. G., Girard, M., Hardy, M., et al. (2010). *Arcy-sur-Cure (Yonne), Grotte du Bison, Rapport 2006–2010*. Dijon: Quadrennial Report to the Service Régionale d'Archéologie de Bourgogne.

Discamp, E., Jaubert, J., & Bachellerie, F. (2011). Human choices and environmental constraints: Deciphering the variability of large game procurement from Mousterian to Aurignacian times (MIS 5–3) in Southwestern France. *Quaternary Science Reviews, 30*, 2755–2775.

Enloe, J. G. (1999). Hunting specialization: Single-species focus and human adaptation. In J. P. Brugal, F. David, J. G. Enloe, & J. Jaubert (Eds.), *Le Bison: Gibier et Moyen de Subsistance des Hommes du Paléolithique aux Paléoindiens des Grandes Plaines* (pp. 501–509). Antibes: Éditions APDCA.

Enloe, J. G. (2001). Investigating human adaptations in the early upper Paleolithic. In M. A. Hayes & P. Thacker (Eds.), *Questioning the answers: Re-solving fundamental problems of the early upper Paleolithic* (pp. 21–26). BAR International Series S1005. Oxford: Archaeopress.

Enloe, J. G. (2003a). Food sharing past and present: Archaeological evidence for economic and social interaction. *Before Farming, 2003/1*(1), 1–23.

Enloe, J. G (2003b). Acquisition and processing of reindeer in the Paris Basin. In S. Costamagno & V. Laroulandie (Eds.), *Zooarchaeological insights into Magdalenian lifeways* (pp. 23–31). BAR International Series S1144. Oxford: British Archaeological Reports.

Enloe, J. G. (2004). Equifinality, assemblage integrity and behavioral inferences at Verberie. *Journal of Taphonomy, 2*(3), 147–165.

Enloe, J. G. (2007). L'écologie du renne et les adaptations des chasseurs: un aperçu du Paléolithique. In S. Beyries & V. Vaté (Éds.), *Les Civilisations du Renne d'Hier et d'Aujourd'hui: Approches Ethnohistoriques, Archéologiques et Anthropologiques.* XXVIIe Rencontres Internationales d'Archéologie et d'Histoire d'Antibes (pp. 209–221). Antibes: Éditions APDCA.

Enloe, J. G. (2011). Middle Paleolithic spatial analysis in caves: Discerning humans from hyenas at Arcy-sur-Cure, France. *International Journal of Osteoarchaelogy*. doi:10.1002/oa.1276.

Enloe, J. G., & David, F. (2010) *Châtelperronian at the Grotte du Bison, Arcy-sur-Cure: Stratigraphic correlations and spatial organization.* Paper Presented at the 19th Annual Meeting of the Paleoanthropology Society, St. Louis, MO, April 12–13, 2010.

Enloe, J. G., David, F., & Baryshnikov, G. (2000). Hyenas and hunters: Zooarchaeological investigations at Prolom II Cave, Crimea. *International Journal of Osteoarchaelogy, 10*, 310–324.

Farizy, C. (1990). Du Moustérien au Châtelperronien à Arcy-sur-Cure: un état de la question. In C. Farizy (Ed.), *Paléolithique Moyen et Paléolithique Supérieur Ancien en Europe* (pp. 281–189). Mémoires 3. Nemours: Musée de Préhistoire d'Île-de-France.

Grayson, D. K. (1984). Quantitative Zooarchaeology: Topics in the Analysis of Archaeological Faunas. Orlando: Academic Press.

Hublin, J.-J., Spoor, F., Braun, M., Zonneveld, F., & Condemi, S. (1996). A late Neandertal associated with upper Palaeolithic artefacts. *Nature, 381*, 224–226.

Lam, Y. M., Chen, X., & Pearson, O. M. (1999). Intertaxonomic variability in patterns of bone density and the differential representation of bovid, cervid, and equid elements in the archaeological record. *American Antiquity, 64*, 343–362.

Leroi-Gourhan, A. (1961). Les fouilles d'Arcy-sur-Cure (Yonne). *Gallia Préhistoire, 4*, 3–16.

Leroi-Gourhan, A. (1988). Le passage Moustérien-Châtelperronien à Arcy-sur-Cure. *Bulletin de la Société Préhistoire Française, 85*, 102–104.

Leroi-Gourhan, A., & Leroi-Gourhan, A. (1964). Chronologie des Grottes d'Arcy-sur-Cure (Yonne). *Gallia Préhistoire, 7*, 1–64.

Lhomme, V., David, F., & Thiebaut, C. (2004). Les industries de la fin du Paléolithique Moyen de la Grotte du Bison à Arcy-sur-Cure (Yonne). In N. Molines, M.-H. Moncel, & J.-L. Monnier (Eds.), *Les Premiers Peuplements en Europe* (pp. 379–499). BAR International Series S1364, Oxford: Hadrian Books.

Lyman, R. L. (1994). Vertebrate Taphonomy. Cambridge: Cambridge University Press.

Morin, E. (2007). Fat composition, skeletal part selection, and Nunamiut decision making. *Journal of Archaeological Science, 34*, 69–82.

Neruda, P., & Nerudová, Z. (2011). The middle-upper palaeolithic transition in Moravia in the context of the middle Danube region. *Quaternary International*. doi:10.1016/j.quaint.2011.08.035.

Perisani, M. (2012). Fifty thousand years of flint knapping and tool shaping across the Mousterian and Uluzzian sequence of Fumane cave. *Quaternary International, 247*, 125–150.

Perisani, M., Chrzavzez, J., Danti, A., DeMarch, M., Duches, R., Gurioli, F., et al. (2011). Fire-places, frequentations and the environmental setting of the final Mousterian at Grotta di Fumane: a report from the 2006–2008 research. *Quatär, 58*, 131–151.

Speth, J. D., & Spielmann, K. A. (1983). Energy source, protein metabolism, and hunter-gatherer subsistence strategies. *Journal of Anthropological Archaeology, 2*, 1–31.

Stiner, M. C. (1994). *Honor among thieves: A zooarchaeological study of Neandertal ecology*. Princeton: Princeton University Press.

Stiner, M. C., Arsebük, G., & Howell, F. C. (1996). Cave bears and Paleolithic artifacts in Yarimburgaz Cave, Turkey: Dissecting a palimpsest. *Geoarchaeology, 11*, 279–327.

Vaquero, M., & Pastó, I. (2001). The definition of spatial units in middle Paleolithic sites: The hearth-related assemblages. *Journal of Archaeological Science, 28*, 1209–1220.

Vaquero, M., Vallverdú, J., Rosell, J., Pastó, I., & Allué, E. (2001). Neandertal behavior at the middle Paleolithic site of Abric Romaní, Capellades, Spain. *Journal of Field Archaeology, 28*, 93–114.

Chapter 11
The Importance of Fish, Fowl and Small Mammals in the Paleolithic Diet of the Swabian Jura, Southwestern Germany

Nicholas J. Conard, Keiko Kitagawa, Petra Krönneck, Madelaine Böhme, and Susanne C. Münzel

Introduction

The development and spread of modern humans and the extinction of archaic humans represents one of the most important processes in the history of our genus. The Swabian Jura of southwestern Germany preserves a rich and unique record of the behavioral patterns of Neanderthals and early modern humans. Since the 1860s archaeologists and Quaternary scientists have studied the caves of the Swabian Jura to gain information on region's archaeology and hominin adaptations during the Late Pleistocene (Müller-Beck 1983; Scheer 1994; Conard and Bolus 2006).

N. J. Conard (✉)
Abteilung Ältere Urgeschichte und Quartärökologie, Institut für Ur- und Frühgeschichte und Archäologie des Mittelalters, Tübingen, Germany
e-mail: nicholas.conard@uni-tuebingen.de

and

Senckenberg Center for Human Evolution and Paleoecology Schloss Hohentübingen, Eberhard-Karls-Universität Tübingen, Tübingen, Germany

K. Kitagawa · P. Krönneck · S. C. Münzel
Institut für Naturwissenschaftliche Archäologie, Eberhard-Karls-Universität Tübingen, Tübingen, Germany
e-mail: keiko.kitagawa@ifu.uni-tuebingen.de

P. Krönneck
e-mail: petra.kroenneck@uni-tuebingen.de

S. C. Münzel
e-mail: susanne.muenzel@uni-tuebingen.de

M. Böhme
Abteilung Paläobiologie, Institut für Geowissenschaften, Eberhard-Karls-Universität Tübingen, Tübingen, Germany
e-mail: m.boehme@ifg.uni-tuebingen.de

and

Senckenberg Center for Human Evolution and Paleoecology, Eberhard-Karls-Universität Tübingen, Tübingen, Germany

One of the many important aspects of the archaeology of the Swabian caves is the abundance of well-preserved faunal assemblages (Weinstock 1999; Münzel and Conard 2004a, b; Niven 2006; Krönneck 2012). Most of the key assemblages come from the caves of the Ach and Lone valleys, but lesser known areas such as the Lauchert Valley also contain caves with rich faunal material (Fig. 11.1). Unfortunately, open-air localities with stratified faunal material are rare in the Swabian Jura. These faunal assemblages from the many caves, however, provide an ideal starting point for reconstructing past subsistence practices. Here we consider new information from the caves of the Swabian Jura that provides insights into the changing subsistence practices of the Paleolithic hunters and gatherers of the region.

Until now, nearly all of the faunal studies from the Swabian caves have addressed issues related to the large mammalian fauna (Lehmann 1954; Gamble 1979; Münzel et al. 1994; Weinstock 1999; Münzel and Conard 2004a, b; Niven 2006; Krönneck 2012; Münzel in press). These studies went a long way toward reconstructing how the region's Middle and Upper Paleolithic peoples made a living and fed themselves. Although different sites and different strata showed a degree of variation in the abundances of game taxa, the mammalian archaeofauna was mainly characterized by continuity (Münzel and Conard 2004a), and most work up to now has stressed the broad similarities between diet during the Middle and Upper Paleolithic. While the specific geographic setting of sites can affect the availability of game species and helps, for example, to explain the relative abundance of ibex (*Capra ibex*) at Große Grotte and Geißenklösterle near the high cliffs of the Ach Valley, most assemblages from both periods are dominated by horse (*Equus* sp.) and reindeer (*Rangifer tarandus*) (Tables 11.1, 11.2; Figs. 11.2, 11.3) (Krönneck et al. 2004; Münzel and Conard 2004a).

The faunal remains of cave bear (*Ursus spelaeus*) and mammoth (*Mammuthus primigenius*) are often very numerous and require special attention since they can

Jamie L. Clark and John D. Speth (eds.), *Zooarchaeology and Modern Human Origins: Human Hunting Behavior during the Later Pleistocene*, Vertebrate Paleobiology and Paleoanthropology, DOI: 10.1007/978-94-007-6766-9_11,

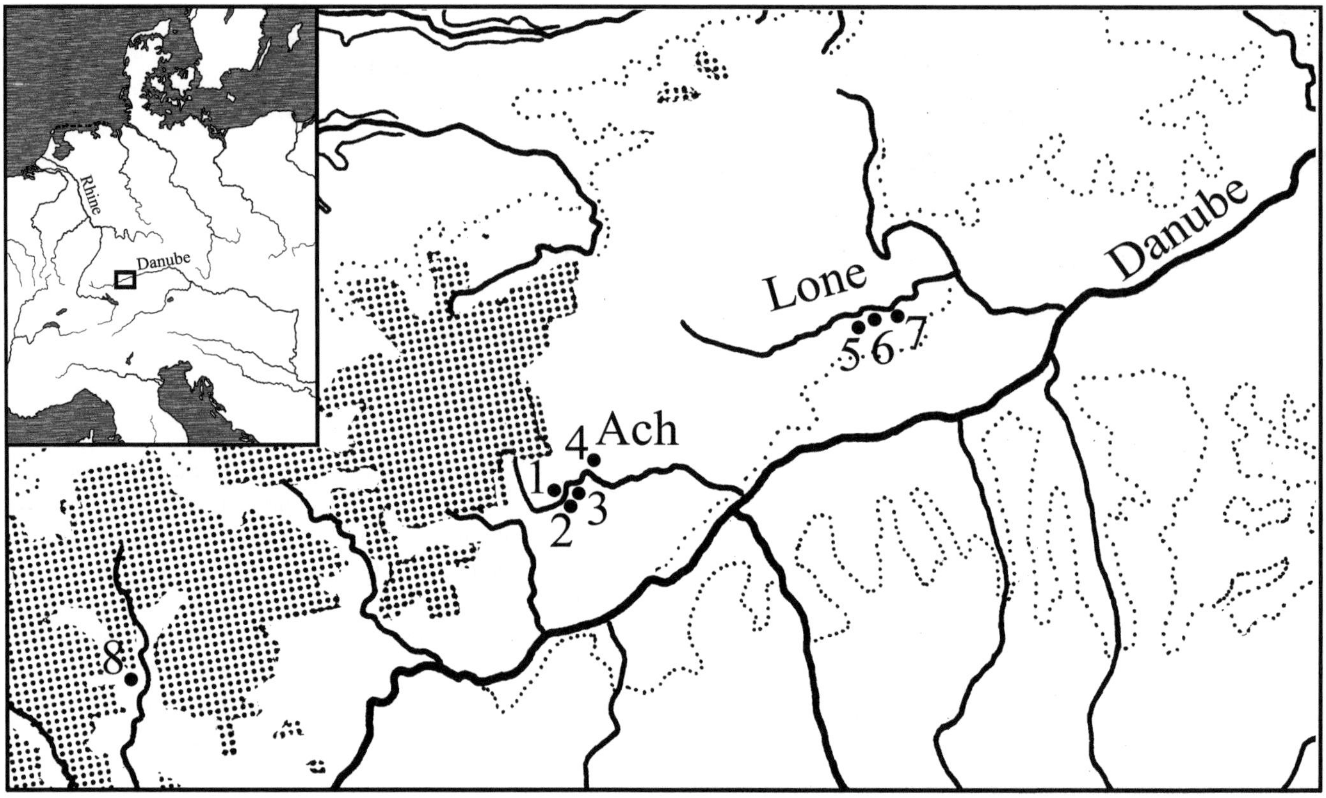

Fig. 11.1 Map of southwestern Germany with the principal sites mentioned in the text. Ach Valley: (*1*) Sirgenstein, (*2*) Hohle Fels, (*3*) Geißenklösterle, (*4*) Brillenhöhle; Lone Valley: (*5*) Bockstein, (*6*) Hohlenstein-Stadel and Hohlenstein-Bärenhöhle, (*7*) Vogelherd; Lauchert Valley: (*8*) Göpfelsteinhöhle

Table 11.1 Geißenklösterle and Hohle Fels

Taxa	Geißenklösterle			Hohle Fels			
	Middle Paleolithic	Aurignacian	Gravettian	Middle Paleolithic	Aurignacian	Gravettian/ Aurignacian	Gravettian
Mammuthus primigenius	3	220	51	0	21	2	40
Coelodonta antiquitatis	8	62	4	2	4	0	1
Equus sp.	21	483	115	10	126	34	336
Cervidae	64	508	186	9	237	65	315
Small ruminants	89	385	155	7	81	25	106
Ursus spelaeus and *U. arctos*	586	2,972	1,419	534	1,021	946	2,273
Canis lupus and other carnivores	39	160	30	7	44	11	54
V. vulpes and *A. lagopus*	26	159	109	4	34	14	73
Lepus sp.	8	209	240	1	52	25	474
Total	844	5,158	2,309	574	1,620	1,122	3,672

NISP of mammalian fauna by taxa without ivory. The main species in *Cervidae* are reindeer and red deer, with reindeer always dominant. The main species for small ruminants are ibex and chamois, with ibex typically being better represented

swamp the faunal signals with material that is not related to past human diets (Tables 11.1, 11.2; Figs. 11.2, 11.3). Throughout the Middle Paleolithic, Aurignacian and Gravettian, remains of cave bear are abundant at Geißenklösterle and Hohle Fels. A high proportion of the cave bear remains accumulated through natural mortality and has no importance for reconstructing human diet. This being said, both during the Middle and Upper Paleolithic, cave bears

Table 11.2 Geißenklösterle and Hohle Fels

Taxa	Geißenklösterle			Hohle Fels			
	Middle Paleolithic	Aurignacian	Gravettian	Middle Paleolithic	Aurignacian	Gravettian/ Aurignacian	Gravettian
Mammuthus primigenius	25	7,261	1,725	0	1,457	41	2,300
Coelodonta antiquitatis	129	869	21	11	71	0	21
Equus sp.	317	6,452	1,113	264	1,920	586	4,673
Cervidae	1,432	3,134	1,386	84	2,123	574	2,100
Small ruminants	530	1161	494	34	617	347	894
Ursus spelaeus and *U. arctos*	2,945	12,687	4,443	4,804	10,622	11,840	25,940
Canis lupus and other carnivores	162	492	50	48	252	792	169
V. vulpes and A. lagopus	35	119	87	7	55	23	106
Lepus sp.	9	222	239	0	44	40	655
Total	5,583	32,396	9,558	5,253	17,161	14,243	36,858

Bone weight (g) of mammalian fauna by taxa without ivory. The main species in *Cervidae* are reindeer and red deer, with reindeer always dominant. The main species for small ruminants are ibex and chamois, with ibex typically being better represented

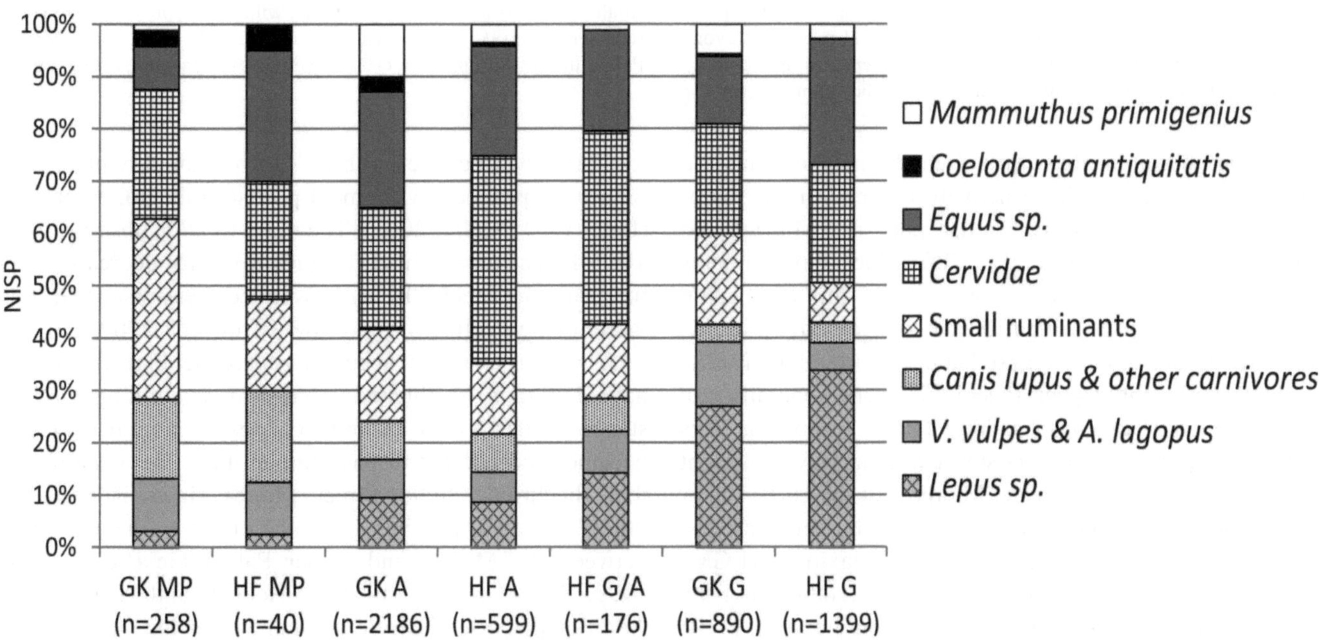

Fig. 11.2 Geißenklösterle and Hohle Fels. NISP% of mammalian fauna by taxa excluding cave bear and mammoth ivory to facilitate comparisons. The main species in *Cervidae* are reindeer and red deer, with reindeer always dominant. The main species for small ruminants are ibex and chamois, with ibex typically being better represented. (*GK* Geißenklösterle, *HF* Hohle Fels, *MP* Middle Paleolithic, *A* Aurignacian, *G/A* Gravettian/Aurignacian, *G* Gravettian)

were occasionally hunted and butchered (Münzel and Conard 2004b). Mammoth is abundant at some sites and particularly frequent during the Aurignacian and to a lesser extent during the Gravettian. This pattern has usually been attributed to the frequent use of mammoth bone and ivory as raw material, rather than a high abundance of mammoth in the diet (Münzel 2001, 2005). Yet, like in other regions (Cavarretta et al. 2001), considerable uncertainty surrounds the question of whether or not early Upper Paleolithic peoples of the Swabian Jura hunted significantly more mammoth than the hominins of the Middle Paleolithic (Niven 2006).

Other trends suggest that woolly rhinoceros (*Coelodonta antiquitatis*) may have been exploited more consistently during the Middle Paleolithic than the Upper Paleolithic. One could also point to other aspects of variability in the large mammalian fauna and their inferred contributions to

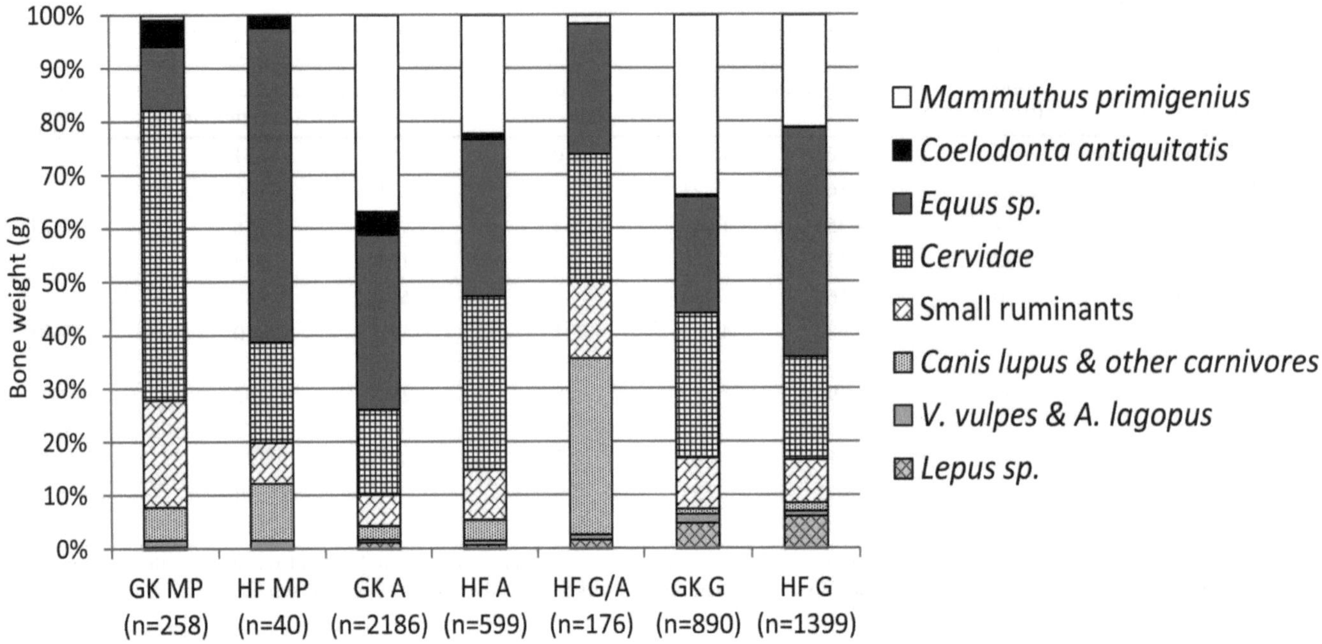

Fig. 11.3 Geißenklösterle and Hohle Fels. Bone weight % of mammalian fauna by taxa excluding cave bear and mammoth ivory to facilitate comparisons. The main species in *Cervidae* are reindeer and red deer, with reindeer always dominant. The main species for small ruminants are ibex and chamois, with ibex typically being better represented. (*GK* Geißenklösterle, *HF* Hohle Fels, *MP* Middle Paleolithic, *A* Aurignacian, *G/A* Gravettian/Aurignacian, *G* Gravettian)

Pleistocene hominin diets. Still, Neanderthals and modern humans both seem to have actively hunted similar mammalian species in broadly similar patterns (Münzel and Conard 2004a; Conard et al. 2006). Since environments were broadly comparable in both periods, this conclusion comes as no surprise (Miller 2009; Krönneck 2012). In this context, our faunal data suggest that the environmental shifts of the Middle and Upper Paleolithic were generally not extreme enough to cause sudden faunal turnovers. This circumstance allowed the persistence of relatively consistent faunal community through the Middle and the early Upper Paleolithic. Only with the end of the local Gravettian around 26 ka BP and the onset of the Last Glacial Maximum (LGM) do we observe a very clear decline in vegetation and the mammalian faunal community. As one would expect, in the absence of rich botanical and faunal communities, human occupation of the region declined radically until the beginning of the Swabian Magdalenian around 13.5–12.5 ^{14}C kBP (Weniger 1987; Hahn 1995; Terberger 2001).

Palynological investigations in southern Germany showed that the vegetation in the last glacial cycle was characterized by steppe-like botanical communities (Müller 2001). During MIS 5d-a, the proportion of trees was higher than later, but there were never dense forests compared to interglacial periods. The arboreal species decreased in MIS 4 and vanished at the end of MIS 3, just before the LGM (MIS 2). From the Middle Paleolithic to the Aurignacian, hominins lived in steppic conditions with thin woods and localized riverine forests. In the Gravettian, only plants that reflect steppe-like environment persist in the pollen profiles (Lang 1994; Müller 2001). The same trend is observed with avian fauna, in which the species diversity decreases over time, and the species living in wooded areas disappear with the climatic decline of the LGM (Krönneck 2009, 2012).

In recent years, advances in zooarchaeological research have led to a heightened interest in examining circumstances under which different classes of animal resources became incorporated in past human diets. For example, in their diachronic studies Stiner, Munro, Haws, Hockett and others pursued possible explanations for dietary shifts between the Middle and Upper Paleolithic (Stiner et al. 1999; Stiner and Munro 2002; Hockett and Haws 2003). Additionally, Klein, Steele and colleagues have argued that the intensity of the exploitation of small game and molluscan resources can be used as a proxy for resource stress and changing population densities (Klein and Cruz-Uribe 2000; Steele and Klein 2005/2006).

In this broad context, Hockett and Haws (2003, 2005) speculated that researchers would find more evidence for the use of small game including small mammals, fish, and birds in the Swabian Aurignacian in comparison to the Middle Paleolithic. This hypothesis implies that modern humans were able to outcompete Neanderthals as a result of incorporating and exploiting nutritional resources that Neanderthals did not use. The expansion of diet breath, all else being equal, would then allow modern humans to maintain larger populations within the occupied areas and outcompete Neanderthals, who were perhaps culturally conservative and highly focused upon exploiting large mammalian game (Conard et al. 2006; Conard 2011). Until now, little data were available to test these ideas based on abundances of small mammals, birds and fish from the caves of the Swabian Jura.

This paper reports the results of the first systematic attempt to present diachronic and numerical data on the abundance of small mammals, birds and fish in the diets of the Paleolithic inhabitants of the Ach Valley in the Swabian Jura. While the analyses are still ongoing, the current data from the Middle Paleolithic and early Upper Paleolithic reflects continuity in most areas accompanied by increases in the use of small mammals, birds and fish.

Cultural Stratigraphy

The Paleolithic cultural stratigraphy and chronostratigraphy of the Swabian caves have been well studied and document reliable records of behavioral evolution. Hohle Fels and Geißenklösterle, the key sites examined in this paper, have been excavated with great care over decades and contain long sequences of Middle and Upper Paleolithic deposits that serve as ideal case studies for examining past patterns of subsistence.

In general, the Middle Paleolithic deposits belong to the Late Pleistocene, but very few of the find horizons have been dated by reliable means. The radiocarbon dates for the Middle Paleolithic often represent minimum ages, and the absolute date for the late phase of the Middle Paleolithic from Geißenklösterle based on ESR dating on tooth enamel falls around 43 ka BP (Richter et al. 2000).

Following what appears to be an occupational hiatus at the end of the Middle Paleolithic (Conard and Bolus 2006; Conard et al. 2006), the Upper Paleolithic begins with the Aurignacian. The Aurignacian appears abruptly in a developed form with no convincing indications for cultural continuity between the Middle and the early Upper Paleolithic. Only in exceptional circumstances, such as at Haldenstein Cave, are leaf points or *Blattspitzen* assemblages found in stratified contexts (Riek 1938). Thermoluminescence dates on heated flint artifacts place the beginning of the Aurignacian around 41 ka BP (Richter et al. 2000), which is consistent with the calibrated ages of the radiocarbon measurements from the early Aurignacian of the region (Conard and Bolus 2003, 2008).

The caves of the Swabian Jura contain rich Aurignacian deposits, which using radiocarbon dating typically date between 40 and 30 ^{14}C kBP. These Aurignacian deposits are usually much richer in artifact finds than the preceding Middle Paleolithic horizons. The period of 30–27 ^{14}C kBP documents the development of the Swabian Gravettian (Moreau 2009). Like the Aurignacian, the dates for the early Gravettian are old in comparison with most regions and points to the central and upper Danube region as a key area of cultural innovation during the early part of the Upper Paleolithic (Conard and Bolus 2003). Many find horizons also point to intense periods of occupation during the Gravettian, particularly in the Ach Valley (Scheer 2000; Conard and Moreau 2004; Moreau 2009). Together with the unfavorable climatic condition of the LGM, the local Gravettian populations declined and eventually left the region. The next significant phase of occupation in the Swabian Jura is the Magdalenian, which appears around 13.5–12.5 ^{14}C kBP and lasts for about a millennium before the Late Paleolithic of the terminal Pleistocene begins. The Middle Paleolithic, Aurignacian and Gravettian form the cultural stratigraphic units of the current study.

Taphonomic Biases

In order to examine how subsistence practices developed during the Middle and Upper Paleolithic, one must consider the extent to which differential preservation of faunal remains or other taphonomic biases affect the archaeological record. The faunal assemblages of the Swabian caves are well preserved, justifying a meaningful comparison of the assemblages over time. The preservation is generally good in terms of morphology, surface preservation and biomolecules. Numerous extractions of collagen for radiocarbon dating and stable isotope studies have shown consistently high yields, and publications on ancient DNA from these cave fauna document excellent preservation (Hofreiter et al. 2002, 2007; Bocherens 2011; Münzel et al. 2011).

The only notable exceptions to this pattern are bones from a number of Middle Paleolithic layers, for example, at Geißenklösterle that show signs of mechanical and chemical surface weathering. These weathered bones are still well preserved in terms of their biomolecular signatures, but they are poorly suited for studying surface modifications. If we quantify anthropogenic versus natural modifications, including carnivore damage, we observe a decrease in carnivore activities and mechanical and chemical weathering from the Middle Paleolithic towards the Upper Paleolithic layers, while the amount of anthropogenic modifications clearly increases (Fig. 11.4). In the current context, we are particularly concerned with the possibility that bones of small mammals, birds and fish may be less well preserved in the deposits of the Middle Paleolithic versus the Upper Paleolithic (Münzel 2009). One argument to suggest the contrary is the presence of microfauna in the deposits from both periods.

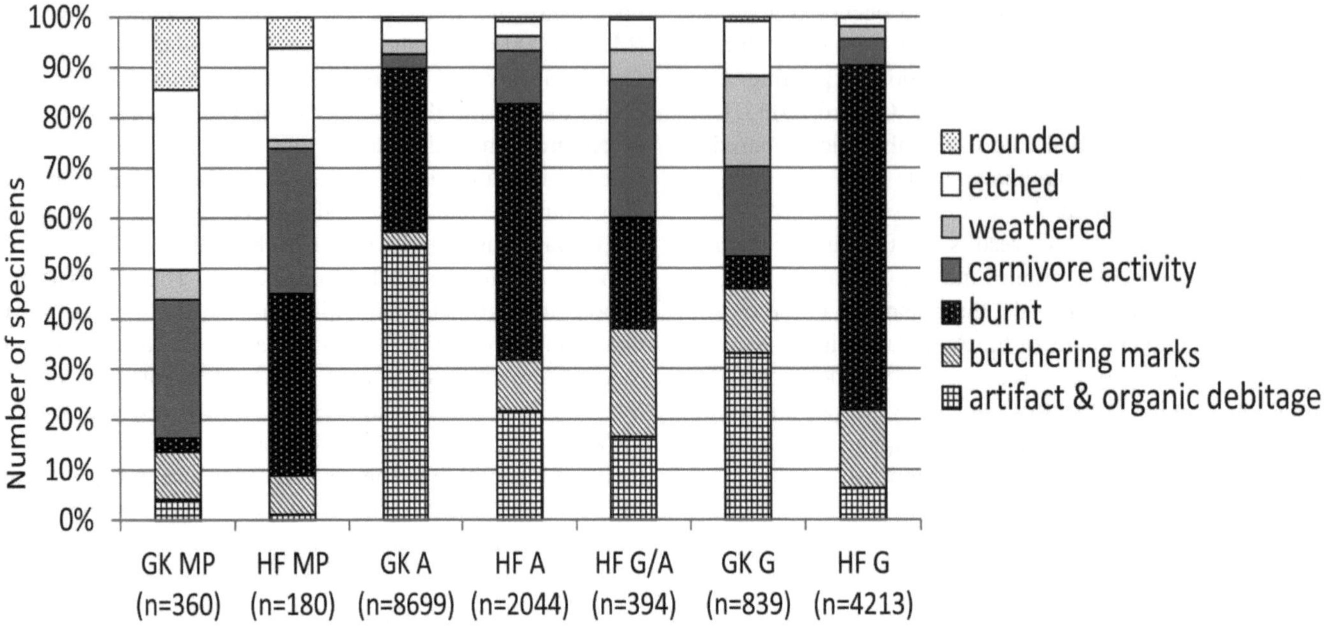

Fig. 11.4 Geißenklösterle and Hohle Fels. Frequency of natural and anthropogenic modifications of bones. To avoid double counting of bones, in the relatively few cases in which bones have multiple modifications, the anthropogenic modification takes precedent over the natural one. Unlike the other diagrams, this includes mammoth ivory. (*GK* Geißenklösterle, *HF* Hohle Fels, *MP* Middle Paleolithic, *A* Aurignacian, *G/A* Gravettian/Aurignacian, *G* Gravettian)

Since the Middle and Upper Paleolithic deposits both contain microfauna, we assume for now that there is no fundamental bias affecting the preservation of bones of small mammals, birds and fish. The question of possible biases in preservation needs to be considered in greater detail as we collect more data on the taphonomy of bones under study.

Another issue is determining the contribution of non-human agents to the accumulation of the faunal assemblages. We see no indications that cave bears brought bones of fish, birds or small mammals into the caves. Work by Bocherens and colleagues (Bocherens et al. 1994, 2006; Münzel et al. 2011) using stable isotopes indicates that the cave bears were herbivores. These conclusions are consistent with studies on tooth morphology, which suggest that cave bears were not habitual carnivores (Kurtén 1976). Another key point is that the small bones and fish scales show no signs of surface weathering and etching that are associated with the digestion of bones by carnivores and birds of prey (Krönneck in press). Gnawed bones exist in both the Middle and Upper Paleolithic horizons, with higher numbers of carnivore damage in the Middle Paleolithic deposits. Another argument against a strong taphonomic bias is the presence of anthropogenically modified bones of small mammals and birds in both the Middle and Upper Paleolithic (Böttcher et al. 2000; Krönneck 2009; Münzel 2009). At this stage of analysis, we have yet to identify taphonomic features of the sites in question that would significantly bias the preservation of small bones in archaeological strata under study. Given the absence of obvious taphonomic bias, we assume that the assemblages of small fauna examined here are broadly comparable.

Sampling

Geißenklösterle and Hohle Fels have been excavated over many seasons using careful methods and systematic water-screening of all the deposits, making them suitable for analyzing small animal exploitation. Fieldwork at Geißenklösterle ran from 1973 to 1991 with a few interruptions. The dig was initially led by Eberhard Wagner and continued starting in 1974 under Joachim Hahn's (1988) direction. Between 2000 and 2002 archaeologists from the University of Tübingen conducted the most recent phase of excavation (Conard and Malina 2003). The modern phase of excavations at Hohle Fels began in 1977 under Hahn's lead and has continued nearly every year since, with the last 15 seasons of the ongoing excavations under Conard's direction.

All archaeological sediments from these excavations have been water-screened with mesh down to 2 mm to recover remains of birds, fish, small mammals and other classes of small materials and artifacts. For each quarter meter of the excavations, we have continuous sequences of uninterrupted finely water-screened and sorted samples. We have processed ca. 12,000 buckets from Geißenklösterle

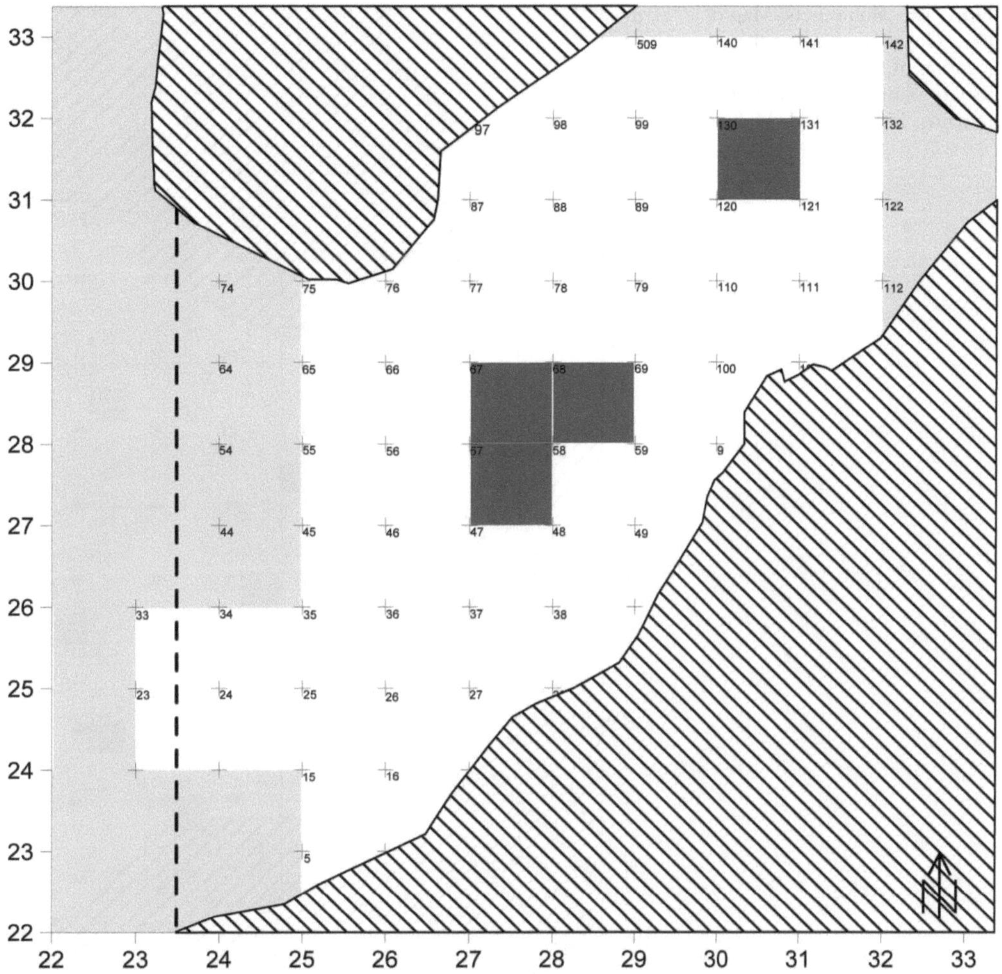

Fig. 11.5 Geißenklösterle. Map of the excavation showing the location of the four square meters sampled for small faunal remains. Figure by M. Malina

and ca. 17,000 buckets from Hohle Fels. Since this work, while productive for a wide range of reasons, is extremely time consuming, we have defined a number of archaeological units for our samples. At Geißenklösterle, the main sample for bird and fish bones originates from 960 washed and sorted buckets of sediment ranging between 1 and 10 l, while most of the small mammalian faunal remains come from the piece-plotted finds. At Hohle Fels, 1,005 water-screened samples provided the majority of the avian and fish material, and the great majority of the small mammalian faunal remains originate from the piece-plotted finds.

These water-screened samples are relatively unimportant for documenting the contribution of large and medium sized game to diets of Neanderthals and modern humans, because the bones of these species are larger and are usually recovered along with the ca. 22,000 and over 60,000 piece-plotted finds from Geißenklösterle and Hohle Fels.

This pilot study makes no attempt to undertake a complete analysis of all of the bones of small mammals, birds and fish recovered during water-screening. Instead, we define samples of the sediments from Geißenklösterle and from Hohle Fels. For the periods of the Upper Paleolithic, the percentage of water-screened samples that have been analyzed is provided in the tables for birds and fish. The sample from Geißenklösterle represents the full column of sediment available from four square meters, while the sample from Hohle Fels represents the full columns available from nine quarter meters (Figs. 11.5, 11.6). Two of the nine columns from Hohle Fels span the entire stratigraphic sequence from the Middle Paleolithic to the Magdalenian, while three of the four columns sampled at Geißenklösterle extend through the Paleolithic sequence from the Middle Paleolithic to the Magdalenian. At Geißenklösterle, the central part of the excavation has mostly been dug to the bedrock. At Hohle Fels, only the central part of the excavation has reached the Middle Paleolithic find horizons, and nowhere has the bedrock yet been reached.

This situation with relatively few sampled columns extending into the Middle Paleolithic leads to a smaller amount of material from this period, and particularly affected the study of fish remains. This bias is exacerbated by the low find density of the Middle Paleolithic materials from Geißenklösterle and Hohle Fels, representing only roughly one-tenth that of the Upper Paleolithic find

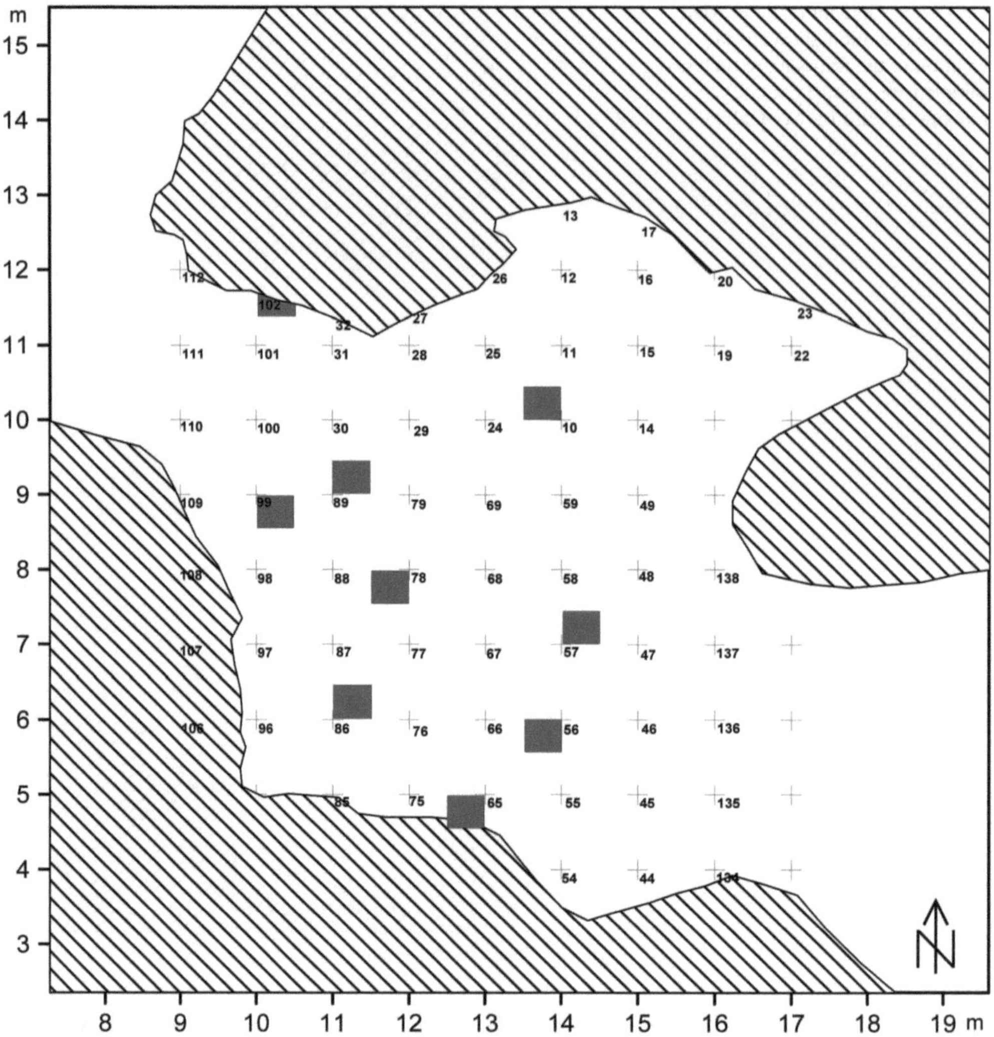

Fig. 11.6 Hohle Fels. Map of the excavation showing the location of the nine quarter meters sampled for small faunal remains. Figure by M. Malina

horizons (Conard et al. 2006; Conard 2011). This means that for the Middle Paleolithic, we need a sample of roughly ten times the volume of the Upper Paleolithic strata to have a comparable sample. To help alleviate this problem, we have collected the fish remains from all the available samples from the Middle Paleolithic of Hohle Fels. We greatly increased the sample used for recovering fish bones, because our initial sample produced no remains of fish at all. At this stage of work, we are looking for the first indications of diachronic trends, and assume that more work will be needed to assess sample bias. The percentage of sampled buckets for each period is indicated in the table that presents the fish data (Table 11.7). While we acknowledge that a range of potential sampling biases as well as variable taphonomic processes and site uses exists, the available samples should give us a first reliable look at how small mammals, birds and fish contributed to hominin diet during the Middle and Upper Paleolithic.

Results

Small Mammals

At Geißenklösterle and Hohle Fels, the bones of red fox (*Vulpes vulpes*), arctic fox (*Alopex lagopus*), and especially hare (*Lepus europaeus* or *L. timidus*) occasionally preserve cutmarks documenting that a portion of the bones from these small species at these sites are the result of hominin predation (Table 11.3; Fig. 11.7). Both foxes are present in all of the periods under consideration. We grouped red fox and arctic fox as well as two hare species together since they are difficult to distinguish for fragmented specimens. Small mammals including hare and foxes form a lower portion of the faunal assemblages of Middle Paleolithic strata at Geißenklösterle than in the overlying Aurignacian and Gravettian deposits. This trend is difficult to confirm at

Table 11.3 Geißenklösterle and Hohle Fels

Taxa/body size	Geißenklösterle			Hohle Fels			
	Middle Paleolithic	Aurignacian	Gravettian	Middle Paleolithic	Aurignacian	Gravettian/ Aurignacian	Gravettian
Lepus sp.	8	209	240	1	52	25	474
V. vulpes and A. lagopus	26	159	109	4	34	14	73
Indet., hare/fox size	5	75	113	10	28	13	197
Total	39	443	462	15	114	52	744

NISP of small mammals by taxa and body size

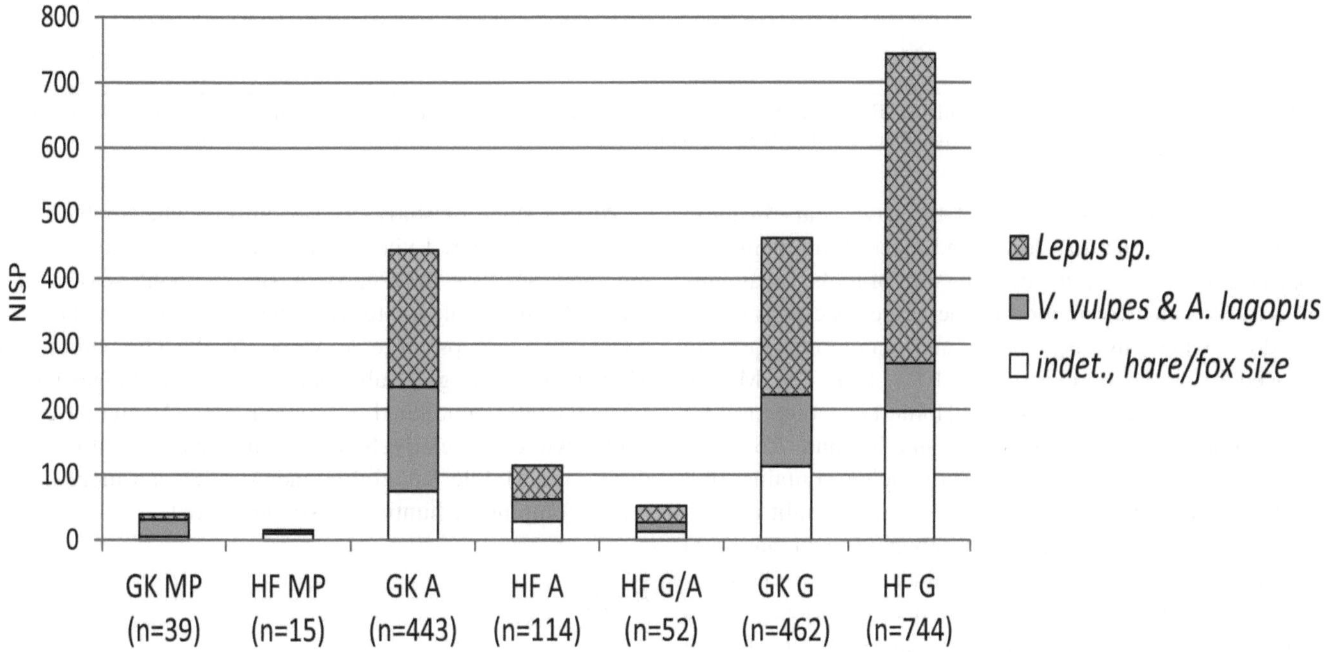

Fig. 11.7 Geißenklösterle and Hohle Fels. NISP of small mammals by taxa and body size. (*GK* Geißenklösterle, *HF* Hohle Fels, *MP* Middle Paleolithic, *A* Aurignacian, *G/A* Gravettian/Aurignacian, *G* Gravettian)

Hohle Fels because of the small size of the available Middle Paleolithic assemblage. At Geißenklösterle, where the sample size is larger, this trend is clear both in terms of NISP and bone weight.

Since the weight of edible animal resources correlates with bone weight, this parameter is the easiest means of approximating the relative contributions of species to the human diet (Uerpmann 1972, 1973). Many scholars quantify faunal remains using additional methods, such as MNI and MAU (Binford 1978), but since Brain's (1967, 1969) work demonstrated that these methods of quantification do not overcome taphonomic biases, which can depend upon biological factors such as the age at death and other post-depositional processes, we use bone weight in addition to NISP for quantifying the bones. Other attempts to quantify bone fragments cannot fully overcome issues of taphonomy (Münzel 1988; Lyman 1994; Marean et al. 2001). Furthermore, Grayson (1984) and Lyman (1994) have shown that quantifying MNI depends on a number of assumptions, particularly on the use of carcasses by past hunters and gatherers and depends upon how assemblages are aggregated based on stratigraphic units. The parameters of NISP and bone weight can be used to approximate biomass exploited by foragers in the past. In the case of Paleolithic sites, however, the importance of faunal raw material should not be underestimated. At Geißenklösterle, we recognize a close relationship between specific skeletal elements represented in the assemblage and their use as raw materials to make tools, demonstrating that the assemblage underwent a strong selection by the site's inhabitants (Münzel 2001; Barth 2007; Barth et al. 2009).

Looking at these variables, NISP and bone weight, we see that small-sized game comprise a greater proportion of the assemblages of the Upper Paleolithic fauna than the

Table 11.4 Geißenklösterle and Hohle Fels

Taxa	Geißenklösterle			Hohle Fels			
	Middle Paleolithic	Aurignacian	Gravettian	Middle Paleolithic	Aurignacian	Gravettian/ Aurignacian	Gravettian
Waterfowls	0	18	48	0	6	2	1
Birds of prey and owls	0	5	18	0	7	3	3
Galliforms	2	61	121	2	111	15	27
Songbirds	0	36	26	7	117	9	31
Others	0	16	7	0	5	0	4
Birds, ind.	0	12	28	1	74	12	39
Total	2	148	248	10	320	41	105
% of samples analyzed	Na	Na	Na	5.7	7.1	5.7	4.3

NISP of avian fauna by taxa. *Na* not available. For Hohle Fels, the sample is derived from 38 buckets from Middle Paleolithic, 143 from the Aurignacian, 78 from the Gravettian/Aurignacian, and 112 from the Gravettian

Middle Paleolithic. Since most of the cave bears in these deposits represent natural deaths and most of the mammoth ivory is associated with the manufacture of tools, ornaments and artworks, the distribution of the game species is easier to address when cave bear bones and mammoth ivory are removed from the figures (Figs. 11.2, 11.3). The Middle Paleolithic of Geißenklösterle and Hohle Fels contains 13.2 and 12.5 % small game by NISP and 0.7 and 1.6 % by weight respectively, while the Aurignacian contains 16.8 and 14.4 % by NISP and 3.1 and 4.1 % by weight respectively. The greatest change in the abundance of hare and foxes is in the Gravettian, where values go up to 39 % by NISP and 3.9–6.4 % by weight. Both Geißenklösterle and Hohle Fels show a remarkably similar signature for all periods that have sufficiently large samples (Fig. 11.7).

If we consider the pattern of faunal exploitation, we see a trend toward a greater use of hare during the Gravettian than the earlier periods. Overall, the data show that modern humans of the Upper Paleolithic exploited small mammals at higher levels than did Middle Paleolithic Neanderthals, and the increase between the Middle Paleolithic and the Aurignacian is more remarkable than the increase between the Aurignacian and the Gravettian. From this point of view, one could view this shift as a clear trend rather than a fundamental difference. Given the presence of cut marks on phalanges and metapodia, we assume that hare and foxes were often trapped for their pelts, but they were probably eaten with regularity as well. The multiple uses of small game are also demonstrated by the frequent presence of perforated fox canines as ornaments in both the Aurignacian and Gravettian of the Swabian Jura (Conard 2003). Tooth pendants are much more frequent in the Gravettian and ivory pendants are predominant in the Aurignacian.

At this stage of study, we are unsure whether hare and foxes were procured via trapping, netting, hunting or driving, and whether they carry a signal about seasonality. Similarly, it is difficult to say whether lone individuals or cooperative groups were at work. In the Gravettian and Magdalenian, the great abundance of hare is suggestive of systematically conducted group activities (Napierala 2008, 2009), while the relatively small number of faunal remains during the Middle Paleolithic and Aurignacian more likely reflect trapping or hunting of single animals.

Birds

Although most zooarchaeological work in the Swabian Jura has focused on mammalian taxa, Krönneck (2009, in press) and others (Götz 1949; Boessneck and von den Driesch 1973; Mourer-Chauviré 1983; Böttcher et al. 2000; Rathgeber 2004) have compiled and published data on the importance of avian fauna for Paleolithic subsistence. In the context of this study, Krönneck examined additional samples from Geißenklösterle and Hohle Fels to augment earlier data from Brillenhöhle and elsewhere. These studies indicate that game birds including galliforms, as well as other birds were hunted or trapped during both the Middle and Upper Paleolithic.

We draw our inferences on the role of birds in Paleolithic diet mainly from the study of avian faunal remains as well as from the abundance of gastroliths, because gizzard stones are abundant in fowl, especially galliforms. The fact that both of these lines of evidence point toward the same conclusions attests to the validity of the temporal trends we address here. The use of birds is also documented by feather

Table 11.5 Geißenklösterle and Hohle Fels

Taxa	Geißenklösterle			Hohle Fels			
	Middle Paleolithic	Aurignacian	Gravettian	Middle Paleolithic	Aurignacian	Gravettian/ Aurignacian	Gravettian
Grey partridge, *Perdix perdix*	0	0	2	0	16	2	4
Partridge, *Perdix*/ *Alectoris*	0	0	0	0	0	0	1
Partridge, *Perdicinae*	0	1	1	0	2	0	1
Quail, *Coturnix coturnix*	0	0	0	0	1	1	0
Hazel Grouse, *Tetrastes bonasia*	0	0	5	0	3	1	1
Ptarmigan, *Lagopus* sp.	2	49	87	0	19	2	5
Black grouse, *Lyrurus tetrix*	0	2	6	0	0	0	0
Capercaillie, *Tetrao urogallus*	0	1	2	0	0	0	0
Galliform, *Phasianidae*	0	8	18	2	70	9	15
Total	2	61	121	2	111	15	27

NISP of galliforms by taxa

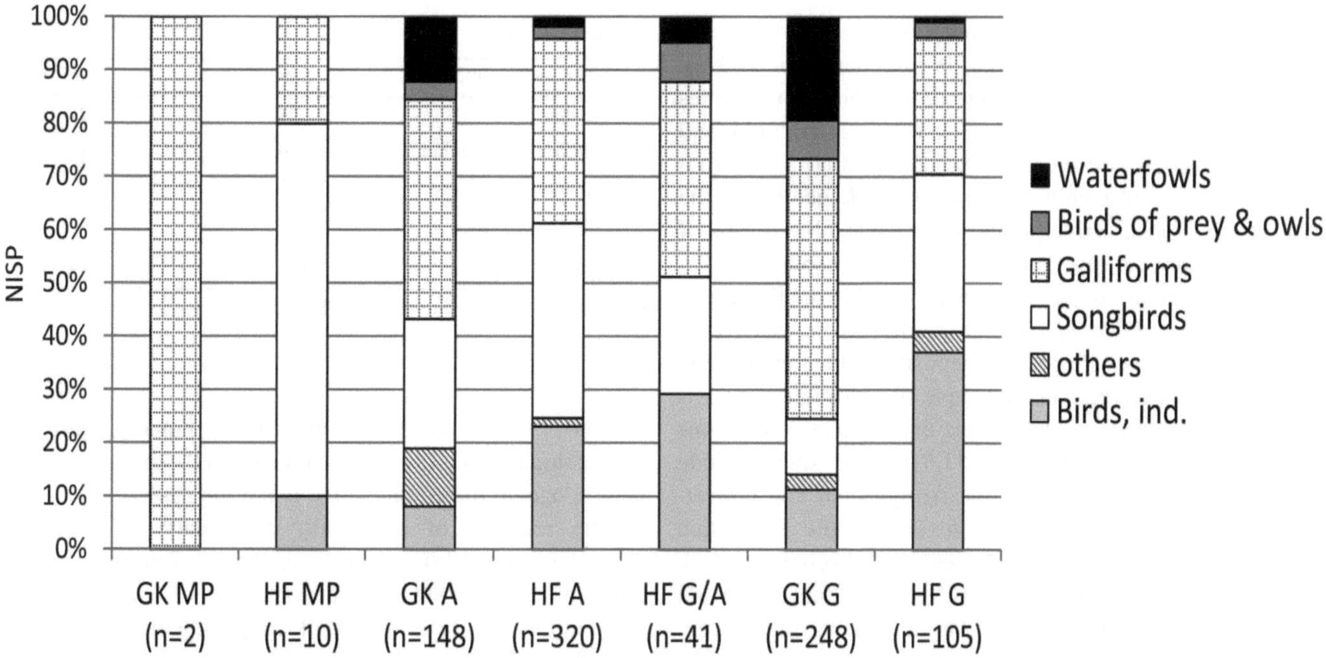

Fig. 11.8 Geißenklösterle and Hohle Fels. NISP% of avian fauna by taxa. (*GK* Geißenklösterle, *HF* Hohle Fels, *MP* Middle Paleolithic, *A* Aurignacian, *G/A* Gravettian/Aurignacian, *G* Gravettian)

fragments on stone tools in the Aurignacian context at Hohle Fels (Hardy et al. 2008).

Hohle Fels is a deep cave unsuited for owls or other roosting birds whose pellets in other settings could contribute to the avian fauna. At Geißenklösterle, the situation is less clear, but the surface preservation of the bones and occasional anthropogenic modifications also suggest that a portion of the avian fauna is the result of activities related to human subsistence and the procurement of resources including bones for flutes and feathers.

Remains of bone flutes have been recovered from Aurignacian contexts at Vogelherd in the Lone Valley and at

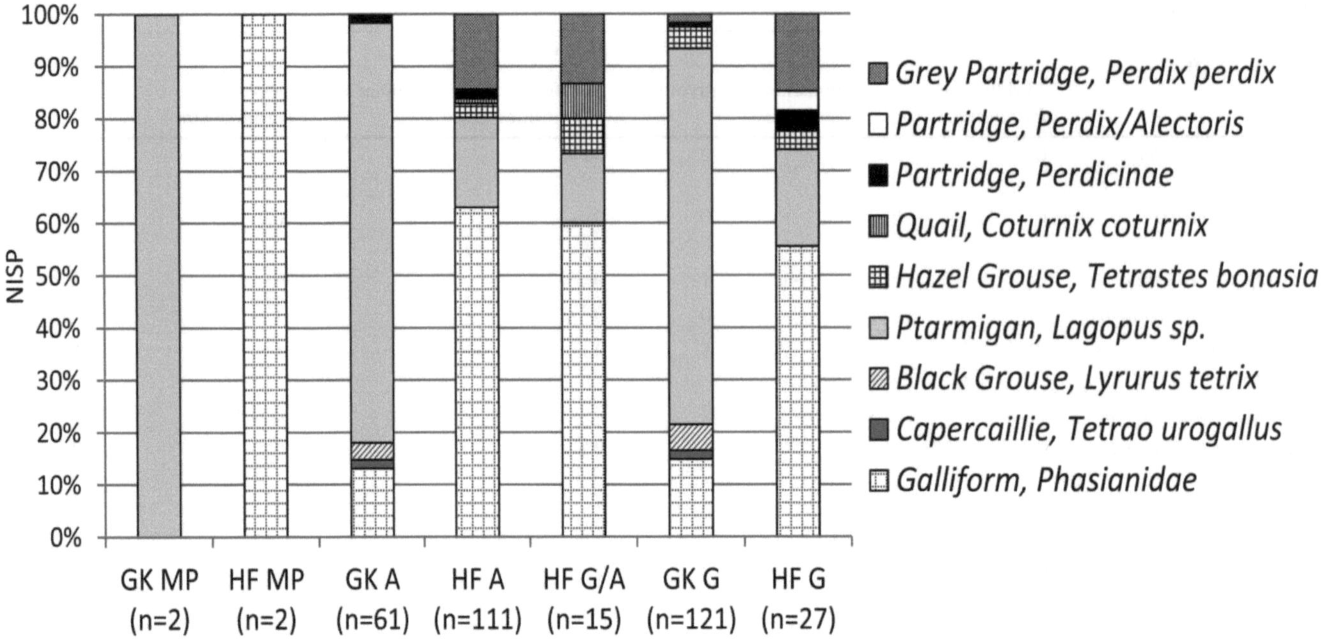

Fig. 11.9 Geißenklösterle and Hohle Fels. NISP% of galliforms by taxa. (*GK* Geißenklösterle, *HF* Hohle Fels, *MP* Middle Paleolithic, *A* Aurignacian, *G/A* Gravettian/Aurignacian, *G* Gravettian)

Geißenklösterle and Hohle Fels in the Ach Valley (Hahn and Münzel 1995; Conard and Malina 2006; Conard et al. 2009a). At Geißenklösterle and Vogelherd, the bird bone flutes appear to have been made from swan (*Cygnus* sp.) radii, and at Hohle Fels one well-preserved bone flute was carved from the radius of a griffon vulture (*Gyps fulvus*). Especially in the case of the swans, active predation is a plausible means of obtaining bones for raw material and edible resources.

The most important game bird was probably the ptarmigan (*Lagopus mutus/L. lagopus*). The ptarmigan is well represented in the Upper Paleolithic deposits of Geißenklösterle and Hohle Fels where a good sample of ptarmigan bones is available, allowing us to draw some quantitative conclusions (Tables 11.4, 11.5; Figs. 11.8, 11.9). Both sites include numerous ptarmigan bones in the Aurignacian and Gravettian. This pattern is in contrast to the Middle Paleolithic period where only few bird bones have been recovered, including two ptarmigan specimens from Geißenklösterle and two galliforms, which are slightly larger than ptarmigan, from Hohle Fels. At Hohle Fels, where gastroliths were systematically recovered, we can see that the abundance of ptarmigan bones and gastroliths co-vary with the highest values in the Aurignacian (Table 11.6; Fig. 11.10). Our data show that gastroliths are most numerous in the site's Aurignacian deposits, with the fewest present in the Middle Paleolithic deposits. The close correlation between the abundance of lithic and faunal artifacts with the presence of gastroliths also indicates that gizzard stones arrived at the sites as a result of human activities. If the gastroliths occurred in these deposits naturally, we would expect them to be equally present in archaeological find horizons and in strata lacking and poor in anthropogenic materials, which is not the case.

Table 11.6 Hohle Fels

Number of specimens	Gastroliths	Galliforms
Gravettian	21	27
Gravettian/Aurignacian	14	15
Aurignacian	65	111
Middle Paleolithic	2	2
Total	102	155

Frequencies of gastroliths and galliforms

The remains of songbirds, particularly the jackdaw (*Coloeus monedula*), are also suggestive of human activity. One could speculate that the presence of these bones in large numbers results from collecting the young birds from their nests. This practice is well documented ethnographically (Hölzinger 1987; Fisher 1997), and possibly has contributed to the accumulation of jackdaw and other species of small songbirds. For comparison, a 250 g jackdaw is much smaller and less meaty than a 400–500 g ptarmigan or a 2000–3000 g capercaillie (*Tetrao urogallus*), or large swans that weigh 7,000–12,000 g. Taking young birds out of their nests is more like harvesting or gathering than hunting, possibly making this practice productive despite their small size.

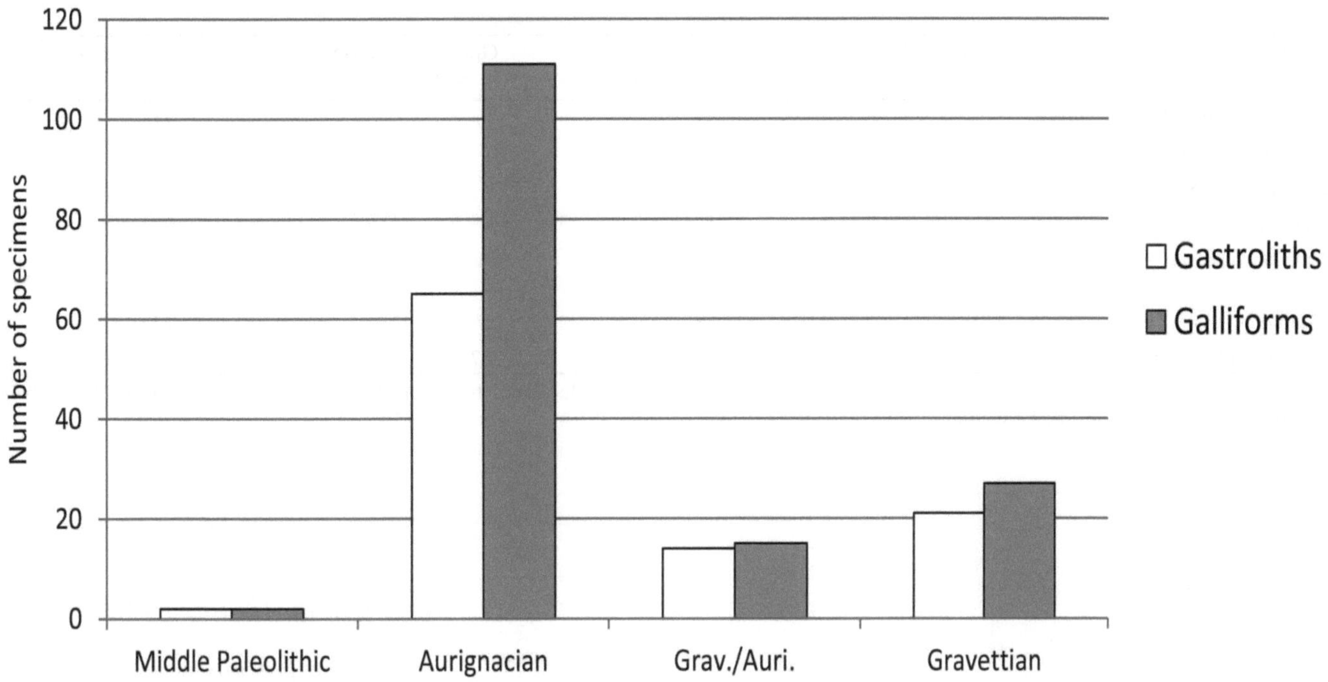

Fig. 11.10 Hohle Fels. Frequencies of gastroliths and galliforms

Bird remains are sometimes found in Middle Paleolithic sites, but are more frequent in the Upper Paleolithic (Mourer-Chauviré 1974; Laroulandie 2004; Krönneck 2009). The method used to obtain the birds is seldom clear and how the birds were used is uncertain. Birds have been documented from the Middle Paleolithic in Italy, in Fumane Cave (Peresani et al. 2011), and in France, Grotte de l'Hortus, Herault (Laroulandie 2004). The Upper Paleolithic provides more evidence of cutmarks, use of feathers, and the production of tools and flutes. Such traces are especially common in Magdalenian contexts (Laroulandie 2004; Krönneck 2009).

In summary, the avian fauna is suggestive of a shift toward more intensive use of ptarmigan and other species beginning with the Aurignacian. This pattern of exploitation is consistent with an expansion of diet breadth in connection with increasing population densities starting at the beginning of the Upper Paleolithic.

Fish

The assemblages of fish bone provide less clear-cut evidence for a shift to new resources in connection with the colonization of southwestern Germany by modern humans and the local extinction of Neanderthals. Work on the fish from Geißenklösterle by Torke (1981) and G. Böhme and from Hohle Fels by M. Böhme has led to the identification of five species of fish from the Danube and its tributaries such as the Ach from the Middle Paleolithic to the Gravettian (Table 11.7; Fig. 11.11). These include: carp (*Cyprinidae*), Danube salmon (*Hucho hucho*) ca. 10–15 kg, brown trout (*Salmo trutta*) ca. 500 g, grayling (*Thymallus thymallus*) ca. 500 g and burbot (*Lota lota*) ca. 500 g. Of these, all but the last have been identified by M. Böhme at Hohle Fels. The overall pattern from both sites shows similarities with common occurrences of grayling, burbot and brown trout, but a detailed comparison must await the publication of the fish remains from Geißenklösterle (Böhme in press).

During fieldwork at Hohle Fels, excavators occasionally identified fish bones and fish scales in the Aurignacian and Gravettian deposits. This, however, was not the case in the Middle Paleolithic deposits, where finds from all classes of anthropogenic material are less frequent. Studies of the archaeological assemblages from the caves of the Swabian Jura have demonstrated the occupation intensity during the Middle Paleolithic represents approximately one-tenth that of the Aurignacian (Conard 2011; Conard et al. 2012). To increase the likelihood of identifying the use of fish during the Middle Paleolithic, we collected the fish remains from the entire volume of the Middle Paleolithic and not just the sample columns. In this manner, we increased our sample by more than a factor of 10, making the sample roughly comparable to those from the Aurignacian and Gravettian. The abundance of the different taxa at Hohle Fels varies greatly with the common occurrence of grayling and burbot in the Aurignacian. Grayling dominates the Gravettian assemblage

Table 11.7 Hohle Fels

Taxa	Middle Paleolithic (n = 19)	Aurignacian (n = 78)	Gravettian /Aurignacian (n = 27)	Gravettian (n = 47)
Grayling, *Thymallus thymallus*	3	35	20	43
Burbot, *Lota lota*	13	39	6	2
Brown trout, *Salmo trutta*	0	3	1	2
Danube salmon, *Hucho hucho*	0	1	0	0
Carp, *Cyprinidae*	3	0	0	0
Total	19	78	27	47
% of samples analyzed	100	7.1	5.7	4.3

NISP of fish by taxa. The sample is derived from 672 buckets from Middle Paleolithic, 143 from the Aurignacian, 78 from the Gravettian/ Aurignacian, and 112 from the Gravettian

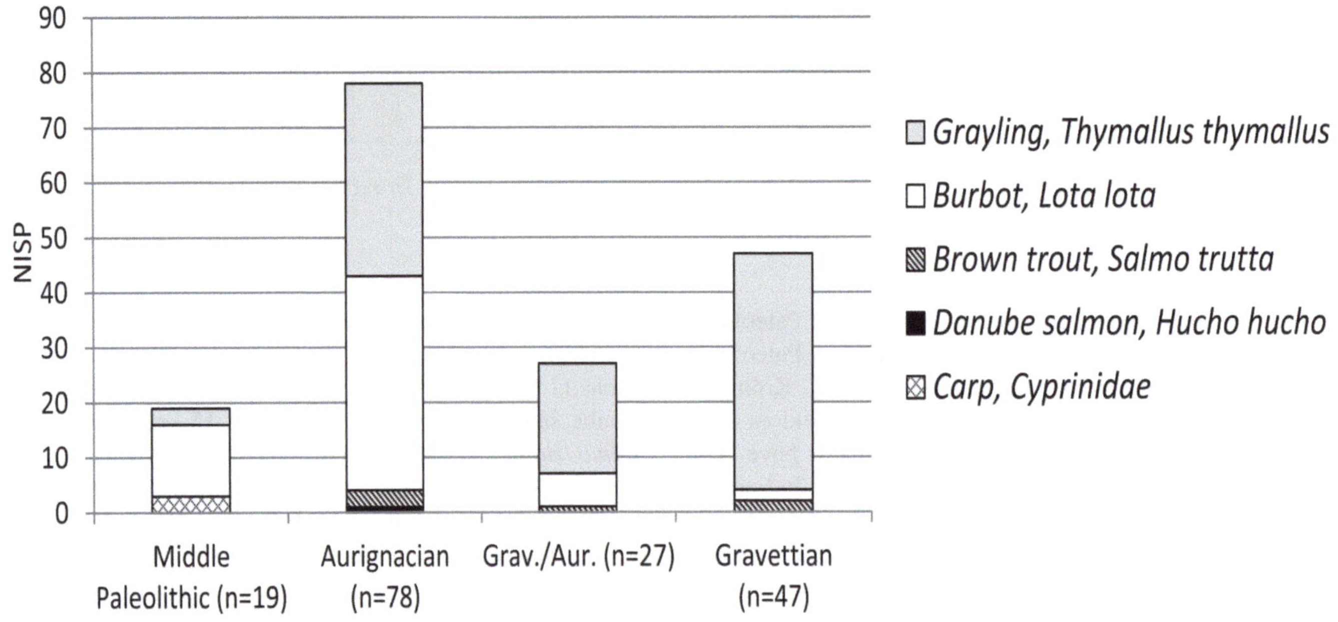

Fig. 11.11 Hohle Fels. NISP of fish by taxa

Fig. 11.12 Vogelherd. Aurignacian-aged depiction of a fish carved from mammoth ivory. Dimensions: 70.3 × 30.8 × 9.2 mm; Photo J. Lipták, copyright University of Tübingen

(Table 11.7; Fig. 11.11). The Middle Paleolithic contains only few fish remains with the majority being burbot and a few specimens of grayling and carp. The Danube salmon, which was fished in late spring (Torke 1981), is documented by only a single specimen in the Aurignacian. In addition to the taphonomic argument discussed above, the relatively large size of the species of fish present attests to the exploitation of freshwater resources by humans and not by other predators such as birds.

The overall abundance of fish rises sharply with the start of the Aurignacian, but the intensity of exploitation of fish appears to drop during the Gravettian. The Gravettian assemblage also contains a smaller number of species. In this context, one of the recently discovered ivory figurines from Vogelherd, which depicts a fish that resembles a trout, is noteworthy and provides an indication of the importance of fish for early modern humans (Conard et al. 2009b)

(Fig. 11.12). The rise in the abundance of several species of fish and the increased number of species in the diet of modern humans at the start of the Upper Paleolithic represents a change in comparison with the Middle Paleolithic, but, as with the use of small game, this change seems to reflect a more intense exploitation of fish, but not a fundamental behavioral shift.

Conclusions

The results presented here on small mammals, birds and fish complement our understanding of large mammalian game species in the past. As we have argued earlier (Münzel and Conard 2004a; Conard 2011), the use of larger mammalian species reflects continuity between the subsistence practices of Neanderthals and modern humans. In all Paleolithic periods, large- and medium-sized mammalian game, particularly horse and reindeer provided the bulk of the calories. We can now document that the start of the Upper Paleolithic during the Swabian Aurignacian saw an increased exploitation of small game. These results augment a growing body of data that documents the variability of faunal exploitation during both the Middle and the Upper Paleolithic and the diversity of the dietary shifts between the Middle and Upper Paleolithic in different regions of western Eurasia (Stiner et al. 1999; Stringer et al. 2008; Peresani et al. 2011; Blasco and Fernández Peris 2012).

We see this dietary shift in Swabia as reflecting new adaptations that allowed modern humans to maintain higher population densities. This being said, the change in diet and the use of resources is gradual when we consider the small mammalian data and the samples of fish from Hohle Fels. We see a clear intensification from both sites, but not a radical shift in prey selection. The break is seen in the leap in the exploitation at the start of the Aurignacian, but in the case of small mammals, a second and perhaps more significant shift occurs during the Gravettian. The changes reflect a mixture of continuous use of the main game species, a moderate shift in the use of small mammals and fish, and a sharp increase in the exploitation of birds during the Aurignacian.

Despite our expectations to the contrary, we have no evidence that modern humans and Neanderthals met in the Swabian Jura, and most of the sites in the region document an occupational hiatus between the late Middle Paleolithic and the Aurignacian (Schmidt 1912; Riek 1934; Hahn 1988; Conard and Malina 2003; Conard et al. 2006; Miller 2009). Still, modern humans appear to have quickly developed new practices of subsistence in the Swabian Jura that required the investment of additional effort to outcompete the indigenous Neanderthals by extracting more calories from their environment. This increased investment in food procurement strategies allowed modern humans to feed larger groups of people and to maintain a higher population density relative to the culturally more conservative Neanderthals (Conard 2011). At this time, we also see numerous innovative lithic and organic tools and weapons that have no counterparts in the material culture of the Middle Paleolithic. Furthermore, a vast array of innovations in the symbolic realm, including figurative depictions, mythical imagery, personal ornaments with three dimensional forms and musical instruments also emerge (Conard 2008, 2009; Conard et al. 2009a). While these important innovations cannot be explained solely in terms of dietary and nutritional variables, the evidence for expanding diet breadth and intensified extraction of animal resources from the environment likely contributed to the demographic success of modern humans and the ultimate extinction of Neanderthals.

Acknowledgments We thank Jamie Clark and John Speth for their invitation to participate in their symposium at the ICAZ congress in 2010 in Paris. We thank all of the organizations that have funded this work including the Deutsche Forschungsgemeinschaft, the Heidelberger Akademie der Wissenschaften, the Universität Tübingen, Heidelberger Cement, the Voith Corporation, Alb-Donau-Kreis, Kreis Heidenheim, the Gesellschaft für Urgeschichte, the Museumsgesellschaft Schelklingen, the Landesamt für Denkmalpflege Baden-Württemberg, the towns of Blaubeuren and Niederstotzingen, and the Förderverein Eiszeitkunst im Lonetal. We are grateful to Hervé Bocherens for his helpful advice on the diet and behavior of cave bears. We thank Maria Malina for her important support during excavations and labwork, and we also thank the hundreds of excavators, water-screeners and sorters who have contributed to this research.

References

Barth, M. M. (2007). *Familienbande? Die gravettienzeitlichen Knochen- und Geweihgeräte des Achtals (Schwäbische Alb). [Family ties? The Gravettien bone and antler tools from the Ach valley (Swabian Alb)]*. Tübinger Arbeiten zur Urgeschichte 4. Rahden: Marie Leidorf.

Barth, M. M., Conard, N. J., & Münzel, S. C. (2009). Palaeolithic Subsistence and Organic Technology in the Swabian Jura. In L. Fontana, F.-X. Chauvière & A. Bridault (Eds.), *In search of Total Animal Exploitation. Cases Studies in Upper Palaeolithic and Mesolithic*. Proceedings of the XVth UISPP World Congress, Lisbon, 4-9 September 2006, Session C 61, vol. 42, Oxford, J. & E. Hedges (BAR International Series 2040), 5–20.

Binford, L. R. (1978). *Nunamiut ethnoarchaeology*. New York: Academic Press.

Blasco, R., & Fernández Peris, J. (2012). A uniquely broad spectrum diet during the Middle Pleistocene at Bolomor Cave (Valencia, Spain). *Quaternary International, 252*, 16–31.

Bocherens, H. (2011). Diet and ecology of Neanderthals: Implications from C and N isotopes. In N. J. Conard & J. Richter (Eds.), *Neanderthal lifeways, subsistence and technology* (pp. 73–86). Heidelberg: Springer.

Bocherens, H., Fizet, M., & Mariotti, A. (1994). Diet, physiology and ecology of fossil mammals as inferred by stable carbon and nitrogen isotopes biogeochemistry: Implications for Pleistocene bears. *Palaeogeography, Palaeoclimatology, Palaeoecology, 107*, 213–225.

Bocherens, H., Drucker, D. G., Billiou, D., Geneste, J.-M., & van der Plicht, J. (2006). Bears and Humans in Chauvet Cave (Vallon-Pont-d'Arc, Ardèche, France): Insights from stable isotopes and radiocarbon dating of bone collagen. *Journal of Human Evolution, 50*, 370–376.

Boessneck, J. & von den Driesch, A. (1973). Die jungpleistozänen Tierknochenfunde aus der Brillenhöhle. In G. Riek (Ed.), *Das Paläolithikum der Brillenhöhle bei Blaubeuren (Schwäbische Alb). Teil II.* Forschungen und Berichte zur Vor- und Frühgeschichte in Baden-Württemberg 4/II. Stuttgart: Theiss Verlag.

Böhme, G. (in press). Die Amphibien, Reptilien und Fische In N. J. Conard, M. Bolus, & S. Münzel (Eds.), *Geißenklösterle II. Fauna, Flora und Umweltverhältnisse im Mittel- und Jungpaläolithikum.* Tübingen: Kerns Verlag.

Böttcher, R., Çep, B., Kind, C.-J., Mörike, D., Pawlik, A., Rähle, W., et al. (2000). Kogelstein - eine mittelpaläolithische Fundstelle bei Schelklingen-Schmiechen. *Fundberichte aus Baden-Württemberg, 24*, 7–176.

Brain, C. K. (1967). Hottentot food remains and their bearing on the interpretation of fossil bone assemblages. *Scientific papers of the Namib Desert Research Station, 32*, 1–11.

Brain, C. K. (1969). The contribution of the Namib Desert Hottentots to an understanding of Australopithecine bone accumulations. *Scientific papers of the Namib Desert Research Station, 39*, 13–22.

Cavarretta, G., Giola, P., Mussi, M., & Palombo, M. R. (Eds.). (2001). *The World of elephants: Proceedings of the 1st International Congress.* Rome: Consiglio Nazionale della Ricerche.

Conard, N. J. (2003). Eiszeitlicher Schmuck auf der Schwäbischen Alb. In S. Kölbl & N. J. Conard (Eds.), *Eiszeitschmuck- Status und Schönheit* (pp. 15–49). Urgeschichtliches Museum Blaubeuren, Museumsheft 6. Blaubeuren: Urgeschichtliches Museum.

Conard, N. J. (2008). Die Neugrabungen am Vogelherd. In N. J. Conard & E. Seidl (Eds.), *Das Mammut vom Vogelherd: Tübinger Funde der ältesten erhaltenen Kunstwerke* (pp. 34–43). Tübingen: Museum der Universität Tübingen.

Conard, N. J. (2009). A female figurine from the basal Aurignacian of Hohle Fels Cave in southwestern Germany. *Nature, 459*, 248–252.

Conard, N. J. (2011). The demise of the Neanderthal cultural niche and the beginning of the Upper Paleolithic in Southwestern Germany. In N. J. Conard & J. Richter (Eds.), *Neanderthal lifeways, subsistence and technology* (pp. 223–240). Heidelberg: Springer.

Conard, N. J., & Bolus, M. (2003). Radiocarbon dating the appearance of modern humans and timing of cultural innovations in Europe: New results and new challenges. *Journal of Human Evolution, 44*, 331–371.

Conard, N. J. & Bolus, M. (2006). The Swabian Aurignacian and its place in European prehistory. In O. Bar-Yosef & J. Zilhao (Eds.), *Toward a definition of the Aurignacian* (pp. 209–237). Lisboa: Trabalhos de Arqueologia 45.

Conard, N. J., & Bolus, M. (2008). Radiocarbon dating the late Middle Paleolithic and the Aurignacian of the Swabian Jura. *Journal of Human Evolution, 55*, 886–897.

Conard, N. J. & Malina, M. (2003). Abschließende Ausgrabungen im Geißenklösterle bei Blaubeuren, Alb-Donau-Kreis. In Landesdenkmalamt Baden- Württemberg (Ed.), *Archäologische Ausgrabungen in Baden-Württemberg 2002* (pp. 17–21). Stuttgart: Theiss Verlag.

Conard, N. J. & Malina, M. (2006). Schmuck und vielleicht auch Musik am Vogelherd bei Niederstotzingen-Stetten ob Lontal, Kreis Heidenheim. In Landesdenkmalamt Baden- Württemberg (Ed.), *Archäologische Ausgrabungen in Baden-Württemberg 2005* (pp. 21–25). Stuttgart: Theiss Verlag.

Conard, N. J., & Moreau, L. (2004). Current research on the Gravettian of the Swabian Jura. *Mittelungen der Gesellschaft für Urgeschichte, 13*, 29–59.

Conard, N. J., Bolus, M., Goldberg, P., & Münzel, S. C. (2006). The last Neanderthals and first modern humans in the Swabian Jura. In N. J. Conard (Ed.), *When Neanderthals and modern humans met* (pp. 305–341). Tübingen: Kerns Verlag.

Conard, N. J., Malina, M., & Münzel, S. C. (2009a). New flutes document the earliest musical tradition in southwestern Germany. *Nature, 460*, 737–740.

Conard, N. J., Malina, M., & Verrept, T. (2009b). Weitere Belege für eiszeitliche Kunst und Musik aus den Nachgrabungen 2008 am Vogelherd bei Niederstotzingen-Stetten ob Lontal, Kreis Heidenheim. In Landesdenkmalamt Baden- Württemberg (Ed.), *Archäologische Ausgrabungen Baden-Württemberg 2008* (pp. 23–26). Stuttgart: Theiss Verlag.

Conard, N. J., Bolus, M., & Münzel, S. C. (2012). Middle Paleolithic land use, spatial organization and settlement intensity in the Swabian Jura, southwestern Germany. *Quaternary International, 247*, 236–245.

Fisher, C. T. (1997). Past human exploitation of birds on the Isle of Man. *International Journal of Osteoarchaeology, 7*, 292–297.

Gamble, C. S. (1979). Hunting strategies in the central European Palaeolithic. *Proceedings of the Prehistoric Society, 45*, 35–52.

Götz, W. H. J. (1949). Eine altsteinzeitliche Vogelfauna aus der Schafstallhöhle bei Veringenstadt (Hohenzollern). In E. Lindner, M. Bräuhäuser, & H. Rauther (Eds.), *Jahreshefte des Vereins für vaterländische Naturkunde in Württemberg*, 97–101, Jahrgang 1941–1945 (pp. 101–114). Stuttgart: Schweizerbartsche Verlagsbuchhandlung.

Grayson, D. K. (1984). *Quantitative zooarchaeology: Topics in the analysis of archaeological faunas.* Orlando: Academic Press.

Hahn, J. (1988). *Die Geißenklösterle-Höhle im Achtal bei Blaubeuren I. Fundhorizontbildung und Besiedlung im Mittelpaläolithikum und im Aurignacien.* Stuttgart: Theiss Verlag.

Hahn, J. (1995). Neue Beschleuniger-^{14}C-Daten zum Jungpaläolithikum in Südwestdeutschland. *Eiszeitalter und Gegenwart, 45*, 86–92.

Hahn, J., & Münzel, S. C. (1995). Knochenflöten aus dem Aurignacien des Geißenklösterle bei Blaubeuren, Alb-Donau-Kreis. *Fundberichte aus Baden-Württemberg, 20*, 1–12.

Hardy, B. L., Bolus, M., & Conard, N. J. (2008). Hammer or crescent wrench? Stone-tool form and function in the Aurignacian of southwest Germany. *Journal of Human Evolution, 54*, 648–662.

Hockett, B. S., & Haws, J. A. (2003). Nutritional ecology and diachronic trends in Paleolithic diet and health. *Evolutionary Anthropology, 12*, 211–215.

Hockett, B. S., & Haws, J. A. (2005). Nutritional ecology and the human demography of Neandertal extinction. *Quaternary International, 137*, 21–34.

Hofreiter, M., Capelli, C., Krings, M., Waits, L., Conard, N. J., Münzel, S., et al. (2002). Ancient DNA analyses reveal high mitochondrial DNA sequence diversity and parallel morphological evolution of late Pleistocene Cave bears. *Molecular Biology and Evolution, 19*, 1244–1250.

Hofreiter, M., Münzel, S., Conard, N. J., Pollack, J., Slatkin, M., Weiss, G., et al. (2007). Sudden replacement of cave bear mitochondrial DNA in the late Pleistocene. *Current Biology, 17*, 122–123.

Hölzinger, J. (Ed.) (1987). *Die Vögel Baden-Württembergs* (Vol. 1.2). Stuttgart: Verlag Eugen Ulmer.

Klein, R. G., & Cruz-Uribe, K. (2000). Middle and Later Stone Age large mammal and tortoise remains from Die Kelders Cave 1, Western Cape Province, South Africa. *Journal of Human Evolution, 38*, 169–195.

Krönneck, P. (2009). Vögel—Bonusmaterial in der Archäozoologie. Jungpleistozäne Inventare als Auswertungsbeispiele. In R. de Beauclair, S. Münzel, & H. Napierala (Eds.), *Knochen pflastern ihren Weg: Festschrift für Margarethe und Hans-Peter Uerpmann* (pp. 133–141). Rahden: Marie Leidorf.

Krönneck, P. (2012). *Die pleistozäne Makrofauna des Bocksteins (Lonetal—Schwäbische Alb). Ein neuer Ansatz zur Rekonstruktion der Paläoumwelt*. Ph. D. Dissertation, University of Tübingen.

Krönneck, P. (in press). Die Vogelknochen vom Geißenklösterle. In N. J. Conard, M. Bolus, & S. Münzel (Eds.), *Geißenklösterle II. Fauna, Flora und Umweltverhältnisse im Mittel- und Jungpaläolithikum*. Tübingen: Kerns Verlag.

Krönneck, P., Niven, L., & Uerpmann, H.-P. (2004). Middle Paleolithic subsistence in the Lone Valley (Swabian Alb, Southern Germany). *International Journal of Osteoarchaeology, 14*, 212–224.

Kurtén, B. (1976). *The cave bear story: Life and death of a vanished animal*. New York: Columbia University Press.

Lang, G. (1994). *Quartäre Vegetationsgeschichte Europas*. Stuttgart: Fischer Verlag.

Laroulandie, V. (2004). Exploitation des ressources aviaires durant le Paléolithique en France: bilan critique et perspectives. In J.-P. Brugal & J. Desse (Eds.), *Petits animaux et sociétés Humaines du complément alimentaire aux ressources utilitaires XXXIVe rencontres internationales d'archéologie et d'histoires d'Antibes* (pp. 163–172). Antibes: Editions APDCA.

Lehmann, U. (1954). Vogelherd und Bocksteinschmiede im Lonetal. *Eiszeitalter und Gegenwart, 4*(5), 142–146.

Lyman, R. L. (1994). *Vertebrate taphonomy*. Cambridge: Cambridge University Press.

Marean, C. W., Abe, Y., Nilssen, P. J., & Stone, E. C. (2001). Estimating the minimum number of skeletal elements (MNE) in zooarchaeology: A review and a new image-analysis GIS approach. *American Antiquity, 66*, 333–348.

Miller, C. (2009). *Formation processes, paleoenvironments and settlement dynamics of the Paleolithic cave sites of Hohle Fels and Geißenklösterle: A geoarchaeological and micromorphological perspective*. Ph. D. Dissertation, University of Tübingen.

Moreau, L. (2009). *Geißenklösterle. Das Gravettien der Schwäbischen Alb im europäischen Kontext*. Tübingen: Kerns Verlag.

Mourer-Chauviré, C. (1974). *Les oiseaux du Pléistocène moyen et superieur de France*. Lyon: Documents des Laboratoires de Gèologie de la Facultè des Science de Lyon 64.

Mourer-Chauviré, C. (1983). Die Vogelknochen aus dem magdalénienzeitlichen Fundplatz Petersfels, Grabungen 1974–1976. In G. Albrecht, H. Berke, & F. Poplin (Eds.), *Naturwissenschaftliche Untersuchungen an Magdalénien-Inventaren vom Petersfels, Grabungen 1974–1976* (pp. 132–229). Tübingen: Archaeologica Venatoria.

Müller, U. (2001). *Die Vegetations- und Klimaentwicklung im jüngeren Quartär anhand ausgewählter Profile aus dem südwestdeutschen Alpenvorland*. Tübingen: Tübinger Geowissenschaftliche Arbeiten, Series D Vol 7.

Müller-Beck, H. (1983). *Urgeschichte in Baden-Württemberg*. Stuttgart: Theiss Verlag.

Münzel, S. C. (1988). Quantitative analysis and archaeological site interpretation. *Archaeozoologia, 2*, 93–109.

Münzel, S. (2001). The production of Upper Palaeolithic mammoth bone artifacts from southwestern Germany. In G. Cavarretta, P. Giola, M. Mussi, & M. R. Palombo (Eds.), *The World of elephants: Proceedings of the 1st international congress* (pp. 448–454). Rome: Consiglio Nazionale della Ricerche.

Münzel, S. (2005). Mammoth remains in the Upper and Middle Paleolithic layers of Geißenklösterle Cave (Ach Valley, Swabian Jura, Southwestern Germany). Hunting season, acquisition of raw material and tool production at Geißenklösterle Cave. In D. Vialou, J. Renault-Miskovsky, & M. Patou-Mathis (Eds.), *Comportements des hommes du Paléolithique moyen et supérrieur en Europe: territoires et milieux* (pp. 39–49). Liège: Eraul 111.

Münzel, S. (2009). Taphonomische Anmerkungen zum Subsistenzverhalten von Neandertalern und modernen Menschen am Beispiel der Geißenklösterle-Höhle bei Blaubeuren, Schwäbische Alb. In R. de Beauclair, S. Münzel, & H. Napierala (Eds.), *Knochen pflastern ihren Weg: Festschrift für Margarethe und Hans-Peter Uerpmann* (pp. 169–179). Rahden: Marie Leidorf.

Münzel, S. (in press). Die jungpleistozäne Großsäugerfauna aus dem Geißenklösterle. In N.J. Conard, M. Bolus, & S. Münzel (Eds.), *Geißenklösterle II. Fauna, Flora und Umweltverhältnisse im Mittel- und Jungpaläolithikum*. Tübingen: Kerns Verlag.

Münzel, S., & Conard, N. J. (2004a). Change and continuity in subsistence during the Middle and Upper Palaeolithic in the Ach Valley of Swabia (south-west Germany). *International Journal of Osteoarchaeology, 14*, 225–243.

Münzel, S., & Conard, N. J. (2004b). Cave bear hunting in Hohle Fels Cave in the Ach Valley of the Swabian Alb. *Revue de Paléobiologie, 23*, 877–885.

Münzel, S., Morel, P., & Hahn, J. (1994). Jungpleistozäne Tierreste aus der Geißenklösterle-Höhle bei Blaubeuren. *Fundberichte aus Baden-Württemberg, 19*, 63–93.

Münzel, S., Stiller, M., Hofreiter, M., Mittnik, A., Conard, N. J., & Bocherens, H. (2011). Pleistocene bears in the Swabian Jura (Germany): Genetic replacement, ecological displacement, extinctions and survival. *Quaternary International, 245*, 225–237.

Napierala, H. (2008). *Die Tierknochen aus dem Kesslerloch. Neubearbeitung der paläolithischen Fauna*. Beiträge zur Schaffhauser Archäologie 2.

Napierala, H. (2009). Reindeer and seasonality: An overview of methods and their applicability to archaeological materials, as exemplified by the Kesslerloch Cave faunal assemblage. In R. de Beauclair, S. Münzel, & H. Napierala (Eds.), *Knochen pflastern ihren Weg: Festschrift für Margarethe und Hans-Peter Uerpmann* (pp. 181–192). Rahden: Marie Leidorf.

Niven, L. (2006). *The Paleolithic occupation of Vogelherd Cave: Implications for the subsistence behavior of late Neanderthals and early Modern Humans*. Tübingen: Kerns Verlag.

Peresani, M., Romandini, M., Fiore, I., Gala, M., & Tagliacozzo, A. (2011). Late Neandertals and the intentional removal of feathers as evidenced from bird bone taphonomy at Fumane Cave 44 ky B.P., Italy. *Proceedings of the National Academy of Sciences of the United States of America, 108*, 3888–3893.

Rathgeber, T. (2004). Die quartäre Tierwelt der Höhlen um Veringenstadt (Schwäbische Alb). *Laichinger Höhlenfreund, 39*, 207–228.

Richter, D., Waiblinger, J., Rink, W. J., & Wagner, G. A. (2000). Thermoluminescence, electron spin resonance and ^{14}C-dating of the Late Middle and Early Upper Palaeolithic site of Geißenklösterle Cave in Southern Germany. *Journal of Archaeological Science, 27*, 71–89.

Riek, G. (1934). *Die Eiszeitjägerstation am Vogelherd im Lonetal: Die Kulturen*. Tübingen: Akademische Buchhandlung Franz F. Heine.

Riek, G. (1938). Festellung des Hochsolutréen in Württemberg. *Forschungen und Fortschritte, 14*, 147–148.

Scheer, A. (Ed.). (1994). *Höhlenarchäologie im Urdonautal bei Blaubeuren. Museumshefte 1*. Blaubeuren: Urgeschichtliches Museum Blaubeuren.

Scheer, A. (2000). The Gravettian in southwest Germany: Stylistic features, raw material resources and settlement patterns. In W. Roebroeks, M. Mussi, J. Svoboda, & K. Fennema (Eds.), *Hunters of the golden age* (pp. 257–270). Leiden: University of Leiden.

Schmidt, R. R. (1912). *Die diluviale Vorzeit Deutschlands* (3 Vols.). Stuttgart: Schweizerbartsche Verlagsbuchhandlung.

Steele, T. E. & Klein R. E. (2005/2006). Mollusk and tortoise size as proxies for Stone Age population density in South Africa: Implications for the evolution of human cultural capacity (Homenaje a Jesús Altuna). *Munibe, 57*, 221–237.

Stiner, M. C., & Munro, N. D. (2002). Approaches to prehistoric diet breadth, demography, and prey ranking systems in time and space. *Journal of Archaeological Method and Theory, 9*, 175–208.

Stiner, M. C., Munro, N. D., Surovell, T. A., Tchernov, E., & Bar-Yosef, O. (1999). Paleolithic population growth pulses evidenced by small animal exploitation. *Science, 283*, 190–194.

Stringer, C. B., Finlayson, J. C., Barton, R. N. E., Fernándo-Jalvo, Y., Cáceres, I., Sabin, R. C., et al. (2008). Neanderthal exploitation of marine mammals in Gibraltar. *Proceedings of the National Academy of Sciences of the United States of America, 105*, 14319–14324.

Terberger, T. (2001). *Vom Gravettien zum Magdalénein in Mitteleuropa. Aspekte der menschlichen Besiedlungsgeschichte in der Zeit um das zweite Kältemaximum der letzten Eiszeit.* Habilitationsschrift, anders trennen: University of Greifswald.

Torke, W. (1981). *Fischreste als Quellen der Ökologie und Ökonomie in der Steinzeit Südwest-Deutschlands*. Urgeschichtliche Materialhefte 4. Tübingen: Archaeologica Venatoria.

Uerpmann, H.-P. (1972). Tierknochenfunde und Wirtschaftsarchäologie. Eine kritische Studie der Methoden der Osteo-Archäologie. *Archäologische Informationen, 1*, 9–27.

Uerpmann, H.-P. (1973). Ein Beitrag zur Methodik der wirtschaftshistorischen Auswertung von Tierfunden aus Siedlungen. In J. Matolcsi (Ed.), *Domestikationsforschung und Geschichte der Haustiere* (pp. 391–395). Budapest: Akademiai Kiado.

Weinstock, J. (1999). *The Upper Pleistocene mammalian fauna from the Große Grotte near Blaubeuren (southwestern Germany).* Stuttgarter Beiträge zur Naturkunde, Serie B (Geologie und Paläontologie) 277. Stuttgart: Staatliches Museum für Naturkunde.

Weniger, G. C. (1987). Magdalenian settlement and subsistence in southwest Germany. *Proceedings of the Prehistoric Society, 59*, 293–307.

Chapter 12
Neanderthals and Cave Hyenas: Co-existence, Competition or Conflict?

Gerrit L. Dusseldorp

Introduction

It is hypothesized that the Early Pleistocene hominin colonization of Europe was hampered by the presence of a large carnivore guild. During the Middle Pleistocene a turnover took place, with a replacement of a number of large carnivores by smaller animals. This turnover is often dubbed the *Crocuta crocuta* event, due to the replacement of the giant hyena (*Pachycrocuta brevirostris*) by spotted or cave hyena (*Crocuta crocuta spelaea*) (Turner 1992; Madurell-Malapeira et al. 2010; Martínez-Navarro 2010; Palombo 2010). The latter species is abundant in the fossil record of Europe from the Middle Pleistocene onwards. After the *Crocuta* event, a substantial increase in hominin presence in Europe is also recorded (see review in Roebroeks 2006). During the Middle Pleistocene, European hominins developed into the Neanderthals (*Homo neanderthalensis*), which functioned as top-predators (e.g. Bocherens 2009; Richards and Trinkaus 2009). Competition with carnivores likely influenced the Neanderthal foraging niche (e.g. Brantingham 1998; García and Virgós 2007). The exact influence that the co-existence of Neanderthals with large carnivores such as cave hyena had on their foraging strategies is not yet clear.

Neanderthals and cave hyenas were predators of roughly similar size. Both preyed on large ungulates. Since the size of a predator is an important determining factor in the size of its prey (Radloff and Du Toit 2004), the potential dietary overlap between the two is large. Taphonomic studies show that many Pleistocene bone assemblages contain evidence for the activities of both agents (e.g. Armand 1998; Villa et al. 2004; Miracle 2005). Moreover, Neanderthal fossils are sometimes present in hyena dens (e.g. Tournepiche 1994; Beauval et al. 2005; Enloe 2012). This suggests that the species co-existed in many areas. However, prey deposits uniquely accumulated by cave hyenas are also known (e.g. Brugal et al. 1997; Fosse 1997; Bartram and Villa 1998; Fosse et al. 1998; Diedrich and Žák 2006).

Ecologically, if two species occupy a similar niche, competitive exclusion is expected to result in the local extinction of one of the two species. However, subtle differences between niches can allow the co-existence of similar organisms. In addition, the co-existence of similar animals can result in character displacement enabling their continued sympatry (e.g. Hardin 1960; Davies et al. 2007). In order to illuminate the degree of niche overlap between Neanderthals and cave hyenas, a comparison of the representation of prey animals in both Neanderthal and cave hyena accumulated bone assemblages is required.

Here I compare the species composition of a large sample of Neanderthal bone accumulations with those of a number of hyena dens. Both the Neanderthal and hyena assemblages date to Marine Isotope Stage (MIS) 4 and 3, while geographically, the sample is limited to modern-day France. In terms of species representation, a large degree of overlap between Neanderthal and cave hyena prey deposits is apparent, with bovids, equids and cervids being the most important animal groups in both Neanderthal and cave hyena assemblages. However, Neanderthals appear to be preferentially associated with cervids, while hyena deposits contain more bovids, equids and megafauna.

I discuss possible taphonomic and behavioral causes of the differences in representation of animal groups between both predators' prey deposits. I suggest that taphonomic causes alone cannot explain the differences and some degree of niche separation between cave hyenas and Neanderthals was in place.

G. L. Dusseldorp (✉)
Institute for Human Evolution, University of the Witwatersrand, PO 2050 WITS, Johannesburg, South Africa
e-mail: g.dusseldorp@gmail.com

and

Faculty of Archaeology, Leiden University, PO Box 9515, 2300 RA, Leiden, The Netherlands

Jamie L. Clark and John D. Speth (eds.), *Zooarchaeology and Modern Human Origins: Human Hunting Behavior during the Later Pleistocene*, Vertebrate Paleobiology and Paleoanthropology, DOI: 10.1007/978-94-007-6766-9_12,

Neanderthal and Cave Hyena Foraging Strategies

Neanderthals

During the 1980s, Binford hypothesized that Neanderthals were mainly marginal scavengers and not, as had often been assumed, effective hunters of large mammals (Binford 1985, 1988). This led to a proliferation of meticulous studies of Middle Palaeolithic bone assemblages. Hence, the foraging methods practiced by Neanderthals can be well characterized due to a wealth of taphonomic and zooarchaeological studies that have been performed on Neanderthal bone assemblages.

The studies stemming from the hunting versus scavenging debate have shown that at many sites, Neanderthals effectively hunted large animals, often preying on prime-aged ungulates, such as reindeer (*Rangifer tarandus*), horse (*Equus caballus*), aurochs (*Bos primigenius*) and bison (*Bison priscus*) (e.g. Jaubert et al. 1990; Farizy et al. 1994; Auguste 1995; Gaudzinski 1995, 1996; Gaudzinski and Roebroeks 2000; Schild et al. 2000; Adler et al. 2006; Costamagno et al. 2006). Their subsistence repertoire was not static; Neanderthals were able to effectively exploit herbivore communities in a host of different environmental settings. In the Near East for example, they successfully hunted gazelle (*Gazella gazella*), fallow deer (*Dama mesopotamica*) wild goats (*Capra aegagrus*) and camels (*Camelus* sp.) (Speth and Tchernov 1998, 2001; Griggo 2004). The predecessors of Neanderthals, *Homo heidelbergensis*, were also successfully hunting large ungulates, as is dramatically illustrated by the site of Schöningen in Germany. Here eight wooden spears were found together with the remains of around 20 butchered horses that were secondarily exploited by carnivores (Thieme 1997; Voormolen 2009).

Neanderthal foraging strategies are characterized by a lower diet breadth than later anatomically modern human populations living in Europe (e.g. Richards et al. 2001; Stiner and Kuhn 2006). Neanderthals are also thought to have lacked projectile weapons, such as spearthrowers or bow and arrow (Shea and Sisk 2010). Furthermore, the paucity of small, fast-moving prey at most Middle Palaeolithic sites suggests that they may not have used snares or traps (e.g. Stiner et al. 1999, 2000). Finally, the evidence for Neanderthal exploitation of aquatic resources such as fish or shellfish is scarce. Although at some localities they definitely did exploit aquatic resources (Stiner 1994; Barton 2000; Stringer et al. 2008), their importance appears to be far smaller than at contemporaneous Middle Stone Age sites in Africa (Klein and Steele 2008). In short, at most sites we know, during their long existence over the large area they occupied, Neanderthals appear to have been specialized in hunting large, preferably prime-aged, ungulates.

Cave Hyenas

Cave hyenas are anatomically similar to modern spotted hyenas (*Crocuta crocuta*). They were substantially larger, however, especially during cold periods (Klein and Scott 1989). Genetically, cave hyenas are indistinguishable from modern East African spotted hyenas (Rohland et al. 2005).

Although hyenas are often characterized as scavengers, spotted hyenas are proficient hunters, for whom scavenging appears to be a foraging strategy of secondary importance (e.g. Kruuk 1972; Mills 1990; Holekamp et al. 1997; Cooper et al. 1999; Hayward 2006). Since cave hyenas are anatomically and genetically similar to spotted hyenas, it is likely that hunting was also the main subsistence strategy of Pleistocene cave hyenas. There is a relationship between predator and prey size (Radloff and Du Toit 2004), leading to the expectation that Pleistocene European hyenas were able to tackle larger prey than modern African spotted hyenas.

Spotted hyenas are social carnivores, living in clans averaging 25 individuals, but sometimes many more (Kruuk 1972; Mills and Bearder 2006; Watts and Holekamp 2007). Hyenas are cursorial predators, pursuing their prey for large distances until it is exhausted. 30–35 % of hunts appear to be successful (Holekamp et al. 1997; Hayward 2006). Spotted hyena prey-size varies, especially depending on hunting group size (e.g. Drea and Frank 2003; Trinkel 2010). Most spotted hyena hunts are done solitarily (Watts and Holekamp 2007). Alone, spotted hyenas are able to hunt prey up to about 250 kg, and can hunt animals such as blue wildebeest (*Connochaetes taurinus*) and gemsbok (*Oryx gazella*) (e.g. Cooper et al. 1999; Mills and Bearder 2006). When hunting in groups, spotted hyenas' hunting success increases (Holekamp et al. 1997). In addition, they are also able to tackle larger prey, such as zebra (*Equus quagga*) or kudu (*Tragelaphus strepsiceros*) and even buffalo (*Syncerus caffer*). In exceptional cases giraffes (*Giraffa camelopardalis*) and even juvenile elephants (*Loxodonta africana*) are taken (e.g. Cooper et al. 1999; Salnicki et al. 2001; Trinkel 2010). The importance of scavenging depends on the abundance of prey in an area. In arid, poor environments, scavenging generally accounts for a larger proportion of consumed calories than in rich environments. However, hunted meat always accounts for the majority of consumed calories (Cooper et al. 1999; Mills and Bearder 2006). Since the mammoth steppe likely contained abundant herbivore biomass (Delpech 1999), hunting was in all probability cave hyenas' primary foraging strategy.

Spotted hyena prey selection reflects the relative abundance of the available herbivore species. In contrast to other large carnivores, who usually preferentially prey on three to five species, they do not appear to significantly avoid species that are present, except for the largest ones (Hayward 2006). This means that they have a wide diet breadth. Hyenids accumulate prey bones in their dens. This behavior is well documented for striped hyenas (*Hyaena hyaena*) and brown hyenas (*Parahyaena brunnea*). Spotted hyenas collect bones at a slower rate (Lam 1992; Brugal et al. 1997; Pokines and Peterhans 2007; Lansing et al. 2009). A number of cave hyena dens in northwestern Europe has been excavated (e.g. Brugal et al. 1997; Fosse 1997; Fosse et al. 1998; Diedrich and Žák 2006). This allows the study of Pleistocene hyena accumulated bone assemblages to determine the prey choice of cave hyenas.

Spotted hyenas are characterized by a variety of foraging strategies, solitary hunting, group hunting and scavenging. Prey selection is varied and since they are cursorial hunters and also scavenge, weak individuals are generally overrepresented among the consumed animals. In Pleistocene Europe, the same ungulates that are important at many Neanderthal sites occur at most cave hyena dens.

Materials and Methods

I have compared the composition of the bone assemblages of a sample hyena dens with the composition of a large number of Neanderthal-accumulated assemblages. I use 12 published bone assemblages from hyena dens from France dating to MIS 3 and 4, listed in Table 12.1. Hyena dens are identified by an abundance of traces of carnivore activity, such as gnaw marks and traces of digestion on bone surfaces and the absence or rarity of traces of human activity. Further, the presence of juvenile hyena remains and coprolites are indicative of hyena dens. Juvenile hyenas are vulnerable and do not leave the den until 9–12 months old (Lansing et al. 2009). Hence, signs of denning suggest that the influence of other accumulating agents were negligible, since hyenas are not expected to den in places where their young can be disturbed (e.g. Bar-Yosef et al. 1992; Speth and Tchernov 1998).

The cave hyena bone assemblages are compared to the bone assemblages of roughly contemporaneous Neanderthal sites. For this, I have used the database of Mousterian assemblages published by Grayson and Delpech (2006). I did not include their Châtelperronian sample, since there is no consensus on the authorship of this technocomplex (Bar-Yosef and Bordes 2010). Furthermore, Grayson and Delpech (2006) use assemblages with a minimum Number of Identified specimens (NISP) of 20, while the hyena assemblages under consideration are larger. Hence, I only considered the Mousterian assemblages with $\sum$NISP $\geq$ 100. Moreover, I only considered the assemblages from MIS 3 and 4. The Mousterian assemblages under consideration are listed in Table 12.2.

To illuminate prey preferences of hyenas and Neanderthals, I compare the importance of different prey across the assemblages. To aid analysis, I focus on the importance of large bovids (*Bos* and *Bison*), Equids (mostly horse; small

Table 12.1 The hyena bone accumulations under consideration

Site	$\sum$NISP herbivores	NTAXA herbivores	Carnivore NISP	Date	References
Les Auzières 2	109	9	64	MIS 4	Marchal et al. (2009)
Bois Roche	853	6	81	MIS 4	Villa et al. (2004, 2010)
Camiac	939	10	92	MIS 3	Guadelli et al. (1988), Guadelli (1989)
Morancourt	368	10	59	Upper Pleistocene[a]	Fosse (1997); Mouton and Joffroy (1948)
Les Plumettes, ass. Inférieur	2,721	10	433	MIS 3	Beauval and Morin (2010)
Les Plumettes, ass. Supérieur	2,019	10	469	MIS 3	Beauval and Morin (2010)
Rochers-de-Villeneuve, ens. Intermediaire	492	9	119	MIS 3	Beauval and Morin (2010)
Rochers-de-Villeneuve, ens. Supérieur	2,413	10	302	MIS 3	Beauval and Morin (2010)
Tournal B/C	883	10	316	MIS 3	Magniez (2009)
Tournal D	519	9	127	MIS 3	Magniez (2009)
Unikote I (7–9)	102	10	218	MIS 3	Michel (2005)
Unikote I (10–12)	74	11	344	MIS 3	Michel (2005)

For the number of herbivore taxa, the "large bovid" category was counted as 1 taxon, unless the presence of both aurochs and bison are explicitly mentioned. Specimens determined to "indet herbivore" and herbivore size classes were incorporated in the $\sum$NISP herbivores

[a] The presence of cold-loving taxa, such as woolly rhinoceros, mammoth and especially reindeer suggests a cold phase

Table 12.2 The Mousterian assemblage listed by Grayson and Delpech (2006) used in the analysis

Site	∑NISP herbivores	NTAXA herbivores
Beauvais 1	224[a]	5
Beauvais 2	409	5
Chez Pinaud 10	459	4
Chez Pinaud 14	414	3
Chez Pinaud 18	189	3
Chez Pinaud 20	162	3
Chez Pinaud 22	2,135	5
Chez Pinaud 24	374	3
Combe-Grenal 11	158	8
Combe-Grenal 13	171	6
Combe-Grenal 14	537	8
Combe-Grenal 17	148	7
Combe-Grenal 19	104	5
Combe-Grenal 20	190	6
Combe-Grenal 21	270	8
Combe-Grenal 22	982	7
Combe-Grenal 23	1,101	6
Combe-Grenal 24	285	6
Combe-Grenal 25	284	6
Combe-Grenal 26	105	4
Combe-Grenal 27	296	8
Combe-Grenal 28	122	5
Combe-Grenal 29	224	8
Combe-Grenal 32	153	7
Combe-Grenal 35	331	9
Espagnac IV A	125	3
Genay	1,408	8
La Baume de Gigny XIX	145	8
Hortus 10A3B	143	3
Hortus 11C	112	2
Hortus 13A-14	486	5
Hortus 15	164	2
Hortus 17-21B	206	5
Hortus 26B	110	3
Hortus 28A	108	1
Hortus 29	252	2
Hyene IVb6	5,168	8
Mauran	4,192	3
Moula Guercy	160	12
Mutzig I C5	297	7
Mutzig I C6	289	6
Pech de l'Azé Ib4	327	8
Pech de l'Azé Ib6	412	8
Pech de l'Azé II 3	173	11
Pech de l'Azé II4B4	315	10
Pech de l'Azé II 4C 2	492	8
Pech de l'Azé IV H2	102	5
Pech de l'Azé IV I1	240	5
Pech de l'Azé IV I2	539	7
Pech de l'Azé IV J3	292	10
Pech de l'Azé J3a	262	8
Pech de l'Azé IV J3b	519	8
Pech de l'Azé IV 3Jc	210	8
Pech de l'Azé IV J4	185	6
Salpêtre de Pompignan IX	240	5
Portel Ouest F1	516	7
Portel Ouest F2	5,448	8
Portel Ouest F3	2,369	7
Portel Ouest G	166	6
La Quina 2b	437	4
La Quina 4b	885	4
La Quina 6a	972	5
La Quina 6c	271	3
La Quina 6d	152	5
La Quina 8	2,560	5
Regourdou 2	416	7
Regourdou 4	121	6
Grotte du Renne XI	1,351	7
Grotte du Renne XII	780	7
Roc-en-Pail	2,430	4
Saint-Césaire EGPF	1,205	8
Saint-Marcel G	321	6
Saint-Marcel H	365	7
Saint-Marcel I	332	7
Grotte Tournal IA	108	4
Vaufrey I	140	6
Grotte XVIC	554	10

If no bovids determined to taxon were listed, the category Bos/Bison was counted as one taxon

[a] Grayson and Delpech (2006) list a ∑NISP of 234, yet adding up the numbers in their table yields a ∑NISP of 224

numbers of *Equus hydruntinus* are present in some assemblages), cervids (*Cervus elaphus*; *Cervus simplicidens*; *Dama dama*; *Capreolus capreolus*; *Megaloceros giganteus* and *Rangifer tarandus*), caprids (*Capra ibex*, *Capra pyrenaica* and *Rupicapra rupicapra*), megafauna (proboscideans and rhinocerotids) and wild boar (*Sus scrofa*) in the assemblages.

Assemblage diversity is analyzed by comparing the total ∑NISP in an assemblage with the number of identified non-overlapping taxa (NTAXA) (cf. Grayson 1991; Grayson and

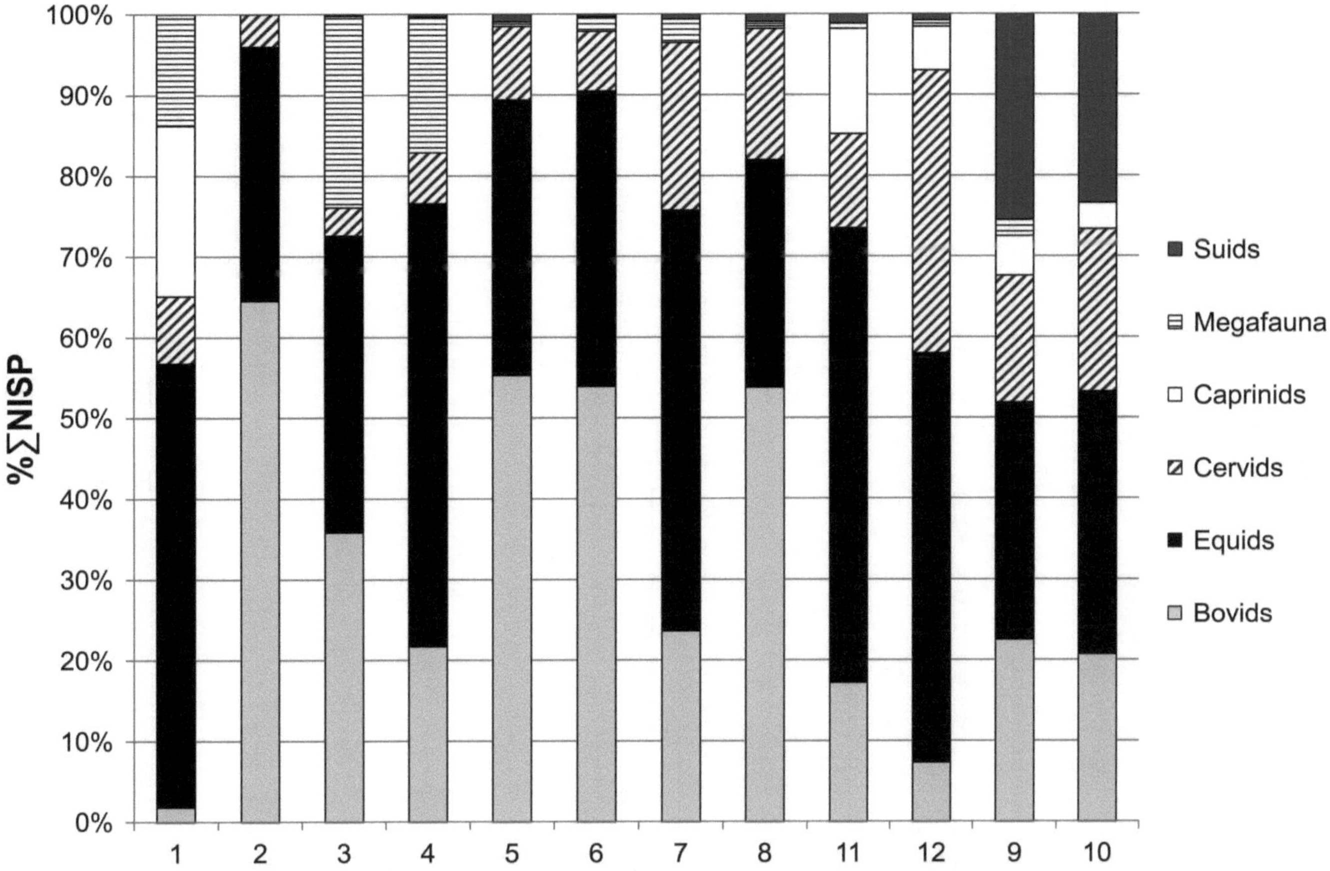

Fig. 12.1 The importance of the different groups of herbivores in the hyena dens under consideration. The NISP determined to size class only were divided among the species belonging to that size class present in the assemblage in the ratio of their presence among the NISP determined to taxon. *1* Les Auzières 2; *2* Bois Roche; *3* Camiac; *4* Morancourt; *5* Les Plumettes ass. inférieur; *6* Les Plumettes ass. supérieur; *7* Les Rochers-de-Villeneuve ens intermédiaire; *8* Les Rochers-de-Villeneuve ens supérieur; *9* Tournal B/C; *10* Tournal D; *11* Unikoté I, *7–9*; *12* Unikoté I *10–12*

Delpech 1998). This measure assumes that, although NTAXA is dependent on assemblage size, assemblages representing foraging strategies exploiting a broader set of resources will show more NTAXA for a similar assemblage size than assemblages representing foraging strategies where only a narrow set of resources is (Grayson 1991; Grayson and Delpech 1998).[1]

Results and Discussion

The importance of the prey animal groups in the hyena den assemblages is shown in Fig. 12.1. The difference in the average importance of the animal groups (bovids, equids, cervids, caprids, megafauna and suids) between Neanderthal sites and hyena dens is illustrated in Fig. 12.2. It is immediately apparent from the illustration that bovids, equids and cervids are the most important groups in both Neanderthal and hyena bone assemblages. However, the importance of the categories differs between them. To determine the statistical significance of these differences, the %∑NISP data of all animal groups were tested for normality using a Kolmogorov–Smirnov test in SPSS 19.0. The datasets were found to significantly deviate from the normal distribution. Hence, whether the differences in the importance of the groups between Mousterian bone assemblages and hyena dens are statistically significant was tested using a Mann–Whitney U test in SPSS 19.0. The results of these tests are listed in Table 12.3.

In terms of assemblage diversity, hyena-accumulated assemblages appear to be more diverse than assemblages accumulated by Neanderthals. As shown in Fig. 12.3, there are many very large Mousterian assemblages containing fewer herbivore taxa than even the smallest hyena assemblages.

[1] For the hyena assemblages from Bois Roche, Les Plumettes and Rochers-de-Villeneuve, bones identified to herbivore size classes were included in the ∑NISP; this may lead to a slight underestimation of hyena assemblage diversity relative to the Mousterian assemblages.

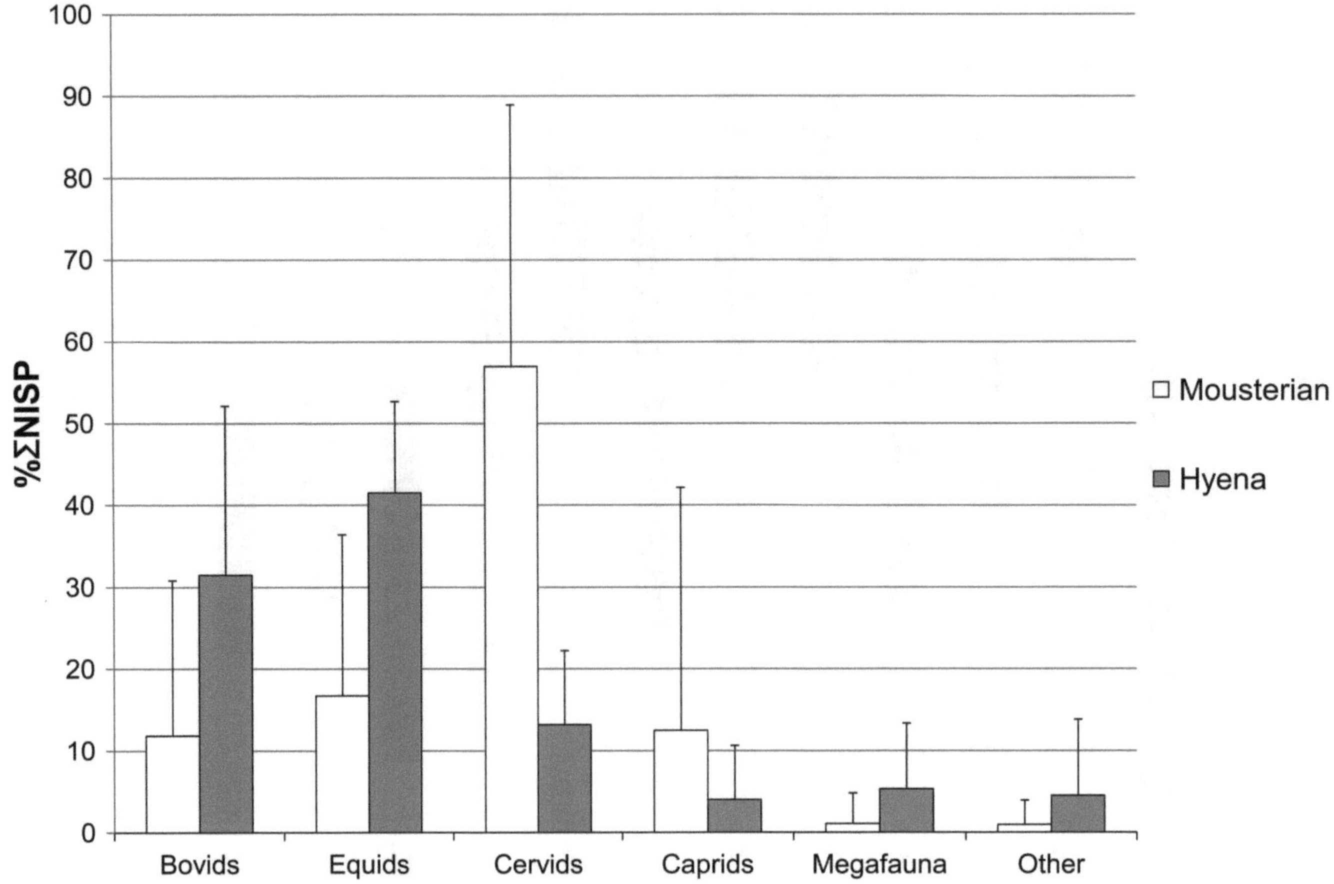

Fig. 12.2 Comparison of the average proportion the groups of species make up of hyena and Mousterian bone assemblages. The *error bars* indicate the standard deviations

Table 12.3 Results of the Mann–Whitney U tests comparing the representation of animal groups between Mousterian and hyena bone assemblages

	Outcome	U value	Significance	Rank Mousterian	Rank hyena
Bovids	Different	772	<0.001	42	71
Equids	Different	826	<0.001	42	75
Cervids	Different	153	<0.001	50	19
Caprids	Not different	439	0.672	46	43
Megafauna	Different	755	<0.001	42	69
Suids	Different	725	0.001	43	70

As shown, the differences in importance of the animal groups between Neanderthal and hyena assemblages are, except in the case of caprids, highly significant. Bovids and equids are better represented in hyena assemblages, while cervids are better represented in Mousterian assemblages. Interestingly, the amount of megafauna at hyena dens is significantly higher than it is in Mousterian assemblages. Moreover, in terms of the number of species of herbivores present, hyena assemblages appear to be richer than Mousterian assemblages. This suggests that hyenas and Neanderthals were preferentially targeting different herbivore groups and that hyenas were targeting a wider suite of species than Neanderthals. However, in order to accept this hypothesis, a number of confounding factors that may also have resulted in the differential representation of species in hyena and Neanderthal bone assemblages need to be examined.

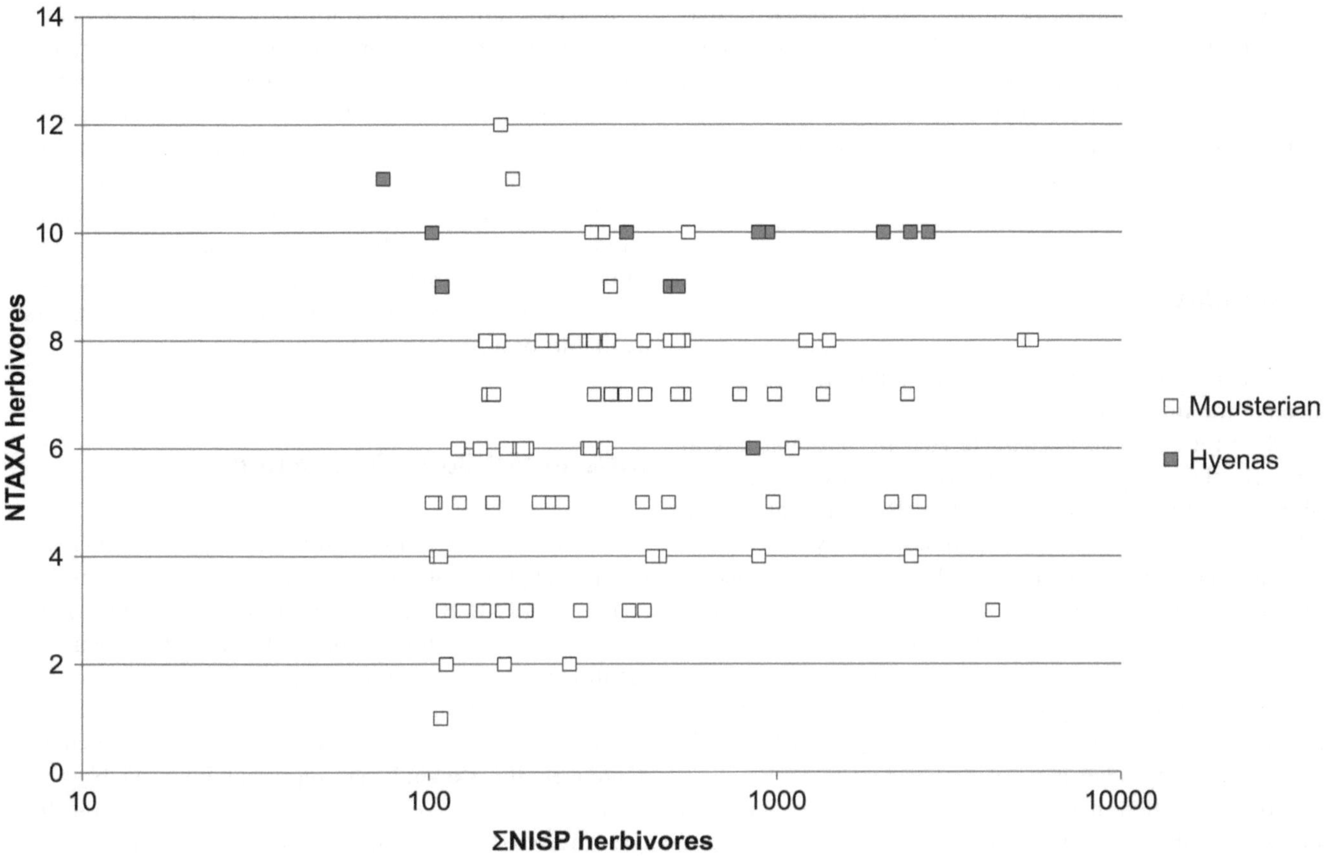

Fig. 12.3 Comparison of the relationship between the number of identified herbivore taxa and the $\sum$NISP of herbivore species between Mousterian and hyena assemblages

Availability

The most important factor determining whether species are present in faunal assemblages is whether they are present in the surroundings of the sites. In the sample under discussion, this issue is particularly important with regard to caprids and suids.

The absence of caprids in many assemblages is an important reason why the difference in representation of this group between Neanderthal and cave hyena assemblages is not significant. However, ibex (*Capra ibex*) and chamois (*Rupicapra rupicapra*) are characteristic of mountainous habitats (e.g. Delpech 1999). Their absence at most sites simply reflects the sites' location in unsuitable environments.

Modern spotted hyenas are usually associated with savannahs, open plains and open woodlands but, they occur in mountainous areas up to altitudes of 4,000 m (Skinner and Chimimba 2005). Nevertheless, their method of hunting, pursuing prey over large distances appears unsuited to mountainous, steep terrain. Ibex, especially mothers with young, use steep slopes as an anti-predator strategy (e.g. Villaret et al. 1997). This may have been a more effective strategy against spotted hyenas than the anti-predator strategies of other ungulates in the same area. Hence the poorer representation of caprids in hyena dens may reflect a real phenomenon, since hyenas may be expected to preferably hunt prey in the flatter parts of the landscape. On the other hand, the assemblages of Les Auzières 2 and Tournal show that caprids were certainly not immune to hyena exploitation.

In suitable environments, Neanderthals often exploited caprids intensively. In a number of assemblages, most notably from l'Hortus, caprids account for the majority of identified specimens, up to 98 % (Grayson and Delpech 2006). Their rarity in the hyena dens is at least partly due to the fact that only two sites, Les Auzières 2 and Tournal, apparently sample environmental conditions in which caprids thrived. Research on more hyena dens from mountainous environments may indicate that caprids were exploited more frequently than suggested by Fig. 12.2.

Suids are rare in most assemblages under consideration with the exception of Unikoté. Wild boar has a relatively narrow habitat tolerance, occurring only in forested environments (Delpech 1999). It would be expected to be more common during warmer periods. The low frequency of wild boar at most sites can thus not be used to come to conclusions on its role in either Neanderthal or cave hyena subsistence strategies.

Seasonality

Seasonal migrations, either of prey or predator may also account for differential representation of species at sites. Reindeer for example is characterized by seasonal migrations. Stable isotope analysis shows that this species was migratory during the Pleistocene too (Britton et al. 2011). Other species, such as bison and horse may have been present year-round (Britton et al. 2011: 183). If Neanderthals or hyenas were territorial, a migratory species may only have been available in their territory during a certain part of the year. If sites were seasonally occupied, some sites may lack migratory species, even though they played an important role in the subsistence strategies of the site's occupants during other parts of the year.

Modern spotted hyenas often occupy permanent and relatively stable territories. However, it appears that, if resident prey are rare and herbivore concentrations migratory, spotted hyenas will either forage in adjacent territories on a seasonal basis, or use a territory that is seasonally increased in size. Only if resident prey is common enough will territories be exclusively used by one clan (Hofer and East 1993; Trinkel et al. 2004; Höner et al. 2005).

The situation during MIS 3 and 4 is difficult to assess. The mammoth steppe that covered large parts of France during this period was probably high in ungulate biomass (Delpech 1999). However, especially during the coldest part of MIS 4, conditions may have been harsh. In these circumstances, resident herbivores could have been sparsely distributed and reindeer may have been one of the most important species on the landscape and extra-territorial foraging would become important for hyenas. Studies on modern spotted hyena suggest that migratory prey tends to be overrepresented at den sites compared to resident prey (Lansing et al. 2009). In addition, analysis of the seasonality of den occupation suggests that den placement is geared towards following migratory prey (Boydston et al. 2006: 83). Since spotted hyena prey deposits are time-averaged palimpsests accumulated over long periods of time (Lansing et al. 2009), the species representation of accumulated prey is expected to reflect the long-term importance of species; seasonal variations will likely be averaged out. These factors combined suggest that the migratory nature of prey species would not lead to their underrepresentation at hyena den sites.

Reindeer are an important prey in many Neanderthal assemblages. They account for part of the difference in the importance of cervids between hyena and Neanderthal assemblages. However, other cervids, especially the non-migratory red deer account for much of the cervids in Mousterian assemblages. This species is also poorly represented in the hyena assemblages under consideration. In conclusion, the higher frequency of cervids in Mousterian assemblages does not seem related to different responses of Neanderthals and hyenas to seasonally available prey.

Differential Bone Destruction

Hyena dentition enables them to destroy bones, this allows them to exploit marrow and grease contained in the bones (Sutcliffe 1970; Blumenschine 1987; Lam 1992; Ferretti 2007). The destruction of bones by hyenas influences the identifiability of the assemblages. This may lead to an underrepresentation of small species in bone assemblages, as their bones are more vulnerable to fragmentation. Moreover, the skeletal part representation can be biased, with high-density elements and the cranial skeleton being less affected than low-density elements. As a result, long bones at hyena dens are generally represented by the high-density epiphyses, while the diaphyses have been destroyed by gnawing (e.g. Cruz-Uribe 1991; Fosse 1996; Tournepiche and Couture 1999; Pickering 2002). Cranial parts, especially of smaller species are often overrepresented, since postcrania are preferentially destroyed. In larger species, postcrania are better represented (e.g. Fosse 1996; Brugal et al. 1997; Kuhn et al. 2010).

Pleistocene cave hyena dens show that horse, large bovid and even woolly rhinoceros postcrania did not escape the destructive activities of cave hyenas. In many cases the epiphyses of horse and large bovid bones were completely destroyed (e.g. Guadelli 1989; Villa and Bartram 1996; Diedrich and Žák 2006). At some dens, even the amount of identifiable large ungulate postcrania of is small. At Les Plumettes for example, teeth represent between one-third and three-quarters of identified remains for ungulate species (e.g. Beauval and Morin 2010). Furthermore, at Les Plumettes and Rochers-de-Villeneuve, a clear difference in the kind of traces of hyena activities that were present on bones could be observed. Large bones were heavily gnawed, while smaller elements primarily showed traces of digestion (Beauval and Morin 2010).

The preferential destruction of cervid and caprid bones could account for the underrepresentation of cervid and caprid bones in hyena dens. However, there is a number of arguments to suggest that although the destruction of elements surely influenced the representation of these species, this cannot account for the large difference that is observed between cave hyena and Mousterian assemblages. First, a number of Pleistocene hyena dens containing large amounts of cervids is known (e.g. Lunel Viel, Cueva de las Hienas, Bucca della Iena, Grotta dei Moscerini level 5, Grotta Guattari level 0–1) (Stiner 1992, 2004; Fosse 1996, 1997). Secondly, in modern spotted hyena assemblages from Africa small ungulates are well represented (Skinner et al. 1986; Lam 1992; Pokines and Peterhans 2007; Kuhn et al. 2010). Thirdly, for Bois Roche, Les Plumettes and Rochers-de-Villeneuve, bones not identifiable to species level have been determined to ungulate size classes (Villa et al. 2004; Beauval and Morin 2010). In all assemblages, this resulted in a large number of specimens being assigned to a size 3/4 ungulate (i.e. equid/bovid sized). Only in Rochers-de-Villeneuve was a small number of size 2 specimens identified (Beauval and Morin 2010). If small ungulates formed an important part of the diet, but their bones had been preferentially destroyed resulting in an underrepresentation, one would expect that more ungulate size 2 category specimens would have been identified. Hence, if hyenas were exploiting cervids in a similar degree to Neanderthals, one would expect this to be more apparent from the dens.

Nevertheless, it is possible that the destruction of smaller species was more intense at the sites under consideration than at sites from other periods and regions that do contain large numbers of cervids. Faith and Behrensmeyer (2006) observed that increased intensity of conspecific spotted hyena competition results in increased intensity of bone destruction and this is expected to affect the elements of small-sized species more heavily than larger-sized species. However, the transition of MIS 4 to 3 appears to show an increase in hyena population size in France (Discamps 2010). The population size of African predators in general, and spotted hyenas specifically, appears to be limited by the availability of food (East 1984; Grange and Duncan 2006). This suggests that ungulate prey was relatively abundant. Hence the intensity of competition was not dramatic.

In all, although the preferential destruction of skeletal elements of smaller species might lead to a degree of underrepresentation of smaller species, dense elements and elements poor in marrow and grease are expected to survive. The assemblages of Les Auzières 2, Tournal and the upper assemblage of Les Plumettes suggest that if caprids and cervids formed an important part of the hyena diet, they would be recorded in the bone assemblages. In all, for the sample under consideration, cervids really do seem to have been exploited less intensively by cave hyenas than by Neanderthals.

The substantial amount of megafaunal remains at some hyena dens has been ascribed to their higher resistance to destructive processes than the bones of smaller species (e.g. Guadelli 1989; Diedrich and Žák 2006). However, a taphonomic analysis of the hyena den of Camiac, rich in woolly rhinoceros, suggests that the importance of the species here cannot be explained by the preferential destruction of bovids and equids (Discamps 2011). The regular occurrence of woolly rhinoceros, also in hyena accumulations outside France suggests that this species was a structural ingredient of the cave hyena menu (e.g. Discamps 2011; Dusseldorp 2011).

Transport

The representation of animal species at archaeological sites and hyena dens is also affected by transport decisions. Hominins preferentially process larger carcasses in the field, hence the quantity of skeletal elements that is transported back to a central place will decrease as the size of prey carcasses increases (e.g. O'Connell et al. 1990; Monahan 1998; Lupo 2006; Faith et al. 2009). It is generally predicted that when larger species are processed, those body parts that are most rewarding in nutritional value will be transported back to the site. However, this prediction is sometimes violated in anthropological studies (e.g. O'Connell et al. 1990; Domínguez-Rodrigo 2002). The violation of the predictions regarding the preferential transport of skeletal parts appear to be caused by the fact that many, sometimes competing economic indices are taken into account in hunter-gatherer transport decisions (Lupo 2006: 56–57).

The apparent dominance of cervids in Neanderthal bone assemblages may thus be caused by more intensive processing of the carcasses of larger species. The magnitude of this effect is difficult to gauge. However, looking at the skeletal part representation of a number of Palaeolithic sites in France and Italy, it appears that transport decisions are similar across red deer, ibex, horse and aurochs. Only the smallest ungulates such as roe deer are transported more completely (Valensi and Psathi 2004). Roe deer is rare or absent in most of the Mousterian assemblages under consideration. In only 7 assemblages in the sample does it constitute more than 10 % of the $\sum$NISP (Grayson and Delpech 2006). This suggests that similar parts of the carcass were transported to archaeological sites, regardless of carcass size. Other researchers have also documented transport decisions being similar across bovids, equids and cervids (e.g. Morin 2004).

Transport of megafauna by Neanderthals was probably selective. Their remains are generally rare at Middle Palaeolithic sites. Some sites, geographically and temporally outside of the study area, do contain large amounts of megafauna, such as Biache-Saint-Vaast, Mont-Dol, Taubach and La Cotte de Saint-Brélade (Scott 1980;

Auguste 1995; Bratlund 1999). These sites appear to represent locations in the close vicinity of the place where the animals were killed. However, postcranial elements were probably transported away from La Cotte de Saint- Brélade (Scott 1980). Similarly, it has been proposed that meat bearing parts were preferentially transported away from the mammoth accumulation of Lynford (Schreve 2006). This scenario has been contested though (Smith 2008). If the representation of megafauna is scrutinized in terms of Minimum Number of Individuals (MNI), the impact of preferential transport in terms of species representation is expected to be diminished. A comparison of the megafauna representation at the hyena den of Camiac and the Neanderthal site of Saint-Césaire in terms of MNI still suggests that cave hyenas were exploiting megafauna in larger numbers than Neanderthals (Dusseldorp 2011).

In the case of spotted hyenas, selective transport of energy rich parts appears to be in evidence (e.g. Henschel et al. 1979). As a whole, hyena bone assemblages appear to reflect the composition of the herbivore guild surrounding the site (Skinner et al. 1986; Stiner 1992; Lansing et al. 2009). The smaller species are underrepresented relative to their natural abundance, but this appears to reflect the hyenas' preferential consumption of medium to large-sized ungulates (Henschel et al. 1979; Brugal et al. 1997). Since hyenas appear to hunt prey in proportion to their presence on the landscape (Hayward 2006), hyena dens give a good insight in their prey choice.

The mobility strategy of Neanderthals may also have influenced the representation of different animal groups at archaeological sites. Spotted hyenas move their den on average once every 30 days (Boydston et al. 2006). This suggests that the residues of foraging activities during the period of den occupation will present a representative picture of the foraging activities that were performed. Neanderthals on the other hand may have been characterized by a system of very high mobility, perhaps staying no more than a single night at a time at most sites (cf. Langbroek 2001, 2012). Their mobility may have been routed between magnet locations (cf. Binford 1984), lacking "central places" (cf. Binford 1984, 1987; Sept 1992; Kolen 1999). In this kind of system, the archaeological substrate of subsistence activities may be reflected in a diffuse scatter of archaeological features, a "veil of stones", across the landscape (e.g. Roebroeks et al. 1992b; Langbroek 2012). The archaeological sites we excavate may represent overnight camps where the toolkit was maintained. Remains of subsistence activities, especially remains of large animals, would be unlikely to be transported to such locations.

This scenario is difficult to test. However, circumstantial evidence suggests that Neanderthals utilized some form of central places where the spoils of exploiting the foraging radius were transported to for some amount of time. Neanderthals subsisted on hunted meat to an important degree, especially in cold periods. Hunting is a knowledge intensive foraging strategy (cf. Kaplan et al. 2000). In this situation, juveniles would not be able to provide their own sustenance and would thus need to be provisioned. This appears to be the case for pregnant women too, whose energetic demands would be high (cf. Mussi 2007; Ponce de Léon et al. 2008). Even if no division of labor was in place (cf. Kuhn and Stiner 2006), it is unlikely that pregnant women participated in hunting large mammals. Similarly, wounded and invalid Neanderthals sometimes survived for considerable periods of time (e.g. Trinkaus and Zimmerman 1982; Berger and Trinkaus 1995; Zollikofer et al. 2002). It thus appears likely that some kind of provisioning of people that did not participate in hunting activities was in place. Hypermobility can then likely not account for the structural underrepresentation of particular classes of animal at a large sample of Middle Palaelithic sites.

A fuller discussion of this issue is beyond the scope of this paper. However, it is clear that different mobility strategies can influence the representation of different resources. Although a kind of central place model is often implicitly assumed for Neanderthals, this need not apply in all regions and periods occupied by Neanderthals.

Comparing Archaeozoological and Isotopic Results

Stable isotope analysis can also be used to study dietary habits of Pleistocene predators. It has been applied to bone assemblages of two Neanderthal sites, as well as a hyena den and a sinkhole assemblage dating to MIS 4 and 3 from southwestern France. The results have been used to propose a differentiation of dietary niches between Neanderthals and cave hyenas (Bocherens et al. 2005; Richards et al. 2008b). Interestingly the interpretation of the stable isotope results appears to contradict archaeozoological analysis (Dusseldorp 2011). It was suggested, based on ^{13}C and ^{15}N signatures of hyenas and Neanderthals, that hyenas exploited reindeer much more intensively than Neanderthals and that Neanderthals preferentially focused on megafauna (Bocherens et al. 2005) and on large bovids and horse (Richards et al. 2008b).

Some factors discussed above, such as seasonal occupation of sites, differential transport and differential destruction of bones may have caused these apparently contradictory results. First, if one or both species had a fixed seasonal "round", it is possible that resources whose consumption does show up in isotopic values were deposited at as yet unexcavated sites. However, with regards to cervids and megafauna, the pattern apparent in the isotopically sampled sites is similar to the pattern of the much larger sample of sites analyzed here. Bovids and horse are well represented in both the isotopically sampled hyena den and Middle Palaeolithic sites (Dusseldorp 2011). This interpretation does not furnish an explanation for the observed pattern.

Second, underrepresentation of small species in hyena assemblages due to their preferential destruction may have played a role. Reindeer are virtually absent in the isotopically sampled hyena bone assemblages, this may be because their remains were preferentially destroyed. However, a number of dens with substantial amounts of cervids is known (e.g. Stiner 1992, 1994; Fosse 1996), suggesting this factor alone cannot explain the observed pattern. The presence of substantial amounts of reindeer at Tournal (Magniez 2009) suggests that hyenas did, at least occasionally, consume large quantities of reindeer. However, it also suggests that this would remain visible in the archaeozoological record.

Third, field processing by Neanderthals could account for an underrepresentation of megafauna at Mousterian sites. In view of the existence of Neanderthal sites containing large numbers of megafaunal remains, this factor too cannot account fully for the observed pattern. Moreover, economics of transport would suggest that large megafaunal elements would also be less likely to be transported over large distances by cave hyenas than smaller elements.

Stable isotope analysis informs on the average isotopic composition of the diet in the long-term. Reindeer is characterized by high ^{13}C values compared to other herbivores. Its consumption would be expected to show up in the form of elevated ^{13}C values in sampled predators. However, Neanderthal reindeer consumption, evidenced at many sites, could be masked by the consumption of resources with isotopic signatures characterized by low ^{13}C values. One of the resources that could have this effect is freshwater fish (Balter and Simon 2006). Evidence for the consumption of fish is rare in the Middle Palaeolithic archaeological record. However, there are some arguments to suggest that it may have been practiced. First, fish remains are taphonomically much more vulnerable than bones of other animal groups (Wheeler and Jones 1989). Second, in older excavations, the presence of fish remains was likely missed if sieving was not practiced (cf. Van Niekerk 2011). Third, there is some evidence showing that fish consumption was practiced at least occasionally (e.g. Hardy and Moncel 2011). Furthermore, fish consumption need not be practiced directly by Neanderthals in order to show up in their isotope values. Fish consumers, such as brown bear (*Ursus arctos*), and in some cases the largely herbivorous cave bear (*Ursus spelaeus*) are characterized by similar isotopic values (e.g. Richards et al. 2008a, b; Peigné et al. 2009). The consumption large numbers of bears has been observed at a number of Neanderthal sites (Auguste 2003).

The consumption of predators and/or fish also furnishes an alternative explanation for the very high ^{15}N values characterizing sampled Neanderthals. These high values have been attributed to the consumption of megafauna (Bocherens et al. 2005). Since consumption of predators such as brown bear may explain both the high ^{13}C and ^{15}N values characterizing Neanderthals, and there is more substantial archaeological evidence for their consumption than for the regular consumption of megafauna by Neanderthals, this may be the most parsimonious explanation for the pattern. However, further research into this problem is needed.

Differentiated Dietary Niches

Both hyenas and Neanderthals relied to an important extent on three main groups of prey during MIS 3 and 4: bovids, equids and cervids. However, there is a significant difference in the importance of the groups between the two accumulators. Neanderthals appear to rely more heavily on cervids, while hyenas show a higher reliance on equids and bovids. In addition, megafauna is better represented in the hyena sample (Discamps 2011; Dusseldorp 2011). Some additional factors may also have contributed to the differentiation of the niches of both predators.

First, the dietary importance of carnivores may have differed between Neanderthals and cave hyenas. Hyena dens are generally much richer in carnivore remains than Neanderthal sites (see Table 12.1 for the total number of carnivore remains per assemblage). This has been proposed to be a diagnostic criterion for the identification of hyena dens (Cruz-Uribe 1991; Pickering 2002). Therefore, if cave hyenas consumed larger amounts of carnivores, their diet breadth could be broader than that of Neanderthals.

Actualistic research suggests that, while striped and brown hyenas accumulate many small carnivores, spotted hyena dens often contain few carnivore remains (Kuhn et al. 2010). The only carnivore that is often present in large numbers in spotted hyena dens is spotted hyena itself. Here it usually concerns young individuals, which are often killed by their siblings (e.g. Frank et al. 1991; Pickering 2002; Drea and Frank 2003). As shown in Fig. 12.4, the majority of carnivores at most of the hyena dens under consideration are cave hyenas. The two assemblages from Tournal form an exception. Here, large amounts of cave bear bones account for the majority of carnivore remains.

The presence of cave bears in deposits need not be related to hyena foraging strategies. Cave bears experienced attritional mortality during hibernation (e.g. Stiner 1998). Since cave bears would not use dens regularly used by predators, the use of Tournal Cave by cave bears and cave hyenas was likely separated in time. Moreover, if hyenas occasionally did exploit cave bears either by hunting or scavenging, this does not necessarily differentiate them

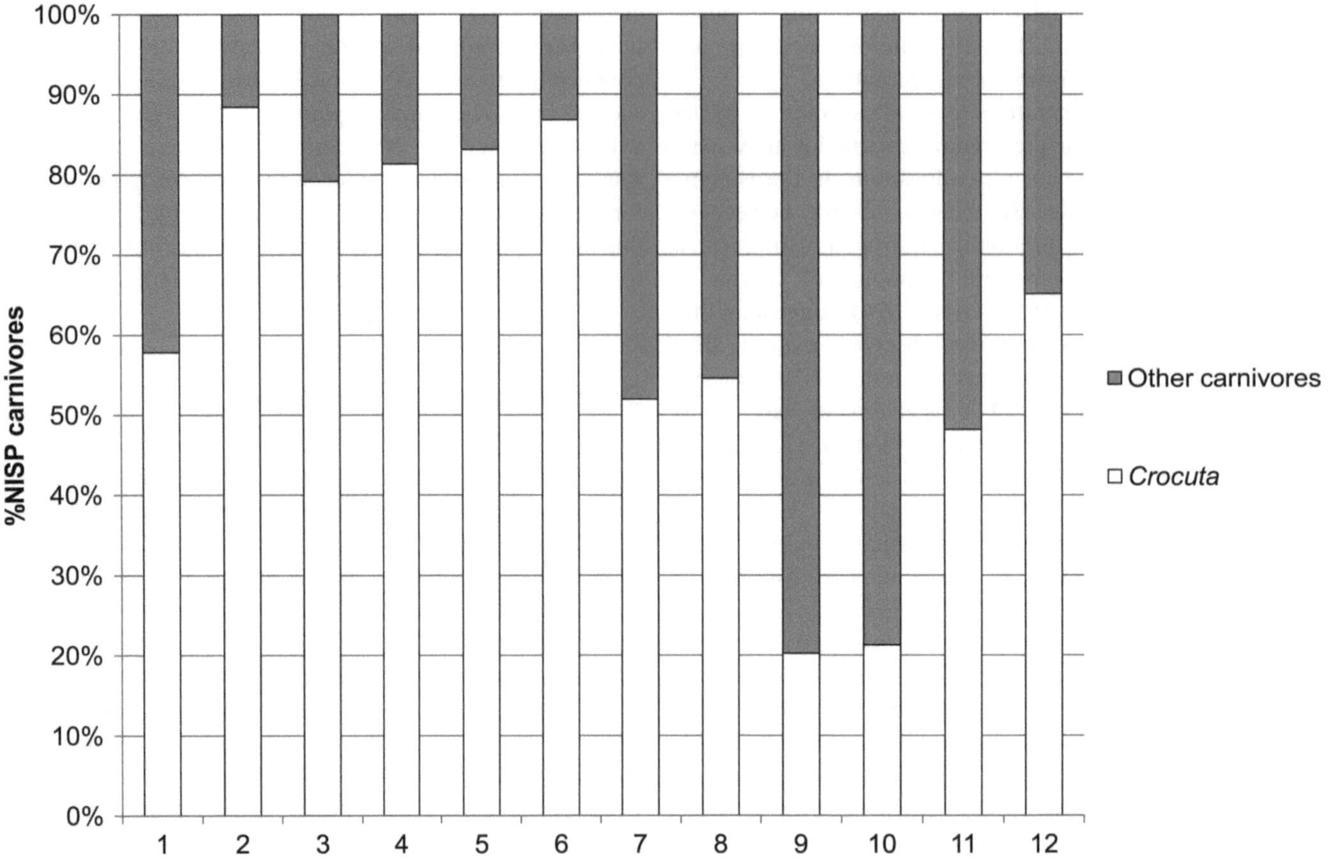

Fig. 12.4 The proportion of cave hyenas in the ∑NISP of carnivores of the hyena dens under consideration. *1* Les Auzières 2; *2* Bois Roche; *3* Camiac; *4* Morancourt; *5* Les Plumettes ass. inférieur; *6* Les Plumettes ass. supérieur; *7* Les Rochers-de-Villeneuve ens intermédiaire; *8* Les Rochers-de-Villeneuve ens supérieur; *9* Tournal B/C; *10* Tournal D; *11* Unikoté I, *7–9*; *12* Unikoté I *10–12*

from Neanderthals, since they are also known to have consumed cave bear at some sites, such as Arcy-sure-Cure and Balve Cave (David 2002; Kindler 2012). Therefore, although the consumption of carnivores may have been a factor in the diversification of Neanderthal and cave hyena dietary niches, its role was probably small.

In addition to the choice of prey species, differentiation of resource exploitation can be effected when different predators exploit different categories of individuals from a species. Since hyenas are cursorial hunters, weak individuals (old and young animals) tend to be overrepresented among their prey. This would lead one to expect prey species in hyena dens to be characterized by an attritional mortality profile (Cruz-Uribe 1991; Stiner 2004). As discussed above, many Neanderthal bone assemblages dominated by prime-aged individuals are known. Hence, this may be a factor of importance in the diversification of their dietary niches.

However, an emphasis on juvenile and aged individuals is not apparent in all Pleistocene hyena dens. At the hyena den of Fouvent, the mortality profile of horse is undistinguishable from those from the Neanderthal sites of Bau de l'Aubesier and Combe-Grenal (Fernandez et al. 2006). Here, the age-profile does not appear to be unduly influenced by differential preservation (Fernandez et al. 2006: 179). At the sites under consideration, coarse information on the age-class distribution is available for Camiac (Fosse 1996: 78). Here it appears that all animal groups, except for proboscideans are represented mainly by adult remains. However, no distinction between adult and senile is provided. An underrepresentation of juveniles may be caused by differential preservation and transport, which can change the mortality profile of remains found in dens from the expected profile based on the categories of prey that are killed (Pickering 2002). It thus appears that hyenas may not have focused on weak individuals as strongly as expected. Studying the proportion of senile individuals in bone assemblages may further clarify this issue.

The differential targeting of prey age groups by Neanderthals and hyenas may have contributed to the differentiation of hyena and Neanderthal dietary niches. Surprisingly however, Pleistocene cave hyenas in France also managed to target prime-age equids. Hence, the difference in age-classes exploited by both predators may not have been very large.

Neanderthals, being primates, were omnivorous. As suggested by Hardy (2010), even during cold periods, Neanderthals could have procured significant amounts of food from vegetable resources. However, modeling suggests that European hunter-gatherers would usually procure less than 50 % of their consumed calories from plant foods (Binford 2007). Especially at latitudes in excess of the threshold of 40°N and S, the importance of animal foods increases dramatically in modern hunter-gatherers. During the Weichselian, the harsh conditions and reduction of vegetable productivity in the winter would result in meat being the most important ingredient of the Neanderthal diet (Roebroeks et al. 1992a; Roebroeks 2003; Mussi 2007).

Reconstructing the actual importance of Neanderthal plant food consumption is problematic, since the remains of vegetable foods themselves are generally not preserved. The presence of starch microfossils from calculus of Neanderthal teeth and microwear analysis confirms that plant foods played a significant role in the Neanderthal diet (El Zaatari et al. 2011; Henry et al. 2011). How important plant foods were exactly is difficult to determine. Microwear analysis suggests that none of the analyzed fossils (some of which are from the Levant and Mediterranean and some from warm periods) consumed more than 50 % plant foods (El Zaatari et al. 2011). Unfortunately, since the nutrients used to build bone collagen are preferentially drawn from the protein component of the diet, the consumption of plant foods in a species subsisting for over 50 % on meat cannot be traced using stable isotope analysis (Bocherens 2009). Since plant foods are expected not to play a role of any significance in spotted hyena diets (although fruit is occasionally consumed) (Lam 1992), the consumption of plant foods by Neanderthals will have lowered the degree of overlap between Neanderthal and cave hyena dietary niches.

Another mechanism of niche differentiation is the temporal separation of activities of competing species (Hayward and Slotow 2009). This mechanism may have significantly influenced Neanderthal—cave hyena niche overlap. Neanderthals are expected to have been diurnal foragers. Spotted hyenas are nocturnal or crepuscular (Hayward and Slotow 2009). Hence, it is likely that cave hyenas were nocturnal or crepuscular too. If this were the case, this would contribute to the degree of differentiation of their niches.

Spotted hyenas, although primarily hunters, can and do scavenge, especially in situations where food is scarce (Cooper et al. 1999; Mills and Bearder 2006). This foraging strategy may also have contributed to the differentiation of cave hyena and Neanderthal niches. However, scavenging is not completely unknown from the Neanderthal archaeological record (e.g. Gaudzinski 2004).

Conclusion

The representation of bovids, equids, cervids and megafauna differs significantly across Neanderthal and hyena accumulated bone assemblages from MIS 4 and 3 from France. Bovids, equids and megafauna are better represented at hyena dens, while cervids are better represented at Mousterian sites. This representation is opposite to the expectations based on stable isotope analyses of the Neanderthal sites of Jonzac and Saint-Césaire and the hyena den of Camiac.

Moreover, the different representation of the animal groups between Neanderthal and hyena-accumulated bone assemblages is highly significant, except in the case of caprids. As discussed, I deem it unlikely that the combination of more intense destruction of smaller animals by hyenas and the preferential transport of smaller animals by Neanderthals can explain the difference. Hence, I propose that the difference in prey representation illustrated here was most likely caused by differing prey choice between Neanderthals and hyenas. In the period and region under consideration, hyenas preferentially preyed on larger animals than Neanderthals. The patterns of prey choice documented for Neanderthals and hyenas during MIS 4 and 3 in France were not immutable. In other periods and areas, hyenas did prey intensively on cervids, while Neanderthals structurally preyed on rhinoceros at the sites of Taubach and Biache-Saint-Vaast for example.

By far the most important cervids in Neanderthal assemblages are reindeer and red deer. Reindeer are a migratory species, living in large herds. Their specialized exploitation may point to a high degree familiarity with the animals, aiding Neanderthals to successfully plan encounters with herds to hunt this species. Red deer live dispersed and are resident in a certain area. Their exploitation may reflect an encounter hunting strategy, although in the exploitation of this species, knowledge of its habits would help in manipulating the encounter rate with it. Since there is an inverse relationship between population density and body size in herbivores, cervids may have represented a larger amount of biomass in a given area than bovids and equids and their targeting may have allowed a more reliable supply of food than targeting larger species.

Hyenas apparently preferentially targeted larger animals when they were available. This is likely a result of their evolution as specialized predator-scavengers. They are able to consume very large amounts of meat and bone in a short time, which may sustain them for a while. Moreover, since they can successfully exploit most of the carcass, including the bone, and they can scavenge, they may have been better able than Neanderthals to deal with a less reliable frequency of encounters with their preferred prey.

Acknowledgments This research was supported by the Netherlands Organization for Scientific Research (NWO; grant no. 360-60-050) and the National Research Foundation, through a South African National Research Foundation postdoctoral fellowship granted by Professor Christopher Henshilwood. I thank Jamie Clark and John Speth for inviting me to contribute to the ICAZ session "Variability in human hunting behavior during Oxygen Isotope Stages (OIS) 4/3: implications for understanding modern human origins". Jerome Reynard provided help regarding the construction of graphs. Kristian Carlson advised on the statistics. Eugene Morin and Michel Lenoir kindly clarified some issues regarding a number of hyena sites. Emmanuel Discamps cordially provided me with advice and work that had not been published at the time of writing. Arne Wossink, Aurore Val, Geeske Langejans and Luc Amkreutz helped me get hold of important literature. Marco Langbroek and two anonymous reviewers provided important comments on how to improve the manuscript.

References

Adler, D. S., Bar-Oz, G., Belfer-Cohen, A., & Bar-Yosef, O. (2006). Ahead of the game, Middle and Upper Palaeolithic hunting behaviors in the Southern Caucasus. *Current Anthropology, 47*, 89–118.

Armand, D. (1998). La faune de la grotte Bourgeois-Delaunay (Commune de la Chaise de Vouthon, Charente). Résultats préliminaires. *Paléo, 10*, 77–86.

Auguste, P. (1995). Chasse et charognage au Paléolithique moyen: L'apport du gisement de Biache-Saint-Vaast (Pas-de-Calais). *Bulletin de la Société Préhistorique Française, 92*, 155–167.

Auguste, P. (2003). La chasse à l'ours au Paléolithique moyen: Mythes, réalités et état de la question. In M. Pathou-Mathis & H. Bocherens (Eds.), *Le rôle de l'environnement dans les comportements des chasseurs-cueilleurs préhistoriques* (pp. 135–142). Oxford: Archeopress.

Balter, V., & Simon, L. (2006). Diet and behavior of the Saint-Cesaire Neanderthal inferred from biogeochemical data inversion. *Journal of Human Evolution, 51*, 329–338.

Barton, N. (2000). Mousterian hearths and shellfish: Late Neanderthal activities on Gibraltar. In C. B. Stringer, R. N. E. Barton, & J. C. Finlayson (Eds.), *Neanderthals on the edge, papers from a conference marking the 150th anniversary of the Forbes' Quarry discovery, Gibraltar* (pp. 211–220). Oxford: Oxbow Books.

Bartram, L. E., & Villa, P. (1998). The archaeological excavation of prehistoric hyena dens: Why bother? In J. P. Brugal, L. Meignen, & M. Pathou-Mathis (Eds.), *Économie préhistorique: les comportements de subsistance au paléolithique actes des rencontres 23-24-25 octobre 1997* (pp. 15–29). Sophia Antipolis: APDCA.

Bar-Yosef, O., & Bordes, J.-G. (2010). Who were the makers of the Châtelperronian culture. *Journal of Human Evolution, 59*, 586–593.

Bar-Yosef, O., Vandermeersch, B., Arensburg, B., Belfer-Cohen, A., Goldberg, P., Laville, H., Meignen, L., et al. (1992). The excavations in Kebara Cave, Mt. Carmel. *Current Anthropology, 33*, 497–550.

Beauval, C., & Morin, E. (2010). Les repaires d'hyènes du Lussacois (Lussac-les-Châteaux, Vienne, France): Apport des sites des Plumettes et des Rochers-de-Villeneuve. In J. Buisson-Catil & J. Primault (Eds.), *Préhistoire entre Vienne et Charente. Hommes et sociétés du Paléolithique*. Chauvigny: Association des Publications Chauvinoises.

Beauval, C., Maureille, B., Lacrampe-Cuyaubère, F., Serre, D., Peressinotto, D., Bordes, J.-G., et al. (2005). A late Neandertal femur from Les Rochers-de-Villeneuve, France. *Proceedings of the National Academy of Sciences, 102*, 7085–7090.

Berger, T. D., & Trinkaus, E. (1995). Patterns of trauma among Neanderthals. *Journal of Archaeological Science, 22*, 841–852.

Binford, L. R. (1984). *Faunal remains from Klasies River Mouth*. New York: Academic Press.

Binford, L. R. (1985). Human ancestors: Changing views of their behavior. *Journal of Anthropological Archaeology, 4*, 292–327.

Binford, L. R. (1987). Searching for camps and missing the evidence? Another look at the Lower Palaeolithic. In O. Soffer (Ed.), *The Pleistocene old world, regional perspectives* (pp. 17–31). New York: Plenum Press.

Binford, L. R. (1988). Étude taphonomique des restes fauniques de la grotte Vaufrey, couche VIII. In J.-P. Rigaud (Ed.), *La grotte Vaufrey, Paléoenvironnement, chronologie, activités humaines* (pp. 525–563). Chalons-sur-Marne: Paquez et fils.

Binford, L. R. (2007). The diet of early hominids: Some things we need to know before "reading" the menu from the archaeological record. In W. Roebroeks (Ed.), *Guts and brains: An integrative approach to the hominin record* (pp. 185–222). Leiden: Leiden University press.

Blumenschine, R. J. (1987). Characteristics of an early hominid scavenging niche. *Current Anthropology, 28*, 383–407.

Bocherens, H. (2009). Neanderthal dietary habits: Review of the isotopic evidence. In J.-J. Hublin & M. P. Richards (Eds.), *The evolution of hominin diets: Integrating approaches to the study of Palaeolithic subsistence* (pp. 241–250). Dordrecht: Springer.

Bocherens, H., Drucker, D. G., Biliou, D., Pathou-Mathis, M., & Vandermeersch, B. (2005). Isotopic evidence for the diet and subsistence pattern of the Saint-Césaire Neanderthal: Review and use of a multi-source mixing model. *Journal of Human Evolution, 49*, 71–87.

Boydston, E. E., Kapheim, K. M., & Holekamp, K. E. (2006). Patterns of den occupation by the spotted hyaena (*Crocuta crocuta*). *African Journal of Ecology, 44*, 77–86.

Brantingham, P. J. (1998). Hominid-carnivore coevolution and invasion of the predatory guild. *Journal of Anthropological Archaeology, 17*, 327–353.

Bratlund, B. (1999). Taubach revisited. *Jahrbuch des Römisch-Germanischen Zentralmuseums, 46*, 61–174.

Britton, K., Grimes, V., Niven, L., Steele, T. E., McPherron, S., Soressi, M., et al. (2011). Strontium isotope evidence for migration in late Pleistocene Rangifer: Implications for Neanderthal hunting strategies at the Middle Palaeolithic site of Jonzac, France. *Journal of Human Evolution, 61*, 176–185.

Brugal, J.-P., Fosse, P., & Guadelli, J.-L. (1997). Comparative study of bone assemblages made by recent and Pleistocene hyenids. In L. A. Hannus, L. Rossum, & R. P. Winham (Eds.), *Proceedings of the 1993 bone conference hot springs, South Dakota* (pp. 158–187). Sioux Falls: Archaeology Laboratory Augustana Collega.

Cooper, S. M., Holekamp, K. E., & Smale, L. (1999). A seasonal feast: Long-term analysis of feeding behaviour in the spotted hyaena (*Crocuta crocuta*). *African Journal of Ecology, 37*, 149–160.

Costamagno, S., Meignen, L., Beauval, C., Vandermeersch, B., & Maureille, B. (2006). Les Pradelles (Marillac-de-Franc, France): A Mousterian reindeer hunting camp? *Journal of Anthropological Archaeology, 25*, 466–484.

Cruz-Uribe, K. (1991). Distinguishing hyena from hominid bone accumulations. *Journal of Field Archaeology, 18*, 467–486.

David, F. (2002). Les ours du Châtelperronien de la Grotte du Renne à Arcy-sur-Cure (Yonne). In T. Tillet & L. R. Binford (Eds.), *L'ours et l'homme* (pp. 185–192). Liège: Université de Liège.

Davies, T. J., Meiri, S., Barraclough, T. G., & Gittleman, J. L. (2007). Species co-existence and character divergence across carnivores. *Ecology Letters, 10*, 146–152.

Delpech, F. (1999). Biomasse d'ongulés au Paléolithique et inférences sur la demographie. *Paléo, 11*, 19–42.

Diedrich, C. J., & Žák, K. (2006). Prey deposits and den sites of the Upper Pleistocene hyena *Crocuta crocuta spelaea* (Goldfuss, 1823) in horizontal and vertical caves of the Bohemian Karst (Czech Republic). *Bulletin of Geosciences, 81*, 237–276.

Discamps, E. (2010). A "hyena event" at the Middle-to-Upper Palaeolithic transition? Preliminary results from the South West of France. *Zona Arqueológica, 13*, 510–516.

Discamps, E. (2011). La place du rhinocéros dans le régime alimentaire des hyènes à Camiac (Gironde, France) et ses implications pour la compétition avec les derniers néanderthaliens. In J.-P. Brugal, A. Gardeisen, & A. Zucker (Eds.), *Prédateurs dans touts leurs états, Actes des XXXIe rencontres internationales d'archéologie et d'histoire d'Antibes* (pp. 33–48). Antibes: Éditions APDCA.

Domínguez-Rodrigo, M. (2002). Hunting and scavenging by early humans, the state of the debate. *Journal of World Prehistory, 16*, 1–54.

Drea, C. M., & Frank, L. G. (2003). The social complexity of spotted hyenas. In F. B. M. De Waal & P. L. Tyack (Eds.), *Animal social complexity, intelligence, culture and individualized societies* (pp. 121–148). Cambridge: Harvard University Press.

Dusseldorp, G. L. (2011). Studying Pleistocene Neanderthal and cave hyena dietary habits: Combining isotopic and archaeozoological analyses. *Journal of Archaeological Method and Theory, 18*, 224–255.

East, R. (1984). Rainfall, soil nutrient status and biomass of large African savanna mammals. *African Journal of Ecology, 22*, 245–270.

El Zaatari, S., Grine, F. E., Ungar, P. S., & Hublin, J.-J. (2011). Ecogeographic variation in Neandertal dietary habits: Evidence from occlusal molar microwear texture analysis. *Journal of Human Evolution, 61*, 411–424.

Enloe, J. G. (2012). Middle Palaeolithic cave taphonomy: Discerning humans from hyenas at Arcy-sur-Cure, France. *International Journal of Osteoarchaeology, 22*, 591–602.

Faith, J. T., & Behrensmeyer, A. K. (2006). Changing patterns of carnivore modification in a landscape bone assemblage, Amboseli Park, Kenya. *Journal of Archaeological Science, 33*, 1718–1733.

Faith, J. T., Domínguez-Rodrigo, M., & Gordon, A. D. (2009). Long-distance carcass transport at Olduvai Gorge? A quantitative examination of Bed I skeletal element abundances. *Journal of Human Evolution, 56*, 247–256.

Farizy, C., David, F., Jaubert, J., & Leclerc, J. (1994). Fonctionnement du site: Hommes et bisons. In C. Farizy, F. David, & J. Jaubert (Eds.), *Hommes et bisons du paléolithique moyen à Mauran (Haute-Garonne)* (pp. 239–245). Paris: CNRS Éditions.

Fernandez, P., Guadelli, J.-L., & Fosse, P. (2006). Applying dynamics and comparing life tables for Pleistocene Equidae in anthropic (Bau de l'Aubesier, Combe-Grenal) and carnivore (Fouvent) contexts with modern feral horse populations (Akagera, Pryor Mountain). *Journal of Archaeological Science, 33*, 176–184.

Ferretti, M. P. (2007). Evolution of bone-cracking adaptations in hyaenids (Mammalia, Carnivora). *Swiss Journal of Geosciences, 100*, 41–52.

Fosse, P. (1996). La grotte n°1 de Lunel-Viel (Hérault, France): Repaire d'hyènes du Pleistocène Moyen. *Paléo, 8*, 47–81.

Fosse, P. (1997). Variabilité des assemblages osseux créés par l'hyène des cavernes. *Paléo, 9*, 15–54.

Fosse, P., Brugal, J.-P., Guadelli, J.-L., Michel, P., & Tournepiche, J.-F. (1998). Les repaires d'hyènes des cavernes en Europe occidentale: Présentation et comparaison de quelques assemblages osseux. In J. P. Brugal, L. Meignen, & M. Pathou-Mathis (Eds.), *Économie préhistorique: les comportements de subsistance au paléolithique actes des rencontres 23-24-25 octobre 1997* (pp. 43–61). Sophia Antipolis: APDCA.

Frank, L. G., Glickman, S. E., & Licht, P. (1991). Fatal sibling aggression, precocial development, and androgens in neonatal spotted hyenas. *Science, 252*, 702–704.

García, N., & Virgós, E. (2007). Evolution of community composition in several carnivore palaeoguilds from the European Pleistocene: The role of interspecific competition. *Lethaia, 40*, 33–44.

Gaudzinski, S. (1995). Wallertheim revisited: A re-analysis of the fauna from the Middle Palaeolithic site of Wallertheim. *Journal of Archaeological Science, 22*, 51–66.

Gaudzinski, S. (1996). On bovid assemblages and their consequences for the knowledge of subsistence patterns in the Middle Palaeolithic. *Proceedings of the Prehistoric Society, 62*, 19–39.

Gaudzinski, S. (2004). A matter of high resolution? The Eemian interglacial (OIS 5e) in North Central Europe and Middle Palaeolithic subsistence. *International Journal of Osteoarchaeology, 14*, 201–211.

Gaudzinski, S., & Roebroeks, W. (2000). Adults only. Reindeer hunting at the Middle Palaeolithic site of Salzgitter-Lebenstedt. *Journal of Human Evolution, 38*, 497–521.

Grange, S., & Duncan, P. (2006). Bottom-up and top-down processes in African ungulate communities: Resources and predation acting on the relative abundance of zebra and grazing bovids. *Ecography, 29*, 899–907.

Grayson, D. K. (1991). Alpine faunas from the White Mountains, California: Adaptive change in the Late Prehistoric Great Basin? *Journal of Archaeological Science, 18*, 483–506.

Grayson, D. K., & Delpech, F. (1998). Changing diet breadth in the Early Upper Palaeolithic of Southwestern France. *Journal of Archaeological Science, 25*, 1119–1129.

Grayson, D. K., & Delpech, F. (2006). Was there increasing dietary specialization across the Middle-to-Upper Palaeolithic transition in France? In N. J. Conard (Ed.), *When Neanderthals and Modern Humans met* (pp. 377–417). Tübingen: Kerns Verlag.

Griggo, C. (2004). Mousterian fauna from Dederiyeh Cave and comparisons with fauna from Umm El Tlel and Douara cave. *Paléorient, 30*, 149–162.

Guadelli, J. L. (1989). Étude taphonomique du repaire d'hyènes de Camiac (Gironde, France), Éléments de comparaison entre un site naturel et un gisement préhistorique. *Bulletin de l'Association Française pour l'Étude du Quaternaire, 2*, 91–100.

Guadelli, J.-L., Lenoir, M., Marambat, L., & Paquero, M.-M. (1988). Un gisement de l'interstade Würmien en Gironde: Le gisement de Camiac à Camiac et Saint-Denis. In L. Binford & J.-P. Rigaud (Eds.), *L'Homme de Neandertal, actes de la colloque International de Liège (4–7 Décembre 1986), Volume 4: La Technique* (pp. 59–60). Liège: Université de Liège.

Hardin, G. (1960). The competitive exclusion principle. *Science, 131*, 1292–1297.

Hardy, B. L. (2010). Climatic variability and plant food distribution in Pleistocene Europe: Implications for Neanderthal diet and subsistence. *Quaternary Science Reviews, 29*, 662–679.

Hardy, B. L., & Moncel, M.-H. (2011). Neanderthal use of fish, mammals, birds, starchy plants and wood 125–250,000 years ago. *PLoS One, 6*, e23768.

Hayward, M. W. (2006). Prey preferences of the spotted hyena (*Crocuta crocuta*) and degree of dietary overlap with the lion (*Panthera leo*). *Journal of Zoology, 270*, 606–614.

Hayward, M. W., & Slotow, R. (2009). Temporal partitioning of activity in large African carnivores: Tests of multiple hypotheses. *South African Journal of Wildlife Research, 39*, 109–125.

Henry, A. G., Brooks, A. S., & Piperno, D. R. (2011). Microfossils in calculus demonstrate consumption of plants and cooked foods in Neanderthal diets (Shanidar III, Iraq; Spy I and II, Belgium). *Proceedings of the National Academy of Sciences, 108*, 486–491.

Henschel, J. R., Tilson, R., & Von Blottnitz, F. (1979). Implications of a spotted hyena bone assemblage in the Namib Desert. *South African Archaeological Bulletin, 34*, 127–131.

Hofer, H., & East, M. L. (1993). The commuting system of Serengeti spotted hyaenas: How a predator copes with migratory prey. I. Social organization. *Animal behaviour, 46*, 547–557.

Holekamp, K. E., Smale, L., Berg, R., & Cooper, S. M. (1997). Hunting rates and hunting success in the spotted hyena (*Crocuta crocuta*). *Journal of Zoology, 242*, 1–15.

Höner, O. P., Wachter, B., East, M. L., Runyoro, V. A., & Hofer, H. (2005). The effect of prey abundance and foraging tactics on the population dynamics of a social, territorial carnivore, the spotted hyena. *Oikos, 108*, 544–554.

Jaubert, J., Lorblanchet, M., Laville, H., Slot-Moller, R., Turq, A., & Brugal, J.-P. (1990). Les chasseurs d'Aurochs de La Borde. Paris: Editions de la maison des sciences de l'homme.

Kaplan, H., Hill, K., Lancaster, J., & Hurtado, A. M. (2000). A theory of human life-history evolution: Diet intelligence and longevity. *Evolutionary Anthropology, 9*, 156–185.

Kindler, L. (2012). *Die Rolle von Raubtieren in der Einnischung und Subsistenz jungpleistozäner Neandertaler, Archäozoologie und Taphonomie der mittelpaläolithischen Fauna aus der Balver Höhle (Westfalen)*. Mainz: Verlag des Römisch-Germanischen Zentralmuseums.

Klein, R. G., & Scott, K. (1989). Glacial/interglacial size variation in fossil spotted hyenas (*Crocuta crocuta*) from Britain. *Quaternary Research, 32*, 88–95.

Klein, R. G., & Steele, T. E. (2008). Gibraltar data are too sparse to inform on Neanderthal exploitation of coastal resources. *Proceedings of the National Academy of Sciences, 105*, E115.

Kolen, J. (1999). Hominids without homes: On the nature of Middle Palaeolithic settlement in Europe. In W. Roebroeks & C. Gamble (Eds.), *The Middle Palaeolithic of Europe* (pp. 139–175). Leiden: University of Leiden.

Kruuk, H. (1972). *The spotted hyena, A study of predation and social behavior*. Chicago: Chicago University Press.

Kuhn, S. L., & Stiner, M. C. (2006). What's a mother to do? The division of labor among Neanderthals and Modern Humans in Eurasia. *Current Anthropology, 47*, 953–980.

Kuhn, B. F., Berger, L. R., & Skinner, J. D. (2010). Examining criteria for identifying and differentiating fossil faunal assemblages accumulated by hyenas and hominins using extant hyenid accumulations. *International Journal of Osteoarchaeology, 20*, 15–35.

Lam, Y. M. (1992). Variability in the behaviour of spotted hyaenas as taphonomic agents. *Journal of Archaeological Science, 19*, 389–406.

Langbroek, M. (2001). The trouble with Neanderthals. *Archaeological Dialogues, 8*, 123–151.

Langbroek, M. (2012). Trees and ladders: A critique of the theory of human cognitive and behavioural evolution in Palaeolithic archaeology. *Quaternary International, 270*, 4–14.

Lansing, S. W., Cooper, S. M., Boydston, E. E., & Holekamp, K. E. (2009). Taphonomic and zooarchaeological implications of spotted hyena (*Crocuta crocuta*) bone accumulations in Kenya: A modern behavioral ecological approach. *Paleobiology, 35*, 289–309.

Lupo, K. D. (2006). What explains the carcass field processing and transport decisions of contemporary hunter-gatherers? Measures of economic anatomy and zooarchaeological skeletal part representation. *Journal of Archaeological Method and Theory, 13*, 19–65.

Madurell-Malapeira, J., Minwer-Barakat, R., Alba, D. M., Garcés, M., Gómez, M., Aurell-Garrido, J., et al. (2010). The Vallparadís section (Terrassa, Iberian Peninsula) and the latest Villafranchian faunas of Europe. *Quaternary Science Reviews, 29*, 3972–3982.

Magniez, P. (2009). Taphonomic study of the Middle and Upper Palaeolithic large mammal assemblage of Tournal Cave (Bize-Minervois, France). *Journal of Taphonomy, 7*, 203–233.

Marchal, F., Monchot, H., Coussot, C., Desclaux, E., Deschamp, P., Thiébaut, C., et al. (2009). Neandertals paleoenvironment in Western Provence: The contribution of Les Auzières 2 (Méthamis, Vaucluse, France). *Comptes Rendus Palevol, 8*, 493–502.

Martínez-Navarro, B. (2010). Early Pleistocene faunas of Eurasia and hominin dispersals. In J. G. Fleagle, J. J. Shea, F. E. Grine, A. L. Baden, & R. E. Leakey (Eds.), *Out of Africa I: The first hominin colonization of Eurasia* (pp. 207–224). Dordrecht: Springer.

Michel, P. (2005). Un repaire Würmien d'hyènes des cavernes: La Grotte d'Unikoté (Iholdy, Pyrénées-Atlantiques, France). In J. A. Lasheras & R. Montes (Eds.), *Actas de la Reunión Científica "Neandertales Cantábricos, estado de la cuestión"*. Museo de Altamira: Santander.

Mills, M. G. L. (1990). *Kalahari Hyaenas: Comparative behavioural ecology of two species*. London: Unwin Hyman.

Mills, M. G. L., & Bearder, S. K. (2006). Hyena Family. In D. W. MacDonald (Ed.), *The Encyclopedia of mammals, New edition*. (pp. 620–625). Oxford: Oxford University Press.

Miracle, P. (2005). Late Mousterian subsistence and cave use in Dalmatia: The zooarchaeology of Mujina Pećina, Croatia. *International Journal of Osteoarchaeology, 15*, 84–105.

Monahan, C. M. (1998). The Hadza carcass transport debate revisited and its archaeological implications. *Journal of Archaeological Science, 25*, 405–424.

Morin, E. (2004). *Late Pleistocene population interaction in western Europe and modern human origins: New insights based on the faunal remains from Saint-Césaire, southwestern France*. Unpublished PhD, University of Michigan, Ann Arbor.

Mouton, P., & Joffroy, R. (1948). Paléolithique Moyen et repaire d'hyènes au "Perthuis de Roche" de Morancourt. *Haute-Marne Bulletin de la Société préhistorique française, 45*, 256–259.

Mussi, M. (2007). Women of the middle latitudes. The earliest peopling of Europe from a female perspective. In W. Roebroeks (Ed.), *Guts and brains: An integrative approach to the hominin record* (pp. 165–183). Leiden: Leiden University Press.

O'Connell, J. F., Hawkes, K., & Blurton-Jones, N. (1990). Reanalysis of large mammal body part transport among the Hadza. *Journal of Archaeological Science, 17*, 301–316.

Palombo, M. R. (2010). A scenario of human dispersal in the northwestern Mediterranean throughout the Early to Middle Pleistocene. *Quaternary International, 223–224*, 179–194.

Peigné, S., Goillot, C., Germonpré, M., Blondel, C., Bignon, O., & Merceron, G. (2009). Predormancy omnivory in European cave bears evidenced by a dental microwear analysis of Ursus spelaeus from Goyet, Belgium. *Proceedings of the National Academy of Sciences, 106*, 15390–15393.

Pickering, T. R. (2002). Reconsideration of criteria for differentiating faunal assemblages accumulated by hyenas and hominids. *International Journal of Osteoarchaeology, 12*, 127–141.

Pokines, J. T., & Peterhans, J. C. K. (2007). Spotted hyena (*Crocuta crocuta*) den use and taphonomy in the Masai Mara National

Reserve, Kenya. *Journal of Archaeological Science, 34*, 1914–1931.

Ponce de Léon, M. S., Golovanova, L., Doronichev, V., Romanova, G., Akazawa, T., Kondo, O., et al. (2008). Neanderthal brain size at birth provides insights into the evolution of human life history. *Proceedings of the National Academy of Sciences, 105*, 13764–13768.

Radloff, F. G. T., & Du Toit, J. T. (2004). Large predators and their prey in a southern African savanna: A predator's size determines its prey size range. *Journal of Animal Ecology, 73*, 410–423.

Richards, M. P., & Trinkaus, E. (2009). Isotopic evidence for the diets of European Neanderthals and early modern humans. *Proceedings of the National Academy of Sciences, 106*, 16034–16039.

Richards, M. P., Pettitt, P. B., Stiner, M. C., & Trinkaus, E. (2001). Stable isotope evidence for increasing dietary breadth in the European mid-Upper Palaeolithic. *Proceedings of the National Academy of Sciences, 98*, 6528–6532.

Richards, M. P., Pacher, M., Stiller, M., Quiles, J., Hofreiter, M., Constantin, S., et al. (2008a). Isotopic evidence for omnivory among European cave bears: Late Pleistocene Ursus spelaeus from the Pestera cu Oase, Romania. *Proceedings of the National Academy of Sciences, 105*, 600–604.

Richards, M. P., Taylor, G., Steele, T., McPherron, P., Soressi, M., Jaubert, J., et al. (2008b). Isotopic dietary analysis of a Neanderthal and associated fauna from the site of Jonzac (Charente-Maritime), France. *Journal of Human Evolution, 55*, 179–185.

Roebroeks, W. (2003). Landscape learning and the earliest occupation of Europe. In M. Rockman & J. Steele (Eds.), *Colonization of unfamiliar landscapes, The archaeology of adaptation*. London: Routledge.

Roebroeks, W. (2006). The human colonisation of Europe: Where are we? *Journal of Quaternary Science, 21*, 425–435.

Roebroeks, W., Conard, N. J., & Van Kolfschoten, T. (1992a). Dense forests, cold steppes and the Palaeolithic settlement of Northern Europe. *Current Anthropology, 33*, 551–586.

Roebroeks, W., De Loecker, D., Hennekens, P., & Van Ieperen, M. (1992b). A "veil of stones": On the interpretation of an early Middle Palaeolithic low density scatter at Maastricht-Belvedere (The Netherlands). *Analecta Prehistorica Leidensia, 25*, 1–16.

Rohland, N., Pollack, J. L., Nagel, D., Beauval, C., Airvaux, J., Paäbo, S., et al. (2005). The population history of extant and extinct hyenas. *Molecular Biology and Evolution, 22*, 2435–2443.

Salnicki, J., Teichmann, M., Wilson, V. J., & Murindagomo, F. (2001). Spotted hyenas *Crocuta crocuta* prey on new-born elephant calves in Hwange National Park, Zimbabwe. *Koedoe, 44*, 79–83.

Schild, R., Tomaszewski, A. J., Sulgostowska, Z., Gautier, A., Bluszcz, A., Bratlund, B., et al. (2000). The Middle Palaeolithic kill-butchery site of Zwolen, Poland. In A. Ronen & M. Weinstein-Evron (Eds.), *Toward modern humans: The Yabrudian and Micoquian, 400–50 k-years ago: Proceedings of a congress held at the University of Haifa, November 3–9, 1996* (pp. 189–207). Oxford: Archeopress.

Schreve, D. C. (2006). The taphonomy of a Middle Devensian (MIS 3) vertebrate assemblage from Lynford, Norfolk, UK, and its implications for Middle Palaeolithic subsistence strategies. *Journal of Quaternary Science, 21*, 543–556.

Scott, K. (1980). Two hunting episodes of Middle Palaeolithic age at La Cotte de Saint-Brelade, Jersey (Channel Islands). *World Archaeology, 12*, 137–152.

Sept, J. M. (1992). Was there no place like home? A new perspective on early hominid archaeological sites from the mapping of Chimpanzee nests. *Current Anthropology, 33*, 187–207.

Shea, J. J., & Sisk, M. L. (2010). Complex projectile technology and *Homo sapiens* dispersal into western Eurasia. *PalaeoAnthropology, 2010*, 100–122.

Skinner, J. D., & Chimimba, C. T. (2005). *The mammals of the southern African subregion* (3rd ed.). Cambridge: Cambridge University Press.

Skinner, J. D., Henschel, J. R., & Van Jaarsveld, A. S. (1986). Bone-collecting habits of spotted hyenas *Crocuta crocuta* in the Kruger National Park. *South African Journal of Zoology, 21*, 301–308.

Smith, G. M. (2008). *Middle Palaeolithic mammoth hunters? The role of hominins at Lynford, Norfolk, UK*. Palaeolithic Mesolithic Conference. British Museum, London.

Speth, J. D., & Tchernov, E. (1998). The role of hunting and scavenging in Neandertal procurement strategies, new evidence from Kebara. In K. Aoki, T. Akazawa, & O. Bar-Yosef (Eds.), *Neandertals and modern humans in Western Asia* (pp. 223–240). New York: Plenum Press.

Speth, J. D., & Tchernov, E. (2001). Neandertal hunting and meat processing in the Near East, evidence from Kebara cave (Israel). In C. B. Stanford & H. T. Bunn (Eds.), *Meat-eating and human evolution* (pp. 52–72). Oxford: Oxford University Press.

Stiner, M. C. (1992). Overlapping species "choice" by Italian Upper Pleistocene predators. *Current Anthropology, 33*, 433–451.

Stiner, M. C. (1994). *Honor among thieves, a zooarchaeological study of Neandertal ecology*. Princeton: Princeton University Press.

Stiner, M. C. (1998). Mortality analysis of Pleistocene bears and its paleoanthropological relevance. *Journal of Human Evolution, 34*, 303–326.

Stiner, M. C. (2004). Comparative ecology and taphonomy of spotted hyenas, humans and wolves in Pleistocene Italy. *Revue de Paléobiologie de Genève, 23*, 771–785.

Stiner, M. C., & Kuhn, S. L. (2006). Changes in the 'connectedness' and resilience of Paleolithic Societies in Mediterranean ecosystems. *Human Ecology, 34*, 693–712.

Stiner, M. C., Munro, N. D., Surovell, T. A., Tchernov, E., & Bar-Yosef, O. (1999). Paleolithic population growth pulses evidenced by small animal exploitation. *Science, 283*, 190–194.

Stiner, M. C., Munro, N. D., & Surovell, T. A. (2000). The tortoise and the hare: Small game use, the broad-spectrum revolution and Paleolithic demography. *Current Anthropology, 41*, 39–73.

Stringer, C. B., Finlayson, J. C., Barton, R. N. E., Fernández-Jalvo, Y., Cáceres, I., Sabin, R. C., et al. (2008). Neanderthal exploitation of marine mammals in Gibraltar. *Proceedings of the National Academy of Sciences, 105*, 14319–14324.

Sutcliffe, A. J. (1970). Spotted hyaena: Crusher, gnawer, digester and collector of bones. *Nature, 227*, 1110–1113.

Thieme, H. (1997). Lower Palaeolithic hunting spears from Germany. *Nature, 385*, 807–810.

Tournepiche, J.-F. (1994). Un néanderthalien dévoré par des hyènes: La grotte de Rochelot (Saint-Amand de Bonnieure, Charente). *Paléo, 6*, 319–321.

Tournepiche, J.-F., & Couture, C. (1999). The hyena den of Rochelot Cave. In S. Gaudzinski & E. Turner (Eds.), *The role of early humans in the accumulation of European Lower and Middle Palaeolithic bone assemblages, Ergebnisse eines Kolloquiums* (pp. 89–101). Bonn: Verlag des Römisch-Germanischen Zentralmuseums.

Trinkaus, E., & Zimmerman, M. R. (1982). Trauma among the Shanidar Neandertals. *American Journal of Physical Anthropology, 57*, 61–76.

Trinkel, M. (2010). Prey selection and prey preferences of spotted hyenas *Crocuta crocuta* in Etosha National Park, Namibia. *Ecological Research, 25*, 413–417.

Trinkel, M., Fleischmann, P., Steindorfer, A. F., & Kastberger, G. (2004). Spotted hyenas (*Crocuta crocuta*) follow migratory prey. Seasonal expansion of a clan territory in Etosha, Namibia. *Journal of Zoology, 264*, 125–133.

Turner, A. (1992). Large carnivores and earliest European hominids: Changing determinants of resource availability during the Lower and Middle Pleistocene. *Journal of Human Evolution, 22*, 109–126.

Valensi, P., & Psathi, E. (2004). Faunal exploitation during the Middle Palaeolithic in south-eastern France and north-western Italy. *International Journal of Osteoarchaeology, 14*, 256–272.

Van Niekerk, K. L. (2011). *Marine fish exploitation during the Middle and Later Stone Age of South Africa*. Unpublished PhD-thesis, University of Cape Town.

Villa, P., & Bartram, L. (1996). Flaked bone from a hyena den. *Paléo, 8*, 143–159.

Villa, P., Castel, J.-C., Beauval, C., Bourdillat, V., & Goldberg, P. (2004). Human and carnivore sites in the European Middle Palaeolithic: Similarities and differences in the bone modification and fragmentation. *Revue de Paléobiologie, Genève, 23*, 705–730.

Villa, P., Sanchez-Goñi, M. F., Bescós, G. C., Grün, R., Ajas, A., Pimienta, J. C. G., et al. (2010). The archaeology and paleoenvironment of an Upper Pleistocene hyena den: An integrated approach. *Journal of Archaeological Science, 37*, 919–935.

Villaret, J. C., Bon, R., & Rivet, A. (1997). Sexual segregation of habitat by the alpine ibex in the French Alps. *Journal of Mammalogy, 78*, 1273–1281.

Voormolen, B. (2009). Ancient hunters, modern butchers : Schöningen 13II-4, a kill-butchery site dating from the northwest European Lower Palaeolithic. *Journal of Taphonomy, 6*, 71–240.

Watts, H. E., & Holekamp, K. E. (2007). Hyena societies. *Current Biology, 17*, R657–R660.

Wheeler, A., & Jones, A. K. G. (1989). *Fishes*. Cambridge: Cambridge University Press.

Zollikofer, C. P. E., Ponce de Léon, M. S., Vandermeersch, B., & Lévêque, F. (2002). Evidence for interpersonal violence in the St. Césaire Neanderthal. *Proceedings of the National Academy of Sciences, 99*, 6444–6448.

Chapter 13
Bone Grease Rendering in Mousterian Contexts: The Case of Noisetier Cave (Fréchet-Aure, Hautes-Pyrénées, France)

Sandrine Costamagno

Fat is an essential element of the diet of human groups living in subarctic regions (Binford 1978; Saint-Germain 1997, 2005; Outram and Mulville 2005; Karlin and Tchesnokov 2007; Costamagno and David 2009). In addition to contributing vitamins, the consumption of grease can be of crucial importance during certain periods of the year for groups with very high protein diets with a limited carbohydrate content (Binford 1978; Speth and Spielmann 1983; Abe 2005). At the end of winter and beginning of spring, herbivores are generally in poor physical condition. The sequence of fat loss begins with subcutaneous fat depletion, followed by muscular, organ and then marrow depletion. The lower extremities (metapodials and phalanges), mandible and brain retain fat the longest (Speth 1983; Stiner 1991). In environments lacking exploitable vegetal resources during these periods, the ingestion of lean meat with no supplementary lipid source can lead to serious dietary deficiencies, sometimes resulting in death (Speth and Spielmann 1983).

Hunter-gatherers can compensate for these deficiencies by eating carbohydrates, especially dried fruits or nuts. They may also hunt the animals most rich in fat, use fat reserves constituted during other seasons and/or extract greater quantities of grease from the available carcasses (Speth and Spielmann 1983; Saint-Germain 2005). Extracting the fat contained in bone tissues is one of the methods that could have been used to increase the lipid portion of the diet. Boiling is the method often used to render bone grease: the bones are submerged in water and then heated over a slow fire for a varying amount of time (Saint-Germain 1997). Bone grease rendering was commonly practiced by forager groups living in very diverse environments (cold tundra, boreal forest, plains, high desert, etc.) (Denys 1672; Leechman 1951; Vehik 1977; Binford 1978; Grønnow et al. 1983; Brink 1997; Morrison 1998; Abe 2005; Saint-Germain 2005; Costamagno and David 2009). Though it is frequent, this food preparation technique is not systematically observed among modern hunter-gatherers (e.g. Hadza: Bunn et al. 1988; O'Connell et al. 1988; Dassanetch: Gifford-Gonzalez 1989).

The bones used to obtain fat are very diverse, including long bones, carpal bones, tarsal bones, vertebrae, and ribs. The Nunamiut (Binford 1978) were very selective, favoring long bone extremities and sometimes a few carpal and tarsal bones. The same is true of the Hidatsa (Wilson 1924 cited in Logan 1998), who preferred the bones of limb extremities. Nunamiut groups often used the fat of vertebrae and ribs to fabricate a sort of candle, while the skull, antlers and scapulae were never processed for bone grease (Binford 1978). Binford argued that the proportion of mono-unsaturated oleic acid was responsible for the Nunamiut's preference for lower limb fats. According to Brink (1997), grease weight is the most important variable underlying the Nunamiut selection. This debate was recently renewed by Morin (2007) who showed that parts rich in oleic acid would have been selected more often than expected based on grease weight alone. Morin (2007: 81) argued that "(...) taste, texture, time of consumption, sequence of fat mobilization, and ease of extraction are all potentially contributing to the bias favoring the selection of skeletal parts rich in unsaturated fatty acids". This fat preference is not universal among forager groups, however. The Cree (Calling Lake Alberta), the Copper Inuit and the Omaha used the articular extremities of long bones, the coxal bones (except the acetabulum), the glenoid cavity of the scapula, the vertebrae and the ribs (Bonnichsen 1973; Jenness, 1922 cited in Morrison 1998; Dorsey, 1884 cited in Vehik 1977). Some Siberian groups who practice reindeer herding still extract fat from a highly variable range of bones (Evenki: Abe 2005; Koriak: Karlin and Tchesnokov 2007). In all of these examples, with the exception of the Koriak, who

S. Costamagno (✉)
TRACES-UMR 5608, CNRS, PRES Toulouse, Université Toulouse 2 – Le Mirail, 5 Allées A. Machado, 31058 Toulouse Cedex 9, France
e-mail: costamag@univ-tlse2.fr

Jamie L. Clark and John D. Speth (eds.), *Zooarchaeology and Modern Human Origins: Human Hunting Behavior during the Later Pleistocene*, Vertebrate Paleobiology and Paleoanthropology, DOI: 10.1007/978-94-007-6766-9_13,

sometimes used shaft fragments (Karlin, personal communication), only cancellous portions are processed for bone grease rendering. However, Delpech and Rigaud (1974) also note the use of long bone shafts by Nunamiut groups; these fragments were submerged in water to make a beverage that was consumed as is.

Few data describe the procedure used to obtain the articular extremities of long bones. When processing marrow bones in their residential camps, the Nunamiut strike long bones as close as possible to the extremities in order to detach the articular portions with as few remaining shaft portions as possible (Binford 1981: 158); the Cree, on the other hand (Bonnichsen 1973), place the articular extremities of long bones on stones and then strike them one or two times in the middle of the shaft. The marrow is thus collected and the shaft fragments attached to the articular extremities are fractured. In all of the recorded cases, the bones used for fat extraction are intensively fragmented through the use of an anvil and percussor: Abe (2005) speaks of fragments that are 1 cm in length, Leechman (1951) of pieces the size of a fingernail, Delpech and Rigaud (1974) of small fragments of epiphyses with an average length of 2 cm, Karlin and Tchesnokov (2007) of a high degree of crushing, and Denys (1672) of a reduction into powder. Higher fragmentation rates, created by increasing the bone surface exposed to boiling water, could result in a more rapid release of grease in a shorter amount of time (Lupo and Schmitt 1997; Saint-Germain 1997). Based on long bone boiling experiments, Church and Lyman (2003) argued that small fragments are in fact not necessary for efficient grease extraction. Nonetheless, the proportion of grease extracted after one hour of boiling is the greatest for the smallest fragments: after one hour, 4 cm long pieces had released 41 % of their fat, while pieces 1 cm long had released 63 %.

Contrary to marrow extraction, bone fat rendering is particularly difficult to detect in archaeological faunal assemblages. Roberts and collaborators (2002) showed that the physicochemical characteristics of bone are not modified (loss of collagen, increased porosity, increased crystallinity) until the bones have been boiled for at least ten hours, which is much longer than the time generally recorded in the ethnographic examples. According to Koon and collaborators (2003), collagen alteration induced by heating at low temperatures can be observed using Transmission Electron Microscopy (TEM). Koon and collaborators (2010) argue that this visualization technique allows us to discriminate between cooked and uncooked bones in recent archaeological contexts. The efficacy of this technique for older sites and on cancellous specimens has not yet been tested, however. For the Paleolithic, due to the lack of durable vessels that can be directly exposed to flames, many researchers have proposed that hot rocks were dropped in a skin or gut container filled with water (Frison 1978; Stiner 2003; Karlin and Tchesnokov 2007). Examples of hot-rock boiling are widely recorded in the ethnographic record (e.g. Ryder 1966, 1969; Vehik 1977; Binford 1978). According to Stiner (2003), three criteria are necessary to identify bone grease rendering at prehistoric sites: (1) intensively fragmented bones; (2) large quantities of heated rocks; (3) anvils with use traces. Few sites have yielded all of these three factors in combination. In the Gravettian at Vale Boi (Portugal), Stiner (2003) noted the presence of fire-cracked rocks and stone anvils with large depressions associated with an intensive fragmentation of cancellous bones. In level 7 of Le Flageolet, the presence of heated rocks and extremely fragmented articular ends, which were concentrated in well delimited zones, led Delpech and Rigaud (1974) to propose that the Gravettian occupants of this site made a fat bouillon. In level IV0 of Pincevent, Julien and Beyries (in Bodu et al. 2006: 83) noted the presence of three sandstone rocks with impact traces that could have "served to fracture bones to collect the marrow or to crush them to extract the fat", while heated limestone pieces could have been used to boil horse bones (March et al. in Bodu et al. 2006: 116). At sites with bison in North America, pits in the ground have been interpreted as evidence of this activity (Logan 1998). Few sites display this type of structure, however, and it is often based on bone residue analysis that bone fat extraction is proposed (Davis and Fisher 1990; Outram 1999; Costamagno and Fano Martínez 2005; Munro and Bar-Oz 2005; Prince 2007).

Few ethnohistoric sources specify the parts and dimensions of the bones that were crushed but most of the examples cited show an intensive crushing of bones before they are dropped in water and boiled, as well as a preferential use of cancellous portions. Shaft fragments were rarely used for bone grease extraction. However, many different taphonomic processes can lead to the fragmentation of cancellous bones and it is very difficult to distinguish post-depositional taphonomic processes from bone grease rendering techniques. For this reason, the comparison of the percentage of preserved cancellous parts with their fragmentation rate (NISP/MNE) proposed by Munro and Bar-Oz (2005) seems of little use as long as we lack a reference base that would enable us to distinguish fresh from post-depositional breakage of cancellous elements. Conversely, the intensive fragmentation of cancellous parts can lead to a problem of differential determination (Lyman and O'Brien 1987; Outram 1999) and result in an under-representation of some skeletal parts. The identification of long bone articular ends may therefore be influenced by this processing technique. An under-representation of long bone epiphyses relative to shaft portions may thus provide an archaeological signature for bone grease rendering. However, numerous other taphonomic processes can also lead to a scarcity of epiphyses, including post-depositional processes (Brain 1969; Behrensmeyer 1975), carnivore gnawing (Brain 1981; Haynes 1983; Hudson 1993), or the use of bone as fuel (Théry-Parisot et al. 2004; Costamagno et al. 2005). It is thus essential to identify the causes of bone preservation/fragmentation to resolve the problems of equifinality (Outram 2001, 2004). Carnivore actions and bone use as fuel are relatively easy to distinguish (e.g. Richardson 1980; Binford 1981; Lyman 1994; Fisher 1995; Costamagno et al. 1998, 2008b). For density-mediated destruction, this distinction is far more problematic, however. As shown by Lyman (1984) and Morin (2010), a very strong negative correlation exists between long bone density and the percentage of fat. Among artiodactyls, bone density is largely correlated with fat content. Therefore, differential preservation linked with density-mediated destruction may mimic a bone grease rendering pattern. The appearance of the bones, the presence of fetal bones or divergent skeletal representations depending on the species, can thus be used to discriminate the taphonomic processes that may have contributed. The bone assemblage of Noisetier Cave enables us to avoid the problems created by the equifinality of these different processes.

Site Presentation

The Mousterian site of Noisetier Cave was first noted by Viré in 1898. Allard finally excavated test pits there in the late 1980s (Allard 1993). Since 2004, Mourre and Thiébaut have been conducting a full research excavation of this site. Noisetier Cave is a small cavity located in the Pyrenees Mountains. It is situated within a limestone cliff at an altitude of 825 m above sea level. The excavations by Allard mainly concerned the central part of the cave. The current excavations, extending over approximately 30 m^2 (Fig. 13.1), have revealed a relatively complex stratigraphic sequence approximately 3 m thick (Fig. 13.2). All of the levels of this sequence have yielded a Mousterian industry characterized by an association of discoid and Levallois debitage, and some Neanderthal human remains (Mourre et al. 2008). Radiocarbon dates have been obtained from wood charcoal fragments originating from an in situ hearth and bone remains from the US0 and levels 1 and 3 (Table 13.1). These dates must be considered with caution, however, as they are at the limit of the radiocarbon method. Based on the abundant microfauna assemblage, all of the levels can nonetheless be attributed to Isotope Stage 3 (Jeannet 2001). The identified animal species (rodents and large mammals) indicate that the deposits were formed during a relatively temperate climatic period (Mourre et al. 2008).

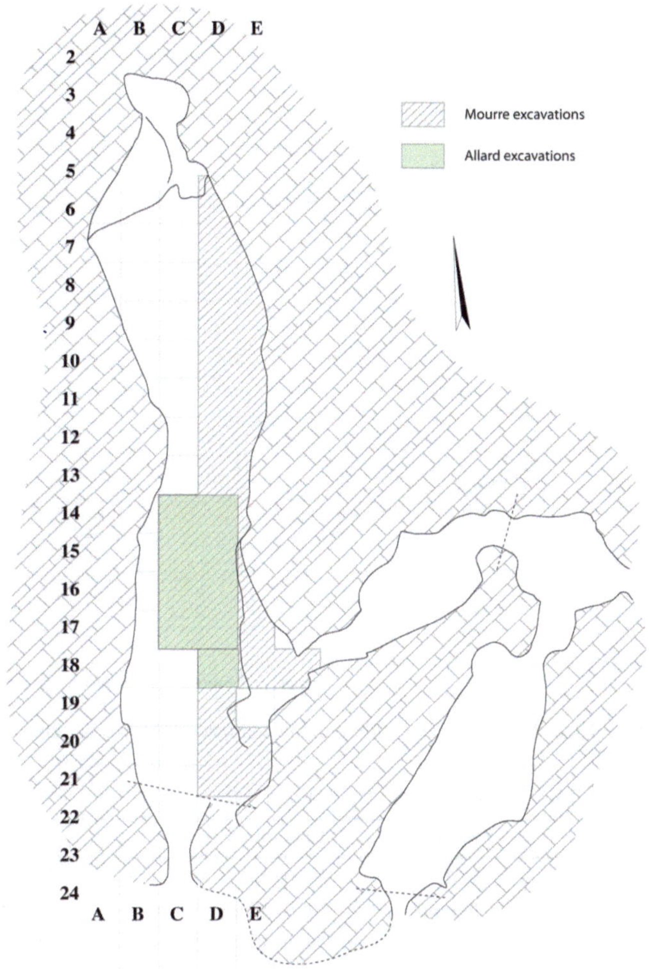

Fig. 13.1 Noisetier cave: zones excavated since 1987 (Mourre *in* Mourre et al. 2010)

The Cartesian coordinates of the archaeological remains collected during both excavations were recorded using the grid installed by Allard. In comparison to the methods employed by Allard, the artifacts recovered during the current excavation are more systematically recorded and the sediments more meticulously sieved. When possible, all of the identifiable bone remains, and all other pieces over 3 cm long, are systematically recorded in three dimensions. All of the sediments are water-sieved through a mesh of 1.6 mm. Once dry, these sieved sediments are sorted to extract the lithic, faunal, microfaunal, malacofaunal and human remains.

Taphonomic analyses have shown that a large portion of the bone assemblage has a non-human origin: the chamois remains (*Rupicapra rupicapra*) are mostly natural, the red deer remains (*Cervus elaphus*) are anthropogenic, and the ibex remains (*Capra ibex* cf. *pyrenaica*) are of a mixed origin (Costamagno et al. 2008a; Mallye et al. 2010).

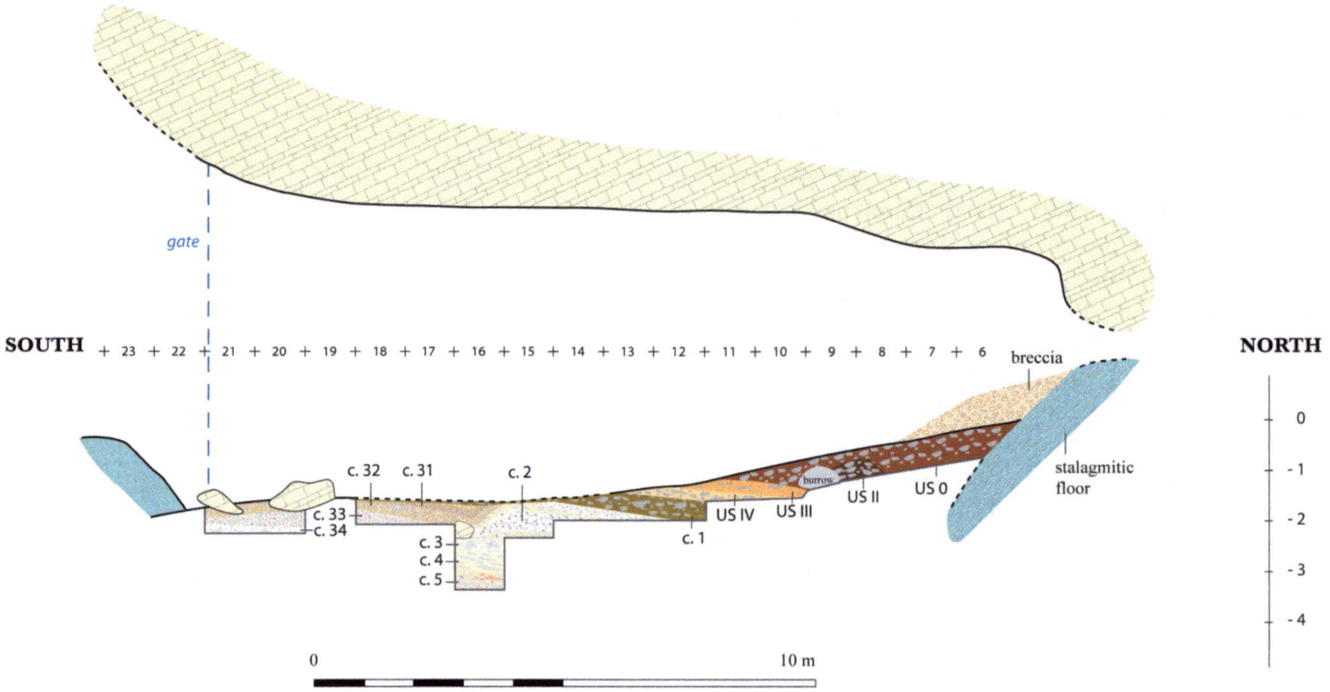

Fig. 13.2 Noisetier cave: stratigraphic sequence (Mourre *in* Mourre et al. 2010)

Table 13.1 Noisetier cave: radiocarbon dates

Layer	Nature	Laboratory number	^{14}C age in years BP	Collagen yield (%)	References
USO	Bone	Poz-14257	29,500 ± 300	4.8	Mourre et al. (2008)
1	Charcoal	Poz-13720	33,700 ± 500	–	Mourre et al. (2008)
1	Charcoal	Poz-13757	31,500 ± 600	–	Mourre et al. (2008)
1	Bone	GIF-7997	42,000 ± 3,100	–	Allard (1993)
3	Bone	Poz-14255	47,000 ± 2,000	0.6	Mourre et al. (2008)

The discovery of dhole[1] (*Cuon alpinicus*) deciduous teeth led us to propose the hypothesis of a cuon den alternating with human occupations (Mallye et al. 2010). This carnivore, rather than the bearded vulture (*Gypaetus barbatus*), as was previously proposed (Costamagno et al. 2008a), would have accumulated the majority of the chamois remains (Mallye et al. 2010). Based on their taphonomic characteristics, these chamois remains could have originated from feces accumulated in the latrine zones, since cuons very rarely bring carcasses into their dens (Fox 1984).

Materials and Methods

Due to a less exhaustive recovery of bone remains, the assemblages originating from the excavations by Allard were not included in this study. The upper levels found in the back of the cave (USO to US IV) were also excluded due to the abundance of cave bear remains (Fig. 13.3). In the test pit, the lower levels (starting with level 4) show a low degree of anthropogenic modification and were thus not included in this analysis. As the other levels displayed clear taphonomic and taxonomic similarities (Costamagno et al. 2010), the bone remains were analyzed with no distinctions between the levels in order to benefit from a sufficiently large sample of red deer.

The assemblage studied in detail is composed of 3,123 remains corresponding to all of the bone remains recorded between 2004 and 2008 (1,558 specimens) and all the taxonomically and anatomically identifiable remains (1,565 specimens[2]) recovered from the sediments sieved during this same period. To make taphonomic and zooarchaeological observations, all the bone surfaces were examined under a low-angled light using a hand lens (magnification: 12×). The criteria used to identify these traces are those

[1] Determination by Boudadi Maligne.

[2] 76 % of these remains are less than 2 cm long.

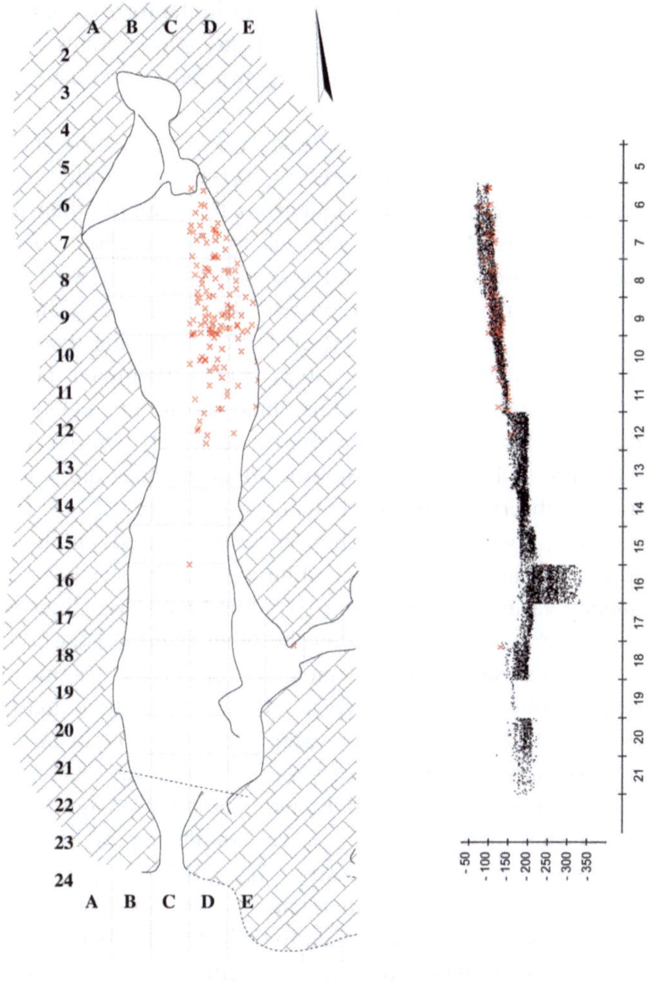

Fig. 13.3 Noisetier cave: spatial distribution of bear remains (*Ursus spelaeus*) (Thiébaut *in* Mourre et al. 2010)

listed by Blumenschine et al. (1996). For the remainder of the sieved materials (i.e. unidentifiable remains), we studied a sample of 20,000 remains. These pieces, nearly 97 % of which are less than 2 cm long, were not subject to a detailed taphonomic analysis. The unidentified bones collected through sieving (20,015 pieces) were sorted into four categories: dental fragments, compact tissue fragments, compact tissue fragments with spongiosa and cancellous bone fragments. The length of these pieces and the presence of burning traces were also recorded. Among the ungulates present, only chamois and red deer are presented in detail in this paper. Ibex, the third most frequent species, was not included in this study due to its mixed origin.

To evaluate the influence of carnivores on the bone assemblage, the percentage of fragments with tooth marks among the recorded remains (number of bone remains with tooth marks/number of bone remains[3] × 100) and the mean percentage of shaft fragments with tooth marks were calculated. The cut marks (number of bone remains with cut marks/number of observable bone remains × 100) and percussion traces (number of bone remains with percussion marks/number of bone remains[4] × 100) present on the recorded remains were also studied. For digestive attacks, all remains were taken into account. To determine whether the fragmentation of long bones occurred on fresh or dry bone, the method defined by Villa and Mahieu (1991) was used. Since burned bones are most abundant among small remains (Stiner et al. 1995; Costamagno et al. 2005), their proportion was estimated based on all the bone specimens (including the fine sieving fraction). The percentage of burned bones (number of burned bones/total number of bone remains × 100) and the percentage of burned cancellous remains (number of burned cancellous remains/number of burned remains × 100) were calculated. To evaluate the relative abundance of skeletal elements or portions rich in cancellous material, the %NNISP,[5] which gives results comparable to the %MAU (Grayson and Frey 2004), was used. I considered only the articular portions of long bones because the discard of vertebrae at a butchery site and their disappearance in association with problems of determination and/or preservation is extremely difficult to distinguish (Marean and Frey 1997; Costamagno 2004; Costamagno and Fano 2005). These skeletal part profiles were subsequently analyzed using bone mineral densities calculated by Computed Tomography (Lam et al. 1999). The correlation coefficients calculated are Spearman's correlation coefficients.

Results

The faunal assemblage of Noisetier Cave is largely dominated by chamois (65.2 %), followed by red deer (12.3 %) and ibex (9 %). Carnivores represent 4.3 % of the assemblage (Table 13.2). The faunal remains are generally very well preserved. Tooth marks are present on 3 % of the remains, and cut marks are present on 11 %. Digestion stigmata were observed on 40 % of the bone remains (Table 13.3). With regards to taphonomic and zooarchaeological observations, there is a clear difference between red deer and chamois. The red deer remains are characterized by a large number of cut marks and low number of digestion

[3] Tooth fragments excluded.

[4] Tooth fragments excluded.

[5] The Normed NISPs (NNISP) represent "the skeletal part NISP values divided by the number of times the relevant part occurs in the skeleton of the animal involved" (Grayson and Frey 2004, p. 31). For %NNISP, the highest NNISP is set to 100 % and the rest of the NNISP values are scaled to that figure.

Table 13.2 Noisetier cave: taxonomic abundance

Taxa	NISP	%NISP
Ursus spelaeus	5	0.3
Panthera pardus	3	0.2
Canis lupus	5	0.3
Cuon alpinus	12	0.7
Vulpes vulpes	33	1.8
Canidae	17	0.9
Putorius putorius	1	0.1
Bovinae	28	1.6
Capra ibex cf. *pyrenaica*	162	9.0
Rupicapra rupicapra	1,170	65.2
Caprinae	32	1.8
Cervus elaphus	221	12.3
Capreolus capreolus	1	0.1
Cervidae	4	0.2
Equus caballus	2	0.1
Lepus sp.	20	1.1
Marmota marmota	78	4.3

marks (Tables 13.3 and 13.4). More than 73 % of the chamois remains, on the other hand, are digested and few display evidence of human activities (2.7 %). Among the chamois middle shaft long bones, 19.6 % have cut marks and 17 % are digested. For chamois, the epiphyses are characterized by a digestion rate varying between 74 and 100 % (Fig. 13.4). Among the long bones, between 6 and 40 % of the middle shaft bones are digested. Table 13.5 shows the percentage of digested bones by size class for chamois and red deer. Even among the small fragments, the red deer bones rarely display traces of digestion.

Carnivore tooth marks are relatively rare on both of these species (Tables 13.3 and 13.4). For red deer, they were recorded on 13 bone specimens mostly composed of rib and middle shaft fragments. Only 1.9 % of the middle shaft bones of red deer, and none of the epiphyses, have tooth marks. For Chamois, the percentages are respectively 5.7 and 2 %. These light bite marks occur in the form of pitting or scoring. No whole red deer or chamois long bones were found and nearly 90 % of the fracture edges indicate the fracturing of fresh bone.

The skeletal profiles of these two species also show clear differences. Fragments of vertebrae, girdles and phalanges are much more numerous for chamois than for red deer (Fig. 13.5). The same is true for compact bone (carpals, patella, tarsals and sesamoids). Along with the phalanges, these pieces represent 46 % of the chamois remains versus 8 % for red deer. For chamois, these pieces nearly always display digestion marks (compact bone: 88 %, phalanges: 83 %). Other than the femur and tibia, both species are characterized by relatively similar frequencies of long bones. Significant differences are nonetheless observable in the skeletal portions (Fig. 13.6). For red deer, regardless of the long bone considered, epiphysis fragments are scarce (depending on the bone, this frequency varies between 2 and 16 %) and are never digested (Fig. 13.6a). For chamois, on the other hand, all the long bones are mostly represented by epiphysis fragments (Fig. 13.6b), 80 % of which display digestion marks (Fig. 13.4). Bone density has been used as a proxy measure of resistance to destructive processes. For red deer, the scatterplot shows a positive and statistically significant relationship between the relative abundance of skeletal portions and their density ($r_s = 0.720$, $p < 0.01$) (Fig. 13.7). The relative abundance of chamois long bone portions, however, is not linked to their density, as is shown by the Spearman correlation coefficient ($r_s = 0.009$) (Fig. 13.8). The spatial analysis of the bone remains recorded during excavation shows an identical spatial dispersion for red deer and chamois (Fig. 13.9).

Among the specimens collected through sieving, compact bones, with or without cancellous tissue, are dominant (90.4 %); cancellous fragments represent less than 9.3 % of the remains and tooth fragments are extremely rare (0.2 %). Less than 8 % of the bone remains are burned and only 7.7 % correspond to cancellous bone fragments (Fig. 13.10).

Table 13.3 Noisetier cave: frequency of tooth-marked, cut-marked, percussion-marked and digested bones

	Tooth-marked		Digested		Cut-marked		Percussion-marked	
Red deer	2.9	(6/210)	3.2	(7/221)	40.6	(76/187)	13.8	(29/210)
Chamois	1.2	(14/1126)	73.2	(857/1,170)	2.7	(14/524)	0.5	(6/1126)

Table 13.4 Noisetier cave: frequency of tooth-marked, cut-marked, percussion-marked and digested bones on red deer and chamois middle shaft fragments

	Tooth-marked		Digested		Cut-marked		Percussion-marked	
Red deer	1.9	(2/105)	0	(49/95)	51.6	(49/95)	20.9	(22/105)
Chamois	5.7	(3/53)	17.0	(9/53)	19.6	(9/46)	9.4	(5/53)

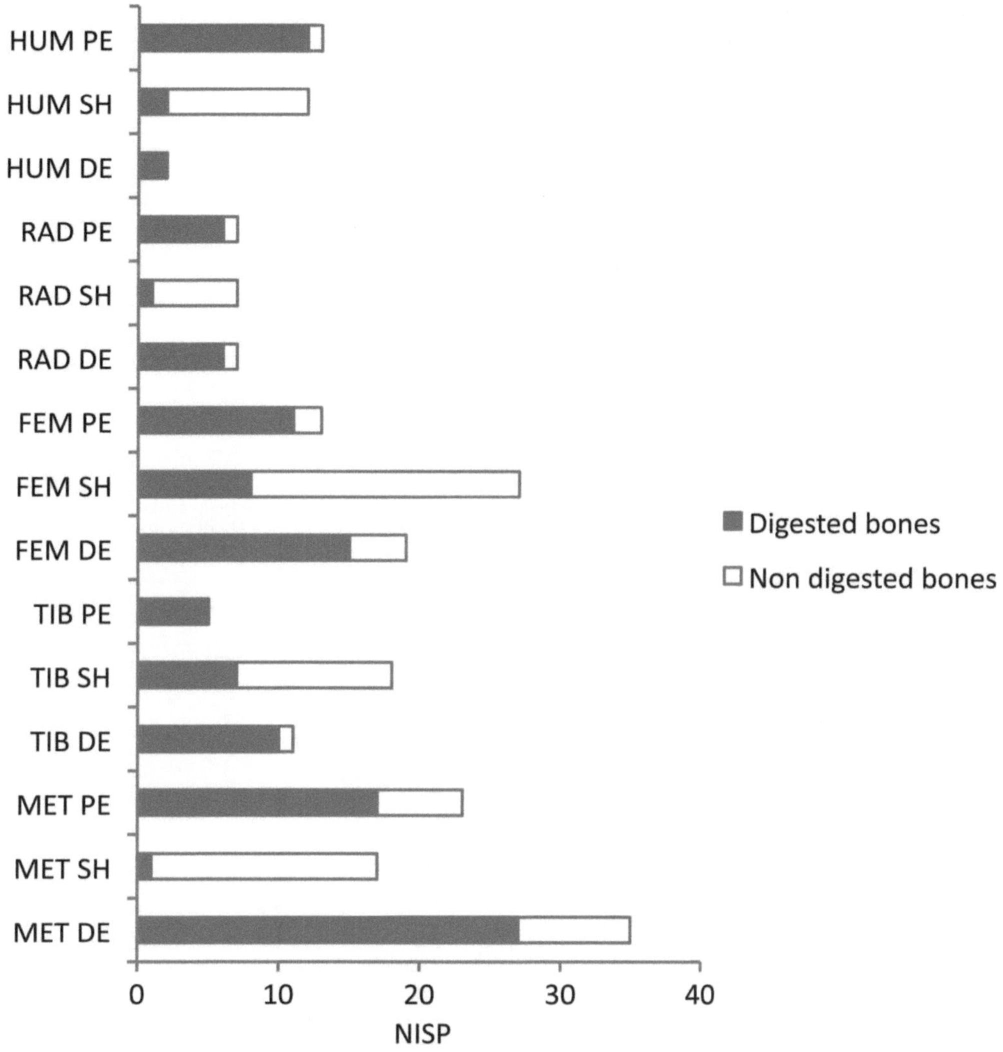

Fig. 13.4 Noisetier cave chamois: digested and non digested bones per long bone portions

Table 13.5 Noisetier cave red deer and chamois: digestion frequency per bone fragment size (length in cm)

	Chamois			Red deer		
	Digested bones	Non digested bones	Digested (%)	Digested bones	Non digested bones	Digested (%)
0–1	90	21	81.1	0	0	0
1–2	598	163	75.6	2	7	22.2
2–3	148	67	68.8	3	16	15.8
3–4	20	21	48.8	2	23	8.0
4–5	1	15	6.3	0	31	0
>5	0	24	0	0	137	0

Discussion

Based on the abundance of cut marks on red deer bones and the high percentage of digested chamois bones, Costamagno et al. (2008a) and Mallye et al. (2010) argued that the red deer specimens were accumulated by humans, whereas most of the chamois specimens were accumulated by the actions of a non human predator.

The scarcity of some skeletal parts suggests that the red deer were hunted by Neanderthals who brought them into the cave in portions, which they then defleshed, as is indicated by the abundance of cut marks. The absence of complete long bones, the presence of impact points on the shaft fragments, and the fracture edges indicating that mostly fresh bone was fractured, are all evidence that long bones were fragmented to extract their marrow. The red

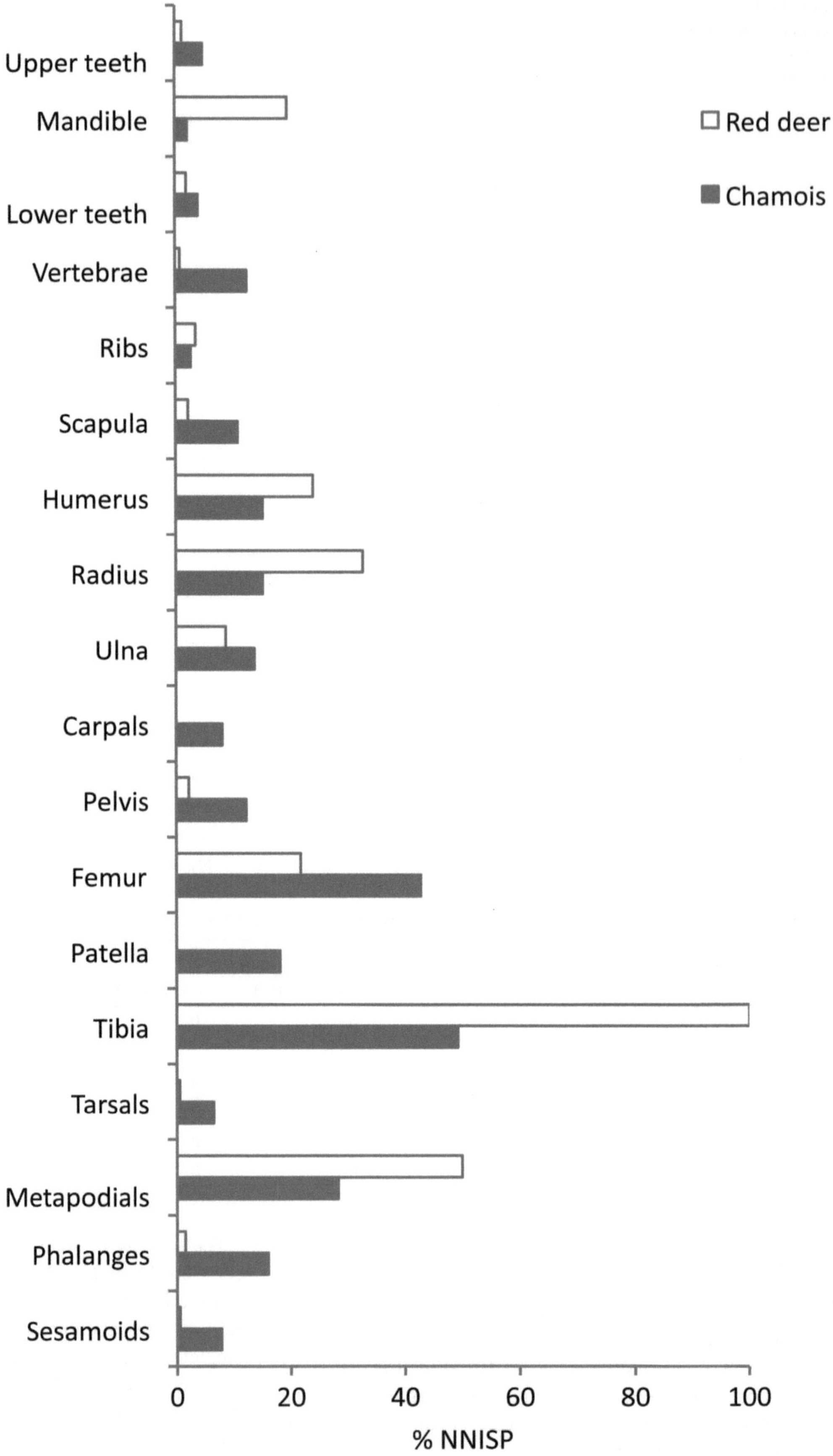

Fig. 13.5 Noisetier cave red deer and chamois: scaled normed NISP (%NNISP) per skeletal part

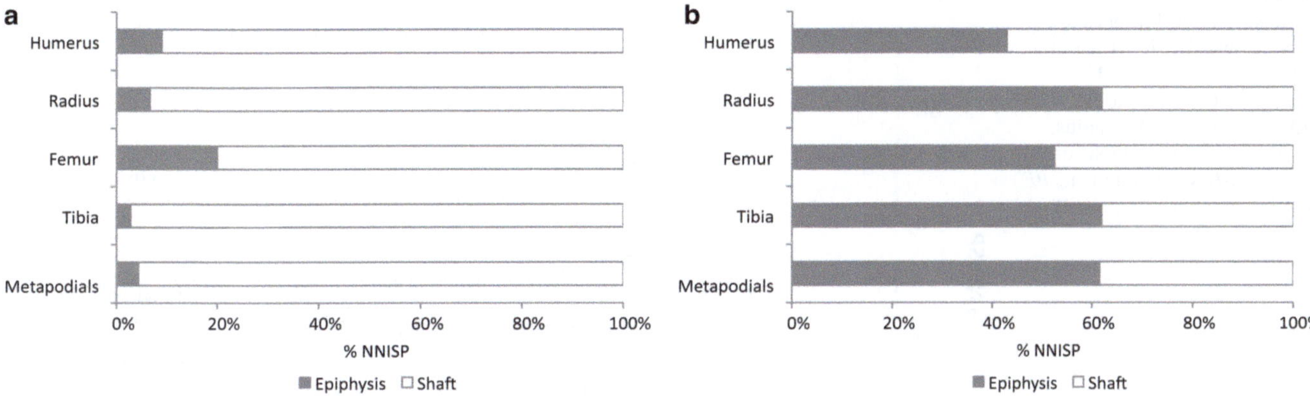

Fig. 13.6 Noisetier cave: relative abundance of epiphysis and shaft fragments per long bone. **a** Red deer. **b** Chamois

Fig. 13.7 Noisetier cave red deer: relationship between %NNISP and bone mineral density (Lam et al. 1999) (*FEM* femur, *HUM* humerus, *MCM* metacarpal, *MTM* metatarsal, *RAD* radius, *TIB* tibia, *DE* distal epiphysis fragment, *SH* shaft fragment, *PE* proximal epiphysis fragment)

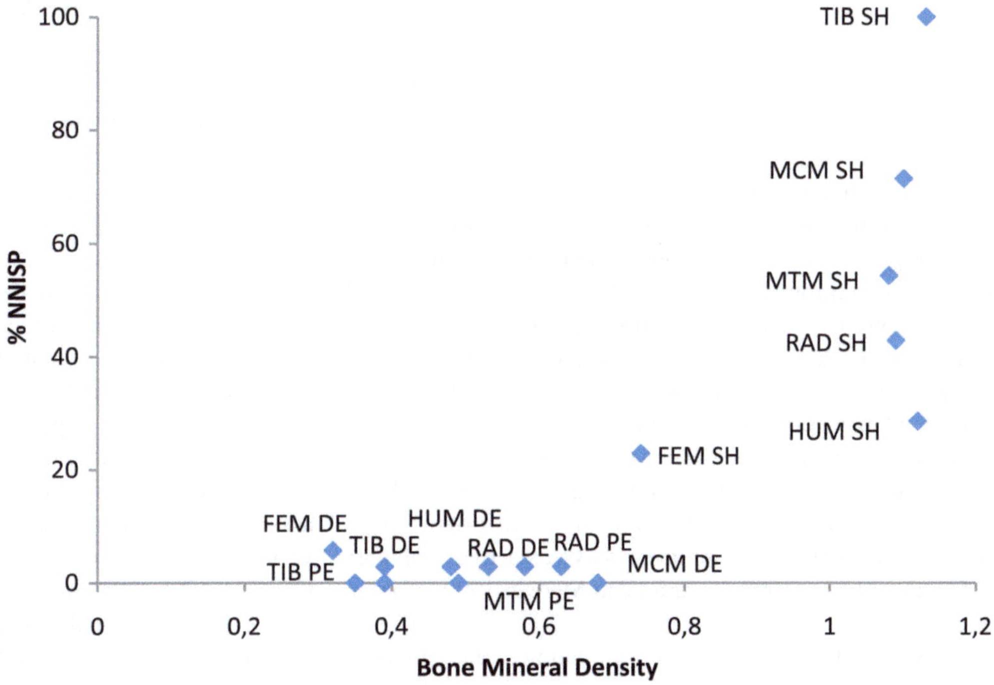

deer remains are also characterized by a clear under-representation of epiphysis portions relative to long bone shafts.

The comparisons made by Mallye et al. (2010) show numerous similarities between the chamois bones accumulated in Noisetier Cave and bones originating from canid scat (Barja Nuñez and Corona-M. 2007; Esteban-Nadal et al. 2010; Mallye et al. unpublished). The percentage of digested bones provides a strong argument for a coprocoenosis. The small size of the bones (97 % of the chamois remains are less than 4 cm long), the low percentage of tooth marks and the identified skeletal elements are also characteristic of bones originating from scat (see Mallye et al. 2010 for a detailed discussion).

Conversely, the hypothesis that the chamois carcasses were scavenged after being processed by Neanderthals is difficult to support based on the available data. If we take into account the remains under 4 cm long—the maximum size of bones generally ingested by canids (Esteban-Nadal 2010)—a clear dichotomy in terms of digestion is still evident between chamois (75.9 %) and red deer (13.2 %) (Table 13.5). In addition, while the under-representation of digested red deer phalanges could be explained by differential transport, the absence of digested red deer long bones indicates clear differences between the skeletal profiles of red deer and chamois. From both a taphonomic and skeletal point of view, the two species show distinct patterns that allow us to exclude the hypothesis of an identical

Fig. 13.8 Noisetier cave chamois: relationship between %NNISP and density (Lam et al. 1999) (*FEM* femur, *HUM* humerus, *MCM* metacarpal, *MTM* metatarsal, *RAD* radius, *TIB* tibia, *DE* distal epiphysis fragment, *SH* shaft fragment, *PE* proximal epiphysis fragment)

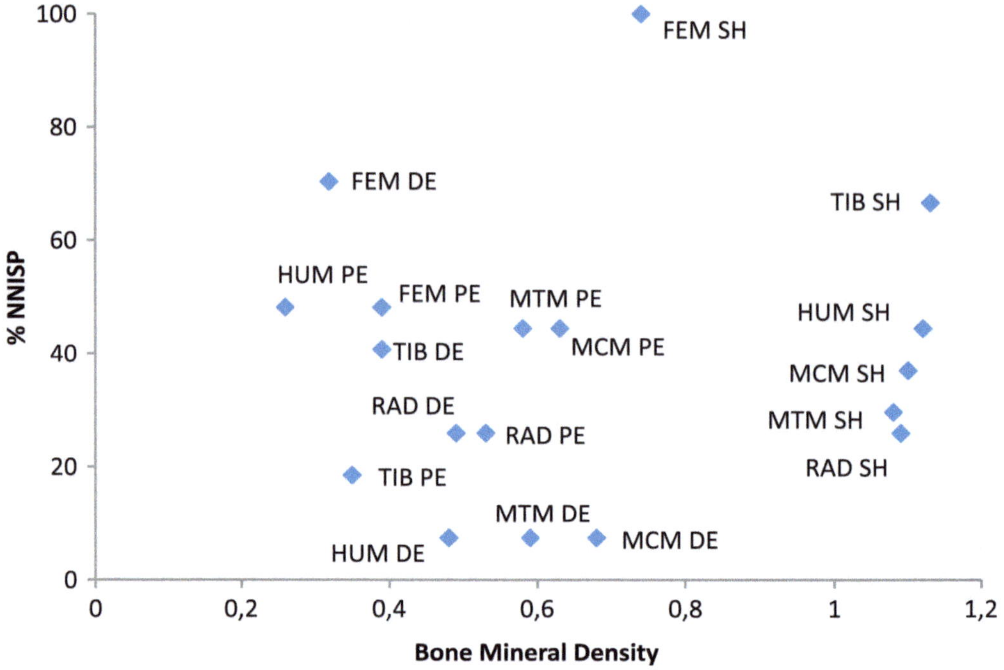

taphonomic history. A few chamois remains with characteristics similar to those of the red deer remains (mainly shaft fragments with or without traces) appear nonetheless to have an anthropogenic origin. The extreme rarity of red deer epiphyses raises interesting questions. Numerous studies have shown that when carnivores had access to carcasses already exploited by humans, they preferentially consumed the epiphyseal fragments due to their high fat content (Blumenschine 1988; Marean and Spencer 1991; Marean et al. 1992; Blumenschine and Marean 1993; Faith et al. 2007). These fragments are therefore often underrepresented in bone assemblages scavenged by carnivores (Marean and Kim 1998; Bartram and Marean 1999; Marean et al. 2000).

Most of the experiments simulating dual-patterned archaeofaunal assemblages (carcass processing by humans followed by scavenging by carnivores) have been conducted on hyenas (Blumenschine 1988; Marean et al. 1992; Capaldo 1998; Domínguez-Rodrigo 1997). Lacking a reference base on canids, I used the data available for hyenas to evaluate the pre-depositional ravaging of red deer bones by carnivores. At Noisetier Cave, tooth marks are present on less than 3 % of the red deer remains. This percentage is well below the values recorded for modern reference collections of carcasses exploited by humans and then scavenged by carnivores (Blumenschine 1988 (tooth-marked on long bones: 15 %); Capaldo 1998 (tooth-marked: 23.2 %)). If we consider the mid-shaft fragments, this percentage (1.9 %) is still very low compared to that recorded for assemblages secondarily scavenged by carnivores, which varies between 6.4 and 15.4 % (Blumenschine 1995; Capaldo 1998; Marean et al. 2000; Lupo and O'Connell 2002). The epiphyseal to shaft fragment ratio of red deer at Noisetier Cave is very low (0.07), corresponding to experiments in which epiphyseal loss is the highest (Seregeti fat-rich experiments, Blumenschine and Marean 1993). At these simulation sites, the proportion of tooth marks on mid-shaft fragments is significantly higher (10 %) than on the

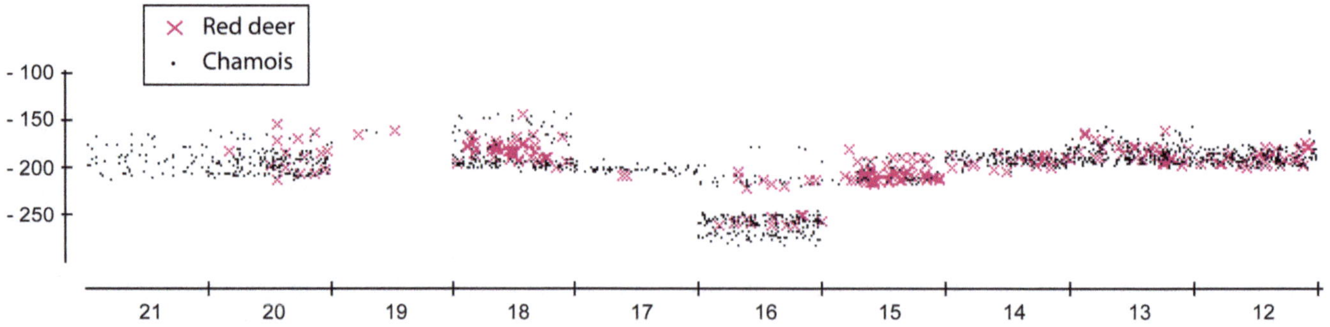

Fig. 13.9 Spatial distribution of red deer and chamois

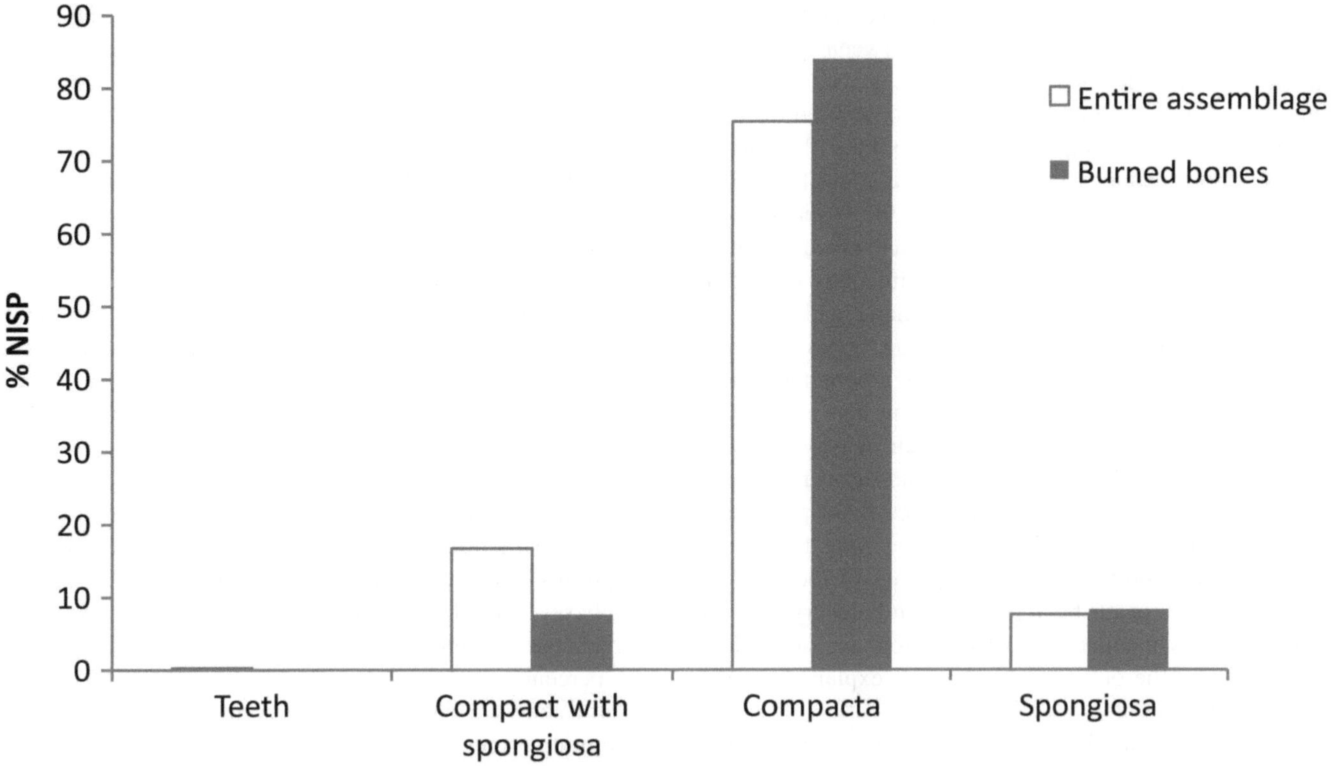

Fig. 13.10 Noisetier cave: relative proportions of bone tissue types for the entire bone assemblage and for the burned bones

red deer bones from Noisetier Cave. In destroying epiphyses, carnivores produce tooth-marked near-epiphyseal fragments in high frequencies (Blumenschine and Marean 1993). In Serengeti fat-rich assemblages, 60 % of the epiphyseal fragments are tooth marked, while at Noisetier Cave less than 5 % display tooth marks. Finally, in actualistic studies, though most of the epiphyses are swallowed by carnivores, the remaining epiphyses are commonly tooth marked, at a rate ranging from 40 to 100 % (Blumenschine and Marean 1993). At Noisetier Cave, none of the recovered red deer epiphyses have such marks. A reference base concerning canid scavenging of human-butchered assemblages would be useful from a comparative point of view, but based on the epiphyseal to shaft fragment ratio, the degree of involvement of canids should have been high for the red deer bones at Noisetier Cave. Even if we can imagine that tooth mark frequencies would be lower in assemblages scavenged by canids, it seems highly unlikely that frequencies this low could result in such a high intensity of epiphyses destruction. The hypothesis that the epiphyseal fragments are due only to ravaging by carnivore scavenging is thus difficult to support.

The use of bone as fuel is another possible cause. Several experiments have shown that: (1) the elements rich in cancellous tissue constitute effective combustion materials, in contrast to the portions composed of compact tissue[6] (Costamagno et al. 1998, 2008b; Villa et al. 2002; Théry-Parisot and Costamagno 2005; Théry-Parisot et al. 2005); and (2) the bone residues originating from intentional burning are always characterized by intensive fragmentation (Théry-Parisot et al. 2004; Costamagno et al. 2005, 2010). Consequently, a preferential use of epiphyses for burning can lead to a bias of long bone portions in favor of shafts (Lyman and O'Brien 1987). At Noisetier Cave, burned bones constitute only a very small proportion of the bone remains. In addition, among the burned bones, there are very few cancellous fragments and their relative abundance per tissue type attests to an accidental burning of remains in contact with fire (Fig. 13.10) (Costamagno et al. 2008b). The hypothesis that an intentional use of bone as fuel is the cause of the under-representation of long bone epiphyses is thus not validated.

Among the other factors that can lead to an under-representation of epiphyses, climatic-edaphic factors (i.e. duration of exposure of the bones before burial, acidity of the surrounding sedimentary matrix or water circulation) must not be neglected (Behrensmeyer 1978; Lyman 1984, 1994; Andrews 1995). As they are difficult to quantify, their

[6] When the marrow is extracted.

action is often supposed when a positive and statistically significant relationship is revealed between the relative abundance of bones and their density. At Noisetier Cave, this type of correlation is observable for the red deer long bones (Fig. 13.7). Nonetheless, if we take the chamois bones into account, it appears that density-mediated destruction played only a minor role in the formation of the bone assemblage. Though the red deer and chamois remains were accumulated within the same sedimentary matrix, and were thus subjected to identical climatic and diagenetic agents, the epiphyses of red deer long bones are nearly absent, while they are clearly dominant among the chamois remains, and there is thus no correlation between the relative abundance of the portions and their density. The over-representation of chamois proximal and distal femur articular extremities is thus significant since these portions are among of the least dense parts of the long bones. It is therefore more probable that the scarcity of red deer long bone epiphyses is attributable to human butchery processes than to density-mediated destruction. Bone grease rendering then becomes the only process that can explain the described pattern.

The Mousterian site of Les Pradelles shows the same pattern as that of Noisetier Cave. Excavated from 1967 to 1980 by Vandermeersch, and since 2001 by Maureille and Mann, Les Pradelles (Marillac-le-Franc, Charente) is a collapsed gallery within a karstic system (Maureille et al. 2007). Its deposits yielded a Quina industry contemporaneous with the last glaciation cycle (Meignen et al. 2007). In the lowest levels, Facies 2, the faunal spectrum is largely dominated by reindeer and has an exclusively anthropogenic origin (Costamagno et al. 2006; Rendu et al. 2012). The analysis of the skeletal part representation shows a high frequency of long bone fragments, which does not seem to be associated with taphonomic problems as the bones are well preserved and the frequency of carnivore modifications is extremely low (2.7 %). The carcasses were thus introduced into the site as incomplete portions, with a preference for marrow-rich long bones (Costamagno et al. 2006; Rendu et al. 2012). As at Noisetier Cave, long bone epiphyses are poorly represented. Shaft fragments represent 95 % of the long bone sample and epiphysis fragments are on average less than 2 cm long (Fig. 13.11). The under-representation of long bone epiphyses cannot be linked to carnivore actions or to density-mediated destruction since the bones are very well preserved. No burned bones have been found at Les Pradelles. Therefore, as at Noisetier Cave, the under-representation of reindeer epiphyses may be attributable to bone grease rendering.

Noisetier Cave and Les Pradelles are not the only Mousterian sites that display this pattern. For level II of Roc de Marsal I, excavated by Dibble and Turq since 2004 (tooth-marked: 0.3 %, no burned bones), Soulier (2007) proposed bone grease rendering to explain the scarcity of reindeer long bone epiphysis (i.e. 4.6 %). Castel personal communication also observed this pattern in the lower levels of this site. According to Deaujard (2008), the red deer in level 7 of Saint-Marcel[7] (i.e. 7.1 %) and Ibex in levels b and c–d of Les Peyrards[8] (i.e. 18.7 and 15.4 %) may have been processed in the same way. All of these assemblages are characterized by a relatively low ratio of long bone epiphyses. Not all Mousterian sites are characterized by this pattern, however, as is shown at the site of Payre, level Fa, which has a high ratio of long bone epiphyses (i.e. 49.2 %).

The data presented in this paper are based on the NISP since the MNE has not yet been calculated for Noisetier Cave and Les Pradelles. One could therefore object that the high frequency of shaft fragments could be associated with marrow extraction. In the context of this activity, long bone diaphyses indeed yield many more fragments than long bone epiphyses. To test this assumption for the available sites, the percentage of epiphyses based on the NISP was compared to the percentage of epiphyses based on the MNE. The scatterplot shows a strong positive correlation between both quantitative units (Fig. 13.12). Sites where epiphysis fragments are scarce are also characterized by an under-representation of MNE epiphyses. The quantitative unit used for the analyses has no influence on the observed pattern.

No fire-cracked rocks have been found at any of the Mousterian sites discussed in this paper. Given the early date of the excavations at La Baume des Peyrards and Saint-Marcel, this absence could be linked to the excavation methods employed. At Les Pradelles, Noisetier Cave and Roc-de-Marsal, on the other hand, this is not the case. It is thus difficult to consider the possibility of hot-rock bone-boiling technologies at these three sites. Ethnological studies have shown that some hunter-gatherers chew long bone ends after marrow extraction (Schaefer and Steckle 1980; Oliver 1993). This activity produces tooth marks, but the epiphyses are not destroyed. Grinding the bones of small prey to ease consumption has also been recorded (Yohe et al. 1991; Reinhard et al. 2007; Sobolik 2008). According to Brugal (personal communication), Maasai peoples may entirely consume smashed bones. Given the need for fat in cold temperate, subarctic, and arctic environments, Marean (2005) proposed the hypothesis that Neanderthals lacking a hot-rock technology swallowed crushed cancellous bones. At Noisetier Cave, cobbles with percussion traces could be evidence of the crushing of cancellous portions. Could the ingestion of fragments of crushed cancellous bones by Neanderthals explain the scarcity of long bone epiphyses at

[7] Excavated by Gilles from 1977 to 1988.

[8] Excavated by de Lumley from 1955 to 1969.

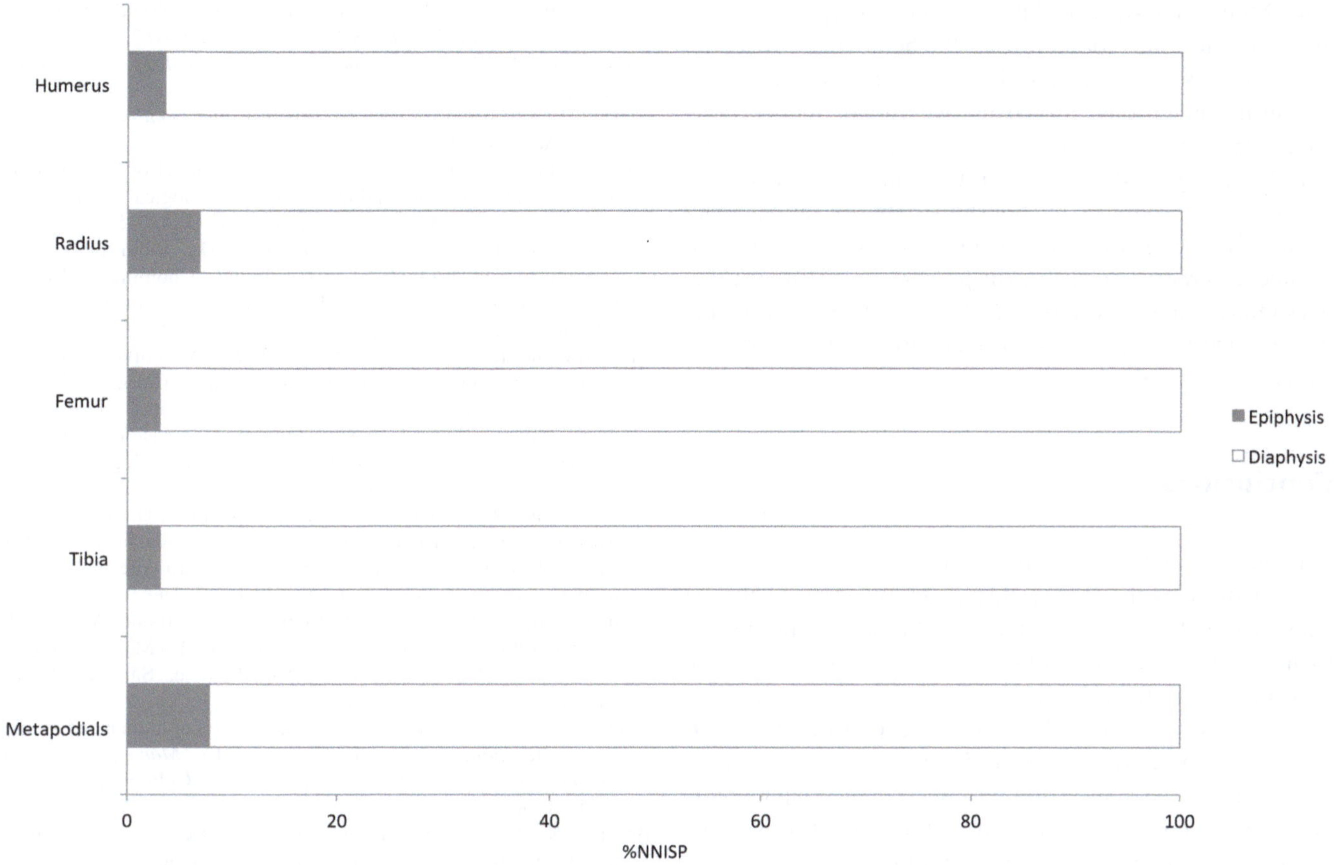

Fig. 13.11 Les Pradelles. Relative abundance of epiphysis and shaft fragments per long bone

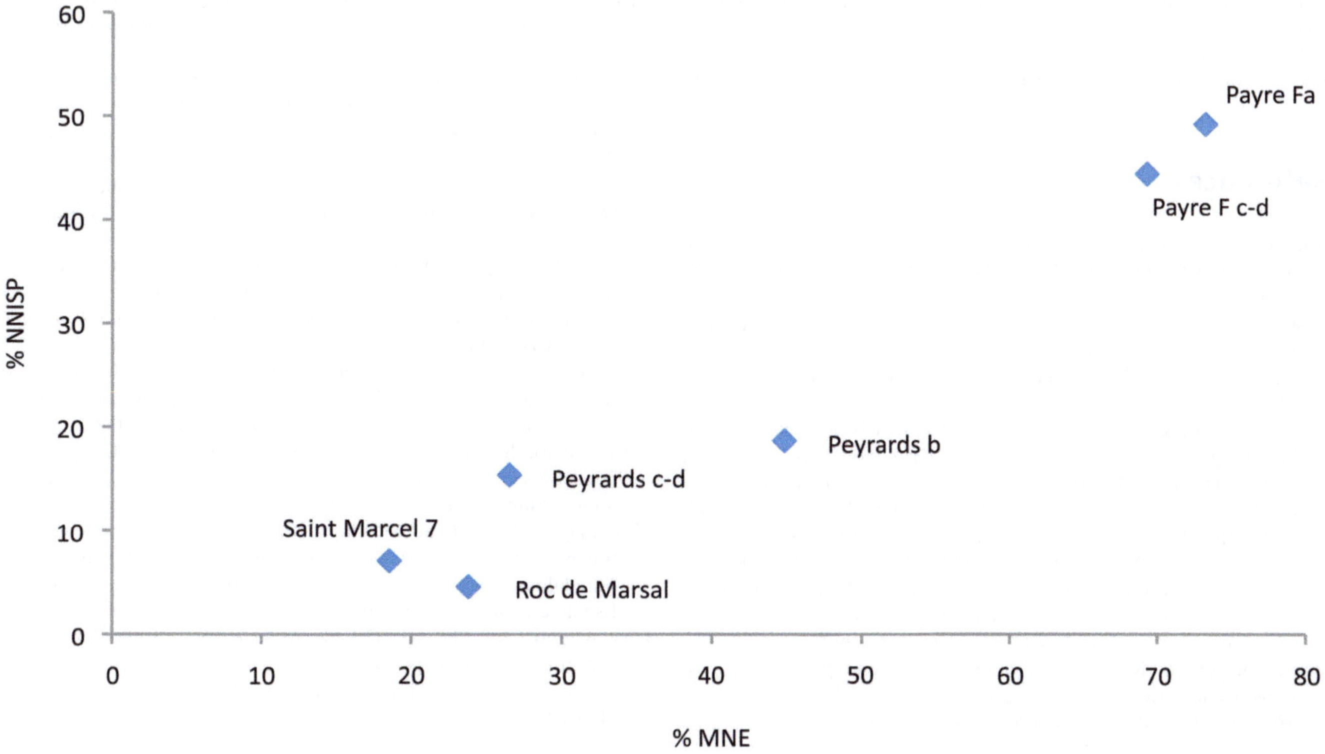

Fig. 13.12 The relationhip between %NNISP and %MNE for the ratio epiphyses/diaphyses (data from Soulier 2007 and Daujeard 2008)

some Mousterian sites? Did the resulting "bony paste" have the same capacities for storage as the bone grease rendered by boiling? Without nutritional studies, it is difficult to draw any firm conclusions. Moreover, we cannot reduce bone grease rendering to seasonal stress or resource intensification. Ethnological data on subarctic hunter-gatherers show that there are complex relationships between humans and animal fat: fat is valuable not just for food, but also for technical purposes such as skin processing, or ritual practices (Karlin and Tchesnokov 2007). It is therefore possible that Neanderthals used this substance for technical, or even cultural, purposes.

Conclusions

Taphonomic studies of the bone remains at Noisetier Cave support the hypothesis that Neanderthals extracted the fat contained within cancellous tissues. This study thus appears to indicate that as early as MIS 3, Neanderthals used technical procedures that enabled them to exploit this substance, whose extraction is known to be difficult. However, since numerous taphonomic processes can result in an under-representation of cancellous elements and/or portions, detailed taphonomic analyses must be made of each site before conclusions can be made concerning this type of exploitation.

Acknowledgments I wish thank to Céline Thiébaut and Vincent Mourre for the spatial distribution figures. The non recorded bone objects were counted by Nadia Cavanhié and Laurianne Streit thanks to financing from the French Ministry of Culture. My sincere gratitude to John Speth for all the constructive discussions that we have had on this subject. The translation of this paper was realized by Magen O'Farrell.

References

Abe, Y. (2005). *Hunting and butchering patterns of the Evenki in the Northern Transbaikalia Russia.* Ph. D., Stony Brook University, New-York.

Allard, M. (1993). Fréchet-Aure, Grotte du Noisetier. *Bilan scientifique de la région Midi-Pyrénées, 1992*, 113–114.

Andrews, P. (1995). Experiments in taphonomy. *Journal of Archaeological Science, 22*, 147–153.

Barja Nuñez, I., & Corona-M. E. (2007). El análisis de excretas desde la etología y la arqueozoología. El caso del lobo ibérico. In E. Corona-M. & J. Arroyo Cabrales (Eds.), *Human and faunal relationships reviewed: An archaeozoological approach* (Vol. 1627, pp. 113–121). Oxford: BAR International Series.

Bartram, L. E., & Marean, C. W. (1999). Explaining the "Klasies patterns": Kua ethnoarchaeology, the Die Kelders Middle Stone Age archaeofauna, long bone fragmentation, and carnivore ravaging. *Journal of Archaeological Science, 26*(1), 9–29.

Behrensmeyer, A. K. (1975). The taphonomy and paleoecology of Pleistocene vertebrate assemblages of Lake Rudolf. *Bulletin of the Museum of Comparative Zoology, 146*, 473–578.

Behrensmeyer, A. K. (1978). Taphonomic and ecologic information from bone weathering. *Paleobiology, 4*(2), 150–162.

Binford, L. R. (1978). *Nunamiut ethnoarchaeology.* New York: Academic Press.

Binford, L. R. (1981). *Bones: Ancient men and modern myths.* New York: Academic Press.

Blumenschine, R. J. (1988). An experimental model of the timing of hominid and carnivore influence on archaeological bone assemblages. *Journal of Archaeological Science, 15*, 483–502.

Blumenschine, R. J. (1995). Percussion marks, tooth marks, and experimental determinations of the timing of hominid and carnivore access to long bones at FLK *Zinjanthropus*, Olduvai Gorge, Tanzania. *Journal of Human Evolution, 29*, 21–51.

Blumenschine, R. J., & Marean, C. W. (1993). A carnivore's view of archaeological bone assemblages. In J. Hudson (Ed.), *From bones to behavior: ethnoarchaeological and experimental contributions to the interpretation of faunal remains* (Vol. Occasional Paper No. 21, pp. 273–300). Carbondale: Center for Archaeological Investigations, Southern Illinois University.

Blumenschine, R. J., Marean, C. W., & Capaldo, S. D. (1996). Blind tests of interanalyst correspondence and accuracy in the identification of cut-marks, percussion marks and carnivore tooth marks on bone surface. *Journal of Archaeological Science, 23*, 493–508.

Bodu, P., Julien, M., Valentin, B., Debout, G., Averbouh, A., Bemilli, C., et al. (2006). Un dernier hiver à Pincevent: les Magdaléniens du niveau IV0 (Pincevent, La Grande Paroisse, Saine-et-Marne). *Gallia Préhistoire, 48*, 1–180.

Bonnichsen, R. (1973). Some operational aspects of human and animal bone alteration. In B. M. Gilbert (Ed.), *Mammalian osteoarchaeology: North America* (pp. 9–24). Columbia: Missouri Archaeological Society.

Brain, C. K. (1969). The contribution of Namib Desert Hottentots to an understanding of australopithecine bone accumulations. *Scientific Papers of the the Namib Desert Research Station, 39*, 13–22.

Brain, C. K. (1981). *The hunters or the hunted? An introduction to African cave taphonomy.* Chicago: University of Chicago Press.

Brink, J. W. (1997). Fat content in leg bones of *Bison bison*, and applications to archaeology. *Journal of Archaeological Science, 24*, 259–274.

Bunn, H. T., Bartram, L. E., & Kroll, E. M. (1988). Variability in bone assemblage formation from Hadza hunting, scavenging, and carcass processing. *Journal of Anthropological Archaeology, 7*(4), 412–457.

Capaldo, S. D. (1998). Simulating the formation of dual-patterned archaeofaunal assemblages with experimental control samples. *Journal of Archaeological Science, 25*, 311–330.

Church, R. R., & Lyman, R. L. (2003). Small fragments make small differences in efficiency when rendering grease from fractured artiodactyl bones by boiling. *Journal of Archaeological Science, 30*(8), 1077–1084.

Costamagno, S. (2004). Facteurs taphonomiques influant sur la représentation différentielle des éléments squelettiques des animaux chassés. *Les Nouvelles de l'Archéologie, 95*(1), 6–9.

Costamagno, S., & David, F. (2009). Comparaison des pratiques bouchères et culinaires de différents groupes sibériens vivant de la renniculture. *Archaeofauna, 19*, 9–25.

Costamagno, S., & Fano Martínez, M. A. (2005). Pratiques cynégétiques et exploitation des ressources animales dans les niveaux du Magdalénien supérieur-final de El Horno (Ramales, Cantabrie, Espagne). *Paleo, 17*, 31–56.

Costamagno, S., Griggo, C., & Mourre, V. (1998). Approche expérimentale d'un problème taphonomique: utilisation de combustible osseux au Paléolithique. *Préhistoire Européenne, 13*, 167–194.

Costamagno, S., Théry-Parisot, I., Brugal, J.-P., & Guibert, R. (2005). Taphonomic consequences of the use of bones as fuel. Experimental data and archaeological applications. In T. O'Connor (Ed.), *Biosphere to lithopshere. New studies in vertebrate taphonomy* (pp. 51–62). Oxford: Oxbow Books, Actes du 9° Colloque de l'ICAZ, Durham, août 2002.

Costamagno, S., Meignen, L., Beauval, C., Vandermeersch, B., & Maureille, B. (2006). Les Pradelles (Marillac-le-Franc, France): A Mousterian reindeer hunting camp? *Journal of Anthropological Archaeology, 25*, 466–484.

Costamagno, S., Robert, I., Laroulandie, V., Mourre, V., & Thiébaut, C. (2008a). Rôle du gypaète barbu (*Gypaetus barbatus*) dans la constitution de l'assemblage osseux de la grotte du Noisetier (Fréchet-Aure, Hautes-Pyrénées, France). *Annales de Paléontologie, 94*(2), 245–265.

Costamagno, S., Théry-Parisot, I., Castel, J.-C., & Brugal, J.-P. (2008b). Combustible ou non? analyse multifactorielle et modèles explicatifs sur les ossements brûlés paléolithiques. In I. Théry-Parisot, S. Costamagno, & A. Henry, (Eds.), *Gestion des combustibles au Paléolithique et au Mésolithique: nouveaux outils, nouvelles interprétations. Actes du XVème Congrès de l'IUSPP, Lisbonne, septembre 4–9, 2006* (pp. 69–84). Oxford: BAR International Series.

Costamagno, S., Théry-Parisot, I., Kuntz, D., Bon, F., & Mensan, R. (2010). Impact taphonomique d'une combustion prolongée sur des ossements utilisés comme combustible. *P@lethnologie, 2*, 173–187.

Daujeard, C. (2008). *Exploitation du milieu animal par les Néandertaliens dans le Sud-Est de la France* (Vol. 1867). Oxford: BAR International Series.

Davis, K. L., & Fisher, J. W. (1990). A late prehistoric model for communal utilization of pronghorn antelope in the Northwestern Plains region, North America. In L. B. Davis & B. O. K. Reeves (Eds.), *Hunters of the recent past* (pp. 241–276). London: Unwin Hyman.

Delpech, F., & Rigaud, J.-P. (1974). Etude de la fragmentation et de la répartition des restes osseux dans un niveau d'habitat paléolithique. In H. Camps-Fabrer (Ed.), *L'Industrie de l'Os dans la Préhistoire* (pp. 47–55). Marseille: Université de Provence.

Denys, N. (1672). *Histoire naturelle des peuples, des animaux, des arbres et plantes de l'Amérique septentrionale et de ses divers climats* (Vol. 2). Paris: Claude Barrin.

Domínguez-Rodrigo, M. (1997). Meat-eating by early hominids at the FLK 22 (*Zinjanthropus*), Olduvai Gorge, Tanzania: an experimental approach using cut-mark data. *Journal of Human Evolution, 33*, 669–690.

Esteban-Nadal, M., Cáceres, I., & Fosse, P. (2010). Characterization of a current coprogenic sample originated by *Canis lupus* as a tool for identifying a taphonomic agent. *Journal of Archaeological Science, 37*(12), 2959–2970.

Faith, J. T., Marean, C. W., & Behrensmeyer, A. K. (2007). Carnivore competition, bone destruction, and bone density. *Journal of Archaeological Science, 34*(12), 2003–2011.

Fisher, J. W. (1995). Bone surface modifications in zooarchaeology. *Journal of Archaeological Method and Theory, 2*(1), 7–68.

Fox, M. W. (1984). *The whistling hunters: Field studies of the Asiatic wild dog (Cuon alpinus)*. Albany: State University of New York Press.

Frison, G. C. (1978). *Prehistoric hunters of the high plains*. New York: Academic Press.

Gifford-Gonzalez, D. (1989). Modern analogues: Developing an interpretative framework. In R. Bonnichsen & M. Sorg (Eds.), *Bone modification* (pp. 43–52). Orono, Maine: Center for the Study of First Americans.

Grayson, D. K., & Frey, C. J. (2004). Measuring skeletal part representation in archaeological faunas. *Journal of Taphonomy, 2*, 27–42.

Gronnow, B., Meldgaard, M., & Berglund Nielsen, J. (1983). *Aasivissuit—the great summer camp. Archaeological, ethnographical and zoo-archaeological studies of a caribou-hunting site of the West Greenland* (Vol. 5). Meddelelser om Gronland: Man and Society.

Haynes, G. (1983). A guide for differentiating mammalian carnivore taxa responsible for gnaw damage to herbivore limb bones. *Paleobiology, 9*(2), 164–172.

Hudson, J. (1993). The impacts of domestic dogs on bone in forager camps; or, the dog-gone bones. In J. Hudson (Ed.), *From bones to behavior: Ethnoarchaeological and experimental contributions to the interpretation offaunal remains* (Vol. Occasional Paper No. 21, pp. 301–323). Carbondale: Center for Archaeological Investigations, Southern Illinois University.

Jeannet, M. (2001). La microfaine et l'environnement de la grotte du noisetier à Fréchet (Hautes-Pyrénées). *Bulletin de la Société Préhistorique Ariège-Pyrénées* (LVI), 83–90.

Karlin, C., & Tchesnokov, Y. (2007). Notes sur quelques procédés de récupération de la graisse du renne: approcje ethnoarchéologique. In S. Beyries & V. Vaté (Eds.), *Les Civilisations du Renne d'hier et d'aujourd'hui. Approches ethnohistoriques, archéologiques et anthropologiques* (pp. 309–323). XVIII° Rencontres Internationales d'Archéologie et d'Histoire d'Antibes. Antibes: APDCA–CNRS.

Koon, H. E. C., Nicholson, R. A., & Collins, M. J. (2003). A practical approach to the identification of low temperature heated bone using TEM. *Journal of Archaeological Science, 30*(11), 1393–1399.

Koon, H. E. C., O'Connor, T. P., & Collins, M. J. (2010). Sorting the butchered from the boiled. *Journal of Archaeological Science, 37*(1), 62–69.

Lam, Y. M., Chen, X., & Pearson, O. M. (1999). Intertaxonomic variability in patterns of bone density and the differential representation of Bovid, Cervid, and Equid elements in the archaeological record. *American Antiquity, 64*, 343–362.

Leechman, D. (1951). Bone grease. *American Antiquity, 16*, 355–356.

Logan, B. (1998). The fat and the land: White Rock phase bison hunting and grease production. *Plains Anthropologists, 43*(166), 349–366.

Lupo, K. D., & O'Connell, J. F. (2002). Cut and tooth mark distributions on large animal bones: Ethnoarchaeological data from the Hadza and their implications for current ideas about early Human carnivory. *Journal of Archaeological Science, 29*(1), 85–109.

Lupo, K. D., & Schmitt, D. N. (1997). Experiments in bone boiling: Nutritional returns and archaeological reflections. *Anthropozoologica, 25*(26), 137–144.

Lyman, R. L. (1984). Bone density and differential survivorship of fossil classes. *Journal of Anthropological Archaeology, 3*, 259–299.

Lyman, R. L. (1994). *Vertebrate taphonomy*. Cambridge: Cambridge University Press.

Lyman, R. L., & O'Brien, M. J. (1987). Plow-zone zooarchaeology: Fragmentation and identifiability. *Journal of Field Archaeology, 14*, 493–498.

Mallye, J.-B., Costamagno, S., Boudadi-Maligne, M., Mourre, V. Prucca, A., Thiébaut, C., et al. (2010). La faune de la grotte du Noisetier (Fréchet-Aure, Hautes-Pyrénées: une histoire taphonomique complexe. Presentation at ICAZ.

Marean, C. W. (2005). From the tropics to the colder climates: contrasting faunal exploitation adaptations of modern humans and Neanderthals. In F. d'Errico & L. Backwell (Eds.), *From tools to symbols. From early hominids to modern humans* (pp. 333–371). Johannesburg: Wits University Press.

Marean, C. W., & Frey, J. (1997). Animal bones from caves to cities: Reverse utility curves as methodological artifacts. *American Antiquity, 62*, 698–711.

Marean, C. W., & Kim, S. Y. (1998). Mousterian large mammal remains from Kobeh Cave (Zagros Mountains, Iran): Behavioral implications for Neanderthals and early modern human. *Current Anthropology, 39*(Supplement), S79–S114.

Marean, C. W., & Spencer, L. (1991). Impact of carnivore ravaging on zooarchaeological measures of element abundance. *American Antiquity, 56*(4), 645–658.

Marean, C. W., Blumenschine, R. J., Spencer, L. M., & Capaldo, S. (1992). Captive hyaena bone choice and destruction, the Schlepp effect, and olduvai archaeofaunas. *Journal of Archaeological Science, 19*, 101–121.

Marean, C. W., Abe, Y., Frey, C. J., & Randall, R. C. (2000). Zooarchaeological and taphonomic analysis of the Die Kelders Cave 1 layers 10 and 11 Middle Stone Age larger mammal fauna. *Journal of Human Evolution, 38*(1), 197–233.

Maureille, B., Mann, A., Beauval, C., Bordes, J.-G., Bourguignon, L., Costamagno, S., et al. (2007). Le gisement moustérien des Pradelles (Marillac-le-Franc, Charente): passé, présent, futur. In J. Evin (Ed.), *Un siècle de construction du discours scientifique en Préhistoire* (pp. 249–261). Actes du XXVIe Congrès préhistorique de France, Avignon, septembre 21–25, 2007. Paris: Société préhistorique française.

Meignen, L., Costamagno, S., Beauval, C., Badré, L., Vandermeersch, B., & Maureille B. (2007). Gestion des ressources lithiques au Paléolithique moyen dans une halte de chasse spécialisée sur le renne: Les Pradelles (Marillac-le-Franc, Charente). In M. H. Moncel, A.-M. Moigne, M. Arzarello, & C. Peretto (Eds.), *Raw material supply areas and food supply areas. Integrated approach of the Behavior* (Vol. 1725, pp. 127–139). Oxford: BAR International Series.

Morin, E. (2007). Fat composition and Nunamiut decision-making: A new look at the marrow and bone grease indices. *Journal of Archaeological Science, 34*(1), 69–82.

Morin, E. (2010). Taphonomic implications of the use of bone as fuel. In I. Théry-Parisot, L. Chabal, & S. Costamagno (Eds.), *The taphonomy of burned organic residues and combustion features in archaeological contexts* (Vol. 2, pp. 209–217). Proceedings of the Workshop, Valbonne, May 27–29, 2008, P@lethnologie.

Morrison, D. (1998). *Caribou hunters in the Western Arctic*. Washington: University of Washington Press.

Mourre, V., Costamagno, S., Thiébaut, C., Allard, M., Bruxelles, L., Colonge, D., et al. (2008). Le site moustérien de la Grotte du Noisetier à Fréchet-Aure (Hautes-Pyrénées): premiers résultats des nouvelles fouilles. In J. Jaubert, J.-G. Bordes, & I. Ortega (Eds.), *Les sociétés du Paléolithique dans un Grand Sud-Ouest de la France: nouveaux gisements, nouveaux résultats, nouvelles méthodes* (Vol. 47, pp. 189–202). Journées scientifiques de la Société préhistorique française, université Bordeaux 1, Talence, novembre 24–25, 2006, Paris: Mémoire de la Société Préhistorique Française.

Mourre, V., Thiébaut, C., & Costamagno, S. (2010). *Le site moustérien de la grotte du Noisetier à Fréchet-Aure (Hautes-Pyrénées)*. Rapport de fouille programmée.

Munro, N. D., & Bar-Oz, G. (2005). Gazelle bone fat processing in the Levantine Epipalaeolithic. *Journal of Archaeological Science, 32*(2), 223–239.

O'Connell, J. F., Hawkes, K., & Blurton-Jones, N. (1988). Hadza hunting, butchering, and bone transport and their archaeological implications. *Journal of Anthropological Research, 44*(2), 113–161.

Oliver, J. S. (1993). Carcass processing by the Hadza: Bone breakage from butchery to consumption. In J. Hudson (Ed.), *From bones to behavior: Ethnoarchaeological and experimental contributions to the interpretation of faunal remains* (Vol. Occasional Paper No. 21, pp. 200–227). Carbondale: Center for Archaeological Investigations, Southern Illinois University.

Outram, A. K. (1999). A comparison of Paleo-Eskimo and medieval norse bone fat exploitation in Western Greenland. *Arctic Anthropology, 36*(1/2), 103–117.

Outram, A. K. (2001). The scapula representation could be the key: A further contribution to the 'Klasies Pattern' debate. *Journal of Archaeological Science, 28*(12), 1259–1263.

Outram, A. (2004). Applied models and indices vs. high-resolution, observed data: Detailed fracture and fragmentation analyses for the investigation of skeletal part abundance patterns. *Journal of Taphonomy, 2*(3), 167–184.

Outram, A. K., & Mulville, J. (2005). The zooarchaeology of fats, oils, milk and dairying: an introduction and overview. In J. Mulville & A. K. Outram (Eds.), *The zooarchaeology of fats, oils, milk and dairying* (Vols. 1–6). Actes du Colloque de l'ICAZ, Durham, août 2002. Oxford: Oxbow Books.

Prince, P. (2007). Determinants and implications of bone grease rendering: A Pacific Northwest example. *North American Archaeologist, 28*(1), 1–28.

Reinhard, K. J., Ambler, J. R., & Szuter, C. R. (2007). Hunter-gatherer use of small animal food resources: Coprolite evidence. *International Journal of Osteoarchaeology, 17*(4), 416–428.

Rendu, W., Costamagno, S., Meignen, L., & Soulier, M.-C. (2012). Monospecific faunal spectra in Mousterian contexts: Implications for social behavior. *Quaternary International, 247*, 50–58.

Richardson, P. R. K. (1980). Carnivore damage to antelope bones and its archaeological implications. *Paleontologia Africana, 23*, 109–125.

Roberts, S. J., Smith, C. I., Millard, A., & Collins, M. J. (2002). The taphonomy of cooked bone: Characterizing boiling and its physico-chemical effects. *Archaeometry, 44*(3), 485–494.

Ryder, M. L. (1966). Can one cook in a skin? *Antiquity, 40*, 225–227.

Ryder, M. L. (1969). Can one cook in a skin? *Antiquity, 43*, 217–220.

Saint-Germain, C. (1997). The production of bone broth: A study in nutritional exploitation. *Anthropozoologica, 25*(26), 153–156.

Saint-Germain, C. (2005). Animal fat in the cultural world of the native peoples of Northeastern America. In J. Mulville & A. K. Outram (Eds.), *The zooarchaeology of fats, oils, milk and dairying* (pp. 107–113). Actes du Colloque de l'ICAZ, Durham, août 2002. Oxford: Oxbow Books.

Schaefer, O., & Steckle, J. (1980). *Dietary habits and nutritional base of native populations of the Northwest Territories*. Yellowknife: Department of Information, Government of the Northwest Territories.

Sobolik, K. D. (2008). Dietary reconstruction as seen in coprolites. *Cambridge Histories Online*, pp. 44–51.

Soulier, M.-C. (2007). *Etude archéozoologique du carré M16 de la couche II du gisement moustérien du roc de Marsal (Dordogne)*. Université Toulouse 2—Le Mirail, Mémoire de Master.

Speth, J. D. (1983). *Bison kills and bone counts: Decision making by Ancient Hunters*. Chicago: University of Chicago Press.

Speth, J. D., & Spielmann, K. A. (1983). Energy source, protein metabolism, and hunter-gatherer subsistence strategies. *Journal of Anthropological Archaeology, 2*, 1–31.

Stiner, M. C. (1991). Food procurement and transport by human and nonhuman predators. *Journal of Archaeological Science, 18*, 455–482.

Stiner, M. C. (2003). Zooarchaeological evidence for resource intensification in Algarve, Southern Portugal. *Promontoria, 1*, 27–61.

Stiner, M. C., Kuhn, S. L., Weiner, S., & Bar-Yosef, O. (1995). Differential burning, recrystallization, and fragmentation of archaeological bones. *Journal of Archaeological Science, 22*, 223–237.

Théry-Parisot, I., & Costamagno, S. (2005). Propriétés combustibles des ossements. Données expérimentales et réflexions archéologiques sur leur emploi dans les sites paléolithiques. *Gallia Préhistoire, 47*, 235–254.

Théry-Parisot, I., Brugal, J.-P., Costamagno, S., & Guilbert, R. (2004). Conséquences taphonomiques de l'utilisation des ossements comme combustible. *Approche expérimentale. Les Nouvelles de l'Archéologie, 95*(1), 19–22.

Théry-Parisot, I., Costamagno, S., Brugal, J.-P., Fosse, P., & Guilbert, R. (2005). The use of bone as fuel during the Palaeolithic, experimental study of bone combustible properties. In J. Mulville & A. K. Outram (Eds.), *The zooarchaeology of fats, oils, milk and dairying* (pp. 50–59). Actes du Colloque de l'ICAZ, Durham, août 2002. Oxford: Oxbow Books.

Vehik, S. C. (1977). Bone fragments and bone grease manufacture: A review of their archaeological use and potential. *Plains Anthropologist, 22*, 169–182.

Villa, P., & Mahieu, E. (1991). Breakage patterns of human long bones. *Journal of Human Evolution, 21*, 27–48.

Villa, P., Bon, F., & Castel, J.-C. (2002). Fuel, fire and fireplaces in the palaeolithic of Western Europe. *The Review of Archaeology, 23*(1), 33–42.

Viré, A. (1898). Les Pyrénées souterraines (1ère campagne, 1897). *Mémoires de la Société de Spéléologie, 14*(III), 59–96.

Yohe, R. M., Newman, M. E., & Schneider, J. S. (1991). Immunological identification of small-mammal proteins on aboriginal milling equipment. *American Antiquity, 56*(4), 659–666.

Chapter 14
Foraging Goals and Transport Decisions in Western Europe during the Paleolithic and Early Holocene

Eugène Morin and Elspeth Ready

Introduction

Human foragers who radiate out from a central place to hunt commonly face significant transport constraints. One of these constraints results from the limited transport capacities of humans relative to the sometimes large body size of their prey (Barlow and Metcalfe 1996; Bettinger et al. 1997). Research has shown that load weights carried by human foragers may at best represent 70–83 % of body mass, as documented for African women (Maloiy et al. 1986) and Nepalese porters (Bastien et al. 2005a, b; Minetti et al. 2006), respectively. Using the range of body masses (40–76 kg) documented in human forager populations (Walker et al. 2006) as proxies, maximum loads transported by individuals could rarely have exceeded 25–60 kg in the past. Even the highest of these last values represents only 1/15–1/30th of the estimated weight of a healthy steppe bison (*Bison priscus*), a prey species that is well represented in anthropogenic assemblages from Europe (Brugal et al. 1999). In Tanzania, O'Connell and colleagues (1988) observed that Hadza foragers usually carry loads of meat weighing ~10–20 kg per person, with ~45 kg being the upper limit when dealing with exceptionally large carcasses (e.g., giraffe). The implication of these limited transport capacities for small groups of people hunting bison and other large prey taxa is that they probably frequently had to abandon parts of the carcass in the field (Binford 1978; Speth 1983; O'Connell et al. 1990; Monahan 1998; Cannon 2003; Burger et al. 2005).

E. Morin (✉)
Department of Anthropology, Trent University, DNA Bldg Block C, 2140 East Bank Drive, Peterborough, ON K9J 7B8, Canada
e-mail: eugenemorin@trentu.ca

E. Ready
Department of Anthropology, Stanford University, Main Quad Building 50, 450 Serra Mall, Stanford, CA 94305-2034, USA
e-mail: eready@stanford.edu

Theoretical modeling of transport decisions suggests that the distance of kill sites from camp is an important factor conditioning culling decisions and residential mobility (Barlow and Metcalfe 1996; Bettinger et al. 1997; Cannon 2003). In mammals, the population density of species—a factor that significantly influences the encounter rate—tends to relate inversely to body size, with large animals typically occurring at low densities (White et al. 2007). From the point of view of the hunter, these trends mean that larger prey taxa are likely to be encountered in more distant patches. This argument is corroborated by actualistic observations, which show that large taxa such as caribou (*Rangifer tarandus*) and zebra (*Equus burchelli*) are commonly killed at significant distances (>5 km) from the residential camp (Binford 1978; Winterhalder 1981; Bunn et al. 1988; O'Connell et al. 1988, 1990). These transport distances are important as they likely influenced the frequency with which camps were moved (Barlow and Metcalfe 1996).

Three decades ago, Binford (1978) showed that, among the Nunamiut (Alaska), the range of animal body parts transported back to a campsite is variable. Depending on the circumstances, the focus of transport decisions may be the acquisition of meat, skins, fat, or a combination of these products. Binford attributed this variation to the interaction of several factors, including the quantity of food available at the camp and/or in caches, the season of procurement, the size and sex of the animal(s), the travel conditions, the size of the hunting party, etc. To examine the selection criteria for parts transported under different procurement goals, Binford (1978) constructed several economic indices. These can be conceived of as scales ranking skeletal parts in terms of their relative meat, marrow, grease, dry meat or overall food "utility." The use of these indices presupposes that hunters are concerned with maximizing foraging efficiency when transporting animal products. Indeed, as noted by Grayson and Cannon (1999: 142–143), Binford's: "analysis of the interrelationships between 'economic anatomy' and body part frequency in bone assemblages assumes that

Jamie L. Clark and John D. Speth (eds.), *Zooarchaeology and Modern Human Origins: Human Hunting Behavior during the Later Pleistocene*, Vertebrate Paleobiology and Paleoanthropology, DOI: 10.1007/978-94-007-6766-9_14,

individuals forage across the differential opportunities provided by the carcass of a large mammal in the same way that evolutionary ecologists maintain that people forage across larger landscapes."

Because they were intended for testing hypotheses about foraging behavior using faunal assemblages, Binford's utility indices received critical attention in the discipline (e.g., Chase 1985; Lyman 1985, 1992; Jones and Metcalfe 1988; Metcalfe and Jones 1988; Grayson 1989; Klein 1989; Emerson 1990, 1993; Bartram 1993; Blumenschine and Madrigal 1993; Brink 1997; Friesen 2001; Morin 2007). In Europe, several prehistoric sites have been interpreted using these analytical tools (e.g., Jaubert and Brugal 1990; David and Farizy 1994; Auguste 1995; Blasco 1997; Bellai 1998; Cho 1998; Boyle 2000; Rosell 2001; Haws 2003; Vannoorenberghe 2004; Costamagno et al. 2006; Faith 2007; Daujeard 2008; Gaudzinski-Windheuser and Niven 2009; Fernández-Laso 2010; Rendu 2010). Yet, despite a relatively wide range of available utility models, several studies of European assemblages have been limited to the examination of the Food Utility Index (FUI, Metcalfe and Jones 1988). This model ranks anatomical parts relative to their total soft tissue weight, taking into account the presence of "riders" (skeletal elements of low to moderate utility that are immediately adjacent to parts of higher utility) in the limbs. This focus on the FUI assumes that meat—by far the most heavily weighted product in this index—was the principal factor mediating the transport of skeletal parts in prehistoric Europe. In this paper, we investigate the validity of this assumption and examine whether it most parsimoniously explains patterns of anatomical representation in this region.

Materials and Methods

Assemblage Selection

The sample used here to examine variations in transport goals in prehistoric Europe consists of 167 faunal assemblages from 63 sites (Table 14.1). Archaeological sequences that contributed several assemblages to the sample include Abri Pataud ($n = 19$, Cho 1998; Sekhr 1998; Vannoorenberghe 2004), Gabasa 1 ($n = 10$, Blasco 1997), Saint-Césaire ($n = 9$, Morin 2012), Rond-du-Barry ($n = 8$, Costamagno 1999), Grotte XVI ($n = 7$, Grayson and Delpech 2003) and Jonzac ($n = 6$, Beauval 2004). Assemblages with indications of a significant contribution by carnivores or with evidence of biased recovery of faunal specimens—a problem common in old excavations—were excluded from the analysis. Sites with abundant bear remains but in which other carnivores are rare, such as Grotte XVI (Grayson and Delpech 2003) and Payre (Daujeard 2008), were included in our sample because bears are not considered to be significant bone accumulators. Unfortunately, many publications did not provide clear information about excavation methods and the analysis of long bone fragments. We excluded assemblages only if there was clear evidence that long bone shaft portions were not kept and/or analyzed.

Assemblages with very poor faunal preservation were also excluded. However, certain anthropogenically-accumulated assemblages in our sample show moderate signs of carnivore overprinting (e.g., Grotte du Bison layer D, Gabasa 1, Gatzarria, Solutré) or significant damage caused by weathering and plant growth (e.g., Etiolles, Pincevent IV0-T125, Champ-Chalatras, Le Sire). Although these phenomena might have altered patterns of skeletal representation in these assemblages, they were retained in the analysis as they are from regions or are associated with contexts for which we had small sample sizes. Because the issue of density-mediated attrition is of critical importance for the interpretation of anatomical profiles, we performed the analysis both with and without low survival elements.

Assemblages with fewer than 75 *postcranial* specimens (by NISP counts) were not included in the present study (Table 14.1). Given the data at hand, this threshold appeared to us as a satisfactory compromise between statistical robustness and a large sample size of assemblages. We focused on postcranial remains to determine sample size because teeth preserve much better than bones, which allowed us to objectively eliminate assemblages that were argued—based on independent evidence (e.g., heavily damaged bones surfaces)—to show poor faunal preservation. This approach also enabled us to exclude paleontologically-oriented studies largely focused on teeth. It should be noted that the small (NISP:63) layer 3 *Capra* assemblage from Pirulejo (Riquelme Cantal 2008) constitutes an exception to the above rule. This last assemblage was kept in the analysis to enlarge the sample size for Iberia.

Composition of the Sample

Geographically, the sample is heavily biased towards France (Table 14.2). Other countries that are represented include Italy, Portugal, Spain and Switzerland. The bias towards France results from two factors. Firstly, France—southwestern France in particular—has a long tradition of Paleolithic research. As a result, taphonomically-oriented faunal studies abound for this region. Secondly, our own regional expertise leads us to a greater familiarity with sites from southwestern France, especially those dating to the

Table 14.1 The sample used for the analysis of transport decisions

Country, taxon	Site and level, species, sample size (in NISP)	Period	MIS	Site type	Unit	References
France						
Cervids	Payre Fc–d, Cel, 153	MP	7	S	NISP	Daujeard (2008)
	Fa, Cel, 381	MP	7	S	"	"
	Lazaret 5 layers, Cel, 2,290	LP/MP	6	S	"	Valensi (2000)
	Saint-Marcel U, Dama, 257	MP	5	S	"	Daujeard (2008)
	c. inférieure, Cel, 93	MP	5	S	"	"
	ens. 7, Cel, 1,296	MP	5 or 3	S	"	"
	ens. 7, Capreolus, 125	MP	5 or 3	S	"	"
	ens. sup, Cel, 131	MP	5 or 3	S	"	"
	Les Canalettes 3, Cel, 525	MP	5	S	"	Patou-Mathis (1993)
	2, Cel, 139	MP	5	S	"	"
	Marillac 10, Rang, 477[a]	MP	4	S	%MAU	Costamagno et al. (2006)
	9, Rang, 800[a]	MP	4	S	"	"
	Jonzac 24, Rang, 250	MP	4	S	NISP	Beauval (2004)
	22, Rang, 1,328	MP	4	S	"	"
	20, Rang, 87	MP	4	S	"	"
	18, Rang, 117	MP	4	S	"	"
	14, Rang, 281	MP	4	S	"	"
	10, Rang, 299	MP	4	S	"	"
	Beauvais 1, Rang, 122	MP	4	O	"	Auguste and Patou-Mathis (1999)
	2, Rang, 260	MP	4	O	"	"
	La Rouquette UA3, Rang, 172	MP	4	O	"	Heriech-Briki (2003)
	UA2, Rang, 245	MP	4–3	O	"	"
	Mutzig I c. 6, Rang, 207	MP	4–3	O	"	Patou-Mathis (1999)
	c. 5, Rang, 159	MP	4–3	O	"	"
	Gr. XVI C, Rang, 108[b]	MP	4–3	S	"	Grayson and Delpech (2003), Faith (2007)
	Gr. du Noisetier 1, Cel, 125	MP	3	S	"	Costamagno et al. (2008)
	Pech de l'Azé I c. 4, Cel, 394[a]	MP	3	S	%MAU	Rendu (2010)
	c. 7, Cel, 90[a]	MP	3	S	"	"
	Gatzarria Cj, Cel, 255	MP	3?	S	NISP	Ready (2010)
	La Quina 8, Rang, 1,647	MP	3	S	"	Chase (1999)
	7, Rang, 347	MP	3	S	"	"

(continued)

Table 14.1 (continued)

Country, taxon	Site and level, species, sample size (in NISP)	Period	MIS	Site type	Unit	References
	Saint-Césaire EGPF, Rang, 197	MP	3	S	"	Morin (2012)
	EJOP inf, Rang, 88	MP	3	S	"	"
	EJOP sup, Rang, 144	UP, D/Chât	3	S	"	"
	EJO sup, Rang, 324	UP, M Auri	3	S	"	"
	EJF, Rang, 2,697	UP, M Auri	3	S	"	"
	EJM, Rang, 550	UP, Ev Auri	3	S	"	"
	EJJ, Rang, 214	UP, Ev Auri	3	S	"	"
	Gr. du Renne Xc, Rang, 1,208[a]	UP, Chât	3	S	%MAU	David (2002)
	VIII, Rang, 580[a]	UP, Chât	3	S	"	David (2002)
	VII, Rang, 1,229	UP, Ear Auri	3	S	NISP	David and Poulain (2002)
	Gr. du Bison D, Rang, 179	UP, Chât	3	S	"	David et al. (2005)
	Piage B, Rang, 112	UP, Auri	3	S	"	Unpublished data
	Abri Pataud 14, Rang, 2,050	UP, Ear Auri	3	S	"	Sekhr (1998)
	13, Rang, 242	UP, Auri	3	S	"	"
	12, Rang, 208	UP, Auri	3	S	"	"
	11, Rang, 1,155	UP, Auri	3	S	"	"
	Gr. XVI B, Rang, 190[b]	UP, Chât	3	S	"	Grayson and Delpech (2003), Faith (2007)
	Aib, Rang, 155[b]	UP, Auri	3	S	"	", "
	Abb, Rang, 555[b]	UP, Auri?	3	S	"	", "
	Abc, Rang, 659[b]	UP, Gravet	3	S	"	", "
	As, Rang, 254[b]	UP, Solut	2	S	"	", "
	0, Rang, 1,189[b]	UP, Magd	2	S	"	", "
	Les Rois C3, Rang, 184	UP, Ear Auri	3	S	"	Renou (2007)
	C2, Rang, 156	UP, Ev Auri	3	S	"	"
	Abri Pataud 5-lower, Rang, 15,158	UP, Gravet	2	S	"	Vannoorenberghe (2004)
	5-Middle, Rang, 3,238	UP, Gravet	2	S	"	"
	5-upper, Rang, 12,647	UP, Gravet	2	S	"	"
	5, Cel, 272	UP, Gravet	2	S	"	"
	4-lower, Rang, 7,113	UP, Gravet	2	S	"	Cho (1998)

(continued)

Table 14.1 (continued)

Country, taxon	Site and level, species, sample size (in NISP)	Period	MIS	Site type	Unit	References
	4-middle, Rang, 11,239	UP, Gravet	2	S	"	"
	4-upper, Rang, 22,287	UP, Gravet	2	S	"	"
	3-4, Rang, 704	UP, Gravet	2	S	"	"
	3-3, Rang, 5,003	UP, Gravet	2	S	"	"
	3-2, Rang, 861	UP, Gravet	2	S	"	"
	3-1, Rang, 1,287	UP, Gravet	2	S	"	"
	2, Rang, 2,650	UP, Proto-Magd	2	S	"	"
	Le Flageolet I c. V, Rang, 1,496	UP, Gravet	2	S	"	Enloe (1993)
	Combe-Saunière IV, Rang, 3,769	UP, Solut	2	S	"	Castel (1999)
	Cuzoul du Vers 31–30, Rang + MUN[c], 233	UP, Solut	2	S	"	"
	29, Rang + MUN[c], 125	UP, Solut	2	S	"	"
	23, Rang, 1,169	UP, Solut	2	S	"	"
	Rond-du-Barry F2, Rang, 446[a]	UP, Magd	2	S	%MAU	Costamagno (1999)
	E, Rang, 285[a]	UP, Magd	2	S	"	"
	Solutré, Rang, 235	UP, Magd	2	O	"	Turner (2002)
	Gazel, Rang, 6,647[a]	UP, Magd	2	S	%PO	Fontana (1999)
	Le Flageolet II c. IX, Rang, 1,548[d]	UP, Magd	2	S	NISP	Deplano (1994)
	Moulin-Neuf site, Rang, 321[a]	UP, Magd	2	S	%MAU	Costamagno (1999)
	Gr. Tournal G, Rang, 5,743[a]	UP, Magd	2	S	"	Magniez (2010)
	H, Rang, 3,184[a]	UP, Magd	2	S	"	"
	Conques, Rang, 270	UP, Magd	2	S	NISP	Moigne (2003)
	La Magdeleine La Plaine 4e–4n, Rang, 554[a]	UP, Magd	2	S	MNE	Kuntz (2006)
	Bois-Ragot 6, Rang, 359[a]	UP, Magd	2	S	"	Griggo (2005)
	5, Rang, 858[a]	UP, Magd	2	S	"	"
	4, Cel, 427[a]	UP, Magd	2	S	"	"
	Grand Canton, Rang, 186	UP, Magd	2	O	NISP	Bridault and Bémilli (1999)
	Verberie II-1, Rang, 3,600	UP, Magd	2	O	"	Enloe (2004)
	Pincevent IV0-T125, Rang, 441	UP, Magd	2	O	"	Bignon et al. (2006)

(continued)

Table 14.1 (continued)

Country, taxon	Site and level, species, sample size (in NISP)	Period	MIS	Site type	Unit	References
	Pincevent 36, Rang, 2,571	UP, Magd	2	O	"	Enloe (1993)
	Arancou B2, Cel, 161	UP, Magd	2	S	"	Fosse (1999)
	Noyen-sur-Seine, Cel, 772[a]	Hol, Meso	1	O	MNE	Vigne (2005)
Bovines	Arago complexe moyen supérieur, Bison, >200	LP	12	S	MAU	Bellai (1998)
	Biache, Bos, 9,771[a]	MP	7	O	%NISP	Auguste (1993, 1995)
	Payre Fa, BB, 117	MP	7	S	NISP	Daujeard (2008)
	La Borde, Bos, 376	MP	Pre-5	O[e]	"	Slott-Moller (1990), Brugal (2001)
	La Rouquette UA4, BB, 194	MP	5	O	"	Heriech-Briki (2003)
	UA3, BB, 336	MP	4	O	"	"
	UA2, BB, 276	MP	4–3	O	"	"
	Mauran, Bison, 4,216	MP	3	O	"	David and Farizy (1994), Rendu (2007)
	Pech de l'Azé I 4, Bison, 127[a]	MP	3	S	%MAU	Rendu (2010)
	Saint-Césaire EGPF, Bison, 300	MP	3	S	NISP	Morin (2012)
	EJOP sup, Bison, 345	UP, Chât	3	S	"	"
	Abri Pataud 5, BB, 155	UP, Gravet	2	S	"	Vannoorenberghe (2004)
	Moulin-Neuf site, BB, 168[a]	UP, Magd	2	S	%MAU	Costamagno (1999)
	Champ-Chalatras loci 1–3, Bos, 158	Hol, Epipal	1	O	NISP	Pasty et al. (2002)
	La Montagne, Bos, 458	Hol, Meso	1	O	"	Helmer and Monchot (2006)
Caprines	Arago sol F, argali, 6,577[a]	LP	12–11	S	%NISP	Rivals (2002)
	Lazaret 5 layers, ibex, 737	LP/MP	6	S	NISP	Valensi (2000)
	Gr. du Noisetier 1, Capra p, 141	MP	3	S	"	Costamagno et al. (2008)
	Abri Pataud 5, ibex, 114	UP, Gravet	2	S	"	Vannoorenberghe (2004)
	Cuzoul du Vers 23, Rupi, 96	UP, Badeg	2	S	"	Castel (1999)
	Rond-du-Barry F2, ibex, 709[a]	UP, Magd	2	S	%MAU	Costamagno (1999)
	E, ibex, 679[a]	UP, Magd	2	S	"	"
	F2, Rupi, 152[a]	UP, Magd	2	S	"	"
	E, Rupi, 138[a]	UP, Magd	2	S	"	"
	Saint-Germain-la-Rivière 4, saiga, 197	UP, Magd	2	S	NISP	"
	3, saiga, 682	UP, Magd	2	S	"	"
	1, saiga, 1,131	UP, Magd	2	S	"	"

(continued)

Table 14.1 (continued)

Country, taxon	Site and level, species, sample size (in NISP)	Period	MIS	Site type	Unit	References
	C, saiga, 370	UP, Magd	2	S	"	"
	Moulin-Neuf site, saiga, 911[a]	UP, Magd	2	S	%MAU	"
	Conques, Capra p, 181	UP, Magd	2	S	NISP	Moigne (2003)
	Belvis, Capra p, 1,512[a]	UP, Magd	2	S	%PO	Fontana (1999)
Horse	Arago G, Cab, 2,670	LP	12	S	NISP	Bellai (1995)
	Payre Fa, Cab, 80	MP	7	S	"	Daujeard (2008)
	Gr. du Renne VII, Cab, 1,670	UP, Proto-Auri	3	S	"	David and Poulain (2002)
	Abri Pataud 11, Cab, 1,003	UP, Ear Auri	3	S	"	Sekhr (1998)
	Solutré M12 Auri, Cab, 1,495[a]	UP, Auri	3	O	MNI	Olsen (1989)
	L13 Gravet, Cab, 751[a]	UP, Gravet	2	O	"	"
	Le Sire, Cab, 163	UP, Gravet	3	O	NISP	Surmely et al. (2003)
	Combe-Saunière IV, Cab, 314	UP, Solut	2	S	"	Castel (1999)
	Rond-du-Barry F2, Cab, 1,556[a]	UP, Magd	2	S	%MAU	Costamagno (1999)
	E, Cab, 1,035[a]	UP, Magd	2	S	"	"
	Moulin-Neuf site, Cab, 502[a]	UP, Magd	2	S	"	"
	Gr. Tournal G, Cab, 358[a]	UP, Magd	2	S	"	Magniez (2010)
	H, Cab, 436[a]	UP, Magd	2	S	"	"
	Etiolles, Cab, 247	UP, Magd	2	O	NISP	Bignon (2003)
	Tureau-des-Gardes, Cab, 3,897	UP, Magd	2	O	"	"
	Grand Canton, Cab, 3,027	UP, Magd	2	O	"	Bridault and Bémilli (1999)
	Le Closeau 46, Cab, 399	Hol, Ear Azil	1	O	"	Bignon (2003)
Italy					"	
Cervids	Notarchirico, Dama c, 155	LP	7	O	"	Cassoli et al. (1999)
Bovines	Isernia 3a, Bison s, 630	LP	15–12	O	"	Anconetani (1999), Coltorti et al. (2005)
Caprines	Riparo Dalmeri 26c, ibex, 2,218	UP, Epigr	2	S	"	Fiore and Tagliacozzo (2006)
Portugal					"	
Cervids	Galeria Pesada C, cervids, 109	LP	7	S	"	Marks et al. (2002)
	Galeria Pesada B2, cervids, 95	LP	7	S	"	"
	Picareiro F, Cel, 131	UP, Magd	2	S	"	Haws (2003)

(continued)

Table 14.1 (continued)

Country, taxon	Site and level, species, sample size (in NISP)	Period	MIS	Site type	Unit	References
Spain					"	
Cervids	Gran Dolina TD10a, cervids, 181	LP	11	S	"	Rosell (2001)
	Cova Bolomor XII, Cel, 325[a]	MP	6	S	MNE	Blasco et al. (2010)
	Gabasa 1 level h, Cel, 159[a]	MP	4–3	S	MAU	Blasco (1997)
	g, Cel, 511[a]	MP	4–3	S	"	"
	f, Cel, 359[a]	MP	4–3	S	"	"
	e, Cel, 348[a]	MP	4–3	S	"	"
	Abric Romani M, cervids, 377	MP	3	S	NISP	Fernández-Laso (2010)
	K, cervids, 292	MP	3	S	"	"
	Ja, cervids, 317	MP	3	S	"	Rosell (2001)
	Herriko Barra, Cel, 922	Meso?	1	O	"	Mariezkurrena and Altuna (1995)
Caprines	Cueva de Nerja X–VIII, Capra p, 945	UP, Solut	2	S	"	Riquelme Cantal et al. (2005)
	La Fragua 4, Capra, 87	UP, Magd	2	S	"	Marín Arroyo and González Morales (2007)
	El Pirulejo 4, Capra p, 174	UP, Magd	2	S	"	Riquelme Canta l (2008), Cortés Sánchez et al. (2008)
	3, Capra p, 89	UP, Magd	2	S	"	"
Horse	Cova Bolomor XII, Cab, 165[a]	MP	6	S	MNE	Blasco et al. (2010)
	Gabasa 1 level h, Cab, 263[a]	MP	4–3	S	MAU	Blasco (1997)
	g, Cab, 380[a]	MP	4–3	S	"	"
	f, Cab, 249[a]	MP	4–3	S	"	"

(continued)

Table 14.1 (continued)

Country, taxon	Site and level, species, sample size (in NISP)	Period	MIS	Site type	Unit	References
	e, Cab, 374[a]	MP	4–3	S	"	"
	d, Cab, 162[a]	MP	4–3	S	"	"
	a–c, Cab, 206[a]	MP	4–3	S	"	"
Switzerland						
Horse	Champréveyres, Cab, 2,396	UP, Magd	2	O	NISP	Bignon (2003)

The assemblages are presented by country and taxonomic group (cervids, bovines, caprines or horse) and are listed in rough chronological order. Assemblages from the same sites were kept together except when they are separated by a major hiatus

[a] Sample size for these assemblages may comprise elements not included in our analysis of body part representation

[b] At Grotte XVI, NISP counts for the cranium and mandible do not include teeth, in contrast to the other assemblages in our sample

[c] These two assemblages from Cuzoul du Vers combine the NISP counts for reindeer and medium-sized ungulates ("MUN," see text for explanation)

[d] The data in Table 6 in Deplano (1994: 18) clearly contains a mistake, as the values for the talus and tibia are inverted, based on the discussion of these parts in the text. The corrected values are used here

[e] Sinkhole assemblage. The La Borde assemblage is grouped with open air sites as it *may* have coincided with the area of capture (see text for explanation)

Taxa: argali = *Ovis ammon antiqua*; BB = *Bos/Bison*; Bison = *Bison priscus* (steppe bison); Bison s = *Bison schoetensacki*; Bos = *Bos primigenius* (aurochs); Capra p = *Capra pyrenaica* (Pyrenean Ibex); Cel = *Cervus elaphus* (red deer); Dama = *Dama dama* (fallow deer); Dama c = *Dama clactoniana*; Capreolus = *Capreolus capreolus* (roe deer); ibex = *Capra ibex* (Alpine Ibex); Rang = *Rangifer tarandus* (reindeer); Rupi = *Rupicapra rupicapra* (chamois); saiga = *Saiga tatarica* (saiga)

Periods: Auri = Aurignacian; Azil = Azilian; Badeg = Badegoulian; Chât = Châtelperronian; D = Denticulate Mousterian; Ear = Early; Epigr = Epigravettian; Epipal = Epipaleolithic; Ev = Evolved; Gravet = Gravettian; Hol = Holocene; LP = Lower Paleolithic; M = Middle; Magd = Magdalenian; Meso = Mesolithic; MP = Middle Paleolithic; Solut = Solutrean

Other: c. = couche (layer); gr. = grotte (cave); MIS = marine isotope stage; O = open air; S = shelter

Table 14.2 Composition of the sample in terms of geographical origin, taxonomy, site type and cultural period

	Lower Paleolithic	Lower/early Middle Paleolithic	Middle Paleolithic	Upper Paleolithic	Azilian/Mesolithic	Total
Country						
France	3	2	43	87	4	139
Spain	1		15	4	1	21
Italy	2			1		3
Portugal	2			1		3
Switzerland				1		1
Total	8	2	58	94	5	167
Species group						
Cervids	4	1	40	58	2	105
Bovines	2		9	3	2	16
Caprines	1	1	1	18		21
Equids	1		8	15	1	25
Total	8	2	58	94	5	167
Site type						
Shelter	6	2	46	82		136
Open air	2		12	12	5	31
Total	8	2	58	94	5	167

Table 14.3 The sample by marine isotope stage (MIS) and country

	Pre-7	7	6	5	4	3	2	1	Unclear	Total
France	3	5	2	5	12	36	62	4	10	139
Spain	1		2			3	4	1	10	21
Italy	1	1					1			3
Portugal		2					1			3
Switzerland							1			1
Total	5	8	4	5	12	39	69	5	20	167

It should be noted that we attributed the early Gravettian and the later Gravettian to MIS 3 and MIS 2, respectively, as suggested by accepted dates for these episodes (Pesesse 2008)

Middle and Upper Paleolithic. However, we made an effort to include faunal assemblages representing other periods and geographic locations (e.g., Bellai 1995, 1998; Mariezkurrena and Altuna 1995; Haws 2003; Riquelme Cantal et al. 2005, Fiore and Tagliacozzo 2006; Riquelme Cantal 2008; Blasco et al. 2010; Fernández-Laso 2010).

Table 14.2 shows that the majority of the assemblages date to the Middle (58/167 or 34.7 %), and more particularly the Upper (94/167 or 56.3 %), Paleolithic. However, the temporal subdivisions classically used in Paleolithic archaeology (i.e., Lower, Middle and Upper Paleolithic and Epipaleolithic/Mesolithic) are not completely satisfactory for our purposes, as they differ widely in duration and encompass episodes of contrasted climatic regimes (glacials and interglacials). Because climate change may have had an impact on encounter rates, transport distances and foraging decisions, whenever possible we ascribed the assemblages to a specific marine isotope stage (MIS), generally using the attributions made by the authors of the original faunal publications of the assemblages (Table 14.3). MIS refer to alternating long-term (multi-millennial) phases of cool and warm temperature reconstructed based on the isotopic analysis of samples from marine cores (Bowen 2009). Although climatic fluctuations were sometimes significant within MIS (Sánchez Goñi et al. 2008), these coarse temporal units are useful as they allow us to compare variations in procurement goals between and within major climatic episodes.

The sample includes assemblages from four taxonomic groups which are frequently well represented at Western European archaeological sites, and which were likely economically important taxa in the region: cervids, large bovines, caprines and equids. In addition, appropriate utility indices for a species in each of these groups were available or could be constructed from published data. These indices are discussed below. Large carnivores (e.g., brown bear

Ursus arctos) were not considered because, to our knowledge, no utility models are available for them.

The cervid group includes assemblages of the following species: reindeer (*Rangifer tarandus*), red deer (*Cervus elaphus*), roe deer (*Capreolus capreolus*), and fallow deer (*Dama* sp.). Large bovines include both bison (*Bison* sp.) and aurochs (*Bos primigenius*). The caprine group includes ibex (*Capra ibex* and *Capra pyrenaica*), chamois (*Rupicapra rupicapra*) and the argali (*Ovis ammon*). Saiga antelope (*Saiga tatarica*), although not a caprine, is also included in this group because of its body size and morphological similarity. Finally, all of the *Equus* assemblages are from horse. Among these taxonomic groups, cervids (105/167 assemblages or 62.9 %)—especially reindeer—are very well represented (Table 14.2). It should be noted that the carcasses of some of the species considered, such as roe deer and chamois, are small enough to have been transported entirely by small groups of hunters or even by single individuals. Consequently, we might expect skeletal part patterns for these species to differ from those of closely-related but larger species.

Counts of Skeletal Elements

To assess patterns of skeletal abundances we favor the number of identified specimens (NISP) because it avoids some of the critical pitfalls that plague the use of the minimum number of elements (MNE) and the element-based minimum number of individuals ($MNI_{element}$). Both of the latter methods are known to inflate the representation of rare elements, a result of the declining rate of increase as sample size increases (Grayson 1984; Lyman 2008). Furthermore, MNE and MNI values tend to be small in the assemblages considered here, which increases the chance of getting erroneous rankings. Moreover, using NISP and %NISP is advantageous in the present case, as it is the most common (122/167 assemblages or 73.1 %) measure in our sample.

Categories of parts ($n = 19$) that are considered in the present analysis are listed in Appendix 14.1 with associated mineral density values. In creating the database, we made an attempt to standardize the data as much as possible by tabulating skeletal part counts in the same way for all assemblages. However, differences between publications in the presentation of anatomical data led to some difficulty in this process. Consistent tabulation of skull remains (the cranium and mandibles) proved to be the most challenging. While we would have preferred to include only bone portions of the cranium and mandible or single categories of teeth (e.g., D4 and worn M3) in order to avoid inflating head counts due to the high identifiability of tooth fragments, this was not possible because most publications presented data only for teeth and bony portions combined.

Consequently, in order to maintain consistency between assemblages, we included bony portions of the skull, as well as all teeth identified as maxillary or mandibular, in the frequencies of the cranium and mandible. Counts of indeterminate head and tooth fragments (e.g., "unidentified teeth") were disregarded. Nevertheless, some discrepancies between assemblages persist in the dataset. For example, for Grotte XVI, the data on the skull considered here includes only bony portions, because counts of teeth identified as cranial or mandibular were not available (Grayson and Delpech 2003; Faith 2007). Antler fragments were excluded from cranial counts for cervids, to avoid inflating head representation for these animals relative to the other species considered. Excluding antler also avoids inappropriate inflation of the representation of the cranium in cases where shed or unshed antlers may have been preferentially transported for non-food purposes, such as tool-making.

Wherever possible, we recorded vertebrae in three categories: cervical, thoracic, or lumbar, while the sacrum was tabulated with the innominate. However, for many assemblages (52/167 or 31.1 %), only total counts for all vertebrae were available, in which case this figure was recorded in a separate category. When counts of different vertebral segments were provided, we combined the atlas and axis with cervical vertebrae 3–7 to obtain a single value for all cervical vertebrae (C1–7). Phalanges presented a similar problem, as some authors did not separate first, second and third phalanges in their tabulations. For those assemblages (42/167 or 25.1 %), we were only able to record the total number of phalanges present in the sample.

In some publications, frequency data for long bones were provided by bone portion (e.g., proximal, mid-shaft, distal). For NISP data, we recorded the total number of specimens of an element as the sum of the NISP for each portion. This procedure might slightly inflate the representation of long bones if single specimens were counted in more than one category of bone portion (e.g., a whole tibia might be counted three times as the proximal, midshaft, and distal portions are all represented). However, because fragmentation is extensive in the majority of the assemblages included in our sample, it is likely that few fragments represent multiple portions of the bone, and therefore we consider this problem to be relatively minor. Similarly, to improve consistency between assemblages, we combined NISP values for the radius, radio-ulna, and ulna when these values were provided separately. For MNE, MNI and MAU data, we simply recorded the highest value for these elements.

Long bone epiphyses are frequently poorly represented at Paleolithic sites in the study area (e.g., David and Poulain 1990; Deplano 1994; Cho 1998; Castel 1999;

Costamagno 1999; David and Poulain 2002; Haws 2003; Vannoorenberghe 2004; David et al. 2005; Daujeard 2008; Morin 2010, 2012; Ready 2010). Because identification of long bone shaft fragments tends to be a challenging and time-consuming endeavor (Marean and Kim 1998), inter-analyst differences in identification procedures may be a source of some error in the estimated abundances of long bones in the assemblages. We suspect that depressed long bone counts could be a cause of weak correlations with the utility indices in some cases.

In several publications used in the creation of the sample, skeletal part data were not presented in tables but only in graphical form, such as bar charts or scatter plots. We used the software *Graph Grabber* (Quintessa Ltd.) to estimate the abundance values in these cases. While these estimates may not be as accurate as the datasets obtained from published tables, the ordinal ranking of element abundances in the assemblages should be preserved.

An additional problem that may affect the skeletal part patterns is that taxonomic identification of elements becomes more difficult when there are several anatomically-similar species poorly separated in size. For example, at Cuzoul du Vers (Castel 1999), the presence of reindeer, ibex and chamois in certain layers made it difficult to assign less diagnostic skeletal parts (such as ribs, vertebrae and some long bone shafts) to one species. Castel (1999) addressed this problem by providing skeletal part frequencies for "medium-sized" ungulates in addition to counts for each species. For these assemblages, we included these counts in the skeletal part frequencies of the most abundant species (reindeer) in order to avoid underestimating the frequency of less diagnostic elements in the assemblage, even though this has the drawback of possibly including specimens that may not belong to reindeer. These modifications to the NISP counts were limited to Cuzoul du Vers because problems of identification affecting particular taxa were, with few exceptions (e.g., Costamagno 1999; Fosse 1999; Marks et al. 2002; Moigne 2003), rarely mentioned in the publications. This made it difficult to evaluate objectively when the data for a given species needed to be combined with those for the same body size class, when provided. While this is a potential issue that should be kept in mind, it may be of less concern at sites such as Abri Pataud, Saint-Césaire, Jonzac, Grotte XVI and many others where the most abundant species are well-separated in body size.

NISP values were normed (NNISP) in order to enable comparisons with utility indices by accounting for the fact that carcasses may include multiple units of the same category of elements (e.g., a complete reindeer carcass comprises 26 ribs but only one atlas). Although NNISP has its own shortcomings, studies have shown it to be a useful measure of abundance (Grayson and Frey 2004). It should be noted that, in order to prevent inflation of head counts relative to limb bones, archaeological values for the cranium were divided by two because the samples are dominated by isolated teeth and short tooth series (see Appendix 14.1 for details of norming procedures). As emphasized by Grayson and Frey (2004), norming cranial counts by two should provide a more accurate assessment of head counts, as isolated teeth are represented twice in a single cranium (as right and left). When NISP data were not available, %NISP, MNE, $MNI_{element}$ or %MAU values were used as alternatives. Additionally, for a small number of assemblages ($n = 3$), skeletal part data were available only as %PO (*Pourcentages de Parties Observées* or percentages of observed parts). The formula used in these publications to derive %PO values suggests that they are equivalent to %MAU values.

Site Function and Density-Mediated Attrition

At all sites included in the dataset, humans appear to have been the primary accumulator of animal remains. Skeletal part patterns at these sites must therefore reflect transport decisions, which are expected to vary depending on whether the assemblage represents a residential camp, a kill site, a lookout, etc. To examine this issue, the assemblages were classified according to their topographic setting, as this factor might have influenced transport strategies. For simplicity, we used a binary opposition: natural shelters (caves, rockshelters, cliff-base deposits) versus open air sites. Because human-occupied caves and rockshelters are unlikely to have frequently functioned as kill sites or to have provided productive scavenging opportunities, it is reasonable to assume that skeletal parts were largely transported there from other locations. In our sample, most assemblages (136/167 or 81.4 %) are from natural shelters (Table 14.2). This bias was expected, given that in the study area artifacts are generally easier to find in karstic formations than in often deeply buried open air locations.

Whereas faunal assemblages in natural shelters are probably composed of skeletal parts transported to the site, this is not necessarily the case for open air sites, which potentially document a broader range of functions, including residential camps, kill or butchery stations, caches, multi-purpose sites (for instance, if a lookout becomes a cache in the following year), etc. Consequently, faunal remains at open air sites may represent the residuals of decisions to transport parts *away from* as well as *to* the site. For kill sites, we would expect a negative correlation with the utility indices, if high-utility parts were being transported away from the site. It should be noted that the sinkhole assemblage from La Borde (Slott-Moller 1990)

Table 14.4 Description of the various utility indices used in this paper

Taxon, carcass product	Index	Source
Cervids		
Overall food	Simplified Meat Utility Index (MUI, caribou)	Data from Metcalfe and Jones (1988: 489, Table 1), with minor modifications
Overall food, adjusted for riders	Corrected Food Utility Index (CFUI, caribou)	This paper, modified from Metcalfe and Jones (1988: 492, Table 2)
Bone grease	Nunamiut spring grease-rendering episode (caribou)	Binford (1978: 36, Table 1.13, col. 6)
Marrow	Unsaturated Marrow Index (UMI, caribou)	Morin (2007: 77, Table 4)
Dried meat	Meat Drying Index (MDI, caribou)	Friesen (2001: 320, Table 2), with minor modifications
Large bovines		
Overall food	Simplified Meat Utility Index (bison)	Emerson (1990: 609–610, Fig. 8.1, pp. 806–808, Appendix C, Table 1), average of four bison
Overall food, adjusted for riders	Corrected Food Utility Index (CFUI, bison)	This paper, Emerson (1990: 609–610, Fig. 8.1, pp. 806–808, Appendix C, Table 1), average of four bison
Bone grease	Averaged Standardized Skeletal Fat model ([S]AVGSKF (bison)	Emerson (1993: 143, Fig. 8.4b), long bone halves averaged in order to obtain whole bone values
Marrow	Unsaturated Marrow Index (UMI, bison)	This paper, data from Emerson (1990: 337, Table 5.25) (average of four bison), and Meng et al. (1969: 189, Table 1)
Dried meat	Meat Drying Index (MDI, bison)	This paper, data from Emerson (1990: 294–296, Table 5.16, p. 338, Table 5.26, pp. 609–610, Fig. 8.1, pp. 806–808, Appendix C, Table 1), average of four bison
Caprines		
Overall food	Simplified Meat Utility Index (MUI, 90-month sheep)	This paper, data from Binford (1978: 16, Table 1.1)
Overall food, adjusted for riders	Corrected Food Utility Index (CFUI, 90-month sheep)	This paper, data modified from Binford (1978: 16, Table 1.1)
Bone grease	Nunamiut spring grease-rendering episode (caribou)	Binford (1978: 36), Table 1.13
Marrow	Unsaturated Marrow Index (UMI, 90-month sheep)	This paper, data from Binford (1978: 26, Table 1.8), and Meng et al. (1969: 189, Table 1)
Dried meat	Meat Drying Index (MDI, 90-month sheep)	This paper, data from Binford (1978: 16, Table 1.1 and p. 104, Table 3.1)
Equids		
Overall food	General Utility Index (GUI, horse)	Outram and Rowley-Conwy (1998: 844, Table 5), mean of three horses
Overall food, adjusted for riders	Corrected Food Utility Index (CFUI, horse)	Modified from Outram and Rowley-Conwy (1998: 845, Table 6), mean of three horses
Marrow	Unsaturated Marrow Index (UMI, horse)	Morin (2012) [data from Outram and Rowley-Conwy (1998: 843, Table 4), mean of three horses, and Meng et al. (1969: 189, Table 1)]

was grouped with open air sites because the former might have functioned as a natural trap, and therefore, would be conceptually equivalent to an open air kill station. Including La Borde, our sample contains a moderate percentage of open air assemblages (31/167 or 18.6 %, Table 14.2).

Given the differences in preservation conditions between most open air and sheltered sites—notably due to variations in exposure to weathering from sun, rain, wind, and other agents—the quality of faunal preservation may often be poorer in open air sites. The effects of this problem might also be very localized, given that preservation of skeletal remains frequently varies within sedimentary sequences, regardless of the type of site. In addition, other density-mediated processes (e.g., carnivore ravaging, bone grease rendering, selective burning of grease-rich elements as fuel) may have a significant impact on patterns of anatomical representation in an assemblage (Morin 2010). Consequently, we attempted to control for the problem of density-mediated attrition by comparing the correlations with utility indices obtained before and after exclusion of low density parts.

We use the phrase "entire set" to refer to the nineteen classes of skeletal elements included in the comparisons with the utility models and "high survival set" (Marean and Cleghorn 2003) to refer to the reduced set of nine high-density elements that are less likely to be affected by density-mediated attrition. However, it should be noted that these figures may be slightly lower for certain assemblages with missing data (e.g., no data are available on cranium representation at Marillac and Le Sire). For reindeer, bovines and horse, we used the mineral density data collected by Lam et al. (1999) for reindeer (*Rangifer tarandus*), wildebeest (*Connochaetes taurinus*) and *Equus* sp. (Burchell's zebra *Equus burchelli* and Przewalski's horse *Equus ferus przewalskii*), respectively, to create the set of high survival elements. The wildebeest should be a valid proxy for *Bos*/*Bison*, given that Lam and his collaborators (1999) observed only minor differences in density patterns among artiodactyl species. For caprines, we again used the wildebeest as a proxy because Symmons's (2005) density study of sheep (*Ovis aries*) does not consider several of the elements included in our sample. The elements that we included in the high survival set comprise the cranium, mandible, scapula, and all six long bones (Appendix 14.1). This set is identical to the set of high survival elements presented by Marean and Cleghorn (2003), except for the scapula, which was excluded by these last authors for reasons of poor identifiability. In our analyses, we included the scapula in the high survival set because this element is relatively dense and does not present insurmountable problems of taxonomic identification in the species examined here.

Utility Models

To examine what criteria underlie transport decisions in our sample of assemblages, we considered five types of utility models, intended to measure the yields of different carcass products: (i) overall food, (ii) overall food, adjusted for riders, (iii) bone grease, (iv) marrow, and (v) dried meat. Metcalfe and Jones (1988), Jones and Metcalfe (1988), Emerson (1990, 1993), and Lyman (1992) have highlighted several problems in the construction of published utility indices. In this paper, we have attempted to solve some of these problems using the existing data.

The utility models used for each taxonomic group are summarized in Table 14.4. The models for cervids are based on caribou anatomical data presented by Binford (1978). Indices for bison were calculated from raw data in Emerson (1990), which are based on the average weights of anatomical portions of four bison. Indices for sheep were constructed from data on the body composition of a 90-month old domestic sheep from Binford (1978). The calculations used in constructing the indices for this species duplicate those for caribou, as do those for the bison data. Finally, the horse models include a general utility index presented by Outram and Rowley-Conwy (1998) and two additional indices generated based on their data. To improve consistency between taxa and to correct errors, we made minor changes to most of the previously published indices we used, such as the caribou and horse general utility indices (Metcalfe and Jones 1988; Outram and Rowley-Conwy 1998).

For each of the taxonomic groups, we first considered a simple index measuring the overall food content—or general utility—of anatomical portions: the simplified MUI, or Meat Utility Index, which is a measurement of the total soft tissue weight associated with an element (including non-meat products, such as grease and marrow). The simplified MUI is calculated by subtracting the dry bone weight from the gross weight of the element (Metcalfe and Jones 1988). In the simplified MUI, we made two changes relative to Metcalfe and Jones' original MUI. In his analysis of caribou and sheep, Binford (1978) provided wet and dry weight values for one foot (i.e., a set of six phalanges) of the hind leg and one foot of the fore leg, and these weights were used by Metcalfe and Jones (1988). However, these values must be divided by two because, in two-toed artiodactyls, each foot comprises two of each type of phalanx. We then multiplied the values for the combined set of three phalanges by 0.45, 0.35 and 0.20, to obtain rough estimates for the values of the first, second and third phalanx, respectively, based on our approximation of the relative utility of artiodactyl phalanges. We also derived MUI values for the entire set of cervicals (C1–7), as well as for the entire vertebral column, because these are categories often used by archaeozoologists.

We also considered a second general utility index, the CFUI, or Corrected Food Utility Index, which we introduce here. The CFUI is based on the same measurements as the simplified MUI (i.e., total soft tissue weight of an element), but is modified for "riding," that is, the tendency for lower-utility bone portions to be transported as a result of their close association with high-utility parts (Binford 1978). We include the CFUI because this index is similar to Binford's MGUI (Modified General Utility Index) and Metcalfe and Jones' FUI, which are the indices most often used in faunal analysis. However, we prefer the simplified MUI over the rider-averaged CFUI and the original FUI for a number of reasons.

First of all, the simplified MUI is easily constructed and provides a very intuitive measure of utility, as it is strictly

Table 14.5 Derivation of the Corrected Whole Bone FUI (CFUI) for caribou. See text for explanations

Element	FUI	Element	CFUI	Derivation of CFUI
Skull (with brain)[a]	469 (235)	Skull (with brain)	937 (469)	Wet weight (g) – dry bone weight (g)
Mandible (tongue)	1,600	Mandible (tongue)	1,600	Wet weight (g) – dry bone weight (g) (=FUI)
Mandible (w/o t.)	590	Mandible (w/o t.)	590	Wet weight (g) – dry bone weight (g) (=FUI)
Atlas	524	Atlas-axis	524	Wet weight (g) – dry bone weight (g)
Axis	524	Cervicals 3–7	1,905	Wet weight (g) – dry bone weight (g) (=FUI)
Cervicals 3–7	1,905	Thoracic	2,433	Wet weight (g) – dry bone weight (g) (=FUI)
Thoracic	2,433	Lumbar	1,706	Wet weight (g) – dry bone weight (g) (=FUI)
Lumbar	1,706	Pelvis + sacrum	2,531	Wet weight (g) – dry bone weight (g) (=FUI)
Pelvis + sacrum	2,531	Ribs	2,650	Wet weight (g) – dry bone weight (g) (=FUI)
Ribs	2,650	Scapula	2,295	Wet weight (g) – dry bone weight (g) (=FUI)
Scapula	2,295	Humerus	2,093	Average of P hum FUI and D hum FUI
P humerus	2,295	Radio-ulna	1,181	Average of P rul FUI and D rul FUI
D humerus	1,891	Carpals	653	FUI value
P radio-ulna	1,323	Metacarpal	413	Average of P mc FUI and D mc FUI
D radio-ulna	1,039	Femur	5,139	Average of P fem FUI and D fem FUI
Carpals	653	Tibia	2,746	Average of P tib FUI and D tib FUI
P metacarpal	461	Tarsals	1,424	FUI value
D metacarpal	364	Metatarsal	897	Average of P mt FUI and D mt FUI
P femur	5,139	Phalanx 1	99.6	Phalanges FUI/2 × 0.45
D femur	5,139	Phalanx 2	77.5	Phalanges FUI/2 × 0.35
P tibia	3,225	Phalanx 3	44.3	Phalanges FUI/2 × 0.20
D tibia	2,267			
Tarsals	1,424	All cervicals	2,429	Atlas-axis CFUI + cervicals 3–7 CFUI
P metatarsal	1,003	All phalanges	221	Phalanges FUI/2
D metatarsal	792			
Phalanx 1	443			
Phalanx 2	443			
Phalanx 3	443			

[a] Following Binford (1978), values in parentheses for the cranium exclude largely unusable cartilage
w/o t. without tongue; *P* proximal; *D* distal; *hum* humerus; *rul* radio-ulna; *mc* metacarpal; *fem* femur; *tib* tibia; *mt* metatarsal.

based on total soft tissue mass. In contrast, the CFUI is not an empirical index, because it is modified to fit a set of assumptions about the disarticulation behaviors of foragers at butchery sites. These assumptions may not always be appropriate. We would argue that "riding" of low-utility elements is a phenomenon that researchers should attempt to identify in their assemblages rather than assume a priori. For instance, according to central place foraging models, riding of low-utility parts should be more common when kills are made at short distances from the residential place. By inflating the rank of low-utility parts, rider-averaged indices such as the FUI and CFUI may obscure the study of transport strategies in such contexts. However, the differences between the simplified MUI and CFUI should not be overstated, as these models provide similar—although not identical—rank-order information about food utility. Despite the problems that we see with the CFUI, we included it in our analysis to enhance comparisons with other datasets that use the FUI as a reference model.

The CFUI (described for caribou in Table 14.5) is slightly different from other rider-averaged indices such as Metcalfe and Jones' (1988) FUI, because we modified the averaging procedure to rectify some problems identified by Emerson (1990, 1993) and ourselves in the derivation of this index (and of Binford's MGUI). In the original derivation of the FUI, the index values of appendicular parts were adjusted by averaging the value of the part with that of the proximally-adjoining skeletal portion (see Table 14.5). This method was used to divide long bone utility values into proximal and distal portions, which reflected Binford's observation that the Nunamiut sometimes butchered (often frozen) carcasses through the long bones.

Table 14.6 Whole-bone utility indices for caribou/reindeer: the simplified Meat Utility Index (MUI, Metcalfe and Jones 1988: 489, Table 1), the Food Utility Index (FUI, Metcalfe and Jones 1988), the Corrected Food Utility Index (CFUI), a proxy for grease utility based on ethnographic data (Nunamiut spring episode of grease rendering, Binford 1978: 36, Table 1.13, col. 6), the Unsaturated Marrow Index (UMI, Morin 2007: 77, Table 4), and the Meat Drying Index (MDI, Friesen 2001: 320, Table 1)

Element	MUI (g)	FUI	CFUI	Grease	UMI (ml)	MDI
Cranium	937 (469)	469 (235)	937 (469)			1.9
Mandible (with tongue)	1,600	1,600	1,600			66.4
Mandible (w/o tongue)	590	590	590			56.2
Atlas-axis	524		524			88.2
Atlas		524				
Axis		524				
Cervicals 3–7	1,905	1,905	1,905			186.7
Thoracic	2,433	2,433	2,433			311.3
Lumbar	1,706	1,706	1,706			205.8
Pelvis + Sacrum	2,531	2,531	2,531			196.8
Ribs	2,650	2,650	2,650			745.4
Scapula	2,295	2,295	2,295	0.0		89.5
Humerus	1,486		2,093	60.0	22.8	18.5
Proximal		2,295				
Distal		1,891				
Radio-Ulna	755		1,181	46.7	26.3	16.4
Proximal		1,323				
Distal		1,039				
Carpals		653	653	0.0	0.9	
Metacarpal	268		413	25.0	19.6	15.5
Proximal		461				
Distal		364				
Femur	5,139		5,139	83.3	34	17
Proximal		5,139				
Distal		5,139				
Tibia	1,310		2,746	62.5	51.1	13
Proximal		3,225				
Distal		2,267				
Tarsals		1,424	1,424	50.0	4.4	
Metatarsal	581		897	50.0	46.5	11.2
Proximal		1,003				
Distal		792				
Phalanx 1	21.0	443	99.6	0.0	3.7	15.1
Phalanx 2	16.3	443	77.5	0.0	1.8	11.8
Phalanx 3	9.3	443	44.3	0.0	0.9	6.7
Phalanges (set of 3)	46.7	443	221	0.0	6.4	33.6
All cervicals	2,429	2,429	2,429			277.3
All vertebrae	6,568		6,568			794.4

The MUI measures the total amount of soft tissues associated with an element (including marrow and grease), not just meat as implied by its name. It should be noted that the MUI, CFUI and MDI values for the phalanges were divided by two to obtain values for single phalanges. In addition, the CFUI values for the cranium and atlas-axis have been corrected (see text for explanations). Following Binford, values in parentheses for the cranium exclude largely unusable cartilage

However, as noted by Emerson (1993), Binford's (1978) procedure exaggerates the utility of long bones because the values for the proximal and distal halves are *both* based on values for the *whole* bone. Furthermore, archaeozoologists frequently tabulate NISP or MNE values for complete long bones rather than for proximal and distal portions. The

Table 14.7 Data used to derive whole-bone utility indices for sheep (*Ovis aries*): the simplified MUI, the CFUI, the UMI and the MDI. As suggested by Friesen (2001), zero values in col. (j) were rounded up to 1 %

Element	Overall food					Marrow			Dry meat			
	Wet weight in g[a]	Dry weight in g[b]	MUI in g[c]	FUI[d]	CFUI[e]	MCV in ml[f]	Prop of UFA[g]	UMI in ml[h]	Brain or marrow weight in g[i]	% Brain-marrow[j] (%)	Meat weight in g[k]	MDI[l]
Cranium	938.05	294.82	643 (322)	643 (322)	643 (322)				158.79	100.00	484.44	1.5
Mandible	1,193.87	167.60	1,026	1,026	1,026				4.1	2.58	1,022.17	55.6
Mandible (w/o tongue)	525.37	167.60	358	358	358				4.1	2.58	353.67	43.7
Atlas-axis	408.24	87.90	320		320				0	1.00	320.34	69.0
Atlas				320								
Axis				320								
Cervicals 3–7	1,088.64	137.40	951	951	951				0	1.00	951.24	120.1
Thoracic	1,758.20	288.58	1,470	1,470	1,470				0	1.00	1,469.62	241.2
Lumbar	871.29	205.35	666	666	666				0	1.00	665.94	157.0
Pelvis + Sacrum	1,623.55	319.80	1,304	1,304	1,304				2.21	1.39	1,301.54	184.2
Ribs	1,995.84	373.04	1,623	1,623	1,623				0	1.00	1,622.8	303.3
Scapula	844.76	75.10	770	770	770				1.5	0.94	768.16	72.3
Humerus	584.86	95.10	490		700	19.2	0.6007	11.5	15.17	9.55	474.59	8.1
Proximal				770								
Distal				630								
Radio-Ulna	324.92	88.50	236		384	14.0	0.7313	10.2	11.06	6.97	225.36	8.8
Proximal				433								
Distal				335								
Carpals				209	209	1.0	0.8776	0.9	0	0.00		
Metacarpal	135.08	51.50	84		131	8.1	0.9320	7.5	6.4	4.03	77.18	7.3
Proximal				146								
Distal				115								
Femur	1,474.20	121.00	1,353		1,353	24.2	0.6546	15.8	19.12	12.04	1,334.08	9.1
Proximal				1,353								
Distal				1,353								
Tibia	498.96	114.00	385		748	23.9	0.7986	19.1	18.89	11.90	366.07	7.0
Proximal				869								
Distal				627								
Tarsals				359	359	3.2	0.8776	2.8	0	0.00		
Metatarsal	149.69	59.49	90		191	9.5	0.9123	8.7	7.51	4.73	82.69	6.9
Proximal				224								
Distal				157								
Phalanx 1			15	111	25	2.0	0.9179	1.8				5.2
Phalanx 2			11	111	19	1.0	0.9184	0.9				4.0
Phalanx 3			6	111	11	1.0	0.9089	0.9				2.3
Phalanges 1–3			32	111	56	4.0	0.9151	3.7	1.58	1.00	30.77	11.5
Cervicals	1,496.88	225.30	1,272	1,272	1,272				0	1.00	1,271.58	191.4
All vertebrae	4,126.37	719.23	3,407	3,407	3,407				0	1.00	3,407.14	593.9

Following Binford (1978), values in parentheses for the cranium exclude largely unusable cartilage

MCV marrow cavity volume; *prop of UFA* proportion of unsaturated fatty acids

[a] Data from Binford (1978: 16, Table 1.1)

[b] Data from Binford (1978: 16, Table 1.1)

[c] Col. (a) – col. (b)

[d] Calculated following Metcalfe and Jones (1988)

[e] Data obtained using the formulas in Table 5

[f] Data from Binford (1978: 24, Table 1.6)

[g] Data from Meng et al. (1969: 169, Table 1)

[h] Col. (e) × col. (f)

[i] Data from Binford (1978: 104, Table 3.1)

[j] Standardized values from col. (i)

[k] Gross weight – (dry bone weight + marrow or brain weight)

[l] Calculated using the formula in Friesen (2001: 319)

Table 14.8 Data used to derive whole-bone utility indices for bison (*Bison bison*): the simplified MUI, the CFUI, the UMI and the MDI. As suggested by Friesen (2001), zero values in col. (j) were rounded up to 1 %

Element	Overall food					Marrow			Dry meat			
	Wet weight in g[a]	Dry weight in g[b]	MUI in g[c]	FUI [d]	CFUI[e]	MCV in ml[f]	Prop of UFA[g]	UMI in ml[h]	Brain or marrow weight in g[i]	Brain-marrow[j] (%)	Meat weight in g[k]	MDI[l]
Cranium		4,307.45										
Mandible		1,358.65				46.75						
Mandible (w/o tongue)		1,358.65				46.75						
Atlas-axis	5,285.83	537.45	4,748		4,748				0.00	1.0	4,748.4	482.8
Atlas				4,748								
Axis				4,748								
Cervicals 3–7	21,150.20	1,067.05	20,083	20,083	20,083				0.00	1.0	20,083.2	1,013.2
Thoracic	26,279.90	2,175.30	24,105	24,105	24,105				0.00	1.0	24,104.6	1,995.2
Lumbar	23,790.98	885.75	22,905	22,905	22,905				0.00	1.0	22,905.2	852.8
Pelvis + Sacrum	19,724.50	2,067.90	17,657	17,657	17,657				6.18	4.1	17,650.4	456.1
Ribs	35,225.45	3,247.25	31,978	31,978	31,978				0.00	1.0	31,978.2	2,947.9
Scapula	14,384.10	1,305.65	13,078	13,078	13,078	3.50			2.40	1.6	13,076.1	752.7
Humerus	15,849.43	1,890.90	13,959		13,959	152.75	0.6007	91.8	124.98	82.1	1,3833.6	20.10
Proximal				13,959								
Distal				13,959								
Radio-Ulna	6,926.03	1,550.35	5,376		8,594	100.13	0.7313	73.2	89.65	58.9	5,286.0	20.09
Proximal				9,667								
Distal				7,521								
Carpals	682.40	267.60	415	4,289	4,289		0.8776			1.0	414.8	162.7
Metacarpal	1,695.68	640.00	1,056		2,268	30.25	0.9320	28.2	25.58	16.8	1,030.1	23.1
Proximal				2,672								
Distal				1,864								
Femur	52,338.00	2,008.45	50,330		50,330	169.13	0.6546	110.7	138.08	90.7	50,191.5	21.2
Proximal				50,330								
Distal				50,330								
Tibia	9,220.10	1,582.60	7,638		23,647	170.00	0.7986	135.8	152.15	100.0	7,485.4	12.9
Proximal				28,984								
Distal				18,311								
Tarsals	1,630.53	807.80	823	9,682	9,682		0.8776			1.0	822.7	407.6

(continued)

Table 14.8 Data used to derive whole-bone utility indices for bison (*Bison bison*): the simplified MUI, the CFUI, the UMI and the MDI. As suggested by Friesen (2001), zero values in col. (j) were rounded up to 1 %

	Overall food					Marrow			Dry meat			
Element	Wet weight in g[a]	Dry weight in g[b]	MUI in g[c]	FUI [d]	CFUI[e]	MCV in ml[f]	Prop of UFA[g]	UMI in ml[h]	Brain or marrow weight in g[i]	Brain-marrow[j](%)	Meat weight in g[k]	MDI[l]
Metatarsal	1,807.33	754.35	1,053		4,289	37.40	0.9123	34.1	31.25	20.5	1,021.7	20.8
Proximal				5,367								
Distal				3,210								
Phalanx 1			170		446	1.24	0.9179	1.1				55.9
Phalanx 2			132		347	0.96	0.9184	0.9				43.5
Phalanx 3			76		198	0.55	0.9089	0.5				24.9
Phalanges 1–3	627.26	249.50	378	1,983	991	2.75	0.9151	2.5	1.83	1.2	375.9	124.3
Cervicals	26,436.03	1,604.50	24,832	24,832	24,832				0.00	1.0	24,831.5	1,507.1
All vertebrae	76,506.90	4,665.55	71,841	71,841	71,841				0.00	1.0	71,841.4	4,381.0

Following Binford (1978), values in parentheses for the cranium excludes largely unusable cartilage. There were no separate data on soft tissue weights of bison crania and mandibles in Emerson (1990), and therefore, we could not calculate a simplified MUI or CFUI value for these portions

MCV marrow cavity volume; *prop of UFA* proportion of unsaturated fatty acids

[a] Data from Emerson (1990: 609–610, Fig. 8.1; pp. 806–808, Appendix C, Table 1; and 294–296, Table 5.6); calculated using Emerson's GWT + dry bone weight

[b] Data from Emerson (1990: 294–296, Table 5.6)

[c] Col. (a) − col. (b) (=Emerson's GWT)

[d] Calculated following Metcalfe and Jones (1988)

[e] Data obtained using the formulas in Table 5

[f] Data from Emerson (1990: 337, Table 5.25)

[g] Data from Meng et al. (1969: 169, Table 1)

[h] Col. (e) × col. (f)

[i] Data from Emerson (1990: 338, Table 5.26)

[j] Standardized values from col. (i)

[k] Gross weight − (dry bone weight + marrow or brain weight)

[l] Calculated using the formula in Friesen (2001: 319)

reason behind this is often due to the difficulty of assigning abundant non-descript shaft portions to the proximal or distal end of the bone. To deal with these problems, in the CFUI we averaged the proximal and distal FUI values for the long bones. Concerning the head, Binford (1978: 74) argued that the value for the skull should be divided by two to account for the presence of largely unusable cartilage. For unknown reasons, the values for this element calculated with and without cartilage ended up being divided by two and four, respectively, in the derivation of the FUI. We correct for this error in the CFUI.

In the FUI, the index values of the atlas and axis were inflated by assigning to each of these vertebrae the value of the two combined. In the CFUI, we have assigned the simplified MUI value to the atlas-axis set to correct for this unnecessary inflation. The same procedures used to calculate values for the phalanges for the simplified MUI were used to derive CFUI values from FUI values for the phalanges. CFUI values for the cervical vertebrae and for the entire vertebral column were also calculated.

For all of the taxa considered in our analysis, the same procedures were used in the derivation of simplified MUI and CFUI indices, except that the values for sets of horse phalanges were not divided by two because, unlike two-toed artiodactyls, phalanges in this species are not present as duplicates in a foot. It should also be noted that, due to a lack of data, the horse simplified MUI and CFUI are constructed slightly differently than those for the other species. The horse MUI and CFUI are based on meat weight plus marrow weight rather than gross weight minus dry bone weight (see Outram and Rowley-Conwy 1998).

The grease index (Tables 14.4, 14.6) consists of data from an ethnographic episode observed by Binford (1978) during which a Nunamiut woman selected caribou parts for grease rendering in spring 1971. To obtain whole bone values for this index, we averaged the proximal and distal long bone values provided by Binford. The same data served as a grease index for cervids, bison, and caprines. However, this index was not used for horse because of the important differences between equids and the other artiodactyls examined here in the bony structure of limb elements (Outram and Rowley-Conwy 1998).

All the assemblages were also compared to a marrow index. The marrow index we used is the Unsaturated Marrow Index (UMI), which measures the total quantity of unsaturated fatty acids in the major marrow-bearing elements of caribou (Morin 2007). To calculate UMI values for the other taxonomic groups, marrow cavity volumes for bison, sheep, and horse, provided by Emerson (1990), Binford (1978), and Outram and Rowley-Conwy (1998), respectively, were multiplied by the corresponding proportions of unsaturated fatty acids in marrow-bearing bones of caribou (Meng et al. 1969). The caribou data should be a reasonable proxy for these species, given strong inter-specific patterning in the fatty acid content of mammalian limb bones (Irving et al. 1957; West and Shaw 1975; Pond 1998; Garvey 2011).

The final index we considered is the Meat Drying Index, or MDI, which was constructed for caribou by Friesen (2001) using data from Binford (1978). The MDI assumes that the suitability of a skeletal portion for drying is related to the quantity of meat in a portion, its surface area, and the absence of brain and marrow which may spoil easily. Consequently, to construct the MDI, Friesen (2001) multiplied the meat weight of a skeletal portion by its ratio of bone weight to gross weight, and then divided the result by the percentage of brain or marrow weight in the element (see Tables 14.7, 14.8). We applied some minor modifications to Friesen's calculations. First, to obtain values for the individual phalanges, we adopted the same procedure used to obtain values for phalanges for the MUI and SMUI: the value for one two-toed artiodactyl foot (six phalanges) was divided by two and then multiplied by 0.45, 0.35, and 0.20, for the first, second, and third phalanges, respectively. We also calculated values for all seven cervicals and for the entire vertebral column. For bison and sheep, we calculated meat drying indices using Friesen's formula and data provided for sheep by Binford (1978) and for bison by Emerson (1990). We did not create a Meat Drying Index for horse because we did not have data on bone weight.

Statistical Analyses

As suggested by Grayson (1984) and Lyman (2008) for comparisons involving archaeozoological data, statistical relationships between the utility models and the faunal assemblages were examined using Spearman's rank-order correlation. In all statistical tests, *p*-values were considered significant when <0.05. It should be noted that certain analyses include fewer categories of elements than others (e.g., comparisons with the high survival set considers at most nine categories of elements versus 19 for the entire set). This means that comparisons with the utility models may reach statistical significance less often in the former analyses. This problem also affects the study of the utility models. For instance, in the entire set, the UMI and the grease rendering episode provide values for only 11 (9 in bovids due to a lack of data for carpals and tarsals) and 12 classes of skeletal elements, respectively, in contrast to ≥ 17 classes for the MUI, CFUI, and MDI. Therefore, in the entire set, the statistical power of the comparisons is likely to be lower for the UMI and the grease rendering model than for the other models. This is also the case for the analyses performed with the high survival set. Although we perform multiple comparisons in our analysis, we did not implement a Bonferroni correction because we considered this approach to be too conservative. Although the Bonferroni correction decreases the probability of Type I error (incorrect rejection of a null hypothesis), it increases the probability of Type II errors (incorrect acceptance of a null hypothesis). Given the low statistical power of several of our comparisons (due to a small number of classes being compared), we preferred not to increase the probability of Type II errors.

To compare how frequently the assemblages from the shelter and open air samples correlate with the various utility models, the equality of percentages was assessed after arcsine-transformation of the data using the *t* statistic (denoted t_s) presented by Sokal and Rohlf (1969: 607–610). Although lesser known in anthropology, this statistic is commonly used in biology to compare proportions (e.g., Bergerud et al. 2007). The same approach was used in our analysis of temporal trends because several expected cells frequencies are less than 5, which prevented the use of Cochran's test of linear trends (see Cannon 2001 for a description of the test).

Results

Our ability to identify foraging goals depends heavily on the discriminatory power of the utility models used in the comparisons. However, these models are often significantly correlated with each other, as shown in Table 14.9 for caribou. The correlations are particularly high between the MUI and CFUI—not surprisingly, as these are variants of the same data—and between the UMI and the grease rendering episode. This last correlation may be explained by physiology, as the anatomical distribution of bone grease is possibly influenced by that of diaphyseal marrow, the distinction between these products being entirely based on differences in modes of extraction rather than in chemical composition or function. In addition, statistically significant correlations were obtained between both of the overall food indices and the grease, marrow and dried meat models. Given these relationships, and assuming that caribou is representative of the other species in the sample, an archaeological assemblage may correlate significantly with more than one utility model. This means that the strength of the correlations—not just their statistical significance—is important when interpreting the data. We will see that there are several instances in our sample of assemblages showing a statistically significant relationship with three or more indices.

To further evaluate the explanatory power of the models, we compared the caribou utility indices with data on bone mineral density. Considering all 19 categories of elements, density is positively and significantly correlated with the grease and UMI models (Table 14.9). These results confirm that density-mediated attrition can confound the analysis of transport decisions (Lyman 1984, 1985; Grayson 1989; Lam et al. 1999). Correlations between the indices and the high survival set were performed in order to gauge the extent of this problem. It is important to reiterate that density values for the shaft portion of long bones were used

Table 14.9 Inter-correlations between the five utility indices used for cervids in this study and bone mineral density

	MUI (overall food)	CFUI (overall food)	Grease (spring episode)	UMI (marrow)	MDI (dry meat)	Mineral density
All parts						
MUI	–	**<0.001**	**<0.05**	**<0.03**	**<0.01**	<0.96
CFUI	**0.90**	–	**<0.02**	**<0.02**	**<0.02**	<0.40
Grease	**0.67**	**0.76**	–	**<0.01**	<0.44	**<0.01**
UMI	**0.78**	**0.76**	**0.86**	–	<0.46	**<0.01**
MDI	**0.82**	**0.64**	0.26	0.27	–	<0.24
Density	0.01	0.21	**0.81**	**0.85**	−0.32	–
High survival						
MUI	–	**<0.02**	<0.34	<0.57	**<0.04**	<0.44
CFUI	**0.88**	–	<0.14	<0.28	<0.16	<0.19
Grease	0.39	0.61	–	<0.18	<0.60	**<0.04**
UMI	0.26	0.49	0.60	–	<0.23	<0.75
MDI	**0.77**	0.50	−0.21	−0.54	–	<0.87
Density	0.32	0.54	**0.86**	0.14	−0.07	–

Statistically significant correlations (at the 0.05 level, in bold in the table) are shown for all 19 categories of elements (*upper*) and for high survival elements only (*lower*). Values below the dashes correspond to the correlation coefficients, while those above the dashes give the *p*-values

Table 14.10 Proportion of assemblages that are significantly correlated with the utility indices in the sample

	Entire set						High survival set					
	Shelter		Open air		Total		Shelter		Open air		Total	
	Neg.	Pos.	Neg.	Pos.	Neg.	Pos.	Neg.	Pos.	Neg.	Pos.	Neg.	Pos.
MUI												
Cervids	1/91	0/91	0/14	0/14	**1/105**	**0/105**	0/91	0/91	1/14	0/14	**1/105**	**0/105**
Bovines	0/7	0/7	0/9	1/9	**0/16**	**1/16**	0/7	0/7	1/9	0/9	**1/16**	**0/16**
Caprines	5/21	0/21			**5/21**	**0/21**	1/21	0/21			**1/21**	**0/21**
Equids	0/17	0/17	1/8	0/8	**1/25**	**0/25**	0/17	0/17	0/8	0/8	**0/25**	**0/25**
Total	**6/136**	**0/136**	**1/31**	**1/31**	**7/167**	**1/167**	**1/136**	**0/136**	**2/31**	**0/31**	**3/167**	**0/167**
CFUI												
Cervids	0/91	1/91	0/14	0/14	**0/105**	**1/105**	3/91	0/91	2/14	0/14	**5/105**	**0/105**
Bovines	0/7	0/7	0/9	1/9	**0/16**	**1/16**	0/7	0/7	0/9	0/9	**0/16**	**0/16**
Caprines	1/21	0/21			**1/21**	**0/21**	1/21	0/21			**1/21**	**0/21**
Equids	0/17	0/17	0/8	0/8	**0/25**	**0/25**	0/17	2/17	0/8	0/8	**0/25**	**2/25**
Total	**1/136**	**1/136**	**0/31**	**1/31**	**1/167**	**2/167**	**4/136**	**2/136**	**2/31**	**0/31**	**6/167**	**2/167**
Grease												
Cervids	0/91	45/91	0/14	9/14	**0/105**	**54/105**	0/91	2/91	0/14	1/14	**0/105**	**3/105**
Bovines	0/7	4/7	0/9	4/9	**0/16**	**8/16**	0/7	0/7	1/9	0/9	**1/16**	**0/16**
Caprines	0/21	8/21			**0/21**	**8/21**	0/21	2/21			**0/21**	**2/21**
Equids												
Total	**0/119**	**57/119**	**0/23**	**13/23**	**0/142**	**70/142**	**0/119**	**4/119**	**1/23**	**1/23**	**1/142**	**5/142**
UMI												
Cervids	0/91	75/91	0/14	12/14	**0/105**	**87/105**	0/91	8/91	0/14	1/14	**0/105**	**9/105**
Bovines	0/7	5/7	0/9	3/9	**0/16**	**8/16**	0/7	0/7	2/9	0/9	**2/16**	**0/16**
Caprines	0/21	11/21			**0/21**	**11/21**	0/21	1/21			**0/21**	**1/21**
Equids	0/17	9/17	0/8	5/8	**0/25**	**14/25**	0/17	2/17	0/8	0/8	**0/25**	**2/25**
Total	**0/136**	**100/136**	**0/31**	**20/31**	**0/167**	**120/167**	**0/136**	**11/136**	**2/31**	**1/31**	**2/167**	**12/167**
MDI												
Cervids	39/91	0/91	4/14	0/14	**43/105**	**0/105**	0/91	0/91	1/14	0/14	**1/105**	**0/105**
Bovines	6/7	0/7	5/9	0/9	**11/16**	**0/16**	2/7	0/7	2/9	0/9	**4/16**	**0/16**
Caprines	9/21	0/21			**9/21**	**0/21**	0/21	0/21			**0/21**	**0/21**
Equids												
Total	**54/119**	**0/119**	**9/23**	**0/23**	**63/142**	**0/142**	**2/119**	**0/119**	**3/23**	**0/23**	**5/142**	**0/142**

in the comparisons, as epiphyses are poorly represented in the archaeological dataset.

When the sample of parts is limited to high survival elements, the grease rendering episode is the only model that is significantly correlated with density, which may indicate that grease-rich epiphyses are in fact associated with shafts that are denser than those attached to grease-poor epiphyses. Because the other utility models are not significantly correlated with density in the high survival set, and given that the elements in this set all have similar density values, analyzing the dataset using only these elements should allow us to assess factors influencing skeletal part transport in assemblages that suffered from differential preservation or other attritional processes, including the use of bone as fuel. However, given that the high survival set is dominated by long bones, the correlations that are based on this set of elements may yield only limited information about transport strategies for the axial skeleton.

Before comparing the assemblages with the utility models, it is useful to highlight some general anatomical trends emerging from the study of the assemblages. Here we limit our comments to qualitative observations, as this issue will be the subject of another publication. A particularly striking pattern in the sample is the low abundance of elements from the spine and pelvis (vertebrae, sacrum, and innominates). Phalanges and ribs are also relatively rare, while many

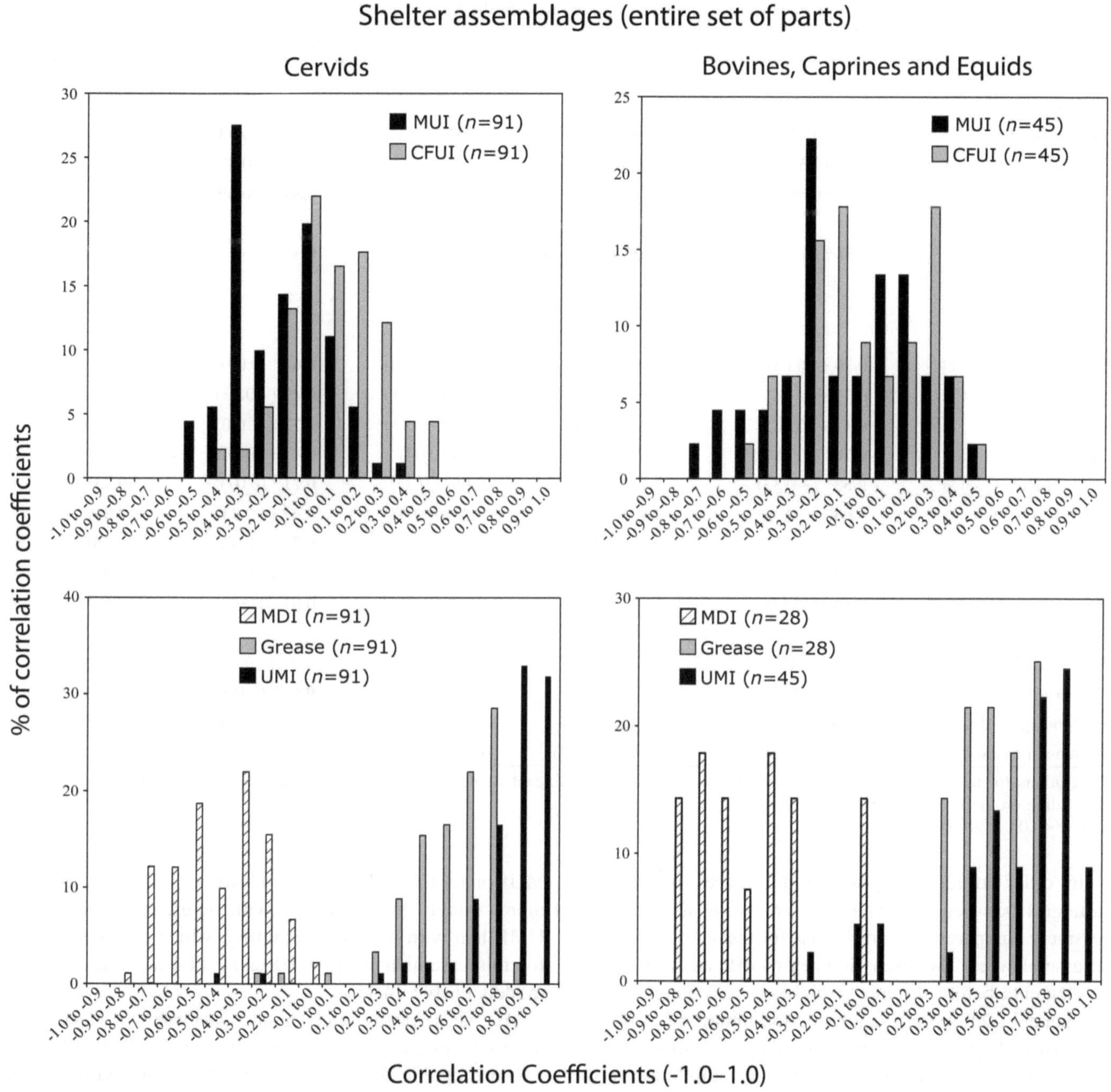

Fig. 14.1 Frequency distributions of correlation coefficients between the utility indices and the entire set of elements in shelter assemblages, shown for cervids (*left*) and for the other three taxonomic groups (bovines, caprines, equids) combined (*right*). *Upper panels*: correlations for food utility indices (FUI and CFUI); *lower panels*: correlations for indices measuring other products (grease, marrow, and dried meat)

assemblages show high frequencies of teeth and long bones. Concerning this last class of elements, a majority of studies emphasize the rarity of long bone epiphyses, which stands in sharp contrast with the abundance of shaft fragments (e.g., David and Poulain 1990; Deplano 1994; Cho 1998; Castel 1999; Costamagno 1999; David and Poulain 2002; Haws 2003; Vannoorenberghe 2004; David et al. 2005; Daujeard 2008; Morin 2010, 2012; Ready 2010). The low representation of vertebrae and long bone epiphyses in the dataset perhaps indicates that various density-mediated processes—including differential preservation, carnivore overprinting, bone grease rendering, and/or selective burning (Morin 2010)—affected the assemblages, an issue to which we return below.

Table 14.11 Assemblages showing a statistically significant relationship with the MUI and/or CFUI

	MUI	CFUI	Grease	UMI	MDI
Positive relationship					
Shelters					
Jonzac 18, Rang	0.28 (0.35)	**0.49*** (0.47)	**0.65*** (0.15)	**0.96**** (**0.93***)	−0.03 (0.10)
Rd-du-Barry F2, Cab	−0.22 (0.52)	−0.03 (**0.83***)		**0.82**** (0.83)	
Rd-du-Barry E, Cab	0.10 (0.59)	0.24 (**0.83***)		**0.84**** (**0.89***)	
Open air sites					
Biache, Bos	**0.56*** (0.14)	**0.53*** (0.43)	**0.81**** (0.68)	**0.88*** (0.66)	−0.46 (**−0.86***)
Negative relationship					
Shelters					
Les Canalettes 3, Cel	−0.39 (−0.68)	−0.30 (**−0.74***)	0.29 (−0.10)	**0.63*** (0.09)	−0.49 (−0.56)
Les Canalettes 2, Cel	−0.37 (−0.58)	−0.39 (**−0.73***)	0.24 (−0.21)	0.49 (−0.13)	−0.40 (−0.33)
Pataud 2, Rang	−0.11 (−0.58)	0.01 (**−0.70***)	0.59 (−0.11)	**0.77*** (−0.03)	−0.34 (−0.40)
Moulin-Neuf, saiga	**−0.70**** (**−0.72***)	**−0.59*** (**−0.74***)	0.35 (0.04)	0.43 (−0.66)	**−0.72**** (−0.62)
St-G.-la-Riv. 4, saiga	**−0.58*** (−0.33)	−0.42 (−0.27)	**0.66*** (0.35)	**0.71*** (0.07)	**−0.83**** (−0.30)
St-G.-la-Riv. 3, saiga	**−0.64*** (−0.55)	−0.47 (−0.50)	0.48 (−0.07)	0.45 (−0.83)	**−0.81**** (−0.23)
St-G.-la-Riv. 1, saiga	**−0.58*** (−0.38)	−0.36 (−0.25)	0.54 (0.18)	**0.75*** (0.14)	**−0.78**** (−0.18)
St-G.-la-Riv. C, saiga	**−0.60*** (−0.22)	−0.47 (−0.07)	0.42 (0.11)	0.57 (−0.26)	**−0.75**** (−0.05)
Picareiro F, Cel	**−0.56** (−0.57)	−0.28 (−0.25)	0.39 (0.19)	0.61 (0.09)	**−0.69**** (−0.50)
Open air sites					
Solutré L13, Cab	**−0.66*** (−0.57)	−0.46 (−0.40)		0.05 (−0.40)	
Verberie II-1, Rang	−0.30 (**−0.87***)	−0.03 (**−0.70***)	0.43 (−0.32)	**0.69*** (−0.49)	−0.45 (−0.55)
La Montagne, Bos	−0.26 (**−0.82***)	−0.26 (−0.75)	0.43 (−0.18)	0.27 (**−0.89***)	**−0.66*** (−0.21)
Noyen-sur-Seine, Cel	−0.01 (−0.53)	0.10 (**−0.70***)	**0.62*** (−0.11)	**0.78*** (−0.37)	−0.21 (−0.41)

Correlations for the high survival set in parentheses. Statistically significant relationships are in bold and marked with one or two asterisks (depending on the level of significance)
Note: *<0.05, **<0.01

In order to identify possible differences in transport strategies between site types, the following analysis successively compares the correlations for the shelter and open air assemblages. The discussion begins with all ≈ 19 classes of elements before turning to high survival elements in a subsequent section that focuses on the impact of attritional processes. Appendix 14.2 presents the correlations for all the assemblages, while Table 14.10 is a compilation of the results broken down by species group and type of site.

Shelters

In the sample of shelter assemblages, the correlations with the MUI and CFUI are generally weak and non-significant, as illustrated by the frequency distributions of the correlation coefficients, which are presented in Fig. 14.1 for cervids and for all other taxa combined. A single assemblage (Jonzac layer 18, 1/136 or 0.7 %) in this sample shows a positive and statistically significant relationship with the MUI or CFUI (Table 14.10). However, this assemblage is more strongly correlated both with the grease model and the UMI than with overall food utility, which means that even for this assemblage the MUI and CFUI are unsatisfactory models (Table 14.11).

At the opposite end of the spectrum, a handful of assemblages (6/136 or 4.4 %)—five of which involve saiga—are significantly, but inversely, correlated with the food utility indices. However, none of these negative correlations are strong (i.e., ≤−0.80). Overall, these results suggest that the transport of skeletal elements to shelters was motivated by goals other than the procurement of meat, or more realistically, that the transport of skeletal parts to these locations was frequently divorced from the transport of muscle masses—presumably through field processing. In contrast, the few assemblages that are negatively correlated with the food utility indices may, in theory, reflect the transport of whole animals to shelters where high utility parts were removed and subsequently transported to another location. Nevertheless, the fact that no assemblage in the dataset fits the expected pattern for this last type of site—

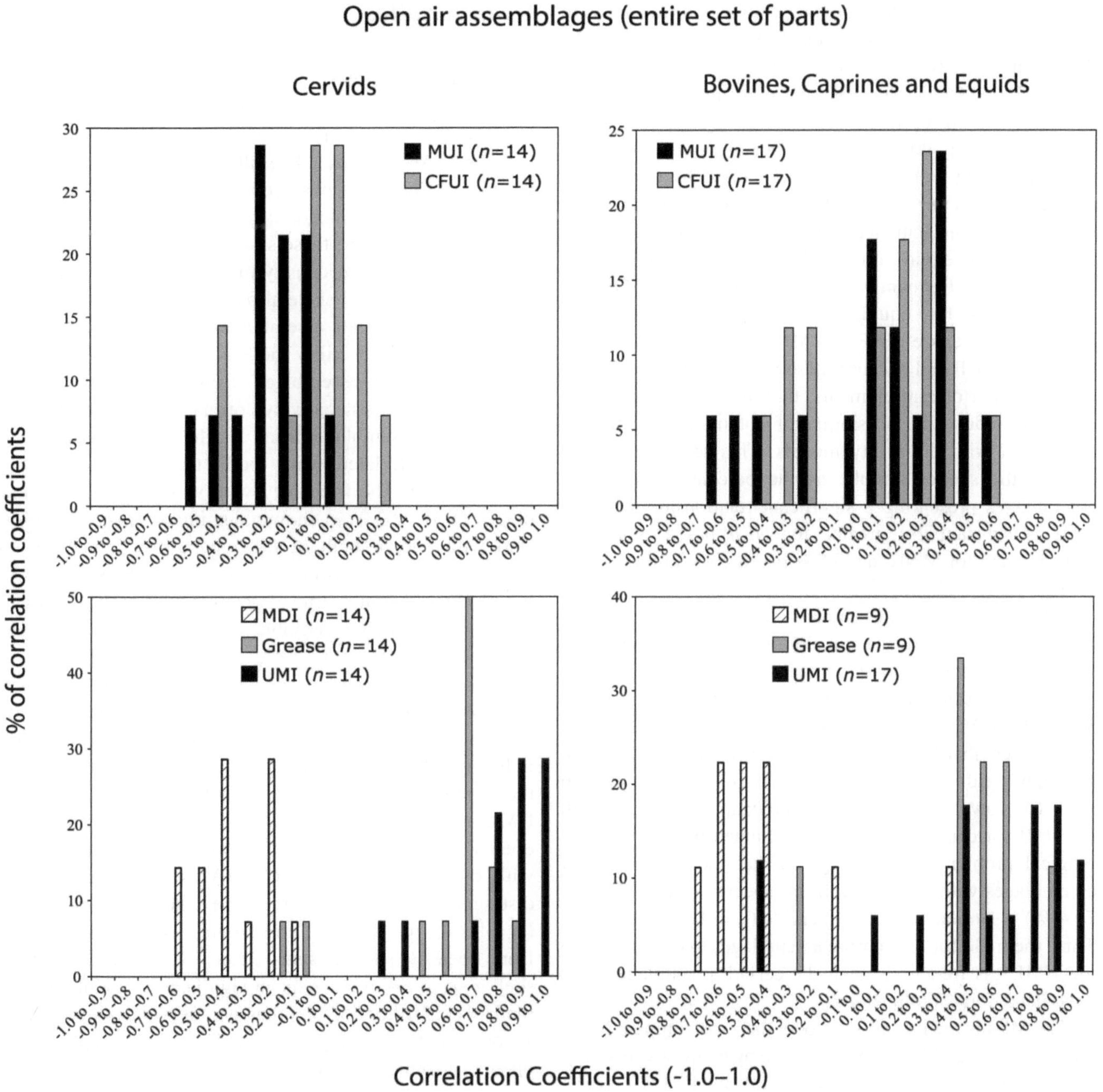

Fig. 14.2 Frequency distributions of correlation coefficients between the utility indices and the entire set of elements in open air assemblages, shown for cervids (*left*) and for the other three taxonomic groups (bovines, caprines, equids) combined (*right*). *Upper panels*: correlations for food utility indices (FUI and CFUI); *lower panels*: correlations for indices measuring other products (grease, marrow, and dried meat)

that they should show a strong positive correlation with overall food utility—argues against this scenario.

While the relationships with the food utility indices are weak, a majority of shelter assemblages are significantly correlated with the UMI (100/136 or 73.5 %) and, to a lesser extent, with the Nunamiut spring episode of grease rendering (57/119 or 47.9 %). The UMI explains variations in the shelter sample very well, given that in 58.1 % (79/136) of the cases, the correlation coefficients are ≥0.80 (Fig. 14.1). The corresponding percentage is significantly lower for the grease model (2/119 or 1.7 %, $t_s = 11.73$, $p < 0.0001$).

Significant relationships with the UMI are observed for all four species groups. However, proportionally more cervid (75/91 or 82.4 %) than caprine (11/21 or 52.4 %, $t_s = 2.71$, $p < 0.01$) and equid (9/17 or 52.9 %, $t_s = 2.45$, $p < 0.02$)

assemblages are significantly and positively correlated with marrow utility, while the difference between cervids and the small sample of bovines is non-significant (5/7 or 71.4 %, $t_s = 0.67, p < 0.51$). These findings may imply that transport strategies varied between the species groups. However, before accepting this conclusion, an alternative explanation—involving a statistical artifact caused by the type of measure used in the correlations—must be considered.

What explains statistical differences between the cervid and the caprine/equid samples? In the shelter sample, NISP data are more frequently available for cervids (70/91 or 76.9 %) than for equids (5/17 or 29.4 %, $t_s = 3.76$, $p < 0.001$). Because MNE and MNI samples in the dataset tend to be small and because these measures exaggerate the representation of rare elements, they may inflate sampling error and reduce the possibility of obtaining a significant correlation with the utility models. This seems to be the case in the shelter sample, as the percentage of cervid assemblages presenting a significant relationship with the UMI is substantially greater for NISP-based (63/70 or 90.0 %) than for MNE- and MNI-based correlations (12/21 or 57.1 %, $t_s = 3.15$, $p < 0.01$).

Although the small number of equid and caprine assemblages precludes a comparable analysis, the equid sample is qualitatively in agreement with the cervid pattern (assemblages statistically correlated with the UMI: NISP, 4/5 or 80.0 %; MNE and MNI, 5/12 or 41.7 %). These limited data suggest that the statistical difference recorded between the equid and cervid assemblages is more apparent than real. The lower frequency of significant correlations with the UMI for caprines cannot be explained by this methodological problem, however, given that the percentage of assemblages with NISP data in this species group (15/21 or 71.4 %) is similar to that for cervids ($t_s = 0.52$, $p < 0.61$). Thus, the trend for the caprine assemblages to correlate with the UMI less frequently than the cervid assemblages may have behavioral implications.

The MDI is the last utility model that was compared with the shelter assemblages. The assemblages are always *negatively*, and often significantly (54/119 or 45.4 %), correlated with the MDI. However, the correlations tend to be weak, as only 4.2 % (5/119) of the assemblages show a correlation coefficient >0.80. Consequently, it seems reasonable to conclude that few of the shelters included in our database were extensively used as caches or for drying meat—unless skeletal parts were dried there and transported elsewhere. As argued above for the correlations with the overall food models, the lack of shelter assemblages in the dataset indicating a significant positive relationship with the MDI is not consistent with this last explanation.

Open Air Sites

As emphasized earlier, more variation is expected in skeletal profiles from open air locations than from shelters because the former may include a wider range of site functions. An unexpected result of the present analysis is the many similarities observed between the two categories of sites when all categories of elements are considered (Fig. 14.2). However, these similarities must be interpreted with caution because density-mediated attrition may have substantially reduced skeletal part variation in the dataset.

Very few open air assemblages are significantly correlated with the food utility models. The exceptions ($n = 2$) include a horse assemblage from Solutré (Gravettian occupation) that is negatively correlated with the MUI and an aurochs assemblage from Biache Saint-Vaast that is positively correlated with both the MUI and CFUI (Table 14.11). The frequency with which open air assemblages (1/31 or 3.2 %) show a significant inverse relationship with one or both of the food utility indices is not statistically different from the value for shelters ($t_s = 0.32$, $p < 0.75$). This is also true of positive relationships with the same models (shelter assemblages: 1/136 or 0.7 %, open air assemblages: 1/31 or 3.2 %; $t_s = 0.97$, $p < 0.34$).

The open air assemblages generally correlate well with the grease model (13/23 or 56.5 %), and especially, the UMI (20/31 or 64.5 %). Both values are statistically indistinguishable from the corresponding percentages for the shelter sample (grease model: $t_s = 0.76$, $p < 0.45$; UMI: $t_s = 0.98$, $p < 0.33$). Likewise, the percentage of open air assemblages showing a strong relationship (≥ 0.80) with the UMI (13/31 or 41.9 %) cannot be statistically distinguished from the percentage calculated for shelters ($t_s = 1.64$, $p < 0.11$). Furthermore, cervid assemblages from open air sites present—again in agreement with the shelter sample—more frequent significant correlations with the UMI (12/14 or 85.7 %) than non-cervid assemblages (8/17 or 47.1, $t_s = 2.36$, $p < 0.02$). The small sample size of open air assemblages does not permit a detailed analysis of the relationship between the type of measure (i.e., NISP vs. MNI- or MNE-based data) and the frequency of significant correlations with different indices.

Comparisons with the MDI show a pattern opposite to that seen with the UMI. Indeed, open air assemblages tend to be inversely—rather than positively—correlated with the former model (9/23 or 39.1 %). This trend is similar to that recorded in the shelter dataset ($t_s = 0.56$, $p < 0.58$). Lastly, no significant positive correlations with the MDI were observed in the sample of open air assemblages and none of the negative correlations are >0.80.

Table 14.12 Summary table showing the percentage of assemblages that are significantly correlated (at the 0.05 level) with different utility models

	MUI or CFUI	Grease	UMI	MDI
%significant (entire set)				
Shelter	0.7 % (4.4 %)	47.9 %	73.5 %	(45.4 %)
Open air	3.2 % (3.2 %)	56.5 %	64.5 %	(39.1 %)
Shelter vs. open air	ns (ns)	ns	ns	(ns)
%significant (high survival set)				
Shelter	1.5 % (2.9 %)	3.4 %	8.1 %	(1.7 %)
Open air	(9.7 %)	4.3 % (4.3 %)	3.2 % (6.4)	(13.0 %)
Shelter vs. open air	ns (ns)	ns (ns)	ns ($t_s = 2.57$, $p < 0.02$)	($t_s = 2.09$, $p < 0.04$)
Other comparisons				
Shelter (entire set) vs. Shelter (high survival set)	ns	$t_s = 8.93$, $p < 0.0001$	$t_s = 12.23$, $p < 0.0001$	$t_s = 9.39$, $p < 0.0001$
Open air (entire set) vs. Open air (high survival set)	ns	$t_s = 4.35$, $p < 0.0001$	$t_s = 5.93$, $p < 0.0001$	$t_s = 2.08$, $p < 0.04$

Negative relationships that are statistically significant are shown in parentheses. Values are not shown when no significant correlations were observed
ns non-significant

In essence, skeletal part patterns are very similar in the shelter and open air datasets. The assemblages frequently show a significant positive relationship with the grease model, but correlations with the UMI are even stronger. In addition, a high proportion of assemblages show a negative relationship with the MDI. However, because the grease model and the UMI are strongly correlated with bone mineral density, the statistical patterns associated with these models may have few behavioral implications. Potentially, these last patterns could be entirely attributable to the action of differential preservation and/or other density-mediated processes. In the following section, we test this possibility by excluding low-survival elements from the analysis.

The Effects of Density-Mediated Processes on the Correlations

Comparisons with the high survival set result in a number of minor and a few more substantial changes in the frequencies with which the assemblages are significantly correlated with the different models. To simplify the discussion, the information on the correlations is summarized in Table 14.12. This table also synthesizes the statistical comparisons performed between the various samples.

For the high survival set, the frequency distributions of the correlation coefficients point to weak relationships with the MUI and CFUI for both the shelter (Fig. 14.3) and open air (Fig. 14.4) samples. In the total sample, only three assemblages (Moulin-Neuf, Verberie, and La Montagne), show a significant—although inverse—relationship with the MUI, the simplest and most intuitive of the food utility indices (Table 14.11). In contrast, when all 19 categories of elements are examined, the relationships with these models are weak for these assemblages, except for Moulin-Neuf. Because the food utility indices are not correlated with bone mineral density (see above), the weak correlations obtained for the entire set of parts suggest that another factor may have mediated skeletal part transport at Verberie and La Montagne, or at least, that overall food utility was not the sole criterion influencing transport decisions at these sites. Therefore, in our sample, the saiga assemblage from Moulin-Neuf is the only case showing a consistently high correlation with the MUI.

A slightly higher number of assemblages ($n = 8$) show a significant correlation with the CFUI (Table 14.12). With the exception of two assemblages from Rond-du-Barry, these significant relationships are all negative. It should be noted that the Rond-du-Barry assemblages also show comparably strong—or even stronger—correlations with the UMI. Consequently, this last model may more parsimoniously explain skeletal patterns at this site.

Correlations between the high survival set and the other utility models are coherent with the trends observed for all 19 parts, although the relationships are generally not as strong. It is important to reiterate that comparisons with the UMI and the grease rendering episode have relatively low statistical power due to the small number of classes of elements (6 for the UMI, 7 for the grease rendering episode) that could be included in the rank order correlations. Consequently, the analysis pays close attention to the shape of the frequency distribution of the correlation coefficients in addition to the significance of individual correlations.

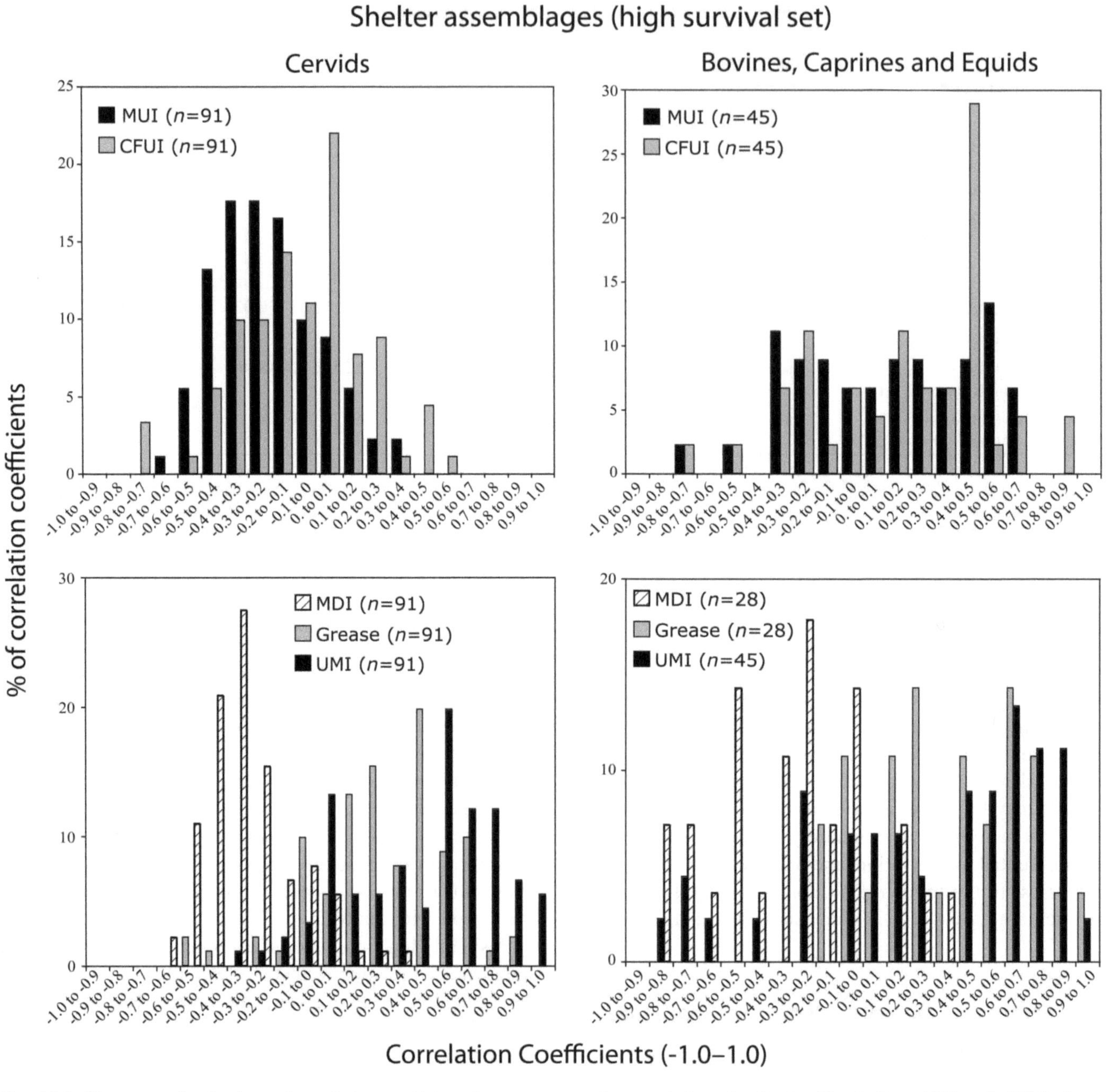

Fig. 14.3 Frequency distributions of correlation coefficients between the utility indices and the high survival set of elements ($n = 8$) in shelter assemblages, shown for cervids (*left*) and for the other three taxonomic groups (bovines, caprines, equids) combined (*right*). *Upper panels*: correlations for food utility indices (FUI and CFUI); *lower panels*: correlations for indices measuring other products (grease, marrow, and dried meat)

In the shelter sample, significant relationships with the grease model were observed in 3.4 % (4/119) of the cases (Table 14.12). This percentage is significantly lower than the value for the entire set of elements ($t_s = 8.93$, $p < 0.0001$). Likewise, the open air sample shows fewer significant correlations (1/23 or 4.3 %) between the grease model and the high survival set than between this model and the entire set ($t_s = 4.35$, $p < 0.0001$).

Despite the small number of skeletal elements that could be examined in the comparisons, the examination of the high survival set indicates that the shelter assemblages are most frequently positively correlated with the UMI (11/136 or 8.1 %). In contrast, few open air assemblages show a significant positive relationship with this model (1/31 or 3.2 %). Nonetheless, the correlation coefficients are often high, particularly in the cervid assemblages (Fig. 14.4),

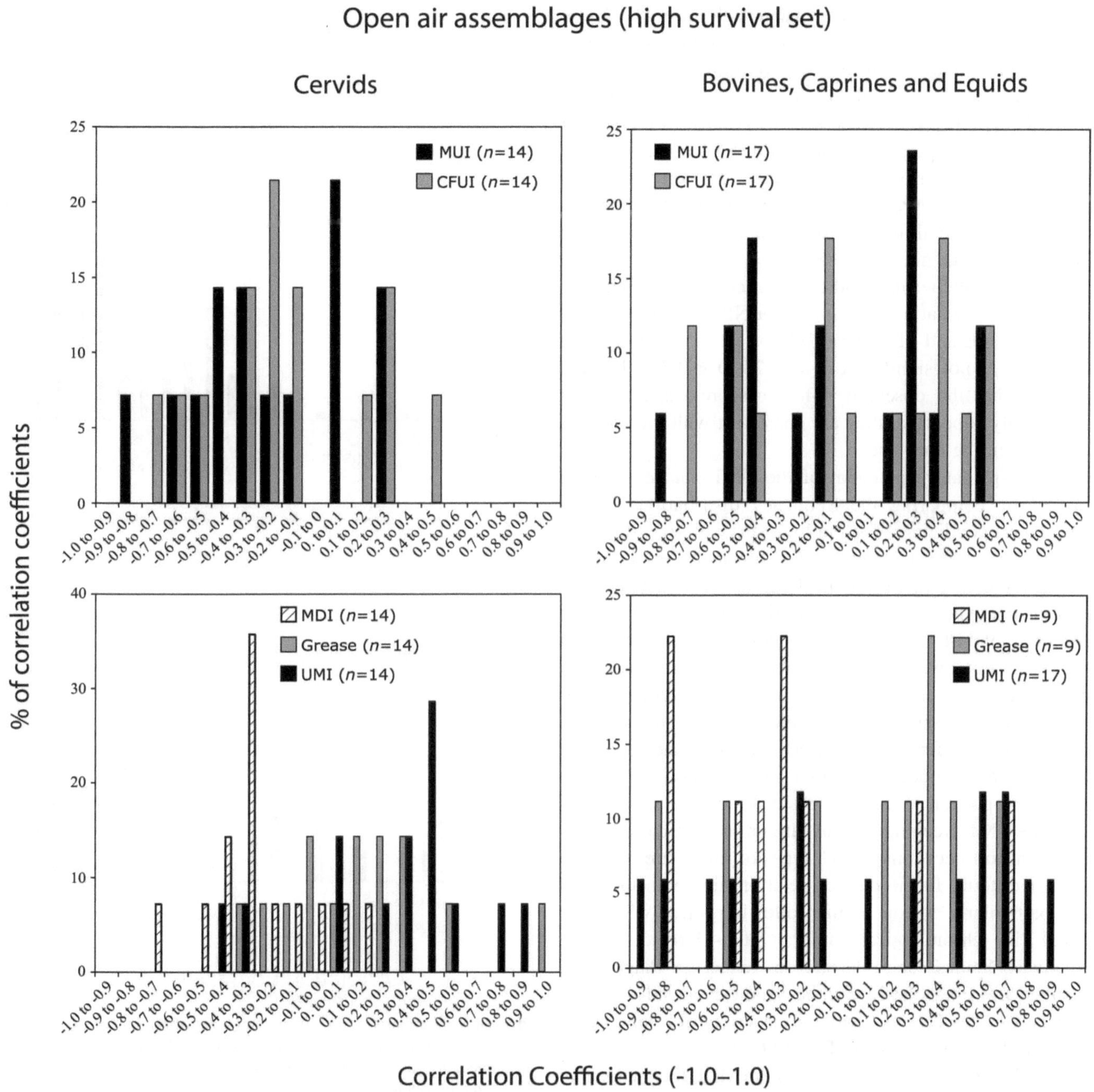

Fig. 14.4 Frequency distributions of correlation coefficients between the utility indices and the high survival set of elements ($n = 8$) in open air assemblages, shown for cervids (*left*) and for the other three taxonomic groups (bovines, caprines, equids) combined (*right*). *Upper panels*: correlations for food utility indices (FUI and CFUI); *lower panels*: correlations for indices measuring other products (grease, marrow, and dried meat)

which suggests that the UMI also best explains patterns of skeletal representation in the open air sample. As was the case for the grease rendering episode, there are fewer significant correlations between the UMI and the high survival set than between this model and the entire set of parts. This pattern characterizes both the shelter ($t_s = 12.23$, $p < 0.0001$) and open air ($t_s = 5.93$, $p < 0.0001$) samples.

Although the UMI is the best predictor of skeletal abundances in our sample, several assemblages—especially those associated with taxa other than cervids—correlate poorly with the UMI. It is unclear at the moment whether these poor correlations are meaningful or simply the consequence of sampling error. As indicated in Table 14.12, two open air bovine assemblages, La Borde and La Montagne, are inversely and significantly correlated with the UMI. These assemblages may attest to the removal and transport of elements of high marrow utility away from the site.

Correlations between the high survival set and the UMI were also compared between species groups. In the shelter sample, a higher percentage of cervid (8/91 or 8.8 %) than non-cervid (3/45 or 6.7 %) assemblages are correlated with this model. However, unlike the situation for the entire set of elements, the difference between the cervids and the other species is not significant ($t_s = 0.45$, $p < 0.66$). In addition, the shelter data indicate that correlations with the UMI are somewhat more common for NISP (8/94 or 8.5 %) than for MNE and MNI data (3/42 or 7.1 %), although the difference is not significant ($t_s = 0.28$, $p < 0.78$). Despite the fact that the observed differences are weak and not statistically supported, they are consistent with the patterns observed in the entire set.

The explanatory potential of the UMI can be assessed further by comparing the frequency distributions of the correlation coefficients for the high survival set with those for the entire set of elements. As shown in Fig. 14.5, the frequency distributions for the UMI indicate that the correlation coefficients are weaker for the high survival set than for all ≈19 parts. This finding suggests that the strength of the correlations with the UMI in the entire set is somewhat exaggerated by density-mediated processes and/or that it is affected by the sample size of elements. In addition, the distributions differ little between shelters and open air assemblages, although the latter sample may comprise a greater proportion of assemblages that are poorly correlated with the UMI. Keeping this caveat in mind, the data suggest that transport decisions at both types of sites were based on similar criteria. In the dataset, relationships with the UMI are particularly strong at the following sites, which are listed in chronological order: Saint-Marcel, Jonzac, Beauvais, Saint-Césaire, Grotte du Renne, Le Piage, Abri Pataud, Le Flageolet I, Combe-Saunière, Rond-du-Barry and Tureau-des-Gardes. Some of the strongest relationships with the UMI are shown in Fig. 14.6.

Turning to the comparisons with the MDI, the analysis of high survival elements indicates that a low percentage of shelter assemblages (2/119 or 1.7 %) and a moderate percentage of open air assemblages (3/23 or 13.0 %) show a significant inverse relationship with this model. Relative to the entire set of parts, both samples of assemblages present fewer correlations that are statistically significant (shelter: $t_s = 9.39$, $p < 0.0001$; open air assemblages: $t_s = 2.08$, $p < 0.04$).

In essence, the examination of the high survival set highlights several important results. When the set of parts is held constant in the comparisons, the shelters and open air sites show few statistical differences in the percentage of assemblages that are significantly correlated with the utility

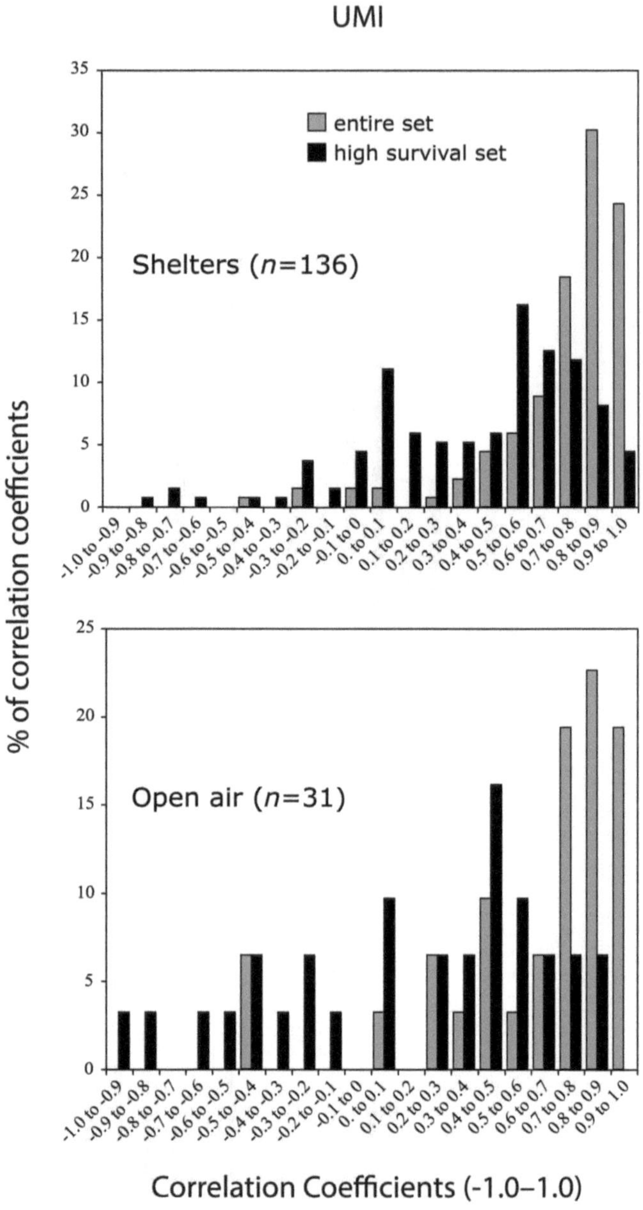

Fig. 14.5 Comparison of frequency distributions of correlation coefficients between the assemblages and the UMI, for the set of all elements and the high survival set. *Upper panel*: shelter assemblages; *lower panel*: open air assemblages

indices. However, we note that with respect to the high survival set, there are significantly more assemblages that are negatively correlated with the UMI and the MDI in the open air sample than in the shelter sample (UMI: $t_s = 2.57$, $p < 0.02$; MDI: $t_s = 2.09$, $p < 0.04$). Another important result is the consistently weak relationships between the assemblages and the MUI or CFUI. These weak relationships contradict the assumption that overall food utility is the most appropriate model for interpreting skeletal profiles in European Paleolithic assemblages. The data also show that when low survival elements are excluded from the

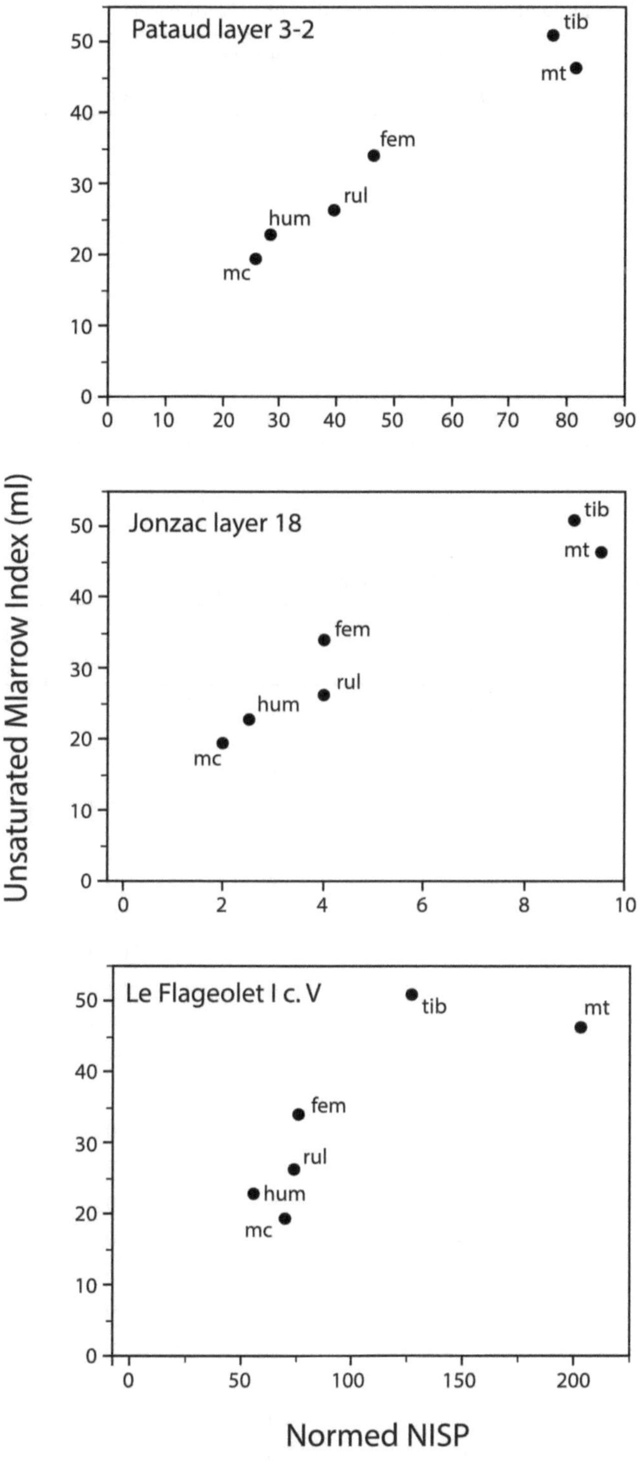

Fig. 14.6 Scatterplots showing the relationship between the Unsaturated Marrow Index and NNISP values (high survival set) for three reindeer assemblages: Abri Pataud (level 3–2), Jonzac (level 18), and Le Flageolet I (couche V)

analysis, there are fewer significant relationships between the archaeological assemblages and utility models (grease model, UMI) that are highly correlated with mineral density. The observed differences may indicate that correlations with these utility models are over-estimated when the entire set of parts is considered, perhaps due to the effects of density-mediated attrition. As discussed above, the differences may also reflect the smaller sample size of elements in the high survival set, which results in lower statistical power for tests performed on this set of parts.

Temporal Trends

Our sample of faunal assemblages covers an important sequence of climatic and cultural events in Western Europe from MIS 15–12 to the beginning of MIS 1. Are there any trends in the sample suggestive of changes in the transport goals or procurement strategies of hominins during this time period? Before addressing this question, it should be noted that in order to increase sample sizes, shelter and open air assemblages were examined together in the analysis of temporal trends, irrespective of species groups and counting methods. However, because all three variables (site type, species group, counting methods) possibly influence the results, it would be useful in future comparisons involving larger samples to analyze the assemblages while holding all three of these variables constant. The entire set of parts was considered in the comparisons, but the analysis focuses more heavily on correlations obtained with the high survival set, in order to counter possible non-random biases caused by density-mediated attrition in our sample. Comparisons were made using a single utility model—the UMI—as this index is best correlated with skeletal patterns in the assemblages. Lastly, we compared "cool" (MIS 6, 4, 2) with "warm" (MIS 7, 5, 3, and 1) climatic stages to explore the effects of general climatic conditions on the correlations. In these last analyses, assemblages pre-dating MIS 7 were excluded from the calculations due to their small numbers and the poorly resolved chronology of some of the sites.

Table 14.13 shows how frequently the assemblages are significantly correlated with the UMI by cultural period (Lower/early Middle, Middle, or Upper Paleolithic). According to the high survival set, none of the comparisons between cultural periods are statistically significant in cervids (Lower/early Middle vs. Middle Paleolithic: $t_s = 1.36$, $p < 0.18$; Lower/early Middle vs. Upper Paleolithic: $t_s = 1.28$, $p < 0.21$; Middle vs. Upper Paleolithic:

Table 14.13 Assemblages showing a statistically significant positive relationship with the UMI as a function of cultural period

	Cervids only			Bovines, caprines, and horse		
	Correlated	Total	Correlated (%)	Correlated	Total	Correlated (%)
Entire set of skeletal elements						
Upper Paleolithic	54	58	93.1	22	36	61.1
Middle Paleolithic	29	40	72.5	8	18	44.4
LP + LP/MP	2	5	40.0	1	5	20.0
High survival set of skeletal elements						
Upper Paleolithic	5	58	8.6	3	36	8.3
Middle Paleolithic	4	40	10.0	0	18	0
LP + LP/MP	0	5	0	0	5	0

"LP/MP" lithic assemblages show a mixture of Lower and (presumably early) Middle Paleolithic features. Because these last assemblages are probably older than the assemblages that are unambiguously associated with the Middle Paleolithic, they were grouped with the Lower Paleolithic sample

Table 14.14 Assemblages showing a statistically significant positive relationship with the UMI as a function of MIS

	Cervids only			Bovines, caprines, and horse		
MIS	Correlated	Total	Correlated (%)	Correlated	Total	Correlated (%)
Entire set of skeletal elements						
1	2	2	100.0	2	3	66.7
2	34	38	89.5	19	31	61.3
3	27	30	90.0	7	9	77.8
4	8	11	72.7	0	1	0
5	3	4	75.0	0	1	0
6	1	2	50.0	1	2	50.0
7	4	5	80.0	3	3	100.0
Total	79	92	85.9	32	50	64.0
"Cool" stages	43	51	84.3	20	34	58.8
"Warm" stages	36	41	87.8	12	16	75.0
High survival set of skeletal elements						
1	0	2	0.0	0	3	0
2	4	38	10.5	3	31	9.7
3	1	30	3.3	0	9	0
4	4	11	36.4	0	1	0
5	0	4	0	0	1	0
6	0	2	0	0	2	0
7	0	5	0	0	3	0
Total	9	92	9.8	3	50	6.0
"Cool" stages	8	51	15.7	3	34	8.8
"Warm" stages	1	41	2.4	0	16	0

"Cool" stages include MIS 6, 4, and 2, whereas "warm" stages include MIS 7, 5, 3 and 1

$t_s = 0.23$, $p < 0.82$). In contrast, the proportion of assemblages that are significantly correlated with the UMI statistically increased in the other taxa between the Middle and the Upper Paleolithic ($t_s = 2.02$, $p < 0.05$), the other comparisons being non-significant (Lower/early Middle vs. Middle Paleolithic: $t_s = 0.00$, $p < 1.00$; Lower/early Middle vs. Upper Paleolithic: $t_s = 1.22$, $p < 0.23$). The entire set of parts indicates some departures from these trends. Although the data confirm that few Lower/early Middle Paleolithic assemblages are significantly correlated with the UMI, the entire set of parts suggests that correlations with this model increased from the Middle to the Upper Paleolithic in *both* the cervids and the other taxa. However, only the trend in cervids is statistically significant (cervids: $t_s = 2.79$, $p < 0.01$; other taxa: $t_s = 1.16$, $p < 0.25$).

Table 14.14 takes a different perspective on the data, presenting them according to MIS rather than cultural period. In the high survival set, none of the cervid assemblages from MIS 7–5 (0/11 or 0 %) correlate significantly with the UMI, while a moderate proportion of later (MIS 4–1, 9/81 or 11.1 %) assemblages do. This pattern of increase is statistically significant ($t_s = 2.11$, $p < 0.04$), whereas that observed in the other taxa is not (MIS 7–5: 0/6 or 0 % vs. MIS 4–1: 3/44 or 6.8 %, $t_s = 1.21$, $p < 0.23$). However, these results are undermined by the fact that they do not find statistical confirmation in the entire set (cervids: 8/11 or 72.7 % vs. 71/81 or 87.7 %, $t_s = 1.19$, $p < 0.24$; other taxa: 4/6 or 66.7 % vs. 28/44 or 63.6 %, $t_s = 0.15$, $p < 0.89$).

Perhaps more revealing are the contrasts between "cool" and "warm" climatic stages. In the high density set, significant correlations with the UMI are more frequent with assemblages attributed to cool than to warm stages, the differences being significant for both the cervids ($t_s = 2.40$, $p < 0.02$) and the other species groups ($t_s = 1.99$, $p < 0.05$). However, once again, the results for the entire set of elements conflicts with those for the high density set and show no significant change between cool and warm stages (cervids: $t_s = 0.48$, $p < 0.64$; other species groups: $t_s = 1.14$, $p < 0.26$).

Altogether, while sample size limitations prevent us from drawing any firm conclusions about temporal trends in the study region, the data nevertheless suggest that, if significant changes in transport decisions occurred during the Pleistocene, they likely preceded, or occurred during, MIS 4. In addition, assemblages seem to correlate with the UMI more frequently during cool than warm isotope stages.

Discussion

In the past two decades, indices measuring overall food utility have frequently been used to assess transport criteria in the Western European Paleolithic. However, in our sample of faunal assemblages, the MUI and CFUI models poorly explain anatomical representation. Our analysis of correlations between Western European Paleolithic assemblages and an expanded set of utility indices indicate that the UMI—which measures marrow utility—shows the highest percentages of significant positive correlations, while significant inverse relationships are sometimes common with the MDI, a proxy for dried meat utility. In contrast, significant correlations with the food utility indices, positive or negative, are rare irrespective of the type of site. This pattern persists for all the taxonomic groups considered: cervids, large bovines, caprines, and equids.

Although we excluded assemblages with very poor preservation, faunal preservation is still less than ideal in certain sites included in the sample. In addition, low-density grease-rich skeletal elements appear to have been frequently burned as fuel at some of the sites examined in this study (e.g., Saint-Césaire, Morin 2010, 2012). For these reasons, it is essential to control for the effect of density-mediated attrition in the assemblages. When low survival parts are omitted from the analysis, correlations with the FUI and CFUI remain weak, the assemblages generally showing a much stronger relationship with the UMI. In our dataset, high-density bone portions containing large quantities of unsaturated marrow fat (e.g., the metatarsal and tibia shafts in cervids) are often more abundant than those associated with large quantities of meat but with only a low or an intermediate ranking in the UMI (e.g., the femur shaft in cervids). Consequently, although we cannot discount the possibility that stripped meat was being returned to sites without the associated skeletal elements, there is clear evidence that marrow content was an important factor in the selection of bones for transport at a large proportion of sites during the Paleolithic of Western Europe.

The marrow-focused pattern documented here seems particularly characteristic of Late Pleistocene assemblages (MIS 4 onwards), although the sample size of pre-MIS 4 sites is small. Assemblages appear to be more frequently significantly correlated with marrow utility during cool isotope stages than during warm ones, which may reflect greater constraints on fat procurement during the former climatic periods. However, because temporal and climatic trends are only weakly supported in our sample, additional data will be needed to verify these inferences.

While nutritional considerations can explain the preferential selection of marrow-rich elements by foragers with high-protein diets (Stefansson 1969; Speth and Spielmann 1983), the selection of elements rich in unsaturated fats over those rich in saturated fats requires additional explanation. In a previous paper, Morin (2007) suggested that possible reasons for preferential selection of marrow rich in unsaturated fat include increased palatability and better taste and texture. In addition, the nutritional condition of ungulates might have also influenced the selection of marrow-bearing elements, as the marrow of distal limb bones—which is rich in unsaturated fatty acids—tends to be mobilized by the animals in poor condition after that contained in more proximal bones.

Differences between open air and shelter sites in the sample are relatively small, although open air assemblages do appear more variable in skeletal composition. Other aspects of the assemblages, however, suggest that the two classes of sites are not always strictly comparable. While whole long bones and anatomical connections are rarely present in human-accumulated shelter assemblages—a nearly universal pattern in Western Europe that contributes to the impression of homogeneity within this class of sites—*some* open air assemblages contain significant proportions of unprocessed long bones *and* moderate to high frequencies of anatomical connections (e.g., Mauran, Solutré). These differences suggest that the latter sites were either functionally different from shelters and other open air sites and/or that they were occupied less intensively, which reduced fragmentation and dispersion of skeletal elements. Concerning this last point, it should be kept in mind that in caves, frequent movement in a confined space may have deleted most anatomical connections. Importantly, none of the open air sites in our sample fits the expected skeletal profile of a kill site. This type of site is probably generally too small and may contain too few artifacts to be highly visible in the archaeological record.

Conclusion

In this analysis of Paleolithic and Epipaleolithic/Mesolithic faunal samples from Western Europe, the strength of the overall patterns of correlations were sometimes weakened by the small number of assemblages for many time periods and for some of the taxonomic groups. Nonetheless, frequent strong correlations with the UMI indicate that unsaturated marrow content was a critical factor mediating the transport of skeletal elements to many locations. A possible implication of our results is that sites where assemblages are strongly correlated with the UMI frequently functioned as a camp, some of which might have been occupied more or less extensively. Despite the key role presumably played by marrow in transport decisions, other factors might have secondarily influenced the selection of skeletal parts. For instance, while overall food utility seems to have been a criterion of lesser importance in most cases, it may have conditioned the transport of axial elements with little to no usable marrow (e.g., the skull, the ribs).

In general, the under-representation of certain classes of elements—including parts with high mineral density (e.g., the cranium as represented by maxillary teeth, the scapula)—in our dataset suggests that transport costs were considerable during the European Pleistocene and that these costs frequently led to substantial field processing. Moreover, several faunal samples in the dataset were significantly correlated with three or more utility models. These findings have significant implications for archaeologists looking to understand assemblage composition, by demonstrating that consideration of a single type of utility model is likely to be an unfruitful strategy. The data also show that density-mediated attrition and the type of counting units used to assess skeletal abundance can impact the correlations with the utility models. We anticipate further exploration of skeletal part patterns on a broader regional and temporal scale, which will explore both the taphonomic and behavioral significance of the patterns reported here.

Acknowledgments We would like to thank Alan Outram and three anonymous reviewers for their critical reading of the manuscript. Their comments significantly improved the text. Our gratitude also goes to Jamie Clark and John Speth for their help and support during the preparation of this paper and for organizing the session.

Appendix

Appendix 14.1 Categories of elements considered in the correlations

#	Element	Value used for norming		Scan site	Mineral density		
		Artiodactyls	Horse		Reindeer	Bovines	Horse
1*	Cranial	2	2				
2*	Mandible	2	2				
3	Cervicals	7	7	CE1	0.45	0.52	0.50
4	Thoracic	13	18	TH1	0.38	0.38	0.32
5	Lumbar	7	6	LU1	0.49	0.58	0.48
6	Rib	26	36	RI3	0.96	1.02	0.50
7*	Scapula	2	2	SP1	1.01	1.02	1.03
8*	Humerus	2	2	HU3	1.12	1.10	1.10
9*	Radio-ulna	2	2	RA3	1.09	1.07	1.08
10	Carpals	12	14	4 bones	0.70	0.73	0.62
11*	Metacarpal	2	2	MC3	1.10	1.15	1.10
12	Sacrum + pelvis	3	3	AC1	0.64	0.64	0.65
13*	Femur	2	2	FE4	1.15	1.16	1.09
14*	Tibia	2	2	TI3	1.13	1.12	1.07
15	Tarsals	10	12	4 bones	0.72	0.74	0.67
16*	Metatarsal	2	2	MR3	1.08	1.14	1.10
17	Phalanx 1	8	4	P1-2	0.92	1.02	1.02
18	Phalanx 2	8	4	P2-2	0.72	0.56	0.59
19	Phalanx 3	8	4	P3-1	0.48	0.53	0.57
	All phalanges	24	12				
	All vertebrae	27	31				

Element numbers with an asterisk identify parts included in the high survival set. Although there is only one cranium in a carcass, this part is counted as two elements because left and right tooth series are generally separated in the archaeological record (see Grayson and Frey 2004 for a similar approach). Density values for reindeer, bovines and horse are from Lam et al. (1999). The density values for bovines are from a relatively closely related species (*Connachaetes taurinus* wildebeest) and for horse, from Burchell's zebra (*Equus burchelli*) and Przewalski's horse (*Equus ferus przewalskii*). Only mid-shaft values are given for the long bones, as epiphyses are rare in the vast majority of the assemblages. Density values for carpals correspond to the average of the lunate, magnum, hamatum, and scaphoid, whereas values for the tarsals correspond to the average of the talus (AS1), calcaneum (CA2), cuneiform and cubo-navicular (NC1 for artiodactyls, navicular for horse)

Appendix 14.2 Correlations obtained for each assemblage with the five utility indices used in our study

	MUI	CFUI	Grease	UMI	MDI
Cervids					
Gr. Dol. TD10a, cervids	−0.45 (−0.33)	−0.31 (−0.39)	0.37 (−0.08)	0.65 (−0.19)	**−0.61*** (−0.09)
Notarchirico, Dama c	−0.14 (−0.45)	0.05 (−0.23)	0.53 (−0.03)	**0.76*** (0.41)	−0.15 (−0.32)
Gal. Pesada C, cervids	−0.25 (0.02)	0.09 (0.06)	0.59 (0.29)	0.67 (−0.09)	**−0.59*** (0.19)
B2, cervids	−0.25 (0.19)	0.17 (0.46)	0.50 (0.39)	**0.92**** (0.79)	**−0.70**** (−0.06)
Payre Fc–d, Cel	0.00 (−0.15)	0.16 (−0.30)	0.45 (−0.58)	**0.86**** (0.19)	−0.21 (0.04)
Fa, Cel	−0.21 (−0.34)	−0.03 (−0.18)	**0.62*** (0.21)	**0.83**** (0.13)	**−0.50*** (−0.43)
Cova Bolomor XII, Cel	−0.26 (0.18)	−0.08 (0.26)	**0.73*** (0.64)	**0.80*** (0.49)	**−0.52*** (0.05)
Lazaret 5 layers, Cel	−0.28 (−0.18)	−0.18 (−0.43)	0.06 (−0.54)	0.67 (−0.03)	**−0.59*** (−0.05)
Saint-Marcel U, Dama	−0.04 (−0.28)	0.12 (−0.18)	0.56 (0.13)	**0.79*** (0.14)	−0.32 (−0.35)
c. inférieure, Cel	0.00 (−0.13)	0.27 (0.24)	**0.79**** (0.71)	**0.92**** (0.74)	−0.31 (−0.45)
ens. 7, Cel	−0.14 (−0.30)	0.10 (0.03)	**0.68*** (0.43)	**0.91**** (0.66)	−0.44 (−0.50)
ens. 7, Capreolus	−0.18 (−0.31)	0.06 (−0.03)	**0.72*** (0.40)	**0.88**** (0.59)	−0.49 (−0.50)
ens. sup, Cel	0.19 (0.27)	0.44 (0.57)	**0.82**** (**0.86***)	**0.86**** (0.37)	−0.13 (0.05)
Les Canalettes 3, Cel	−0.39 (−0.68)	−0.30 (**−0.74***)	0.29 (−0.10)	**0.63*** (0.09)	−0.49 (−0.56)
2, Cel	−0.37 (−0.58)	−0.39 (**−0.73***)	0.24 (−0.21)	0.49 (−0.13)	−0.40 (−0.33)
Marillac 10, Rang	−0.36 (−0.19)	−0.12 (−0.16)	0.24 (−0.08)	0.78 (0.64)	**−0.69*** (−0.15)
9, Rang	−0.41 (0.09)	−0.11 (0.28)	**0.76*** (0.65)	0.44 (0.10)	**−0.77**** (0.01)
Jonzac 24, Rang	−0.03 (−0.13)	0.20 (0.05)	**0.61*** (0.32)	**0.85**** (0.66)	−0.36 (−0.33)
22, Rang	0.18 (−0.02)	0.44 (0.28)	**0.72*** (0.61)	**0.92**** (**0.94***)	−0.16 (−0.35)
20, Rang	0.05 (−0.43)	0.31 (−0.25)	0.55 (−0.03)	**0.84**** (0.60)	−0.15 (−0.43)
18, Rang	0.28 (0.35)	**0.49*** (0.47)	**0.65*** (0.15)	**0.96**** (**0.93***)	−0.03 (0.10)
14, Rang	0.17 (0.02)	0.40 (0.28)	**0.79**** (0.56)	**0.98**** (**0.93***)	−0.14 (−0.28)
10, Rang	0.09 (−0.15)	0.32 (0.10)	0.50 (−0.42)	**0.83**** (0.21)	−0.15 (−0.06)
Beauvais 1, Rang	−0.21 (0.05)	0.06 (0.22)	0.65 (0.54)	**0.90*** (0.77)	**−0.70*** (−0.38)
2, Rang	−0.07 (0.21)	0.19 (0.44)	**0.84*** (**0.92***)	**0.95**** (**0.89***)	**−0.68*** (−0.37)
La Rouqu. UA3, Rang	−0.53 (−0.20)	−0.49 (−0.32)	−0.10 (−0.08)	0.25 (0.03)	−0.50 (−0.15)
UA2, Rang	−0.50 (0.08)	−0.45 (−0.13)	−0.14 (−0.21)	0.39 (0.03)	−0.51 (0.15)
Mutzig I c. 6, Rang	−0.23 (−0.60)	−0.05 (−0.33)	**0.74*** (0.18)	**0.91**** (0.43)	−0.47 (**−0.77***)
c.5, Rang	−0.11 (−0.30)	0.05 (−0.15)	**0.71*** (0.14)	**0.92**** (0.49)	−0.24 (−0.30)
XVI C, Rang	−0.19 (−0.11)	−0.04 (0.11)	0.58 (0.43)	**0.90*** (0.71)	−0.48 (−0.33)
Gabasa 1 level h, Cel	−0.38 (−0.12)	−0.42 (−0.20)	−0.28 (0.21)	−0.49 (0.01)	−0.35 (−0.12)
g, Cel	−0.50 (−0.28)	−0.48 (−0.27)	−0.17 (0.47)	−0.24 (0.81)	**−0.63*** (−0.47)
f, Cel	0.36 (0.09)	0.41 (0.09)	**0.80*** (0.67)	**0.90*** (0.80)	−0.05 (−0.23)
e, Cel	−0.02 (0.19)	0.01 (0.10)	0.34 (0.56)	0.21 (0.36)	−0.27 (−0.01)
Gr. du Noisetier 1, Cel	−0.39 (−0.43)	−0.08 (−0.30)	0.46 (0.11)	**0.71*** (0.03)	**−0.62*** (−0.33)
Pech de l'Azé I c. 4, Cel	−0.30 (−0.03)	−0.12 (0.08)	0.60 (0.40)	0.38 (0.01)	**−0.74**** (−0.12)
c. 7, Cel	−0.48 (0.10)	−0.17 (0.38)	0.71 (0.63)	0.66 (0.51)	**−0.75**** (−0.15)
Gatzarria Cj, Cel	−0.02 (−0.28)	0.23 (−0.10)	**0.68*** (0.40)	**0.91**** (0.59)	−0.28 (−0.40)
A. Romani M, cervids	−0.37 (−0.23)	−0.13 (0.03)	0.42 (0.36)	**0.83*** (0.60)	**−0.75**** (−0.38)
K, cervids	−0.32 (−0.30)	−0.06 (0.03)	0.46 (0.43)	**0.86*** (0.66)	**−0.76**** (−0.50)
Ja, cervids	−0.38 (−0.32)	−0.09 (−0.13)	0.41 (0.11)	0.60 (−0.31)	**−0.67*** (−0.20)
La Quina 8, Rang	−0.35 (−0.37)	−0.11 (−0.09)	**0.75*** (0.24)	**0.90**** (0.59)	**−0.59*** (−0.47)
7, Rang	−0.31 (−0.37)	−0.07 (−0.08)	**0.75*** (0.25)	**0.92**** (0.66)	**−0.60*** (−0.52)
St-Césaire EGPF, Rang	−0.10 (−0.44)	0.13 (−0.20)	0.58 (0.18)	**0.85**** (0.54)	−0.31 (−0.45)

(continued)

Appendix 14.2 (continued)

	MUI	CFUI	Grease	UMI	MDI
EJOP inf, Rang	−0.03 (−0.23)	0.19 (−0.05)	**0.68*** (0.15)	**0.88**** (0.59)	−0.28 (−0.33)
EJOP sup, Rang	0.04 (0.15)	0.23 (0.19)	**0.79**** (0.68)	**0.89**** (0.77)	−0.29 (−0.10)
EJO sup, Rang	−0.14 (−0.26)	0.08 (−0.11)	**0.67*** (0.25)	**0.91**** (0.66)	−0.40 (−0.40)
EJF, Rang	−0.08 (−0.28)	0.14 (−0.10)	**0.67*** (0.25)	**0.93**** (0.66)	−0.36 (−0.40)
EJM, Rang	−0.05 (−0.05)	0.18 (0.10)	**0.73*** (0.46)	**0.94**** (0.77)	−0.37 (−0.30)
EJJ, Rang	0.04 (−0.30)	0.21 (−0.15)	**0.79**** (0.43)	**0.95**** (0.83)	−0.30 (−0.62)
Gr. du Renne Xc, Rang	−0.30 (−0.08)	−0.08 (0.08)	**0.78**** (0.56)	**0.85**** (0.53)	**−0.55*** (−0.23)
VIII, Rang	−0.43 (−0.15)	−0.20 (0.08)	0.56 (0.57)	**0.72*** (0.37)	**−0.62*** (−0.27)
VII, Rang	−0.12 (−0.35)	0.14 (−0.17)	**0.73*** (0.46)	**0.95**** (0.77)	−0.44 (−0.67)
Gr. du Bison D, Rang	−0.09 (−0.09)	0.00 (0.08)	**0.63*** (0.57)	**0.80*** (0.51)	−0.36 (−0.37)
Piage B, Rang	−0.01 (−0.13)	0.17 (−0.10)	**0.76*** (0.46)	**0.91**** (0.70)	−0.33 (−0.31)
Abri Pataud 14, Rang	−0.13 (−0.48)	0.01 (−0.45)	0.48 (−0.07)	**0.76*** (0.09)	−0.27 (−0.32)
13, Rang	−0.13 (−0.30)	−0.01 (−0.35)	0.49 (−0.05)	**0.76*** (0.07)	−0.25 (−0.13)
12, Rang	0.01 (−0.05)	0.09 (0.09)	**0.77*** (0.61)	**0.84**** (0.09)	−0.27 (−0.31)
11, Rang	−0.08 (−0.47)	0.08 (−0.32)	0.55 (0.07)	**0.82**** (0.26)	−0.26 (−0.44)
Gr. XVI B, Rang	−0.38 (−0.50)	−0.06 (−0.33)	0.49 (0.07)	**0.77*** (0.26)	**−0.71*** (−0.47)
Aib, Rang	−0.24 (−0.35)	0.02 (−0.25)	0.42 (−0.06)	**0.75*** (0.20)	−0.59 (−0.26)
Abb, Rang	−0.33 (−0.38)	−0.02 (−0.20)	0.44 (0.21)	**0.85*** (0.54)	**−0.69*** (−0.43)
Abc, Rang	−0.35 (−0.44)	−0.07 (−0.39)	0.47 (0.15)	**0.80*** (0.30)	**−0.68*** (−0.42)
As, Rang	−0.34 (−0.28)	−0.08 (−0.32)	0.40 (0.00)	**0.77*** (0.20)	**−0.59*** (−0.22)
0, Rang	−0.35 (**−**0.45)	−0.15 (−0.44)	0.33 (−0.04)	**0.72*** (0.09)	**−0.64*** (−0.32)
Les Rois C3, Rang	−0.31 (−0.28)	0.07 (0.08)	0.61 (0.50)	**0.97**** (**0.89***)	**−0.79**** (−0.58)
C2, Rang	−0.40 (−0.45)	−0.02 (−0.15)	0.48 (0.18)	**0.87*** (0.54)	**−0.79**** (−0.55)
Pataud 5-lower, Rang	−0.56 (−0.47)	−0.20 (−0.18)	0.47 (0.14)	**0.83*** (0.43)	**−0.82**** (−0.52)
5-Middle, Rang	−0.51 (−0.37)	−0.28 (−0.38)	0.36 (0.04)	**0.72*** (0.09)	**−0.73*** (−0.28)
5-upper, Rang	−0.35 (−0.20)	−0.08 (0.08)	0.60 (0.46)	**0.93**** (0.77)	**−0.75**** (−0.42)
5, Cel	−0.15 (0.10)	0.13 (0.42)	**0.76*** (**0.82***)	**0.98**** (**0.94***)	**−0.63*** (−0.23)
4-lower, Rang	−0.03 (−0.25)	0.18 (−0.07)	**0.65*** (0.29)	**0.91**** (0.60)	−0.29 (−0.33)
4-middle, Rang	−0.01 (−0.12)	0.23 (0.07)	**0.73*** (0.46)	**0.95**** (0.77)	−0.29 (−0.30)
4-upper, Rang	−0.01 (−0.12)	0.22 (0.07)	**0.77*** (0.46)	**0.96**** (0.77)	−0.29 (−0.30)
3-4, Rang	−0.05 (−0.07)	0.20 (0.08)	**0.65*** (0.46)	**0.90**** (0.77)	−0.35 (−0.25)
3-3, Rang	−0.21 (−0.15)	0.05 (0.00)	**0.64*** (0.43)	**0.91**** (0.83)	**−0.52*** (−0.47)
3-2, Rang	−0.20 (0.07)	0.05 (0.22)	**0.74*** (0.68)	**0.96**** (**0.94***)	**−0.51*** (−0.25)
3-1, Rang	0.02 (−0.02)	0.24 (0.13)	**0.71*** (0.54)	**0.94**** (**0.89***)	−0.31 (−0.27)
2, Rang	−0.11 (−0.58)	0.01 (**−0.70***)	0.59 (−0.11)	**0.77*** (−0.03)	−0.34 (−0.40)
Le Flag. I c. V, Rang	0.02 (−0.15)	0.23 (0.13)	**0.72*** (0.57)	**0.93**** (**0.89***)	−0.32 (−0.47)
C.-Saunière IV, Rang	−0.02 (−0.28)	0.23 (−0.10)	**0.67*** (0.25)	**0.93**** (0.66)	−0.30 (−0.40)
C. du Vers 31–30, Rang	0.02 (−0.07)	0.19 (0.05)	**0.77*** (0.36)	**0.89**** (0.54)	−0.32 (−0.32)
29, Rang	−0.04 (−0.13)	0.19 (0.15)	**0.79**** (0.39)	**0.91**** (0.54)	−0.33 (−0.35)
23, Rang	−0.13 (−0.22)	0.08 (0.02)	**0.78**** (0.36)	**0.92**** (0.60)	−0.40 (−0.38)
Picareiro F, Cel	−0.56 (−0.57)	−0.28 (−0.25)	0.39 (0.19)	0.61 (0.09)	**−0.69**** (−0.50)
Rd-du-Barry F2, Rang	0.14 (0.02)	0.32 (0.17)	**0.78**** (0.64)	**0.88**** (0.37)	−0.24 (−0.36)
E, Rang	0.06 (0.36)	0.23 (0.48)	0.56 (0.43)	**0.83**** (0.66)	−0.31 (−0.02)
Solutré, Rang	0.08 (0.25)	0.21 (0.24)	**0.65*** (0.22)	**0.92**** (0.50)	−0.23 (0.09)
Gazel, Rang	−0.36 (−0.30)	−0.26 (−0.40)	0.38 (−0.21)	0.36 (−0.26)	**−0.58*** (−0.17)
Le Flag. II c. IX, Rang	0.14 (−0.42)	0.24 (−0.35)	0.54 (0.18)	**0.83**** (0.54)	−0.12 (−0.48)

(continued)

Appendix 14.2 (continued)

	MUI	CFUI	Grease	UMI	MDI
Moulin-Neuf site, Rang	−0.38 (−0.53)	−0.11 (−0.26)	**0.64*** (0.23)	**0.86**** (0.37)	**−0.59*** (−0.46)
Gr. Tournal G, Rang	−0.35 (−0.37)	−0.18 (−0.18)	0.59 (0.07)	0.58 (0.26)	**−0.54*** (−0.37)
H, Rang	−0.22 (0.20)	0.02 (0.30)	0.57 (0.05)	**0.66*** (0.31)	−0.35 (0.34)
Conques, Rang	−0.06 (−0.25)	−0.08 (−0.42)	0.38 (0.29)	0.62 (0.09)	−0.46 (−0.25)
La Plaine 4e–4n, Rang	−0.25 (0.20)	−0.03 (0.27)	**0.66*** (0.52)	**0.87**** (0.57)	−0.41 (0.22)
Bois-Ragot 6, Rang	−0.35 (−0.46)	−0.09 (−0.11)	**0.68*** (0.23)	**0.87**** (0.37)	**−0.56*** (−0.51)
5, Rang	−0.30 (−0.26)	−0.01 (−0.06)	**0.69*** (0.34)	**0.88**** (0.63)	**−0.53*** (−0.34)
4, Cel	−0.17 (−0.27)	0.10 (−0.07)	**0.61*** (0.25)	**0.87**** (0.50)	−0.35 (−0.25)
Grand Canton, Rang	−0.14 (0.07)	−0.02 (0.12)	**0.66*** (0.39)	**0.76*** (0.37)	−0.29 (0.00)
Verberie II-1, Rang	−0.30 (**−0.87***)	−0.03 (**−0.70***)	0.43 (−0.32)	**0.69*** (−0.49)	−0.45 (−0.55)
Pinc. IV0-T125, Rang	−0.26 (−0.15)	−0.11 (−0.22)	**0.68*** (0.39)	**0.89**** (0.60)	**−0.52*** (−0.28)
36, Rang	−0.26 (−0.37)	−0.03 (−0.28)	**0.63*** (0.21)	**0.87**** (0.37)	**−0.49*** (−0.32)
Arancou B2, Cel	−0.35 (−0.49)	−0.15 (−0.52)	0.57 (0.12)	**0.83**** (0.44)	**−0.59*** (−0.55)
Noyen-sur-Seine, Cel	−0.01 (−0.53)	0.10 (**−0.70***)	**0.62*** (−0.11)	**0.78*** (−0.37)	−0.21 (−0.41)
Herriko Barra, Cel	−0.10 (−0.50)	0.07 (−0.59)	**0.65*** (0.01)	**0.86**** (0.24)	−0.35 (−0.45)
Bovines					
Isernia 3a, Bison s	0.29 (0.21)	0.17 (0.11)	0.55 (0.25)	0.46 (0.26)	−0.45 (−0.54)
Arago c. m. sup, Bison	0.14 (−0.06)	0.26 (0.06)	0.53 (0.49)	**0.70*** (−0.04)	**−0.75*** (−0.19)
Biache, Bos	**0.56*** (0.14)	**0.53*** (0.43)	**0.81**** (0.68)	**0.88*** (0.66)	−0.46 (**−0.86***)
Payre Fa, BB	0.28 (0.21)	0.22 (0.26)	**0.68**** (0.71)	**0.78*** (0.24)	**−0.62*** (−0.56)
La Borde, Bos	−0.01 (−0.40)	−0.23 (−0.72)	−0.23 (**−0.88***)	−0.46 (**−0.90***)	0.37 (0.62)
La Rouquette UA4, BB	0.35 (−0.43)	0.29 (−0.54)	0.43 (−0.57)	0.50 (−0.54)	−0.19 (0.21)
UA3, BB	0.36 (−0.54)	0.30 (−0.50)	**0.60*** (0.11)	0.47 (−0.66)	**−0.59*** (−0.32)
UA2, BB	0.35 (−0.14)	0.30 (−0.11)	**0.61*** (0.32)	0.60 (−0.21)	**−0.57*** (−0.36)
Mauran, Bison	0.09 (−0.21)	0.07 (−0.07)	0.50 (0.36)	**0.72*** (0.09)	**−0.78**** (**−0.86***)
Pech de l'Azé I 4, Bison	−0.38 (0.35)	−0.25 (0.49)	0.49 (0.78)	**0.81*** (0.70)	**−0.87**** (**−0.81***)
St-Césaire EGPF, Bison	0.05 (−0.25)	0.03 (0.18)	**0.50*** (−0.05)	**0.79*** (0.47)	**−0.67**** (−0.50)
EJOP sup, Bison	0.19 (−0.11)	0.21 (0.32)	**0.70**** (0.43)	**0.85*** (0.49)	**−0.61*** (**−0.82***)
Abri Pataud 5, BB	0.45 (0.54)	0.40 (0.68)	**0.63*** (0.61)	0.79 (0.77)	−0.46 (−0.71)
Moulin-Neuf site, BB	−0.30 (0.24)	−0.31 (0.42)	0.33 (0.58)	0.53 (0.66)	**−0.81**** (−0.78)
Ch.-Chalatras 1–3, Bos	0.15 (0.30)	0.19 (0.55)	**0.66**** (0.46)	**0.90*** (0.67)	**−0.63*** (−0.48)
La Montagne, Bos	−0.26 (**−0.82***)	−0.26 (−0.75)	0.43 (−0.18)	0.27 (**−0.89***)	**−0.66*** (−0.21)
Caprines					
Arago sol F, argali	−0.19 (−0.34)	−0.13 (−0.35)	0.56 (−0.12)	0.60 (−0.73)	−0.38 (−0.35)
Lazaret 5 layers, ibex	−0.27 (−0.27)	−0.20 (−0.32)	0.55 (0.21)	0.58 (−0.43)	**−0.65*** (−0.57)
Gr. du Nois. 1, Capra p	−0.16 (0.43)	−0.10 (0.50)	**0.71*** (0.75)	**0.87*** (0.60)	−0.38 (0.32)
Abri Pataud 5, ibex	−0.49 (−0.32)	−0.29 (−0.25)	0.62 (0.21)	0.68 (0.09)	**−0.76**** (−0.26)
C. Ner. X–VIII, Capra p	0.05 (0.18)	0.17 (0.18)	0.35 (−0.11)	**0.70*** (0.20)	−0.05 (0.18)
C. du Vers 23, Rupi	0.16 (0.33)	0.30 (0.48)	**0.76*** (0.63)	**0.92**** (0.60)	−0.02 (0.27)
Rd-du-Barry F2, ibex	−0.22 (−0.08)	−0.04 (0.10)	**0.71*** (0.64)	**0.90**** (0.71)	−0.33 (−0.03)
E, ibex	−0.28 (0.13)	−0.12 (0.20)	**0.76*** (**0.89***)	**0.87**** (**0.89***)	−0.45 (−0.20)
F2, Rupi	−0.37 (−0.18)	−0.28 (−0.17)	**0.78*** (0.50)	**0.80*** (−.20)	**−0.52*** (−0.22)
E, Rupi	−0.25 (−0.11)	−0.16 (−0.25)	**0.77*** (0.62)	**0.80*** (0.19)	−0.48 (−0.52)
St-G.-la-Riv. 4, saiga	**−0.58*** (−0.33)	−0.42 (−0.27)	**0.66*** (0.35)	**0.71*** (0.07)	**−0.83**** (−0.30)

(continued)

Appendix 14.2 (continued)

	MUI	CFUI	Grease	UMI	MDI
3, saiga	**−0.64*** (−0.55)	−0.47 (−0.50)	0.48 (−0.07)	0.45 (−0.83)	**−0.81**** (−0.23)
1, saiga	**−0.58*** (−0.38)	−0.36 (−0.25)	0.54 (0.18)	**0.75*** (0.14)	**−0.78**** (−0.18)
C, saiga	**−0.60*** (−0.22)	−0.47 (−0.07)	0.42 (0.11)	0.57 (−0.26)	**−0.75**** (−0.05)
La Fragua 4, Capra	0.01 (0.17)	0.06 (0.17)	0.50 (0.26)	0.44 (0.09)	−0.02 (0.13)
Moulin-Neuf site, saiga	**−0.70**** (**−0.72***)	**−0.59*** (**−0.74***)	0.35 (0.04)	0.43 (−0.66)	**−0.72**** (−0.62)
El Pirulejo 4, Capra p	−0.29 (−0.31)	−0.23 (−0.35)	0.37 (−0.06)	0.47 (−0.73)	−0.42 (−0.32)
3, Capra p	−0.22 (0.29)	−0.21 (0.25)	0.45 (0.52)	0.56 (0.21)	−0.42 (−0.01)
Conques, Capra p	0.17 (0.03)	0.33 (0.00)	0.57 (0.21)	0.65 (−0.01)	−0.35 (−0.24)
Belvis, Capra p	−0.25 (0.41)	−0.18 (0.44)	**0.73*** (**0.93***)	**0.78*** (0.66)	**−0.50*** (−0.04)
Riparo Dalmeri 26c, ibex	0.19 (−0.23)	0.22 (−0.20)	0.58 (0.18)	**0.78*** (−0.03)	−0.09 (−0.43)
Equids					
Arago G, Cab	0.01 (0.10)	0.22 (0.48)		0.79 (0.66)	
Payre Fa, Cab	−0.27 (−0.03)	−0.14 (0.20)		**0.84**** (0.76)	
Cova Bolomor XII, Cab	−0.09 (0.57)	−0.10 (0.47)		**0.93**** (0.64)	
Gabasa 1 level h, Cab	−0.03 (0.05)	−0.06 (−0.09)		−0.22 (−0.21)	
g, Cab	0.18 (0.65)	0.10 (0.50)		−0.07 (0.67)	
f, Cab	0.31 (0.54)	0.21 (0.34)		0.00 (0.50)	
e, Cab	0.29 (0.47)	0.32 (0.47)		0.33 (0.54)	
d, Cab	0.25 (0.63)	0.16 (0.43)		−0.07 (0.43)	
a–c, Cab	0.35 (0.63)	0.18 (0.33)		0.06 (0.81)	
Gr. du Renne VII, Cab	−0.08 (0.50)	−0.05 (0.52)		**0.93**** (0.71)	
Abri Pataud 11, Cab	0.34 (0.28)	0.41 (0.45)		**0.81*** (0.60)	
Solutré M12 Auri, Cab	−0.44 (−0.15)	−0.33 (−0.18)		−0.47 (−0.24)	
L13 Gravet, Cab	**−0.66*** (−0.57)	−0.46 (−0.40)		0.05 (−0.40)	
Le Sire, Cab	−0.55 (−0.46)	−0.33 (−0.10)		0.68 (−0.10)	
C.-Saunière IV, Cab	0.02 (0.33)	0.13 (0.64)		**0.86**** (0.76)	
Rd-du-Barry F2, Cab	−0.22 (0.52)	−0.03 (**0.83***)		**0.82**** (0.83)	
E, Cab	0.10 (0.59)	0.24 (**0.83***)		**0.84**** (**0.89***)	
Moulin-Neuf site, Cab	−0.29 (0.16)	−0.22 (0.50)		**0.93**** (**0.93***)	
Gr. Tournal G, Cab	−0.41 (−0.10)	−0.31 (0.22)		0.52 (−0.26)	
H, Cab	−0.17 (0.60)	−0.14 (0.42)		**0.90**** (0.84)	
Champréveyres, Cab	0.20 (0.30)	0.26 (0.37)		**0.85**** (0.54)	
Etiolles, Cab	0.45 (0.52)	0.39 (0.30)		**0.71*** (0.43)	
Tureau-des-Gardes, Cab	0.06 (0.37)	0.11 (0.38)		**0.96**** (0.83)	
Grand Canton, Cab	0.00 (0.22)	0.09 (0.38)		**0.79*** (0.54)	
Le Closeau 46, Cab	0.35 (0.57)	0.37 (0.52)		**0.93**** (0.71)	

The correlations are presented for all categories of elements and for high survival elements only (the latter shown in parentheses). The utility models are those listed in Table 14.4, whereas high survival elements are identified in Appendix 14.1

Note $^{*}\alpha < 0.05$, $^{**}\alpha < 0.01$. Abbreviations same as Table 14.1

References

Anconetani, P. (1999). L'assemblage faunique du gisement Paléolithique inférieur d'Isernia-la-Pineta (Molise, Italie) et l'exploitation du bison. In J.-P. Brugal, F. David, J. Enloe, & J. Jaubert (Eds.), *Le bison: Gibier et moyen de subsistance des hommes du Paléolithique aux Paléoindiens des grandes plaines* (pp. 105–120). Antibes: Éditions APDCA.

Auguste, P. (1993). Acquisition et exploitation du gibier au Paléolithique moyen dans le nord de la France. Perspective paléo-écologiques et palethnographiques. In J. Desse & F. Audoin-Rouzeau (Eds.), *Exploitation des animaux sauvages à travers le temps* (pp. 49–62). Juan-les-Pins: Éditions APDCA.

Auguste, P. (1995). Chasse et charognage au Paléolithique moyen: L'apport du gisement de Biache-Saint-Vaast (Pas-de-Calais). *Bulletin de la Societé Préhistorique Française, 92*, 155–167.

Auguste, P., & Patou-Mathis, M. (1999). Beauvais (Oise): Un nouveau gisement du Paléolithique moyen dans le bassin parisien. In S. Gaudzinski & E. Turner (Eds.), *The role of early humans in the accumulations of European Lower and Middle Palaeolithic bone assemblages* (Vol. 42, pp. 343–365). Mainz: Römisch-Germanisches Zentralmuseum.

Barlow, R. K., & Metcalfe, D. (1996). Plant utility indices: Two Great Basin examples. *Journal of Archaeological Science, 23*, 351–371.

Bartram, L. E. (1993). Perspectives on skeletal part profiles and utility curves from eastern Kalahari ethnoarchaeology. In J. Hudson (Ed.), *From bones to behavior: Ethnoarchaeological and experimental contributions to the interpretation of faunal remains* (pp. 115–155). Carbondale: Southern Illinois University, Center for Archaeological Investigation.

Bastien, G. J., Schepens, B., Willems, P. A., & Heglund, N. C. (2005a). Energetics of load carrying in Nepalese porters. *Science, 308*, 1755.

Bastien, G. J., Willems, P. A., Schepens, B., & Heglund, N. C. (2005b). Effect of load and speed on the energetic cost of human walking. *European Journal of Applied Physiology, 94*, 76–83.

Beauval, C. (2004). La faune des niveaux moustériens de "Chez Pinaud" (Jonzac, Charente-maritime, France). Première analyse. *Préhistoire du Sud-Ouest, Supplément no. 8*, 125–156.

Bellai, D. (1995). Techniques d'exploitation du cheval à la Caune de l'Arago (Pyrénées Orientales). *Paléo, 7*, 139–155.

Bellai, D. (1998). Le bison du gisement paléolithique inférieur de la Caune de l'Arago (Tautavel, Pyrénées, France). Nouvelles données archéozoologiques. *Paléo, 10*, 61–76.

Bergerud, A. T., Luttich, S. N., & Lodewijk, C. (2007). *The return of caribou to Ungava*. McGill-Queen's Native and Northern Series, 50. Montreal: McGill-Queen's University Press.

Bettinger, R. L., Malhi, R., & McCarthy, H. (1997). Central place models of acorn and mussel processing. *Journal of Archaeological Science, 24*, 887–899.

Bignon, O. (2003). *Diversité et exploitation des équidés au Tardiglaciaire en Europe occidentale—Implications pour les stratégies de subsistance et les modes de vie au Magdalénien et à l'Azilien ancien du bassin parisien.* Unpublished Ph.D. dissertation, Université Paris X Nanterre.

Bignon, O., Enloe, J. G., & Bémilli, C. (2006). Étude archéozoologique de l'unité T125: Originalité de la chasse des rennes et des chevaux. In P. Bodu, M. Julien, B. Valentin, & G. Debout (Eds.), *Un dernier hiver à Pincevent: Les Magdaléniens du niveau IV0 (Pincevent, La Grande-Paroisse, Seine-et-Marne). Gallia Préhistoire, 48*, 18–35.

Binford, L. R. (1978). *Nunamiut ethnoarchaeology*. New York: Academic Press.

Blasco, M. F. (1997). In the pursuit of game: The Mousterian cave site of Gabasa 1 in the Spanish Pyrenees. *Journal of Anthropological Research, 53*, 177–217.

Blasco, R., Fernández Peris, J., & Rosell, J. (2010). Several different strategies for obtaining animal resources in the late Middle Pleistocene: The case of level XII at Bolomor Cave (Valencia, Spain). *Comptes Rendus Palevol, 9*, 171–184.

Blumenschine, R. J., & Madrigal, T. C. (1993). Variability in long bone marrow yields of east African ungulates and its zooarchaeological implications. *Journal of Archaeological Science, 20*, 555–587.

Bowen, D. Q. (2009). Pleistocene climates. In V. Gornitz (Ed.), *Encyclopedia of paleoclimatology and ancient environments* (pp. 798–804). Dordrecht: Springer.

Boyle, K. V. (2000). Reconstructing Middle Paleolithic subsistence strategies in the south of France. *International Journal of Osteoarchaeology, 10*, 336–356.

Bridault, A., & Bémilli, C. (1999). La chasse et le traitement des animaux. In M. Julien & J.-L. Rieu (Eds.), *Occupations du Paléolithique supérieur dans le sud-est du bassin parisien* (pp. 50–64). Paris: Éditions de la Maison des Sciences de l'Homme.

Brink, J. W. (1997). Fat content in leg bones of *Bison bison*, and applications to archaeology. *Journal of Archaeological Science, 24*, 259–274.

Brugal, J.-P. (2001). Les assemblages fauniques: Paléoenvironnement, taphonomie et archéozoologie. In J.-P. Brugal, P. Chalard, M.-F. Diot, C. Falguères, M. Jarry, J. Jaubert, B. Kervazo, S. Konik, & V. Mourre (Eds.), *Un site moustérien de type Quina dans la vallée du Célé: Pailhes à Espagnac-Sainte-Eulalie (Lot). Gallia Préhistoire, 43*, 33–52.

Brugal, J.-P., David, F., Enloe, J. G., & Jaubert, J. (1999). *Le bison: Gibier et moyen de subsistance des hommes du Paléolithique aux Paléoindiens des grandes plaines*. Antibes: Éditions APDCA.

Bunn, H. H. T., Bartram, L. E., & Kroll, E. M. (1988). Variability in bone assemblage formation from Hadza hunting, scavenging, and carcass processing. *Journal of Anthropological Archaeology, 7*, 412–457

Burger, O., Hamilton, M. J., & Walker, R. (2005). The prey as patch model: Optimal handling of resources with diminishing returns. *Journal of Archaeological Science, 32*, 1147–1158.

Cannon, M. D. (2001). Archaeofaunal relative abundance, sample size and statistical methods. *Journal of Archaeological Science, 28*, 185–195.

Cannon, M. D. (2003). A model of central place forager prey choice and an application to faunal remains from the Mimbres Valley, New Mexico. *Journal of Anthropological Archaeology, 22*, 1–25.

Cassoli, P., Fabbo, G., Fiore, I., Piperno, M., & Tagliacozzo, A. (1999). Taphonomic analysis of some levels of the Lower Palaeolithic site of Notarchirico (Venosa, Basilicata, Italy). In S. Gaudzinski & E. Turner (Eds.), *The role of early humans in the accumulation of European Lower and Middle Palaeolithic bone assemblages* (Vol. 42, pp. 153–172). Mainz: Römisch-Germanisches Zentralmuseum.

Castel, J. C. (1999). *Comportements de subsistance au Solutréen et au Badegoulien d'après les faunes de Combe-Saunière (Dordogne) et du Cuzoul de Vers (Lot).* Unpublished Ph.D. dissertation, Université de Bordeaux I.

Chase, P. G. (1985). On the use of Binford's utility indices in the analysis of archaeological sites. *PACT, 11* (IV.1), 287–302.

Chase, P. G. (1999). Bison in the context of complex utilization of faunal resources: A preliminary report on the Mousterian zooarchaeology of La Quina (Charente, France). In J.-P. Brugal, F. David, J. G. Enloe, & J. Jaubert (Eds.), *Le bison: Gibier et moyen de subsistance des hommes du Paléolithique aux Paléoindiens des grandes plaines* (pp. 159–184). Antibes: Éditions ADPCA.

Cho, T.-S. (1998). Étude archéozoologique de la faune du Périgordien supérieur (couches 2, 3 et 4) de l'Abri Pataud (Les Eyzies, Dordogne): Paléoécologie, taphonomie, paléoéconomie. Unpublished Ph.D. dissertation, Muséum National d'Histoire Naturelle.

Coltorti, M., Feraud, G., Marzoli, A., Peretto, C., Ton-That, T., Voinchet, P., et al. (2005). New $^{40}Ar/^{39}Ar$, stratigraphic and palaeoclimatic data on the Isernia la Pineta Lower Palaeolithic site, Molise, Italy. *Quaternary International, 131*, 11–22.

Cortés Sánchez, M., Jiménez Espejo, F. J., Simón Vallejo, M. D., López Sáez, J. A., Riquelme Cantal, J. A., Fernández Domínguez, E., et al. (2008). La investigación sobre El Pirulejo. Una aproximación interdisciplinar. *Antiquitas, 20*, 213–221.

Costamagno, S. (1999). *Stratégies de chasse et fonction des sites au Magdalénien dans le sud de la France*. Unpublished Ph.D. dissertation, Université de Bordeaux I.

Costamagno, S., Meignen, L., Beauval, C., Vandermeersch, B., & Maureille, B. (2006). Les Pradelles (Marillac-le-Franc, France): A Mousterian reindeer hunting camp? *Journal of Anthropological Archaeology, 25*, 466–484.

Costamagno, S., Robert, I., Laroulandie, V., Mourre, V., & Thiébault, C. (2008). Rôle du gypaète barbu (*Gypaetus barbatus*) dans la constitution de l'assemblage osseux de la grotte du Noisetier (Fréchet-Aure, Hautes-Pyrénées, France). *Annales de Paléontologie, 94*, 245–265.

Daujeard, C. (2008). *Exploitation du milieu animal par les Néanderthaliens dans le sud-est de la France*. Ph.D. dissertation, Université Lumière-Lyon 2.

David, F. (2002). Les ours du Châtelperronien de la grotte du Renne à Arcy-sur-Cure (Yonne). In T. Tillet & L. R. Binford (Eds.), *L'Ours et l'Homme* (pp. 185–192). Liège: ERAUL.

David, F., & Farizy, C. (1994). Les vestiges osseux: Étude archéozoologique. In C. Farizy, F. David, & J. Jaubert (Eds.), *Hommes et bisons du Paléolithique moyen à Mauran (Haute-Garonne)* (pp. 177–234). Paris: Éditions CNRS.

David, F., & Poulain, T. (1990). La faune des grands mammifères des niveaux XI et Xc de la Grotte du Renne à Arcy-sur-Cure (Yonne). In C. Farizy (Ed.), *Paléolithique moyen récent et Paléolithique supérieur ancien en Europe* (pp. 319–323). Nemours: Mémoire du Musée de Préhistoire d'Ile de France 3.

David, F., & Poulain, T. (2002). Les mammifères (herbivores, carnivores, petits mammifères). In B. Schmider (Ed.), *L'Aurignacien de la Grotte du Renne. Les fouilles d'André Leroi-Gourhan à Arcy-sur-Cure (Yonne)* (pp. 49–95). Paris: Éditions CNRS.

David, F., Connet, N., Girard, M., Miskovsky, J.-C., Mourer-Chauviré, C., & Roblin-Jouve, A. (2005). Les niveaux du Paléolithique supérieur à la grotte du Bison (Arcy-sur-Cure, Yonne) couches a à d. *Revue Archéologique de l'Est, 54*, 5–50.

Deplano, S. (1994). Étude de la faune de la couche IX de l'abri du Flageolet II (Dordogne): Approche taphonomique et palethnographique. Master's thesis, Université de Paris I.

Emerson, A. (1990). *Archaeological implications of variability in the economic anatomy of* Bison bison. Unpublished Ph.D. dissertation, Washington State University.

Emerson, A. (1993). The role of body part utility in small-scale hunting under two strategies of carcass recovery. In J. Hudson (Ed.), *From bones to behavior: Ethnoarchaeological and experimental contributions to the interpretation of faunal remains* (pp. 138–155). Carbondale: Center for Archaeological Investigation, Southern Illinois University.

Enloe, J. G. (1993). Subsistence organization in the early Upper Paleolithic: Reindeer hunters of the abri du Flageolet, couche V. In H. Knecht, A. Pike-Tay, & R. White (Eds.), *Before Lascaux: The complex record of the early Upper Paleolithic* (pp. 101–115). Ann Arbor: CRC Press.

Enloe, J. G. (2004). Equifinality, assemblage integrity and behavioral inferences at Verberie. *Journal of Taphonomy, 2*, 147–165.

Faith, J. T. (2007). Changes in reindeer body part representation at Grotte XVI, Dordogne, France. *Journal of Archaeological Science, 34*, 2003–2011.

Fernández-Laso, M. C. (2010). *Remontajes de restos faunísticos y relaciones entre Áreas Domésticas en los niveles K, L y M del Abric Romaní (Capellades, Barcelona, España)*. Unpublished Ph.D. dissertation, Universitat Rovira i Virgili de Tarragona.

Fiore, I., & Tagliacozzo, A. (2006). Lo sfruttamento dello stambecco nel Tardiglaciale di Riparo Dalmeri (TN): Il livello 26c. *Bolzano, 2006*, 59–76.

Fontana, L. (1999). Mobilité et subsistance au Magdalénien dans le bassin de l'Aude. *Bulletin de la Société Préhistorique Française, 96*, 175–190.

Fosse, P. (1999). La grande faune mammalienne: Remarques préliminaires. In C. Chauchat (Ed.), *L'habitat magdalénien de la grotte Bourrouilla à Arancou (Pyrénées-Atlantiques). Gallia Préhistoire, 41*, 98–113.

Friesen, T. M. (2001). A zooarchaeological signature for meat storage: Re-thinking the drying utility index. *American Antiquity, 66*, 315–331.

Garvey, J. (2011). Bennett's wallaby (*Macropus rufogriseus*) marrow quality vs quantity: Evaluating human decision-making and seasonal occupation in late Pleistocene Tasmania. *Journal of Archaeological Science, 38*, 763–783.

Gaudzinski-Windheuser, S., & Niven, L. (2009). Hominin subsistence patterns during the Middle and Late Paleolithic in northwestern Europe. In J.-J. Hublin & M. P. Richards (Eds.), *The evolution of hominin diets: Integrating approaches to the study of Palaeolithic subsistence* (pp. 99–111). Dordrecht: Springer.

Grayson, D. K. (1984). *Quantitative zooarchaeology. Topics in the analysis of archaeological faunas*. New York: Academic Press.

Grayson, D. K. (1989). Bone transport, bone destruction, and reverse utility curves. *Journal of Archaeological Science, 16*, 643–652.

Grayson, D. K., & Cannon, M. D. (1999). Human paleoecology and foraging theory in the Great Basin. In C. Beck (Ed.), *Models for the millennium: Great Basin anthropology today* (pp. 141–151). Salt Lake City: University of Utah Press.

Grayson, D. K., & Delpech, F. (2003). Ungulates and the Middle-to-Upper Paleolithic transition at Grotte XVI (Dordogne, France). *Journal of Archaeological Science, 30*, 1633–1648.

Grayson, D. K., & Frey, C. J. (2004). Measuring skeletal part representation in archaeological faunas. *Journal of Taphonomy, 2*, 27–42.

Griggo, C. (2005). Les grands mammifères de la grotte de Bois-Ragot. In A. Chollet & V. Dujardin (Eds.), *La grotte du Bois-Ragot à Gouex (Vienne)–Magdalénien et Azilien–Essai sur les hommes et leur environnements* (pp. 289–317). Paris: Société Préhistorique Française.

Haws, J. A. (2003). *An investigation of late Upper Paleolithic and Epipaleolithic hunter-gatherer subsistence and settlement patterns in Central Portugal*. Unpublished Ph.D. dissertation, University of Wisconsin-Madison.

Helmer, D., & Monchot, H. (2006). Un site mésolithique de chasse à l'aurochs (La Montagne, Senas, Bouches-du-Rhône). *Anthropozoologica, 41*(2), 215–228.

Heriech-Briki, D. (2003). *Les grands mammifères du gisement de plein air de La Rouquette (Puycelsi, Tarn). Étude paléontologique, archéozoologique. taphonomique, paléoenvironnementale et biostratigraphique*. Unpublished Ph.D. dissertation, Université de Perpignan.

Irving, L., Schmidt-Nielsen, K., & Abrahamsen, S. B. (1957). On the melting points of animal fats in cold climates. *Physiological Zoölogy, 30*, 93–105.

Jaubert, J., & Brugal, J. P. (1990). Contribution à l'étude du mode de vie au Paléolithique moyen: Les chasseurs d'aurochs de La Borde. In J. Jaubert, M. Lorblanchet, H. Laville, R. Slott-Moller, A. Turq, & J. P. Brugal (Eds.), *Les chasseurs d'aurochs de La Borde: Un site du Paléolithique moyen (Livernon, Lot)* (pp. 127–145). Paris: Éditions Maison des Sciences de l'Homme. Documents d'Archéologie Française no. 27.

Jones, K. T., & Metcalfe, D. (1988). Bare bones archaeology: Bone marrow indices and efficiency. *Journal of Archaeological Science, 15*, 415–423.

Klein, R. G. (1989). Why does skeletal part representation differ between smaller and larger bovids at Klasies River Mouth and other archaeological sites? *Journal of Archaeological Science, 6*, 363–381.

Kuntz, D. (2006). Données nouvelles sur les stratégies d'acquisition et de traitement des carcasses au cours du Magdalénien supérieur dans la vallée de l'Aveyron: L'exemple de la Magdeleine la Plaine (Penne, Tarn). *Préhistoire du Sud-Ouest, 13*, 155–165.

Lam, Y. M., Chen, X., & Pearson, O. M. (1999). Intertaxonomic variability in patterns of bone density and the differential representation of bovid, cervid, and equid elements in the archaeological record. *American Antiquity, 64*, 343–362.

Lyman, R. L. (1984). Bone density and differential survivorship of fossil classes. *Journal of Anthropological Archaeology, 3*, 259–299.

Lyman, R. L. (1985). Bone frequencies: Differential transport, in situ destruction, and the MGUI. *Journal of Archaeological Science, 12*, 221–236.

Lyman, R. L. (1992). Anatomical considerations of utility curves in zooarchaeology. *Journal of Archaeological Science, 19*, 7–22.

Lyman, R. L. (2008). *Quantitative paleozoology*. Cambridge: Cambridge University Press.

Magniez, P. (2010). *Étude paléontologique des artiodactyles de la grotte Tournal (Bize-Minervois, Aude, France). Étude taphonomique, archéozoologique et paléoécologique des grands mammifères dans leur cadre biostratigraphique et paléoenvironnemental*. Unpublished Ph.D. dissertation, Université de Perpignan.

Maloiy, G. M. O., Heglund, N. C., Prager, L. M., Cavagna, G. A., & Taylor, C. R. (1986). Energetic cost of carrying loads: Have African women discovered an economic way? *Nature, 319*, 668–669.

Marean, C. W., & Cleghorn, N. (2003). Large mammal skeletal element transport: Applying foraging theory in a complex taphonomic system. *Journal of Taphonomy, 1*, 15–42.

Marean, C. W., & Kim, S. Y. (1998). Mousterian large-mammal remains from Kobeh Cave: Behavioral implications for Neanderthals and early modern humans. *Current Anthropology, 39*, S79–S113.

Mariezkurrena, K., & Altuna, J. (1995). Fauna de mamíferos del yacimiento costero de Herriko Barra. *Munibe, 47*, 23–32.

Marín Arroyo, A. B., & González Morales, M. R. (2007). La Fragua Cave, a seasonal hunting camp in the lower Asón Valley (Cantabria, Spain) at the Pleistocene-Holocene transition. *Anthropozoologica, 42*, 61–84.

Marks, A. E., Brugal, J.-P., Chabai, V. P., Monigal, K., Goldberg, P., Hockett, B., et al. (2002). Le gisement pléistocène moyen de Galeria Pesada (Estrémadure, Portugal): Premiers résultats. *Paléo, 14*, 77–100.

Meng, M. S., West, G. C., & Irving, L. (1969). Fatty acid composition of caribou bone marrow. *Comparative Biochemistry and Physiology, 30*, 187–191.

Metcalfe, D., & Jones, K. T. (1988). A reconsideration of animal body-part utility indices. *American Antiquity, 53*, 486–504.

Minetti, A. E., Formenti, F., & Ardigò, L. P. (2006). Himalayan porter's specialization: Metabolic power, economy, efficiency and skill. *Proceedings of the Royal Society B, 273*, 2791–2797.

Moigne, A.-M. (2003). Les faunes de grands mammifères de la grotte des Conques à Vingrau. In H. Baills (Ed.), *Les Conques. Des chasseurs magdaléniens et leur territoire* (pp. 75–103). Liège: ERAUL.

Monahan, C. M. (1998). The Hadza carcass transport debate revisited and its archaeological implications. *Journal of Archaeological Science, 25*, 405–424.

Morin, E. (2007). Fat composition and Nunamiut decision-making: A new look at the marrow and bone grease indices. *Journal of Archaeological Science, 34*, 69–82.

Morin, E. (2010). Taphonomic implications of the use of bone as fuel. *Palethnologie, 2*, 209–217.

Morin, E. (2012). *Reassessing Paleolithic subsistence: The Neandertal and Modern Human Foragers of Saint-Césaire*. Cambridge: Cambridge University Press.

O'Connell, J. F., Hawkes, K., & Blurton Jones, N. (1988). Hadza hunting, butchering, and bone transport and their archaeological implications. *Journal of Anthropological Research, 44*, 113–161.

O'Connell, J. F., Hawkes, K., & Blurton Jones, N. (1990). Reanalysis of large mammal body part transport among the Hadza. *Journal of Archaeological Science, 17*, 301–316.

Olsen, S. L. (1989). Solutré: A theoretical approach to the reconstitution of Upper Palaeolithic hunting strategies. *Journal of Human Evolution, 18*, 295–327.

Outram, A. K., & Rowley-Conwy, P. (1998). Meat and marrow utility indices for horse (*Equus*). *Journal of Archaeological Science, 25*, 839–849.

Pasty, J. F., Ballut, C., Griggo, C., & Murat, R. (2002). Le gisement épipaléolithique à pointes de Malaurie de Champ-Chalatras (les Martres d'Artière, Puy-de-Dôme). *Paléo, 14*, 101–176.

Patou-Mathis, M. (1993). Étude taphonomique et palethnologique de la faune de l'Abri des Canalettes. In L. Meignen (Ed.), *L'Abri des Canalettes. Un habitat moustérien sur les Grands Causses (Nant, Aveyron)* (pp. 199–237). Paris: Éditions CNRS.

Patou-Mathis, M. (1999). A new Middle Paleolithic site in Alsace: Mutzig I (Bas-Rhin). Subsistence behaviour. In S. Gaudzinski & E. Turner (Eds.), *The role of early humans in the accumulations of European Lower and Middle Palaeolithic bone assemblages* (Vol. 42, pp. 325–341). Maize: Römisch-Germanisches Zentral-museum.

Pesesse, D. (2008). *Les premières sociétés gravettiennes. Analyse comparée des systèmes lithiques de la fin de l'Aurignacien aux débuts du Gravettien*. Unpublished Ph.D. dissertation, Université de Provence.

Pond, C. M. (1998). *The fats of life*. Cambridge: Cambridge University Press.

Ready, E. (2010). *Neandertal foraging adaptations in southwestern France: Implications of Mousterian faunal remains at Gatzarria Cave*. M.A. thesis, Trent University.

Rendu, W. (2007). *Planification des activités de subsistance au sein du territoire des derniers Moustériens. Cémentochronologie et approche archéozoologique de gisements du Paléolithique moyen (Pech-de-l'Azé I, La Quina, Mauran) et Paléolithique supérieur ancien (Isturitz)*. Unpublished Ph.D. dissertation, Université Bordeaux I.

Rendu, W. (2010). Hunting behaviour and Neanderthal adaptability in the Late Pleistocene site of Pech-de-l'Azé I. *Journal of Archaeological Science, 37*, 1798–1810.

Renou, S. (2007). *Les comportements de subsistance dans le Sud-Ouest de la France à l'Aurignacien: Bilan bibliographique et analyse archéozoologique d'un gisement aurignacien, Les Rois (Mouthiers-sur-Boëme, Charente)*. Master's thesis, Université Bordeaux I.

Riquelme Cantal, J. A. (2008). Estudio de los restos oséos de mamíferos de El Pirulejo. Los niveles paleolíticos. *Antiquitas, 20,* 199–212.

Riquelme Cantal, J. A., Simón Vallejo, M. D., & Costés Sánchez, M. (2005). La fauna de mamíferos del Solutrense en la Cueva de Nerja. *Munibe, 57,* 255–263.

Rivals, F. (2002). *Les petits bovidés pléistocènes dans le bassin méditerranéen et le Caucase. Étude paléontologique, biostratigraphique, archéozoologique et paléoécologique.* Unpublished Ph.D. dissertation, Université de Perpignan.

Rosell, J. (2001). *Patrons d'aprofitament de les biomasses animals durant el Pleistocè inferior i mig (Sierra de Atapuerca, Burgos) i superior (Abric Romaní, Barcelona).* Unpublished Ph.D. dissertation, Universitat Rovira i Virgili de Tarragona.

Sánchez Goñi, M. F., Landais, A., Fletcher, W. J., Naughton, F., Desprat, S., & Duprat, J. (2008). Contrasting impacts of Dangaard-Oeschger events over a Western European latitudinal transect modulated by orbital parameters. *Quaternary Science Reviews, 27,* 1136–1151.

Sekhr, A. (1998). *Étude archéozoologique des niveaux aurignaciens (couches 14 à 6) et de la base des niveaux gravettiens (niveaux X à T4) de l'Abri Pataud (Les Eyzies, Dordogne).* Unpublished Ph.D. dissertation, Muséum National d'Histoire Naturelle, Paris.

Slott-Moller, R. (1990). La faune. In J. Jaubert, M. Lorblanchet, H. Laville, A. Turq, & J. P. Brugal (Eds.), *Les chasseurs d'aurochs de la Borde: Un site du Paléolithique moyen (Livernon, Lot)* (pp. 33–68). Paris: Documents d'Archéologie Française 27.

Sokal, R. R., & Rohlf, F. J. (1969). *Biometry: The principles and practice of statistics in biological research.* San Francisco: W. H. Freeman.

Speth, J. D. (1983). *Bison kills and bone counts.* Chicago: Chicago University Press.

Speth, J. D., & Spielmann, K. A. (1983). Energy source, protein metabolism, and hunter-gatherer subsistence strategies. *Journal of Anthropological Archaeology, 2,* 1–31.

Stefansson, V. (1969). *The friendly Arctic. The story of five years in polar regions.* New York: Greenwood Press.

Surmely, F., Alix, P., Costamagno, S., Daniel, P., Murat, R., Renard, R., et al. (2003). Découverte d'un gisement du Gravettien ancien au lieu-dit le Sire (Mirefleurs, Puy-de-Dôme). *Bulletin de la Société Préhistorique Française, 100,* 29–39.

Symmons, R. (2005). New density data for unfused and fused sheep bones, and a preliminary discussion on the modelling of taphonomic bias in archaeofaunal age profiles. *Journal of Archaeological Science, 32,* 1691–1698.

Turner, E. (2002). Solutré. An archaeozoological analysis of the Magdalenian horizon. *Römisch-Germanisches Zentralmuseum Mainz, Monograph 46.*

Valensi, P. (2000). The archaeozoology of Lazaret Cave (Nice, France). *International Journal of Osteoarchaeology, 10,* 357–367.

Vannoorenberghe, A. (2004). *Contribution à la connaissance des comportements de subsistance des Gravettiens du sud-ouest de la France. La faune du Gravettien ancien de l'Abri Pataud (Les Eyzies-de-Tayac, Dordogne).* Unpublished Ph.D. dissertation, Muséum National d'Histoire Naturelle.

Vigne, J.-D. (2005). Découpe du cerf (*Cervus elaphus*) au Mésolithique moyen, à Noyen-sur-Seine (Seine-et-Marne): Analyses tracéologique et expérimentale. *Revue de Paléobiologie, 10,* 69–82.

Walker, R., Gurven, M., Hill, K., Migliano, A., Chagnon, N., de Souza, R., et al. (2006). Growth rates and life histories in twenty-two small-scale societies. *American Journal of Human Biology, 18,* 295–311.

West, G. C., & Shaw, D. L. (1975). Fatty acid composition of Dall sheep bone marrow. *Comparative Biochemistry and Physiology, 50B,* 599–601.

White, E. P., Ernest, S. K. M., Kerkhoff, A. J., & Enquist, B. J. (2007). Relationships between body size and abundance in ecology. *Trends in Ecology and Evolution, 22,* 323–330.

Winterhalder, B. (1981). Foraging strategies in the boreal forest: An analysis of Cree hunting and gathering. In B. Winterhalder & E. A. Smith (Eds.), *Hunter-gatherer foraging strategies: Ethnographic and archaeological analyses* (pp. 66–95). Chicago: University of Chicago Press.

Index

Jamie L. Clark and John D. Speth (eds.), *Zooarchaeology and Modern Human Origins: Human Hunting Behavior during the Later Pleistocene*, Vertebrate Paleobiology and Paleoanthropology, DOI: 10.1007/978-94-007-6766-9,

MIX
Papier aus verantwortungsvollen Quellen
Paper from responsible sources
FSC® C105338

If you have any concerns about our products,
you can contact us on
ProductSafety@springernature.com

In case Publisher is established outside the EU,
the EU authorized representative is:
Springer Nature Customer Service Center GmbH
Europaplatz 3, 69115 Heidelberg, Germany

Printed by Libri Plureos GmbH
in Hamburg, Germany